Mobile Opportunities and Applications for E–Service Innovations

Ada Scupola
Roskilde University, Denmark

Information Science
REFERENCE

Managing Director:	Lindsay Johnston
Editorial Director:	Joel Gamon
Book Production Manager:	Jennifer Yoder
Publishing Systems Analyst:	Adrienne Freeland
Assistant Acquisitions Editor:	Kayla Wolfe
Typesetter:	Henry Ulrich
Cover Design:	Nick Newcomer

Published in the United States of America by
Information Science Reference (an imprint of IGI Global)
701 E. Chocolate Avenue
Hershey PA 17033
Tel: 717-533-8845
Fax: 717-533-8661
E-mail: cust@igi-global.com
Web site: http://www.igi-global.com

Library of Congress Cataloging-in-Publication Data

Mobile opportunities and applications for e-service innovations / Ada Scupola,
editor.
    p. cm.
  Includes bibliographical references and index.
  Summary: "This book brings together different perspectives on the understanding of e-service and mobile communication and their effect on the fields of marketing, management and information systems"--Provided by publisher.
  ISBN 978-1-4666-2654-6 (hbk.) -- ISBN 978-1-4666-2685-0 (ebook) -- ISBN 978-1-4666-2716-1 (print & perpetual access) 1. Information technology. 2. Electronic commerce. 3. Customer services. 4. Management information systems. 5. Mobile communication systems. 6. Mobile computing. I. Scupola, Ada.
  HC79.I55M63 2013
  658.8'72--dc23
                          2012029130

British Cataloguing in Publication Data
A Cataloguing in Publication record for this book is available from the British Library.

The views expressed in this book are those of the authors, but not necessarily of the publisher.

Table of Contents

Section 2
Organizational and Inter-Organizational Issues in the Online Environment

Section 3
Models for Innovative E-Government Services

Detailed Table of Contents

Section 1
Consumers and Innovativeness

Chapter 1

Kaan Varnali, Istanbul Bilgi University, Turkey

Research focusing on consumer behavior in the mobile context is rapidly accumulating. However, the role of personality traits in explaining and predicting users' perceptions regarding mobile services and behavior within the mobile context is conspicuously under-researched. If consumers are considered as dispositional entities, this lack of researcher interest on the role of personality traits on the value creation processes of mobile consumers should be scrutinized. Striving to provide guidance as to why and how to incorporate personality-based variables within prospective research models attempting to explain and predict consumer behavior in the mobile context, this research critically assesses the-state-of-the-art and presents a conceptual discussion regarding related future research avenues.

Chapter 2

JungKun Park, University of Houston, USA
Te-Lin Chung, Purdue University, USA
Won-Moo Hur, Pukyung National University, South Korea

The Internet phone, which was recently introduced to the communication market, is characterized by its lower cost and increased compatibility. However, its popularity and diffusion is relatively low in the North American market compared with Asian or European markets. This study examines the adoption of Internet phone service in Korea to develop a better understanding of consumers' acceptance of the Internet phone service application. The Unified Theory of Acceptance and Use of Technology model (UTAUT) is used as theoretical background with two additional constructs (consumer innovativeness and perceived trust) in the proposed model. Using a mail survey with 437 responses collected in Korea, the results indicate that consumers' trust is the major factor affecting adoption of new phone service. The results also indicate that consumers' innovativeness would influence the effects of trust and facilitating conditions toward intention to use. The performance expectancy shows a dominant effect on trust toward phone services. Although effort expectancy and social influence also significantly contribute to consumers' trust, the effects are relatively minor. The effect of performance expectancy would be stronger on consumers with lower income.

Theodora Zarmpou, University of Macedonia, Greece
Vaggelis Saprikis, University of Macedonia, Greece
Maro Vlachopoulou, University of Macedonia, Greece

This study presents a conceptual model that combines perceived ease of use, perceived usefulness, innovativeness, trust, demographic characteristics and relationship drivers in order to examine their influence on the mobile services' adoption intention. The proposed model is empirically tested using data collected from a survey with questionnaires conducted in Greece. The results are analyzed through factor analysis, stepwise regression analysis, and ANOVAs. The findings show that individuals' innovativeness, their educational level, and the relationship ties between the users and the mobile services are key factors to encourage m-services' adoption. The results provide interesting insights and useful hints to practitioners and researchers.

Jean-Eric Pelet, University of Nantes, France
Panagiota Papadopoulou, University of Athens, Greece

This paper presents the results of an exploratory qualitative study conducted with 26 consumers about their use of computer screen savers. The results show how the use of screen savers remains almost nonexistent. Unknown or taking too long to apply, this feature is not attractive to persons interviewed who do not use it for sustainable development purposes. The paper presents the results of this qualitative study, offering an interpretive analysis of the reasons and factors explaining this type of computer user's behavior. The paper also discusses the potential of using screensaver functionality in e-commerce websites, particularly in the Mediterranean region. In this direction it looks into how this could be provided by the establishment of two elements - a browser and a website extension, which will be tested in a future online experiment.

Chapter 5

Michail N. Giannakos, Ionian University, Greece
Adamantia G. Pateli, Ionian University, Greece
Ilias O. Pappas, Ionian University, Greece

The scope of this paper is to examine the perceptions which induce the Greek customers to purchase over the Internet, testing the direct effect of experience and the moderating effect of satisfaction. A review of research conducted in the Greek online market demonstrates that satisfaction, self-efficacy, and trust keep a prominent role in the Greek customers' shopping behavior. To increase understanding of this behavior, two parameters of the UTAUT model, performance expectancy and effort expectancy, are incorporated. The findings demonstrate that customers' perceptions about all of the parameters do not remain constant, as the experience acquired from past purchases increases. Moreover, the relationship of experience with self-efficacy and intention to repurchase changes, as satisfaction gained from previous purchases increases. The implications of this study are interesting not only for the Greek but also for the Meditterranean researchers and e-retailers, since the Mediterranean ebusiness market shares several cultural similarities with the Greek market.

The rapid growth of mobile technologies and devices makes it possible for the customers of banking services to conduct banking at any place and at any time. Today, most of the banks in the world provide mobile access to its customers for banking as mobile banking systems improve their efficiency and reduce transaction costs. Banks invested heavily in the mobile banking system hoping that its customers would embrace it with open arms. Contrary to the expectation, the lukewarm patronage to mobile banking makes it crucial to understand the factors that contribute to users' intention to use mobile banking. This study extends the applicability of technology acceptance model (TAM) to the mobile banking context. Based on the review of literature, few additional constructs were added to the TAM. Structural equation modeling (SEM) was used to test the casual relationships proposed. Findings of the study support the proposed model's ability of explaining the users' intention to adopt mobile banking.

Section 2
Organizational and Inter-Organizational Issues in the Online Environment

This paper studies the concept of Store Atmosphere in Virtual Commerce (V-Commerce) through the Web in order to empirically define its determinants and investigate their applicability and customization capabilities. A series of in depth interviews with field experts (study #1) along with an online question-naire survey (study #2) served as the data collection mechanisms of the study. The empirical findings suggest that while the social aspect dimension of V-Commerce limits customization capabilities, it provides several innovative options for manipulating Store Atmosphere. Additionally, the results indicate that Store Atmosphere attributes can be grouped in three factors with high average scores concerning the importance users attach to them. Specifically, storefront, store theatrics, colors, music and graphics are grouped in Factor #1 and reflect the "Store's Appeal". Crowding, product display techniques and innovative store atmosphere services are grouped in Factor #2 labeled "Innovative Atmosphere", while store layout constitutes the only attribute included in Factor #3. The paper outlines the theoretical and managerial implications of these research results.

Chapter 8

Logistics Service Providers (3PL) have become important players in supply chain management. In a highly competitive context characterized by "time compression", a successful strategy depends increasingly on the performance of Logistics Service Providers as they play a key integrative role linking different supply chain elements more effectively. However, the role of the information technology capability of these 3PL has not drawn much attention. The research question is: can IT be viewed as a fundamental supply chain management coordination mechanism? If so, does IT capability of third party logistics providers to improve performance in the supply chain and become a bigger factor in a strategic buyer-3PL relationship? By drawing on earlier research on the supply chain management coordination mechanism, the IT capability of third party logistics providers, a case study is conducted.

Section 3
Models for Innovative E-Government Services

Chapter 9

M. Sirajul Islam, Örebro University, Sweden
Ada Scupola, Roskilde University, Denmark

Government 'e-service' as a subfield of the e-government domain has been gaining attention to practitioners and academicians alike due to the growing use of information and communication technologies at the individual, organizational, and societal levels. This paper conducts a thorough literature review to examine the e-service research trends during the period between 2005 and 2009 mostly in terms of research methods, theoretical models, and frameworks employed as well as type of research questions. The results show that there has been a good amount of papers focusing on 'e-Service' within the field of e-government with a good combination of research methods and theories. In particular, findings show that technology acceptance, evaluation and system architecture are the most common themes, which circa half of the studies surveyed focus on the organizational perspective and that the most employed research methods are case studies and surveys, often with a mix of both types of methodologies.

Chapter 10

Tunc D. Medeni, Turksat, Turkey
İ. Tolga Medeni, Turksat, Turkey
Asim Balci, Turksat, Turkey

As an important project for Turkey to achieve Information/Knowledge Society Strategic Goals, the e-Government Gateway currently focuses on the delivery of public services via a single portal on the Internet. In later stages, other channels such as mobile devices will be available for use, underlying a transition towards mobile and ubiquitous government services. In order to provide a supportive base for this transition, the authors develop a modeling of knowledge amphora (@), and link this conceptual model with the e-government gateway. Based on Knowledge Science concepts such as ubiquity, ba (physical, virtual, mental place for relationship-building and knowledge-creation), ma (time-space in-betweenness), reflection and refraction, the modeling of Knowledge Amphora incorporates the interactions @ the Internet and mobile devices that contribute to cross-cultural information transfer and knowledge creation. The paper presents recent electronic and mobile government developments of E-Government Gateway Project in Turkey as an application example of this philosophical and theoretical modeling. The contributed Ubiquitous Participation Platform for Policy Making (UbiPOL) project aims to develop a ubiquitous platform allowing citizens to be involved in policy making processes (PMPs). The resulting work is a practical case study as that develops new m-government operations.

This paper shows the lack of standard procedures to audit e-voting systems and also describes a practical process of auditing an e-voting experience based on a Direct-recording Electronic system (D.R.E). This system has been tested in a real situation, in the city council of Coahuila, Mexico, in November 2008. During the auditing, several things were kept in mind, in particular those critical in complex contexts, as democratic election processes are. The auditing process is divided into three main complementary stages: analysis of voting protocol, analysis of polling station hardware elements, and analysis of the software involved. Each stage contains several items which have to be analyzed at low level with the aim to detect and resolve possible security problems.

Section 4
Interoperability in E-Government and E-Business

This paper introduces the Interoperability Observatory, a structured research effort for measuring interoperability readiness in the regions of South Eastern Europe and the Mediterranean, supported by the Greek Interoperability Centre. The motivation for this effort derives from the fact that, although interoperability is a key element for public administration and enterprises effective operation, and an important enabler for cross-country cooperation, a standard framework for benchmarking interoperability developments at country level is currently not in place. Interoperability-related information is highly fragmented in different ICT, e-Government and e-Business reports. In this context, in the core of the Interoperability Observatory lies the definition of a structured collection of metrics and indicators, associated with the dimension of interoperability-governance, and a mechanism for gathering with regard to the latter suitable information for a number of countries from various sources. The ultimate goal is the use of this information towards the directions of raising awareness on the countries' interoperability status, promoting best practice cases and benchmarking.

This article outlines a business and application architecture for policy-making organisations of public administrations. The focus was placed on the derivation of processes and their IT support on the basis of the policy-cycle concept. The derivation of various (modular) process areas allows for the discussion of generic application support in order to achieve the modular structure of e-government architectures for policy-making organisations of public administrations, as opposed to architectures for operational administration processes by administrations. In addition, further issues and spheres of interest to be addressed in the field of architecture management for policy-making organisations of public administrations will be specified. Different architecture variants are evaluated in the context of a potential application of the architecture design for policy-making organisations of public administrations. This raises questions such as how the issue of interoperability between information systems of independent national, state, and municipal administrations is to be tackled. Further research is needed to establish, for example, the level of enterprise architecture and the depth to which integration in this area must or may extend.

Chapter 14

Elena Sánchez-Nielsen, Universidad de La Laguna, Spain

Daniel González-Morales, Universidad de La Laguna, Spain

Carlos Peña-Dorta, ARTE Consultores Tecnológicos, Spain

Today's Public Administration faces a growing need to share information and collaborate with other agencies and organizations in order to meet their objectives. As agencies and organizations are gradually transforming into "networked organizations," the interoperability problem becomes the main challenge to make possible the vision of seamless interactions across organizational boundaries. Today, diverse architectural engineering guidelines are used to support interoperability at different levels of abstraction. This paper reviews the main guidelines' categories which support aspects of architecture practice in order to develop interoperable software services among networked organizations. The architectural guidelines and practical experiences in the domain of e-Gov employment services for the European Union member state Spain are described. The benefits of the proposed solution and the lessons learned are illustrated.

Section 5
Applications for Innovation in E-Business and E-Government

Chapter 15

Deirdre Lee, National University of Ireland, Ireland

Yojana Priya Menda, National University of Ireland, Ireland

Vassilios Peristeras, National University of Ireland, Ireland

David Price, ThoughtGraph Ltd., UK

The growth of Information and Communication Technologies (ICTs) offers governments advanced methods for providing services and governing their constituency. eGovernment research aims to provide the models, technologies, and tools for more effective and efficient public administration systems as well as more participatory decision processes. In particular, eParticipation opens up greater opportunities for consultation and dialogue between government and citizens. Many governments have embraced eParticipation by setting up websites that allow citizens to contribute and have their say on particular issues. Although these sites make use of some of the latest ICT and Web 2.0 technologies, the uptake and sustained usage by citizens is still relatively low. Additionally, when users do participate, there is the issue of how the numerous contributions can be effectively processed and analysed, to avoid the inevitable information overload created by thousands of unstructured comments. The WAVE platform addresses what the authors see as the main barriers to the uptake of eParticipation websites by adopting a holistic and sustained approach of engaging users to participate in public debates. The WAVE platform incorporates argument visualisation, social networking, and Web 2.0 techniques to facilitate users participating in structured visual debates in a community environment.

Chapter 16

Sebastian Obermeier, ABB Corporate Research, Switzerland
Stefan Böttcher, Universität Paderborn, Germany

A distributed protocol is presented for anonymous and secure voting that is failure-tolerant with respect to malicious behavior of individual participants and that does not rely on a trusted third party. The proposed voting protocol was designed to be executed on a fixed group of N known participants, each of them casting one vote that may be a vote for abstention. Several attack vectors on the protocol are presented, and the detection of malicious behavior like spying, suppressing, inventing, and modifying protocol messages or votes by the protocol is shown. If some participants stop the protocol, a fair information exchange is achieved in the sense that either all votes are guaranteed to be valid and accessible to all participants, or malicious behavior has been detected and the protocol is stopped, but the votes are not disclosed.

Chapter 17

Tyrone Edwards, University of Technology, Jamaica
Suresh Sankaranarayanan, University of West Indies, Jamaica

Access to the correct healthcare facility is a major concern for most people, many of whom gather information about the existing hospitals and healthcare facilities in their locality. After gathering such information, people must do a comparison of the information, make a selection, and then make an appointment with the concerned doctor. The time spent for this purpose would be a major constraint for many individuals. Research is currently underway in this area on incorporating Information and Communication Technology (ICT) to improve the services available in the health industry. This paper proposes an agent based approach to replicate the same search operations as the individual would otherwise do, by employing an intelligent agent. The proposed agent based system has been simulated and also validated through implementation on an individual's smart phone or a PDA using JADE-LEAP agent development kit.

Chapter 18

Ryan Anthony Brown, University of West Indies, Jamaica
Suresh Sankaranarayanan, University of West Indies, Jamaica

Access to the correct healthcare facility is a major concern for most people, many of whom gather information about the existing hospitals and healthcare facilities in their locality. After gathering such information, people must do a comparison of the information, make a selection, and then make an appointment with the concerned doctor. The time spent for this purpose would be a major constraint for many individuals. Research is currently underway in this area on incorporating Information and Communication Technology (ICT) to improve the services available in the health industry. This paper proposes an agent based approach to replicate the same search operations as the individual would otherwise do, by employing an intelligent agent. The proposed agent based system has been simulated and also validated through implementation on an individual's smart phone or a PDA using JADE-LEAP agent development kit.

Preface

This preface to the third book of the series entitled "*Advances in E-Services and Mobile Applications series*" provides an overview of the book and its structure and contextualizes the major theme of the book. As the title of the book "Mobile Opportunities and Applications for E-Service Innovations" indicates, the major overarching theme of the book is innovation and innovativeness.

In fact, the chapters of this book deal with the theme of innovation and innovativeness in different ways. Some chapters deal with behavioural innovativeness at individual level. Most of these chapters deal with the factors that influence the behavioural intention to use and accept e-technologies. Some of these chapters make use of the Technology Acceptance Model or the Unified Theory of Acceptance and Use of Technology model (UTAUT) in their investigations, thus confirming that TAM and its variations are still important and useful models to measure technology acceptance at individual level. However this also shows that there is a need for "innovative" theories and models in this field of research. Such studies focus mainly on the e-services and mobile services within the business context as in the case of Internet Phone Services in the chapter written by Junkun Park or mobile banking services in the chapter written by Selvan *et al*. The two chapters that have conducted an extensive literature review (Varnali in the field of mobile services and Islam and Scupola in the field of e-government services) also show that consumer behaviour and technology acceptance at individual level are recurrent themes in studies dealing with the business and e-government contexts.

However, e-services are important for innovation and innovativeness also at organizational and inter-organizational level. For example, the chapter written by Krasonikolakis *et al*. studies the concept of Store Atmosphere in Virtual Commerce (V-Commerce) and suggest that while the social aspect dimension of V-Commerce limits customization capabilities, it provides several innovative options for manipulating Store Atmosphere in terms of storefront, store theatrics, colours, music and graphics. In addition, the study conducted by Carrus and Pinna in the specific field of logistics service providers shows the importance of innovative IT capabilities for these types of companies in order to improve supply chain performance.

The role of interoperability is in general very important for innovation and innovativeness at organizational and inter-organizational level. This book sees Interoperability as key to e-technologies adoption and diffusion at organizational, inter-organizational and inter-country level, especially in the context of e-government. This is for example showed by the chapter written by Markaki *et al*. .This chapter introduces the Interoperability Observatory, a structured research effort for measuring interoperability readiness in the regions of South Eastern Europe and the Mediterranean region in response to the lack of a standard framework for benchmarking interoperability developments at country level. The other two chapters of this book dealing with interoperability investigate the organizational characteristics that might favour or not e-services acceptance and diffusion such as company policies (Walser and Riedl) and companies' architectures (Sanchez-Nielsen et al). Overall these chapters provide also normative models and recommendations on how to go to address interoperability issues.

Finally the book includes a number of chapters that develop theoretical models, platforms and applications aiming at facilitating such e- and mobile technologies acceptance to innovate and improve e-services both in the business (e.g. Obermeier and Böttcher) and e-government contexts (e.g. Medeni *et al.*; Alaiz-Moreton *et al.*)

The next section describes in detail the structure and contents of the book.

THE CONTENTS OF THE BOOK

The book is structured into five sections. The first section includes six chapters focusing on the factors that impact consumers' level innovativeness in a business context. Two mobile services are particularly analysed in this section: Internet Phone Services (VOIP) and mobile banking. The second section focuses on factors that might impact innovativeness at organizational and inter-organizational level. The third section discusses theoretical models that can be used to innovate e-services and mobile services, while the fourth section focuses on a specific issue: interoperability. Finally the book concludes with a section on applications for e- and mobile services innovation and innovativeness.

A strong emphasis of the book is on the specific geographical region of the Mediterranean Sea. This is due to the fact that several studies investigate problems and issues in the country of Greece or that are of relevance to other Mediterranean countries as well.

THE STRUCTURE OF THE BOOK

Section 1: Consumers and Innovativeness

The first section focuses on innovation and innovativeness for online services related to the consumer market. This section, having a special focus on the individual level, touches upon different issues of innovativeness and acceptance in relation to e-service and mobile services including trust, behavioural intention to use e-technologies as well as the effects of experience and satisfaction on the re-purchasing of e- and mobile services in the online environment. This section includes six chapters.

The first chapter, titled "Personality Traits and Consumer Behavior in the Mobile Context: A Critical Review and Research Agenda," is written by Kaan Varnali, Istanbul Bilgi University, Turkey. The chapter claims that the research focusing on consumer behaviour in the mobile context is rapidly accumulating, while the role of personality traits in explaining and predicting users' perceptions regarding mobile services and behaviour within the mobile context is still under-researched. The author argues that if consumers are considered as dispositional entities, the lack of research interest on the role of personality traits for the value creation process of mobile consumers should be better scrutinized and analysed. In the strive to provide guidance as to why and how to incorporate personality-based variables within prospective research models attempting to explain and predict consumer behaviour in the mobile context, this chapter critically assesses the-state-of-the-art and presents a conceptual discussion regarding related future research avenues.

The second chapter, entitled "The Role of Consumer Innovativeness and Trust for Adopting Internet Phone Services," is written by Jungkun Park, University of Houston, USA. In this chapter the author examines the adoption of Internet phone service in Korea in order to develop a better understanding of consumers' acceptance of such service in this country. The study takes its point of departure in the Unified Theory of Acceptance and Use of Technology model (UTAUT) and adds two new constructs to the model. Such constructs, consumer innovativeness and perceived trust, might help to gain a theoretical understanding of consumers' acceptance of Internet phone service in Korea. The results of the study show indeed that trust in the technology and personal innovativeness are important factors in determining whether or not consumers will use Internet Phone Services. The study also shows that technology performance expectancy is a major factor affecting the trust that consumers have toward Internet phone services.

The third chapter, titled "Examining Behavioral Intention toward Mobile Services: An Empirical Investigation in Greece," is authored by Theodora Zarmpou, Vaggelis Saprikis and Maro Vlachopoulou. This chapter first develops a conceptual model for examining the effects of a number of variables on the intention to use mobile services. The model is then empirically tested using data collected with a survey conducted in Greece. The chapter shows that key factors encouraging m-services' adoption include individuals' innovativeness, the educational level, and the relationship ties between the users and the mobile services in question. The authors argue that what distinguishes this study from previous studies applying multiple behavioral theories and developing conceptual models to identify the different influential factors for the mobile services' usage is that this study adds a marketing perspective to previous investigations. This is specifically done by including a new significant influential factor in the model, called "relationship drivers".

The fourth chapter, titled "Colored vs. Black Screens or How Color Can Favor Green e-Commerce," is written by Jean-Eric Pelet, University of Nantes, France and Panagiota Papadopoulou, University of Athens, Greece. This chapter presents the results of an exploratory study investigating the use of computer screen savers among consumers. The study conducted 26 interviews in order to get empirical insights into the use of computer screen savers among consumers. The results show that the use of screen savers is almost nonexistent.This feature, unknown or taking too long to apply, is not attractive to the persons interviewed who do not use it for sustainable development purposes. The chapter, after presenting the results of the qualitative study, offers an interpretive analysis and discussion of the reasons and factors that explain this type of computer user's behavior. The chapter additionally discusses the potential of using screensaver functionalities in e-commerce websites, particularly in the Mediterranean region. In this direction, the study provides insights into how this could be done through the establishment of two elements - a browser and a website extension. These functionalities are subjects for further research and that can be tested in future online experiments.

The fifth chapter, titled "Identifying the Direct Effect of Experience and the Moderating Effect of Satisfaction in the Greek Online Market," is written by Michail N. Giannakos, Adamantia G. Pateli, and Ilias O. Pappas, all working at the Ionian University in Greece. The scope of this chapter is to examine the perceptions which induce the Greek consumers to purchase over the Internet, by testing the direct effect of experience and the moderating effect of satisfaction. The authors state that a review of research conducted in the Greek online market demonstrates that satisfaction, self-efficacy, and trust have a prominent role in explaining the Greek customers' shopping behaviour. To increase understanding of this behaviour, the authors include two parameters of the UTAUT model, performance expectancy and effort expectancy into their study. The findings demonstrate that customers' perceptions about the investigated parameters do not remain constant, but are a function of the experience accumulated through past purchases. Similarly, the relationship between experience and self-efficacy and intention to repurchase changes, as satisfaction gained from previous purchases increases. The authors expect that the implications of this study are interesting not only for the Greek market but also for the other Mediterranean researchers and e-retailers, since the Mediterranean e-business market shares several cultural similarities with the Greek market.

The sixth chapter, titled, "Behavioral Intention Towards Mobile Banking in India: The Case of State Bank of India (SBI)," is written by N. Thamarai Selvan and B. Senthil Arasu, both at the National Institute of Technology, India and M. Sivagnanasundaram, Kirloskar Institute of Advanced Management Studies, India. The background of the chapter is that the rapid growth of mobile technologies and devices makes it possible for the customers of banking services to conduct banking at any place and at any time. Today, most of the banks in the world provide mobile access to its customers for banking services provisions. One motivation for banks to do this is that mobile banking systems improve banks' efficiency and reduce transaction costs. Banks have invested heavily in the mobile banking systems hoping that their customers would adopt it straight away and without problems. However, contrary to expectations, this has not been all that easy. Therefore, the authors argue for the need to understand the factors that affect users' intention to use mobile banking. The study presented in this chapter makes such an attempt in the Indian mobile banking context. It extends the applicability of the technology acceptance model (TAM) to mobile banking. Based on an extensive review of literature, few additional constructs were added to the TAM model to take into account the case of mobile banking. Structural equation modeling (SEM) was used to test the casual relationships proposed in the modified model. Finally the findings of the study support the proposed model's ability of explaining the users' intention to adopt mobile banking in an Indian context.

Section 2: Organizational and Inter-Organizational Issues in the Online Environment

This section focuses on factors related to innovativeness and innovation in an organizational as well as inter-organizational level and includes two chapters.

The seventh chapter, titled "Defining, Applying and Customizing Store Atmosphere in Virtual Reality Commerce:Back to Basics?" is written by Ioannis G. Krasonikolakis, Adam P. Vrechopoulos, Athanasia Pouloudi, all working at Athens University of Economics and Business, Greece.

This chapter studies the concept of Store Atmosphere in Virtual Commerce (V-Commerce) through the Web in order to empirically define its determinants and investigate their applicability and customization capabilities. The study employees a series of in depth interviews with field experts along with an online questionnaire survey as the data collection mechanisms. The empirical findings suggest that while the social aspect dimension of V-Commerce limits customization capabilities, it provides several innovative options for manipulating Store Atmosphere. Additionally, the results indicate that Store Atmosphere attributes can be grouped in three factors with high average scores concerning the importance users attach to them. Specifically, storefront, store theatrics, colours, music and graphics are grouped in Factor #1 and reflect the "Store's Appeal". Crowding, product display techniques and innovative store atmosphere services are grouped in Factor #2 labelled "Innovative Atmosphere", while store layout constitutes the only attribute included in Factor #3. To conclude the chapter outlines the theoretical and managerial implications of these research results.

The last chapter of the section, titled "Information Technology and Supply Chain Management Co-ordination: The Role of Third Party Logistics Providers," is written by Pier Paolo Carrus and

Roberta Pinna, University of Cagliari, Italy. The chapter argues that Logistics Service Providers (3PL) have become important players in supply chain management and that in a highly competitive context characterized by "time compression", a successful strategy depends increasingly on the performance of Logistics Service Providers as they play a key integrative role linking different supply chain elements more effectively. However, the role of the information technology capability of these 3PL has not drawn much attention, therefore the study investigates whether IT can be viewed as a fundamental supply chain management coordination mechanism and if so how may the IT capability of third party logistics providers improve performance in the supply chain and become a bigger factor in a strategic buyer-3PL relationship. By drawing on earlier research on the supply chain management coordination mechanism, the study conceptualizes the IT capability of third party logistics providers and conducts a case study to illustrate the theory.

Section 3: Models for Innovative E-Government Services

This section provides an overview of the state of the art of e-services within the e-government domain, highlighting the major factors dealing with innovation and innovativeness within this field as well as presents some theoretical models that could be used for innovation in the field of e-government.

The first chapter of the section, titled "E-Service Research Trends in the Domain of E-Government: A Contemporary Study," is written by M. Sirajul Islam, Örebro University, Sweden and Ada Scupola, Roskilde University, Denmark. The chapter's background is that government 'e-service' as a subfield of the e-government domain has been gaining attention to practitioners and academicians alike due to the growing use of information and communication technologies at the individual, organizational, and societal levels. This chapter conducts a thorough literature review to examine the e-service research trends during the period between 2005 and 2009 mostly in terms of research methods, theoretical models, and frameworks employed as well as type of research questions. The results show that there has been a good amount of studies focusing on 'e-Service' within the field of e-government with a good combination of research methods and theories. In particular, the findings show that technology acceptance, evaluation and system architecture are the most common lenses used to investigate innovation and innovativeness, with circa half of the studies surveyed focusing on the organizational level. From a research method perspective, the study shows that the most employed research methods are case studies and surveys, often with a mix of both types of methodologies.

The second chapter, titled "Proposing a Knowledge Amphora Model for Transition Towards Mobile Government," is written by Tunc D. Medeni, Tolga Medeni, Asim Balci, from Turksat, Turkey. The authors explain that the e-Government Gateway in Turkey focuses on the delivery of public services via a single portal on the Internet. The e-Government Gateway is an important project for Turkey to achieve Information/Knowledge Society Strategic Goals. The project has the goals, in later stages, of using other channels such as mobile devices to provide a transition towards mobile and ubiquitous government services. In order to provide a supportive base for this transition, the authors of the chapter develop a modeling of knowledge amphora (@), and link this conceptual model with the e-government gateway. Based on Knowledge Science concepts such as ubiquity, ba (physical, virtual, mental place for relationship-building and knowledge-creation), ma (time-space in-between-ness), reflection and re-fraction, the modeling of Knowledge Amphora incorporates the interactions @ the Internet and mobile devices that contribute to cross-cultural information transfer and knowledge creation. The chapter presents recent electronic and mobile government developments of E-Government Gateway Project in Turkey as an application example of this philosophical and theoretical modeling. The Ubiquitous Participation Platform for Policy Making (UbiPOL) project developed in the chapter aims to develop a ubiquitous platform allowing citizens to be involved in policy making processes (PMPs). The resulting work is the illustration of a practical case study of how that develops new m-government operations.

The third and last chapter of this section, titled "Technical Audit of an Electronic Polling Station: A Case Study," is written by Hector Alaiz-Moreton, Luis Panizo-Alonso, Ramón A. Fernandez-Diaz, And Javier Alfonso-Cendon, Universidad de Leon, Spain. This chapter shows the lack of standard procedures to audit e-voting systems and describes a practical process of auditing an e-voting experience based on a Direct-recording Electronic system (D.R.E). This system has been tested in a real situation, the city council of Coahuila, Mexico, in November 2008. During the auditing, several things were kept in mind, in particular those critical in complex contexts, as democratic election processes are. The auditing process is divided into three main complementary stages: analysis of voting protocols, analysis of polling station hardware elements, and analysis of the software involved. Each stage contains several items which have to be analysed at low level with the aim to detect and resolve possible security problems.

Section 4: Interoperability in E-Government and E-Business

The focus of this section is on interoperability issues in the field of e-government and e-business as essential to innovation and innovativeness at organization, inter-organizational and inter-country level.

The first chapter, titled "Measuring Interoperability Readiness in South Eastern Europe and the Mediterranean: The Interoperability Observatory," is written by Ourania Markaki, National Technical University of Athens, Greece Yannis Charalabidis, University of the Aegean, Greece and Dimitris Askounis, National Technical University of Athens, Greece. This chapter introduces the Interoperability Observatory, a structured research effort for measuring interoperability readiness in the regions of South Eastern Europe and the Mediterranean, supported by the Greek Interoperability Centre. The authors argue that the motivation for this effort derives from the fact that, although interoperability is a key element for public administration and enterprises effective operation, and an important enabler for cross-country cooperation, a standard framework for benchmarking interoperability developments at country level is currently not in place. Interoperability-related information is highly fragmented in different ICT, e-Government and e-Business reports. In this context, the Interoperability Observatory attempts to provide the foundation for a structured collection of metrics and indicators, associated with the dimension of interoperability-governance, and a mechanism for gathering suitable information for a number of countries from various sources. The ultimate goal is the use of this information towards the directions of raising awareness on the countries' interoperability status, promoting best practice cases and benchmarking.

The second chapter of this section, titled "Policy Cycle-Based E-Government Architecture for Policy-Making Organizations of Public Administrations," is written by Konrad Walser and Reinhard Riedl, both at Bern University of Applied Sciences, Switzerland. The chapter outlines a business and application architecture for policy-making organizations in public administrations. The focus of the study is on the derivation of processes and their IT support on the basis of the policy-cycle concept. The derivation of various (modular) process areas allows for the discussion of generic application support in order to achieve the modular structure of e-government architectures for policy-making organizations of public administrations, as opposed to architectures for operational administration processes by administrations. In addition, the chapter identifies further issues and spheres of interest to be addressed in the field of architecture management for policy-making organizations of public administrations. Different architecture variants are evaluated in the context of a potential application of the architecture design for policy-making organizations. This raises questions such as how the issue of interoperability between information systems of independent national, state, and municipal administrations is to be tackled. Further research is needed to establish, for example, the level of enterprise architecture and the depth to which integration in this area must or may extend.

The last chapter of the section, titled "Architectural Guidelines and Practical Experiences in the Realization of E-Gov Employment Services," is written by Elena Sánchez-Nielsen and Daniel González-Morales, Universidad de La Laguna, Spain and Carlos Peña-Dorta, ARTE Consultores Tecnológicos, Spain. The chapter takes its starting point in the fact that today's Public Administration faces a growing need to share information and collaborate with other agencies and organizations in order to meet their objectives. As agencies and organizations are gradually transforming into "networked organizations," the interoperability problem becomes the main challenge to make possible the vision of seamless interactions across organizational boundaries. Today, diverse architectural engineering guidelines are used to support interoperability at different levels of abstraction. This chapter reviews the main guidelines' categories which support aspects of architecture practice in order to develop interoperable software services among networked organizations. The chapter describes the architectural guidelines and practical experiences in the domain of e-Gov employment services for the European Union member state Spain. Finally the authors illustrate the benefits of the proposed solution and the lessons learned.

Section 5: Applications for Innovation in E-Business and E-Government

The last section of this book has the purpose of providing examples of platforms and applications that can be used to make organizations, public and private, more innovative.

The first chapter of this section, titled "The WAVE Platform: Utilizing Argument Visualisation, Social Networking and Web 2.0 Technologies for eParticipation," is written by Deirdre Lee, DERI, National University of Ireland Galway (NUI Galway), Ireland, Yojana Priya Menda and Vassilios Peristeras, National University of Ireland, Ireland, and David Price, ThoughtGraph Ltd., UK. The chapter takes the starting point in the fact that the growth of Information and Communication Technologies (ICTs) offers governments advanced methods for providing services and governing their constituency. eGovernment research aims to provide the models, technologies, and tools for more effective and efficient public administration systems as well as more participatory decision processes. In particular, eParticipation opens up greater opportunities for consultation and dialogue between government and citizens. Many governments have embraced eParticipation by setting up websites that allow citizens to contribute and have their say on particular issues. Although these sites make use of some of the latest ICT and Web 2.0 technologies, the uptake and sustained usage by citizens is still relatively low. Additionally, when users do participate, there is the issue of how the numerous contributions can be effectively processed and analysed, to avoid the inevitable information overload created by thousands of unstructured comments. In this context, the authors propose the WAVE platform to address what the authors see as the main barriers to the uptake of eParticipation websites by adopting a holistic and sustained approach of engaging users to participate in public debates. The WAVE platform incorporates argument visualisation, social networking, and Web 2.0 techniques to facilitate users participating in structured visual debates in a community environment.

The second chapter of this section, titled "Protecting a Distributed Voting Schema for Anonymous and Secure Voting against Attacks of Malicious Partners," is written by Sebastian Obermeier, ABB Corporate Research, Switzerland and Stefan Böttcher, Universität Paderborn, Germany. In the chapter a distributed protocol is presented for anonymous and secure voting that is failure-tolerant with respect to malicious behaviour of individual participants and that does not rely on a trusted third party. The proposed voting protocol was designed to be executed on a fixed group of N known participants, each of them casting one vote that may be a vote for abstention. Several attack vectors on the protocol are presented, and the detection of malicious behaviour like spying, suppressing, inventing, and modifying protocol messages or votes by the protocol is shown. If some participants stop the protocol, a fair information exchange is achieved in the sense that either all votes are guaranteed to be valid and accessible to all participants, or malicious behaviour has been detected and the protocol is stopped, but the votes are not disclosed.

The third chapter of this section, titled "Applications of Intelligent Agents in Hospital Search and Appointment System," is written by Tyrone Edwards, University of Technology, Jamaica and Suresh Sankaranarayanan, University of West Indies, Jamaica. The authors claim that the health sector is an important sector within the e-government domain. Access to the correct healthcare facility is a major concern for most people, many of whom gather information about the existing hospitals and healthcare facilities in their locality. After gathering such information, people must do a comparison of the information, make a selection, and then make an appointment with the doctor. The time spent for this purpose would be a major constraint for many individuals. Research is currently underway in this area on incorporating Information and Communication Technology (ICT) to improve the services available in the health industry. This chapter proposes an agent based approach to replicate the same search operations as the individual would otherwise do, by employing an intelligent agent. The proposed agent based system has been simulated and also validated through implementation on an individual's smart phone or a PDA using JADE-LEAP agent development kit.

The last chapter of this section, titled "Intelligent Store Agent for Mobile Shopping," is written by Ryan Anthony Brown and Suresh Sankaranarayanan, University of West Indies, Jamaica. The chapter deals with the business context and argues that the conventional shopping process involves a human being visiting a designated store and perusing first the items available. A purchase decision is then made based on the information so gathered. However, a number of unique challenges would be faced if potential customer had to execute this process using a mobile device, such as a mobile phone. Taking this aspect into consideration, the authors propose the use of an Intelligent Agent for performing the Mobile Shopping on behalf of customers. In this situation, the agents gather information about the products through the use of 'Store Coordinator Agents' and then use them for comparing the products with the user preferences. The agent based system proposed in the chapter is composed of two agents, viz., a User Agent and Store Coordinator Agent. The implementation of the scheme proposed in the chapter has been done using JADE-LEAP development kit and the performance results are measured and discussed in the chapter.

Ada Scupola
Roskilde University, Denmark

Section 1
Consumers and Innovativeness

Chapter 1
Personality Traits and Consumer Behavior in the Mobile Context:
A Critical Review and Research Agenda

Kaan Varnali
Istanbul Bilgi University, Turkey

ABSTRACT

Research focusing on consumer behavior in the mobile context is rapidly accumulating. However, the role of personality traits in explaining and predicting users' perceptions regarding mobile services and behavior within the mobile context is conspicuously under-researched. If consumers are considered as dispositional entities, this lack of researcher interest on the role of personality traits on the value creation processes of mobile consumers should be scrutinized. Striving to provide guidance as to why and how to incorporate personality-based variables within prospective research models attempting to explain and predict consumer behavior in the mobile context, this research critically assesses the-state-of-the-art and presents a conceptual discussion regarding related future research avenues.

INTRODUCTION

Technological innovations, when they reach a critical threshold level of penetration, may cause tremendous impact on various aspects of daily life. As it was the case in color TV, landline telephone, and PC-based Internet, mobile technology also had a similar effect and caused fundamental shifts on the communication patterns, the temporal and spatial constraints, and the expectations of people (e.g., Balasubramanian, Peterson, & Jarvenpaa, 2002). In fact, the penetration rate of mobile handsets has well passed that of landline phone, PC-based Internet and any other technological

DOI: 10.4018/978-1-4666-2654-6.ch001

devices (Juniper Research, 2008). Eventually, mobile phones have morphed into very capable, constant personal companions that are "always on" and "always connected", enabling companies to establish an ever existing presence alongside their customers through a multitude of interactive mobile applications. The proliferation of the mobile medium and its use for customer interaction represents a discontinuity in the marketplace, and hence mobile marketing phenomena draws mounting interest from both academic and business circles.

Mobile marketing is the use of the mobile medium as a means of marketing communications (Leppäniemi, Sinisalo, & Karjaluoto, 2006). Therefore, the domain of mobile marketing research includes acceptance and use of mobile services, use of mobile applications in consumer service, acceptance and effectiveness of mobile advertising, mobile commerce that involves transactions, and topics related with consumer behavior and policy in the mobile context. More than half of the academic research focusing on mobile marketing consists of articles attempting to explain and predict consumer behavior in the mobile context (Varnali & Toker, 2010). These works have identified a multitude of factors that may have an influence on the acceptance and adoption of mobile services by consumers. Most of these studies have applied extended versions of Technology Acceptance Model, Theory of Reasoned Action, Theory of Planned Behavior, Diffusion of Innovations, and theories from uses and gratifications research. It has been found that content relevance (Heinonen & Strandvik, 2007) and user perceptions regarding various aspects of mobile marketing campaigns and services such as informativeness, entertainment (Bauer et al., 2005), usefulness, ease of use (Hsu & Lu, 2008), expressiveness, behavioral control (Nysveen, Pedersen, & Thorbjornsen, 2005), credibility (Okazaki, 2004), interactivity (Chae et al., 2002), use convenience, connection stability, cost (Park, 2006), riskiness (Chen, 2008), appropriateness of message delivery (Barnes & Scornavacca, 2004;

Kleijnen, Ruyter, & Wetzels, 2007), peer-influence (Hsu & Lu, 2008), and social value (Pihlström, 2007) are the primary predictors of consumer attitudes, intentions, and behavior in the mobile context. Several studies have classified users with respect to their usage and adoption patterns of mobile services and observed that the resulting segments had different psychographic profiles (Kleijnen, Ruyter, & Wetzels, 2004; Marez et al., 2007; Mort & Drennan, 2005). Although such segmentation studies offer strategic guidelines for marketers, they do not provide any explanation for the differences in user perceptions regarding the aforementioned aspects of mobile marketing campaigns and services. Such an insight may be provided by focusing on personality traits. However, very few studies have investigated the effects of personality traits that become salient when an individual is exposed to a mobile marketing message or engaged with a mobile service. This may be due to the fact that researchers, as well as practitioners in the field of mobile marketing, are faced with a bewildering array of personality constructs with little guidance and no rationale at hand.

As an attempt to address these difficulties, this research strives to 1) underpin the importance of including personality traits within prospective research models aiming to explain and predict consumer behavior in the mobile context, 2) identify a variety of context-relevant personality traits, and 3) discuss how these traits are relevant to which specific aspects of the mobile medium. A critical assessment of the state of the art will be presented in order to facilitate future research. The resulting conceptual framework shall provide researchers who are interested in mobile phenomena with a battery of context-relevant personality traits, and shall guide them in forging empirical research agendas. It shall also be beneficial to marketers in developing innovative profiling and segmentation algorithms that would potentially produce increased return rates for their mobile marketing campaigns.

Personality Traits and Consumer Behavior

Learning from experience is not a simple matter of discovering truth; it is a hypothesis testing process in which individuals adapt their beliefs to make sense of new phenomena (Hoch & Deighton, 1989). Perception, a subjective process through which an individual selects, organizes, and interprets stimuli into a meaningful and coherent picture of the world (Schiffman & Kanuk, 2007), is the initial stage of learning from experience, and is highly vulnerable to biases. Two individuals may be exposed to the same stimuli in the same context and still have differing perceptions regarding its utility, riskiness, relevance, emotional appeal and even its appropriateness. Factors that influence differing perceptions include prior knowledge, cultural values, existing stereotypes, cognitive biases, beliefs and intrinsic psychological characteristics. Among these influences, intrinsic psychological characteristics are relatively more enduring and more difficult to change (John & Srivastava, 1999).

There are many accounts explaining the establishment of these intrinsic psychological characteristics, more generally referred to as personality. Freudian account emphasizes the dual influence of heredity and early childhood experiences; while neo-Freudian theories stress broader social and environmental influences (Schiffman & Kanuk, 2007). These theories require measurement of personality with qualitative techniques, and usually view personality as a unified whole. Although it has merits, such an approach is unsuitable for measuring consumer differences in quantitative terms. The alternative is trait theory, which views people as differing in their positions along a continuum on a set of traits. Such an approach enables quantitative measurement of specific individual traits as independent units of analysis (Pervin, 1994). Therefore, most of the literature that investigates the role of interpersonal differences as factors influencing a user's response to certain stimuli in a particular context adopts the trait theory approach (e.g., Dabholkar & Bagozzi, 2002; Kwak, Fox, & Zinkhan, 2002; LaRose & Eastin, 2002).

Study of Personality Traits within the Context of Consumption

While personality traits are being included in consumption related studies since the era of Sigmund Freud, their explanatory power has always been questionable due to the severe discrepancies in empirical findings. Several reasons can be postulated to account for these discrepancies, such as the use of instruments which were originally created to measure gross personality characteristics to diagnose psychological illnesses, the shotgun approach used in many empirical studies without proper theoretical justification, and improper adaptation of scale items (John & Srivastava, 1999; Kassarjian, 1971). Lack of clarity about personality constructs has also been costly to the study of personality in consumer research. The large number of terms has produced some theoretical confusion about both the boundaries of individual constructs and the interrelationships among them (Costa & McCrae, 1995). This theoretical confusion interferes with accumulation of research findings because findings related with a personality construct often remain fragmented under different labels. Kassarjian (1971, p. 416), in his milestone article about the use of personality traits in consumer research, stated that "to expect the influence of personality variables to account for a large portion of the variance is most certainly asking too much." In line with this statement, a great majority of early studies found so weak, if any, relationship between personality and consumer behavior. Following a long period of disparate theories and equivocal findings, the past two decades have seen a revitalization of personality research due to the emergence of better structured integrative personality frameworks (e.g., Baumgartner, 2002; Mowen, 2000). Specifically, Mowen's (2000) hierarchical approach, in which he had identified three levels of personality traits as cardinal, central, and surface, was a very important step towards the conceptualization of the relationship between personality traits and consumer behavior. Such frameworks not only classified personality traits into different abstraction levels but they also applied them in a variety of consumption-related domains. Consequently, more recent studies investigating the relationship between personality traits and users' perceptions, attitudes and behavior in various consumption domains found significant results (e.g., Dabholkar & Bagozzi, 2002; Fraj & Martinez, 2006; Im, Mason, & Houston, 2007; Mooradian, 1996; Mooradian & Olver, 1997), especially in the computer mediated environments (e.g., Donthu & Garcia, 1999; Kwak, Fox, & Zinkhan, 2002; LaRose & Eastin, 2002; Jahng, Jain, & Ramamurthy, 2002; Ross et al., 2009). Findings of these articles provide support for the notion that consumers are dispositional entities, and more importantly that there exists a multitude of domain-specific traits relevant to consumer behavior.

Personality Traits in Mobile Context: The-State-of-the-Art

The mobile medium should not be conceived as an extension to the conventional wired Internet because it has a unique essence of its own. Through distinctive features of mobile devices, the mobile medium not only extends the benefits of the Internet but also presents an original set of value propositions (Clarke, 2001). The use of mobile technologies has relaxed the independent and mutual constraints of space and time, which are among the most valuable resources for consumers. Many activities became completely spatially and temporally flexible in a world with mobile technologies (Balasubramanian, Peterson, & Jarvenpaa, 2002). Further, sensing the location of consumers has never been possible before. Prior literature unanimously agrees on the

significance of the impact of mobile technology on the universe of marketing (Balasubramanian, Peterson, & Jarvenpaa, 2002; Barnes, 2002; Mort & Drennan, 2002; Shugan, 2004; Watson et al., 2002). Therefore, academic research should not entirely rely on models that were developed and validated in the realm of wired Internet to explain and predict consumer behavior in the mobile context. Those models will need to be augmented for sufficient explanatory power, and this research strives to draw attention to the conspicuous lack of academic interest on the effect of personality traits in the domain of mobile marketing, and aims provide the necessary guidance in this respect.

In order to assess the level of researcher attention on personality traits within the mobile context a comprehensive review of mobile marketing research stream was conducted by searching the following online databases: ABI/INFORMS, EBSCOhost, Emerald, IEEE Xplore, Science Direct, and Wiley InterScience. Following the methodology employed by Varnali and Toker (2010) the literature search was limited to peer-reviewed journals and was based on keywords: "mobile marketing", "mobile commerce", "mobile advertising", "mobile consumer", "mobile business" and "mobile services". The resulting list of articles (3116 articles were identified, most of which were not actually related with mobile marketing) was subjected to further filtering in order to identify those articles that focused on consumer side of mobile marketing and include at least one personality trait within their empirical research models. It is important to note that, since the purpose of the review is to identify studies that examine the impact of personality traits in the domain of mobile marketing, articles that focused on adoption of mobile technology were excluded. Adoption of mobile phones is related with the acceptance of the underlying technology of the mobile marketing phenomena, and hence is outside the scope of this research. Table 1 presents the findings related with the personality traits that were subjected to empirical investigation within the mobile context, together with their definitions and reference articles in which their original scales can be found.

The review showed that only 15 personality traits have been subjected to empirical testing within the mobile context. Among those, innovativeness and self efficacy took most of the attention and have appeared in nine and five of a total of 23 articles, respectively. The remaining 13 have been included in research models only once or twice. Therefore, the existing evidence regarding the role of those personality traits in driving behavior in the mobile context is insufficient. It is surprising that nearly a decade has passed since the first article focusing on the adoption of mobile marketing has appeared in 2000, and still researchers have not yet expanded the scopes of their personality-related construct batteries. The ones who have investigated the role of personality traits within the mobile context found that they either influence relationships as moderators, or exert direct influence on user perceptions (see Table 1 for findings).

Given the fact that perceived consumer control (e.g., prior permission, controlling frequency of message delivery, ease of opt-out) is regarded as one of the most important success factors of mobile marketing (Barwise & Strong, 2002; Carroll et al., 2007), it is surprising that any construct related with "sense-of-control" (Rodin, 1990) has not yet been included in empirical studies aiming to explain consumer behavior in the mobile context. Several studies have assessed the impact of perceived behavioral control on the intention to adopt mobile services, either directly or indirectly through perceived ease of use (e.g., Bhatti, 2007; Nysveen, Pedersen, & Thorbjornsen, 2005). However, the potential moderating role of personal sense-of-control on the strength of these relationships has not yet been subjected to empirical testing.

Prior research in computer mediated communications has shown that cognitive constructs are important drivers of user perceptions towards ICT-based services. For instance, need for cognition, which is defined as the tendency to engage in and enjoy effortful cognitive endeavors (Cacioppo & Petty, 1982), is found to be associated with perceptions regarding informational characteristics of a website (Kaynar & Amichai-Hamburger, 2008). Similarly, thinking style, which is defined as the self-consistent mode of functioning that individuals show in their perceptual and intellectual activities (Simon, 1960), is found to be an important predictor of online search intentions (Kao, Lei, & Sun, 2008). Since the mobile medium, at the very basic level, is a new mode of computer mediated communication, the lack of interest on cognitive personality constructs in mobile marketing research, especially within the context of the mobile Internet, deserves critical attention.

CHOOSING RELEVANT PERSONALITY TRAITS

Given the fact that psychology literature offers hundreds of personality traits (Goldberg, 1990), of which many have not even been investigated for their potential effects in any consumption context, choosing appropriate personality traits has always been a challenging task. There exists many personality-based constructs in different abstraction levels (McCrae and Costa, 2003; Mowen, 2000). Some of them share several underlying dimensions (e.g., innovativeness of Midgley and Dowling, 1978, and novelty seeking of Hirschman, 1980). Some seem obviously relevant to the mobile environment, while others require a closer look. These facts bring forth the question: How can researchers identify and select the most relevant personality traits to include in their prospective models aiming to explain behavior in the mobile context? Although it is practically impossible to compile a complete list of personality traits that may be relevant to mobile context, the follow-

Table 1. Personality traits and consumer behavior in the mobile context

Personality Trait	Definition	Findings
Agreeableness	is one of the five arch-personality dimensions, which is characterized by traits such as straightforwardness, altruism, compliance, tender mindedness, and modesty (McCrae & Costa, 2003). Agreeable people possess the types of social graces that make their company desirable (Ross et al., 2009).	Disagreeable people are more likely to report receiving more calls, to report that incoming calls are unwanted and they spend more time on their phone as a display (Butt & Phillips, 2008).
Arousal seeking	is the inherent need to search for higher environmental stimulation. Individuals with higher AS have higher optimum stimulation levels (OSL) and constantly seek new ideas and experiences (Mehrabian & Russell, 1974). It highly overlaps with novelty seeking (Hirschman, 1980)	Novelty seeking positively influences the desire to engage in mobile WOM (Okazaki, 2009).
Cognitive style	is "the characteristic, self-consistent mode of functioning that individuals show in their perceptions and intellectual activities" (Simon, 1960, p. 72). It represents the inherent differences of people in terms of how they acquire and process information while engaged in decision-making or problem solving. These differences can be categorized into certain patterns, such as verbalizers vs. visualizers, wholists vs. analysts (Riding, 2001), active vs. reflective, sensing vs. intuitive, global vs. sequential (Felder & Silverman, 1988), and analytic vs. heuristic processors (Jahng, Jain, & Ramamurthy, 2002).	The frame rate at which multimedia content is displayed influences the levels of information assimilated by visualizers (not verbalizers), and they enjoy presentations in full 24-bit color (Ghinea & Chen, 2008). Consumers who are high on visual orientation find it easier to use handheld devices to access the mobile Internet (Bruner & Kumar, 2005).
Extraversion	is one of the five arch-personality dimensions, which is characterized by venturesomeness, gregariousness, courage, talkativeness, affiliation, positive affectivity, spontaneity, energy, ascendance, and ambition (McCrae & Costa, 2003). Extraverts are predisposed toward positive affect and prefer interpersonal interaction (Mooradian & Olver, 1997).	Extraverts are more likely to spend time calling, and changing ring tone and wallpaper, and less likely to value incoming calls (Butt & Phillips, 2008). Perceptions of excessive SMS use are associated with extroversion (Igarashi et al., 2008).
Neuroticism	is one of the five arch-personality dimensions, which is characterized by traits such as anxiety, angry hostility, depression, vulnerability, self-consciousness, and impulsiveness (McCrae & Costa, 2003).	SMS is more likely to be used by neurotic individuals (Butt & Phillips, 2008). No significant relationship is found between neuroticism and mobile phone use (Bianchi & Phillips, 2005). Neuroticism influences emotional reactions to SMS use and the perceived ability to maintain relationships without SMS (Igarashi et al., 2008).
Innovativeness	is the degree to which an individual is receptive to new ideas and makes innovation decisions independently of the communicated experience of others (Midgley & Dowling, 1978). It drives the desire to engage in new experiences to stimulate one's mind, and covers a multitude of sub dimensions such as venturesomeness, novelty seeking, cosmopolitanism, variety seeking, and information seeking.	Perceived ease of use of 3G mobile value-added services is positively influenced by innovativeness, which is not influential on perceived usefulness (Kuo & Yen, 2009). Innovativeness has a positive direct effect on attitudes toward mobile commerce (Yang, 2007). Innovativeness positively influences perceived ease of use and usefulness of M-commerce (Yang, 2005). Innovativeness has a positive influence on OSL, which in turn has a positive influence on SMS use and m-commerce intention (Mahatanankoon, 2007). Innovativeness has a significant effect on cluster membership with respect to consumers' adoption of 3G wireless services (Kleijnen et al., 2004). Innovativeness is a significant predictor of mobile marketing acceptance in USA but not in Pakistan (Sultan & Rohm, 2008). Innovativeness positively influences knowledge about mobile communications (Bauer et al., 2005). No significant relationship is found between innovativeness and perceived ease of use or perceived usefulness of mobile commerce (Bhatti, 2007). Innovativeness increases intention to use all types of mobile services (Mort & Drennan, 2005).

continued on following page

Table 1. Continued

Personality Trait	Definition	Findings
Impulsiveness	is the tendency to buy things immediately, spontaneously, and unreflectively. It is related to need for stimulation in purchase situations (Menon & Kahn, 1995; Rook & Fisher, 1995).	General impulse buying tendency has been found to influence impulsive use of SMS service (Davis & Sajtos, 2009).
Opinion leadership	is not a fundamental personality trait; rather it is a label for those who tend to give advice to others in terms of product or service purchase decisions. However, it has been found to be a relatively stable personal attribute, and has psychological dimensions, so is usually regarded as an individual-based variable similar to personality traits. Characteristics of an opinion leader include creativity, curiosity, innovativeness, extraversion, inner-directedness, and tendency to perceive less risk in consumption situations (Childers, 1986; Robertson & Myers, 1969).	Opinion leadership positively influences group-level social intention to engage in mWOM (Okazaki, 2009). Early adopters of mobile TV perceive the advantage of early adoption as being confirmed as an opinion leader (Marez et al., 2007).
Optimum stimulation level	is a trait that characterizes an individual in terms of his general response to environmental stimuli. High-OSL individuals require higher environmental stimulation, while low-OSL individuals feel more comfortable with familiar situations and stimuli, and withdraw from new or unusual ones (Raju, 1980).	OSL is associated with SMS use and m-commerce intention (Mahatanankoon, 2007).
Personal attachment to the mobile device	Attachment has been defined as "the extent to which an object...is used by the individual to maintain his or her self-concept" (Ball & Tasaki, 1992, p. 158). Therefore, attachment to a mobile device is the extent to which the mobile phone represents an integral part of a person's self-concept.	Personal attachment is not a significant driver of mobile marketing acceptance (Sultan & Rohm, 2008).
Playfulness	is a situation specific trait that represents a type of intellectual or cognitive tendency to interact spontaneously, inventively, and imaginatively with an object, which is microcomputers in the domain of mobile handsets (Webster & Martocchio, 1992). Playfulness is found to be a stable trait that is unlikely to change over time which significantly contributes to long-term information systems usage (Yager et al., 1997).	Playfulness has a positive influence on OSL, which in turn has a positive influence on SMS use and m-commerce intention (Mahatanankoon, 2007).
Predisposition to trust	is a general inclination to display faith in humanity and to adopt a trusting stance toward others. It is the tendency to believe that better results will be obtained by giving people credit and trusting them, regardless of whether this trust is justified (McKnight, Cummings, & Chervany, 1998).	It positively predicts trust in SMS advertising, which in turn influences the perceived usefulness of and the intention to accept SMS advertising (Zhang & Mao, 2008). It has a significant effect on initial trust in mobile banking, which in turn significantly promotes intention to use related services (Kim, Shin, & Lee, 2009).
Susceptibility to social influence	is the tendency to submit to forces within the social environment when making purchase decisions. It has two dimensions: the informational dimension measures the tendency to obtain information about products or services by observing or directly seeking information from other people. The normative dimension measures an individual's need to use purchases to identify with, or enhance, his or her image in the eyes of significant others and a willingness to conform to the expectations of others in making purchase decisions (Bearden et al., 1989; Silvera, Lavack, & Kropp, 2008).	Susceptibility to social influence is influential on the types of mobile services being used (Mort & Drennan, 2005).

continued on following page

Table 1. Continued

Personality Trait	Definition	Findings
Self-efficacy	is the belief in one's own capability to mobilize the motivation, cognitive resources, and courses of action needed to meet given situational demands (Wood & Bandura, 1989). A measure of self-efficacy with respect to the use of new technologies (Technology Readiness) is developed by Parasuraman (2000) and with respect to the use of computers is developed by Compeau and Higgins (1995).	Self-efficacy amplifies the impact of perceived ease of use on attitude towards mobile payment (Shin, 2009). Self-efficacy has a significant effect on perceived ease of use, which in turn positively influences perceived usefulness, perceived credibility and behavioral intention to use mobile banking (Luarn & Lin, 2005). Mobile Technology Readiness (1) strengthens the impact of perceived usefulness and perceived cost on attitude, (2) attenuates the impact of perceived system quality on attitude, and (3) attenuates the impact of attitude on the intention to use wireless finance (Kleijnen et al., 2004). Self-efficacy has a positive influence on the perceived ease of use of SMS ads (Zhang & Mao, 2008).
Time-consciousness	People differ in terms of their temporal profiles based on the extent to which they are aware of passing time, their need to set and meet deadlines, their tendency to plan time, and the extent to which they engage in several activities simultaneously (Kaufman et al., 1991). Time-consciousness reflects how people innately track and account for time and vary in their sensitivity toward time-critical issues.	The relationship between the five predictor variables (time convenience, user control, risk, cognitive effort) and perceived value of mobile services is moderated by time consciousness (Kleijnen et al., 2007).

ing conceptual discussion provides an organized framework that may be useful to address this issue.

Due to the inherent characteristics of the mobile medium, a comprehensive understanding of consumer behavior in the mobile context requires an integrative approach that views the end-user not only as a technology user but also as a service consumer and a network member (Pedersen et al., 2002). Therefore, when searching for personality traits that may exert an influence on the value creation processes of mobile consumers, it is reasonable to give priority to those that were previously found to be relevant in the domains of technology adoption, service consumption, and networking/ communications. Although this outlook should greatly narrow the domain of the literature to be scanned for, still many traits are likely to be found. Therefore, an additional step is required to further narrow the domain of the literature to be scanned. Since a different set of personality-traits become salient in different usage scenarios, establishing "usage specificity" may serve well in this respect. Since, the ultimate driver of acceptance and classification of all mar-

ket offerings in the eyes of consumers is "value" (Zeithaml, 1988), a taxonomy of mobile services and applications based on value propositions shall be useful both in identifying the relevant individual differences in tendencies to act for specific mobile usage situations, and in presenting the conceptual discussion in an organized account.

In order to create such a taxonomy of mobile services and applications an industry specialist, a frequent user of mobile services, and an independent scholar with prior experience in mobile marketing research were invited to participate in a micro-Delphi panel. Delphi method is an iterative process used for structuring a group communication process to facilitate group problem solving and to structure models using a series of questionnaires interspersed with feedback (Linstone & Turloff, 1975; Rowe & Wright, 1999). When the technique is adapted for use in face-to-face meetings, then it is called micro-Delphi. Three participants were chosen on the basis of their expertise on the subject and they represented different facets of the examined topic. The industry specialist was an experienced mobile marketer who have designed and managed many mobile marketing campaigns that involve various types of mobile applications. The invited scholar has been working on mobile related subjects for more than 4 years and had published several articles on the topic. The other member of the micro-delphi panel was selected among a number of candidate graduate students based on their self-reports of experience with the mobile medium. According to Adler and Ziglio (1996), Delphi participants should meet four "expertise" requirements:

1. Knowledge and experience with the issues under investigation;
2. Capacity and willingness to participate;
3. Sufficient time to participate; and,
4. Effective communication skills.

Participants met all these requirements. The researcher acted as the opinion harvesting facilitator. In the first round, all participants were directly interviewed by the researcher. They provided answers to the following questions:

1. Why do people use a mobile service/application?
2. To your knowledge, what types of mobile services/applications exist?
3. What would you expect from each of those applications/services so that they will satisfy you?

Then, the researcher produced the first synthesis and re-interviewed the participants, probing them about the discrepancies between their answers and others'. At the end of the second round, consensus was achieved and four main value propositions of mobile services and applications were identified which constituted taxonomy. In the third round, a semi-structured focus group was conducted with the micro-Delphi participants and five additional graduate students who were heavy users of mobile applications and services. This session served as a verifica-

tion stage in which the researcher provided the group with a variety of mobile applications/services and asked them to analyze their value propositions using the established taxonomy. All mobile services and applications known to participants have been successfully classified under four categories as follows:

1. **Incentive-Based Participatory Campaigns:** These mobile marketing campaigns require users to take a solicited action, such as sending a premium SMS/MMS, making a voice call, interacting with an automated system, or generating and submitting various types of digital content.
2. **Community-Based Applications:** These mobile applications allow users to interact with each other and establish social relationships (e.g., chat, e-mail, SMS messages, instant messages, social networking sites, and various types of widgets that allow sharing social media content).
3. **Commerce-Based Applications:** These mobile applications allow users to engage in commercial transactions on the move (e.g., mobile shopping, mobile banking, mobile payment, and e-ticketing).
4. **Content-Based Applications:** These mobile applications provide information or entertainment (e.g., downloads, games, mobile TV, news, weather updates, services supporting users in the search for online resources, and other time-sensitive and location-based services).

Several applications involve components that reside in multiple categories of this taxonomy. For instance, a location specific multiplayer game may involve transactions among individuals. Similarly, an incentive-based participatory campaign may involve delivery of digital content to campaign participants. It is important to note that the purpose of this taxonomy is not creating mutually exclusive categories. These categories represent main value drivers of mobile applications and services; and the purpose of this taxonomy is to facilitate the search for relevant personality traits when examining the adoption of specific types of mobile applications and services. Several applications may share a mixture of value drivers, and hence the list of potentially relevant personality traits for such applications would be longer than others.

Incentive-Based Participatory Campaigns

Most of the mobile marketing campaigns take advantage of the interactivity provided by the mobile medium via WAP menus, SMS/MMS participation themes, interactive voice response systems, location-based coupons, user-generated contents, voting, and mobile internet sites. These campaigns typically call users to take a particular action. Although market experience suggests that campaigns that involve an interactive component are more successful in eliciting user response, the question "who participates in such campaigns and why?" has not yet been investigated. Intrinsic factors may be very instrumental in influencing perceptions about the degree to which participation in a mobile marketing campaign is likely to produce desirable outcomes. In other words, the decision to participate (or not) may depend on the degree to which the individual believes he or she can mobilize the motivation, cognitive resources, and courses of action needed to meet given situational demands to produce desirable outcomes through the process. On this basis, personality traits such as self-efficacy (Wood & Bandura, 1989), sense of control (Rodin, 1990; Skinner, 1996), optimism (Scheier & Carver, 1985), Machiavellianism (Hunt & Chonko, 1984), self-confidence (Bearden et al., 2001), and risk-aversiveness (Mitchell & Boustani, 1993), which are all related to one's perception of outcome success, become utterly important for explaining and predicting attitudinal and behavioral differences with respect to interactive mobile campaigns. Also

personality traits such as playfulness (Webster & Martocchio, 1992), arousal seeking (Mehrabian & Russell, 1974), optimum stimulation level (Raju, 1980), inherent innovativeness (Midgley & Dowling, 1978), openness to experience (McCrae & Costa, 2003), and novelty seeking (Hirschman, 1980), which are all related to the engagement process itself, instead of the end-states, may be influential on the degree of hedonic pleasure experienced through participation in interactive mobile campaigns.

Community-Based Applications

Community-based applications allow users to interact with each other and establish social relationships by facilitating social sharing and interaction through the mobile medium. Together with the technological proliferation of the mobile handsets and the underlying infrastructure, interactions within mobile social networks started to evolve from one-to-one messaging towards sophisticated interactions of virtual communities. In most of the cases the value of a social content is directly linked to its timeliness (e.g., a gossip, a critic about a movie, a status update, a hilarious picture taken by a built-in camera). Therefore, the immediacy provided by the mobile handset makes it a perfect tool for sharing social content, especially for the youth. Social sharing and interaction is further facilitated by mobile widgets of social networking giants such as Facebook, Twitter, and MySpace, optimized to work best on devices with small screens and limited usability. These types of applications offer an unprecedented opportunity for marketers to create a strong viral effect. The foremost advantage of the viral effect is that friend-to-friend referrals multiply the reach of the campaign exponentially which significantly increases the campaign's exposure at almost no additional cost for the marketer. Furthermore, it has been found that people prefer to receive mobile ads from another person rather than a company, they are more likely to perceive mobile ads positively if it came from another person than a company, and the risk of damaging brand image is attenuated if mobile ads comes from another person within one's community instead of a company (Wais & Clemons, 2008). Market experience has established best practices to facilitate viral effect such as identifying a topic that is interesting for members of a target community and relating the campaign to it and providing social value in sharing the message. However, questions like "Who will be more willing to forward the mobile digital content that is created to generate a viral effect?", "Who initiates discussions and chat sessions in target mobile social communities?", and "Who gets intrinsic satisfaction from interaction and socialization through mobile platforms?" are still unanswered and offer fruitful research avenues. Individuals to whom such digital content will be delivered should have a positive tendency toward communicating with others, socialization and forwarding. On this basis, relevant personality traits may be social character (Kassarjian, 1965; Riesman, 1950), extraversion (McCrae & Costa, 2003), opinion leadership (Childers, 1986), altruism (Feick, Guskey, & Price, 1995), assertiveness (Richins, 1983), market mavenism (Feick & Price, 1987), susceptibility to social influence (Bearden, Netemeyer, & Teel, 1989), and self-consciousness (Fenigstein et al., 1975). Constructs that are related with an individual's perceived ability to perform the required courses of action to socialize and forward digital content through mobile devices, such as self-efficacy (Wood & Bandura, 1989) and self-confidence (Bearden et al., 2001) may also be instrumental in this respect. Additionally, openness to experience (McCrae & Costa, 2003) may also be relevant in this context, because it may be associated with trying out new methods of communication.

Commerce-Based Applications

Mobile commerce, which is defined as "any transaction with a monetary value – either direct or indirect – that is conducted over a wireless telecommunication network" (Barnes, 2002, p.92), allows people to shop on the move. Due to the recent developments in graphics and input capabilities of smart handsets (e.g., iPhone, Blackberry, PDAs), increased connection speeds, and falling data prices, the number of mobile internet users and their frequency of use have increased, which increased the retailing potential of the mobile medium. Prior TAM and TRA based empirical works have identified that perceptions regarding the usefulness, ease of use, emotional appeal, aesthetics, and enjoyment of web-based marketing tools (e.g., banners, websites, pop-up ads) are associated with success in e-commerce (e.g., Hasan & Mesbah, 2007). Personal characteristics, especially cognitive differences (Kaynar & Amichai-Hamburger, 2008), of users have been found to influence these perceptions in the context of online shopping (Lian & Lin, 2008). Similarly, the valence and magnitude of such perceptions in the context of mobile shopping may also be driven by cognitive personality traits, such as need for cognition (Cacioppo & Petty, 1982), cognitive style (Cox, 1967; Simon, 1960), time-consciousness (Kaufman, Lane, & Lindquist, 1991), tolerance for ambiguity (Budner, 1962), and impulsiveness (Menon & Kahn, 1995; Rook & Fisher, 1995). Also, it has been found that, self-efficacy (Wood & Bandura, 1989), novelty seeking (Hirschman, 1980), need for interaction (Dabholkar, 1996), and self-consciousness (Fenigstein, Scheier, & Buss, 1975) are significant moderators of attitude toward self service technologies (Dabholkar & Baggozzi, 2002). Since mobile commerce is ultimately a self-service technology, these constructs may have an influence on its adoption process.

All commerce-based applications involve virtual transactions at some point. Prior research suggests that users are generally sensitive with regard to services that involve virtual monetary transactions, in which case they perceive a heightened risk for both loss of money and disclosure of personal financial information (Hourahine &

Howard, 2004). Perceived risk is considered as a cost weighed against benefits in value perceptions (Sweeney, Soutar, & Johnson, 1999). Users may feel especially vulnerable to risks in the mobile environment due to the very personal nature of their handheld devices. Therefore, managing risk perceptions regarding the commitment of a virtual service provider in protecting user information from unauthorized use, alteration, disclosure, distribution or access is an important task for mobile marketers. On this basis, personality traits such as risk-aversiveness/tolerance (Mitchell & Boustani, 1993), concern for privacy (Malhotra, Kim, & Agarwal, 2004), optimism/pessimism (Scheier & Carver, 1985), predisposition to trust (McKnight, Cummings, & Chervany, 1998), technology anxiety (Igbaria & Parasuraman, 1989), and computer self-efficacy (Compeau & Higgins, 1995) may be utterly important in understanding consumer behavior involving any kind of virtual transactions. Since self-confidence is related to anxiety and risk reducing strategies (Locander & Hermann, 1979), it is another essential personality trait to consider when examining consumers' handling of risk.

Content-Based Mobile Applications

Content-based mobile applications involve delivery of various types of mobile content, such as branded multimedia, file downloads, games, sponsored information services (e.g., news, traffic, weather or stock updates), services supporting users in the search for online resources, and location-based services. There are two ways of delivering content to mobile handsets: push-type and pull-type (Karjaluoto et al., 2006). In pull-type content delivery the communication is user-initiated; therefore the process through which a user interacts with the digital content is identical to that of commerce-based applications. Therefore, the set of personality traits that are influential in driving user adoption and engagement processes in commerce-based applications also apply in the

context of pull-type content delivery. Push-type delivery, on the other hand, is highly related with the concept of intrusiveness, which refers to the feelings of resentment and irritation as a result of unexpected exposure to advertisements (Godin, 1999). Since intrusiveness is largely related to the utility of the interruption (Li, Edwards, & Lee, 2002), personality traits such as concern for privacy (Malhotra et al., 2004) and personal attachment (Ball & Tasaki, 1992) to the mobile device may be highly influential on consumer perceptions regarding push-type content-based mobile applications.

Since all personality traits identified in this research do not belong to the same abstraction level, some may share several underlying dimensions. For instance, altruism is a sub-dimension of agreeableness, whereas innovativeness and arousal seeking are closely related with each other, and are both characteristics of extraverts (McCrae & Costa, 2003). Similarly, characteristics of an opinion leader include innovativeness, extraversion, inner-directedness, and tendency to perceive less risk in consumption situations (Childers, 1986; Robertson & Myers, 1969). However, none of these constructs adequately capture another's conceptual domain in all mobile usage scenarios. Each may provide a valuable piece of information that may be required in specific types of research questions regarding different aspects of the mobile phenomena.

Implications for Marketers

The fact that a mobile handset is typically used by a sole individual makes the mobile medium the ultimate channel of one-to-one marketing. The practice of one-to-one marketing is driven by the concept of personalization. Personalization is about increasing message relevance, which is identified as one of the most important predictors of consumer attitude toward mobile advertising, and as instrumental in minimization of the intrusiveness of mobile marketing messages (Rettie,

Grandcolas, & Deakins, 2005). In light of the discussion presented throughout the article about how personality traits influence consumer perceptions, it would be reasonable to argue that inclusion of personality traits in user profiling would result in more effective targeting and personalization.

Mobile operators are able to build very rich customer databases with information that would be very difficult to collect via other media. They have easy access to demographics, and can track and record consumers' behavioral patterns, even their locations. This information can collectively be used to derive behavioral profiles of users and group them into segments with similar interests, usage patterns, and personality traits. For instance, an innovative data-mining algorithm may use prior response data, in terms of responsiveness to different types of message content, in order to classify consumers according to their mental processing styles as verbalizers and visualizers. Verbalizers will be those who are more responsive to text-based messages, whereas, visualizers will be those who are more responsive to graphics-based messages. Such a segmentation method will be instrumental in designing messages that will be delivered to target segments. Visualizers will ignore long, complex and unfamiliar texts, whereas verbalizers will find little value in visual content. Other possible ways to derive personality traits of mobile consumers may include identification of 1) social characters of users by monitoring their communication patterns, 2) self-concept, susceptibility to social influence and opinion leadership by monitoring users' activities on mobile social networking platforms, 3) playfulness and technology readiness by monitoring users' interaction with digital content, or 4) concern for privacy by tracking users' responses to different types of mobile marketing campaigns that involve different levels of intrusiveness. Identification of such core personality-based differences will increase marketers' ability to predict how targeted users are likely to react when they are interacting with certain types of mobile services, and when they are

exposed to particular types of mobile ads. While traits cannot be easily changed, various aspects of mobile services and marketing practices that interact with the individual traits can be altered. Hence, incorporation of personality traits in targeting and personalization schemes shall produce improved results in mobile marketing campaigns.

If extensive personalization is at one end of a continuum, the other end would be the absolute inverse strategy, which is designing the campaign components in a way that would minimize the impact of interpersonal differences on campaign outcomes. Instead of accounting for interpersonal differences, mobile marketers may find it more practical to incorporate extrinsic motivators within the message content, such as monetary incentives or social benefits. For instance, if a mobile marketer can find a way to deliver a message via a target user's social network, intentions to open, read, respond and in some cases forward would be greatly increased, even for those who score low on predisposition to trust, risk tolerance, and concern for privacy. Similarly, instant-win themes which reward each participant with pre-paid minutes or discount coupons may also have such an effect. Such incentives may lower the perceived expensiveness and hence increase the willingness to accept and engage in mobile marketing practices. Although minimized in such cases, the effect of interpersonal differences based on core predispositions of individuals may still exert moderating impact on the success of mobile marketing campaigns and may be influential in the adoption of mobile services.

CONCLUSION

The aim of this research is to provide guidance as to why and how to incorporate personality-based variables within prospective research models attempting to explain and predict consumer behavior in the mobile context. In the first part of this article, the importance of including personality traits within prospective research models aiming to explain and predict acceptance and adoption of mobile marketing is underpinned. The second part of this article provided a critical assessment of the state-of-the-art, established a taxonomy of mobile applications and services in order to facilitate the search for context-specific personality traits, and identified a variety of personality traits that are relevant to several specific aspects of mobile applications and services in order to provide guidance for future research.

Relying on literature-based findings, it would be reasonable to encourage researchers to design their theoretical models with using situational variables, in other words success factors of mobile marketing (e.g., design of the interface, timing of delivery, prior permission, communication source, media costs, etc.) as predictors of users' perceptions and responses, and personality traits as intervening variables of the relationships between these predictors and the attitudinal/behavioral outcomes.

It has been commonly acknowledged that the technological proliferation of the mobile medium gave birth to a new type of consumer. Wind and Mahajan (2002) used the metaphor of centaur of the Greek mythology to draw a profile for this new type of consumer. "They are like centaurs, half human and half horse, running with the rapid feet of new technology, yet carrying the same ancient and unpredictable human heart" (Wind & Mahajan, 2002, p. 65). Since marketing is about understanding, communicating with, and delivering value to the consumer constituency, researchers of marketing will have to use all possible ways to increase their knowledge about this new type of consumer. Personality traits constitute an important dimension of the "ancient and unpredictable human heart", but they have been overlooked in mobile marketing research stream. It is hoped that this study would stimulate further research in this domain.

REFERENCES

Adler, M., & Ziglio, E. (1996). *Gazing into the oracle: The Delphi Method and its application to social policy and public health.* London, UK: Jessica Kingsley Publishers.

Balasubramanian, S., Peterson, R. A., & Jarvenpaa, S. L. (2002). Exploring the implications of m-commerce for markets and marketing. *Journal of the Academy of Marketing Science, 30*, 348–361. doi:10.1177/009207002236910

Ball, A. D., & Tasaki, L. (1992). The role and measurement of attachment in consumer behavior. *Journal of Consumer Psychology, 1*, 155–172. doi:10.1207/s15327663jcp0102_04

Barnes, S. J. (2002). The mobile commerce value chain: Analysis and future developments. *International Journal of Information Management, 22*, 91–108. doi:10.1016/S0268-4012(01)00047-0

Barnes, S. J., & Scornavacca, E. (2004). Mobile marketing: The role of permission and acceptance. *International Journal of Mobile Communications, 2*, 128–139. doi:10.1504/IJMC.2004.004663

Barwise, P., & Strong, C. (2002). Permission-based mobile advertising. *Journal of Interactive Marketing, 16*, 14–24. doi:10.1002/dir.10000

Bauer, H. H., Reichardt, T., Barnes, S. J., & Neumann, M. M. (2005). Driving consumer acceptance of mobile marketing: A theoretical framework and empirical study. *Journal of Electronic Commerce Research, 6*, 181–192.

Baumgartner, H. (2002). Toward a personology of the consumer. *The Journal of Consumer Research, 29*, 286–292. doi:10.1086/341578

Bearden, W. O., Hardesty, D. M., & Rose, R. L. (2001). Consumer self-confidence: Refinements in conceptualization and measurement. *The Journal of Consumer Research, 28*, 121–133. doi:10.1086/321951

Bearden, W. O., Netemeyer, R. G., & Teel, J. E. (1989). Measurement of consumer susceptibility to interpersonal influence. *The Journal of Consumer Research*, *15*, 473–479. doi:10.1086/209186

Bhatti, T. (2007). Exploring factors influencing the adoption of mobile commerce. *Journal of Internet Banking and Commerce*, *12*, 2–13.

Bianchi, A., & Phillips, J. G. (2005). Psychological predictors of problem mobile phone use. *Cyberpsychology & Behavior*, *8*, 39–51. doi:10.1089/cpb.2005.8.39

Bruner, G. C., & Kumar, A. (2005). Explaining consumer acceptance of handheld Internet devices. *Journal of Business Research*, *58*, 553–558. doi:10.1016/j.jbusres.2003.08.002

Budner, S. (1962). Intolerance of ambiguity as a personality variable. *Journal of Personality*, *30*, 29–50. doi:10.1111/j.1467-6494.1962.tb02303.x

Butt, S., & Phillips, J. G. (2008). Personality and self reported mobile phone use. *Computers in Human Behavior*, *24*, 346–360. doi:10.1016/j.chb.2007.01.019

Cacioppo, J. T., & Petty, R. E. (1982). The need for cognition. *Journal of Personality and Social Psychology*, *42*, 116–131. doi:10.1037/0022-3514.42.1.116

Carroll, A., Barnes, S. J., Scornavacca, E., & Fletcher, K. (2007). Consumer perceptions and attitudes towards SMS advertising: recent evidence from New Zealand. *International Journal of Advertising*, *26*, 79–98.

Chae, M., Kim, J., Kim, H., & Ryu, H. (2002). Information quality for mobile internet services: a theoretical model with empirical validation. *Electronic Markets*, *12*, 38–46. doi:10.1080/101967802753433254

Chen, L. (2008). A model of consumer acceptance of mobile payment. *International Journal of Mobile Communications*, *6*, 32–52. doi:10.1504/IJMC.2008.015997

Childers, T. L. (1986). Assessment of the psychometric properties of an opinion leadership scale. *JMR, Journal of Marketing Research*, *23*, 184–188. doi:10.2307/3151666

Clarke, I. (2001). Emerging value propositions for m-commerce. *The Journal of Business Strategy*, *18*, 133–149.

Compeau, D. R., & Higgins, C. A. (1995). Computer self-efficacy: Development of a measure and initial test. *Management Information Systems Quarterly*, *19*, 189–211. doi:10.2307/249688

Costa, P. T., & McCrae, R. R. (1990). Primary traits of Eysenck's P-E-N system: Three- and five-factor solutions. *Journal of Personality and Social Psychology*, *69*, 308–317. doi:10.1037/0022-3514.69.2.308

Cox, D. F. (1967). The influence of cognitive needs and styles on information handling in making product evaluations. In Cox, D. F. (Ed.), *Risk taking and information handling in consumer behavior*. Boston, MA: Harvard University Press.

Dabholkar, P. (1996). Consumer evaluations of new technology-based self-service options: An investigation of alternative models of service quality. *International Journal of Research in Marketing*, *13*, 29–51. doi:10.1016/0167-8116(95)00027-5

Dabholkar, P., & Bagozzi, R. P. (2002). An attitudinal model of technology-based self-service: Moderating effects of consumer traits and situational factors. *Journal of the Academy of Marketing Science*, *30*, 184–201.

Davis, R., & Sajtos, L. (2009). Anytime, anywhere: Measuring the ubiquitous consumer's impulse purchase behavior. *International Journal of Mobile Marketing*, *4*, 15–22.

Donthu, N., & Garcia, A. (1999). The Internet shopper. *Journal of Advertising Research, 39,* 52–58.

Feick, L., Guskey, A., & Price, L. (1995). Everyday market helping behavior. *Journal of Public Policy & Marketing, 14,* 255–266.

Feick, L., & Price, L. (1987). The market maven: A diffuser of marketplace information. *Journal of Marketing, 51,* 83–87. doi:10.2307/1251146

Felder, R. M., & Silverman, L. K. (1988). Learning and teaching styles in engineering education. *English Education, 78,* 674–681.

Fenigstein, A., Scheier, M. F., & Buss, A. H. (1975). Public and private self-consciousness: Assessment and theory. *Journal of Consulting and Clinical Psychology, 43,* 522–527. doi:10.1037/h0076760

Fraj, E., & Martinez, E. (2006). Influence of personality on ecological consumer behaviour. *Journal of Consumer Behaviour, 5,* 167–181. doi:10.1002/cb.169

Ghinea, G., & Chen, S. Y. (2008). Measuring quality of perception in distributed multimedia: Verbalizers vs. imagers. *Computers in Human Behavior, 24,* 1317–1329. doi:10.1016/j.chb.2007.07.013

Godin, S. (1999). *Permission marketing: Turning strangers into friends, and friends into customers.* New York, NY: Simon and Schuster.

Goldberg, L. R. (1990). An alternative "description of personality": The big-five factor structure. *Journal of Personality and Social Psychology, 59,* 1216–1229. doi:10.1037/0022-3514.59.6.1216

Hasan, B., & Mesbah, U. A. (2007). Effects of interface style on user perceptions and behavioral intention to use computer systems. *Computers in Human Behavior, 23,* 3025–3037. doi:10.1016/j.chb.2006.08.016

Heinonen, K., & Strandvik, T. (207). Consumer responsiveness to mobile marketing. *International Journal of Mobile Communications, 5,* 603–617. doi:10.1504/IJMC.2007.014177

Hirschman, E. C. (1980). Innovativeness, novelty seeking and consumer creativity. *The Journal of Consumer Research, 7,* 283–295. doi:10.1086/208816

Hoch, S. J., & Deighton, J. (1989). Managing what consumers learn from experience. *Journal of Marketing, 53,* 1–20. doi:10.2307/1251410

Hourahine, B., & Howard, M. (2004). Money on the move: Opportunities for financial service providers in the third space. *Journal of Financial Services Marketing, 9,* 57–67. doi:10.1057/palgrave.fsm.4770141

Hsu, H.-H., & Lu, H.-P. (2008). Multimedia messaging service acceptance of pre- and post- adopters: A sociotechnical perspective. *International Journal of Mobile Communications, 6,* 598–615. doi:10.1504/IJMC.2008.019324

Hunt, S. D., & Chonko, L. B. (1984). Marketing and Machiavellianism. *Journal of Marketing, 48,* 30–42. doi:10.2307/1251327

Igarashi, T., Motoyoshi, T., Takai, J., & Yoshida, T. (2008). No mobile, no life: Self-perception and text-message dependency among Japanese high school students. *Computers in Human Behavior, 24,* 2311–2324. doi:10.1016/j.chb.2007.12.001

Igbaria, M., & Parasuraman, S. (1989). A path analytic study of individual characteristics, computer anxiety and attitudes toward microcomputers. *Journal of Management, 15,* 373–388. doi:10.1177/014920638901500302

Im, S., Mason, C. H., & Houston, M. B. (2007). Does innate consumer innovativeness relate to new product/ service adoption behavior? The intervening role of social learning via vicarious innovativeness. *Journal of the Academy of Marketing Science, 35*, 63–75. doi:10.1007/s11747-006-0007-z

Jahng, J. J., Jain, H., & Ramamurthy, K. (2002). Personality traits and effectiveness of presentation of product information in e-business systems. *European Journal of Information Systems, 11*, 181–195. doi:10.1057/palgrave.ejis.3000431

John, O. P., & Srivastava, S. (1999). The big five trait taxonomy: History, measurement, and theoretical perspectives. In Pervin, L. A., & John, O. P. (Eds.), *Handbook of personality* (pp. 102–138). New York, NY: Guilford Press.

Juniper Research. (2008). *Mobile advertising strategies and forecasts*. Retrieved from http://juniperresearch.com/reports/mobile_advertising

Kao, G. Y., Lei, P., & Sun, C. T. (2008). Thinking style impacts on web search strategies. *Computers in Human Behavior, 24*, 1330–1341. doi:10.1016/j.chb.2007.07.009

Karjaluoto, H., Leppaniemi, M., Standing, C., Kajalo, S., Merisavo, M., Virtanen, V., & Salmenkivi, S. (2006). Individual differences in the use of mobile services among Finnish consumers. *International Journal of Mobile Marketing, 1*, 4–10.

Kassarjian, H. (1965). Social character and differential preference for mass communication. *JMR, Journal of Marketing Research, 2*, 146–154. doi:10.2307/3149978

Kassarjian, H. (1971). Personality and consumer behavior: A review. *JMR, Journal of Marketing Research, 8*, 409–414. doi:10.2307/3150229

Kaufman, C. F., Lane, P. M., & Lindquist, J. D. (1991). Exploring more than 24 hours a day: A preliminary investigation of polychromic time-use. *The Journal of Consumer Research, 18*, 392–401. doi:10.1086/209268

Kaynar, O., & Amichai-Hamburger, Y. (2008). The effects of need for cognition on internet use revisited. *Computers in Human Behavior, 24*, 361–371. doi:10.1016/j.chb.2007.01.033

Kim, G., Shin, B. S., & Lee, H. G. (2009). Understanding dynamics between initial trust and usage intentions of mobile banking. *Information Systems Journal, 19*, 283–311. doi:10.1111/j.1365-2575.2007.00269.x

Kleijnen, M., Ruyter, K., & Wetzels, M. (2004). Consumer adoption of wireless services: Discovering the rules, while playing the game. *Journal of Interactive Marketing, 18*, 51–61. doi:10.1002/dir.20002

Kleijnen, M., Ruyter, K., & Wetzels, M. (2007). An assessment of value creation in mobile service delivery and the moderating role of time consciousness. *Journal of Retailing, 83*, 33–46. doi:10.1016/j.jretai.2006.10.004

Kleijnen, M., Wetzels, M., & Ruyter, K. (2004). Consumer acceptance of wireless finance. *Journal of Financial Services Marketing, 8*, 206–217. doi:10.1057/palgrave.fsm.4770120

Kuo, Y. F., & Yen, S. N. (2009). Towards an understanding of the behavioral intention to use 3G mobile value-added services. *Computers in Human Behavior, 25*, 103–110. doi:10.1016/j.chb.2008.07.007

Kwak, H., Fox, R., & Zinkhan, G. (2002). What products can be successfully promoted and sold via the Internet? *Journal of Advertising Research, 42*, 23–38.

LaRose, R., & Eastin, M. (2002). Is online buying out of control? Electronic commerce and consumer self-regulation. *Journal of Broadcasting & Electronic Media, 46*, 549–564. doi:10.1207/s15506878jobem4604_4

Leppäniemi, M., Sinisalo, J., & Karjaluoto, H. (2006). A review of mobile marketing research. *International Journal of Mobile Marketing, 1*, 30–40.

Li, H. S., Edwards, M., & Lee, J. (2002). Measuring the intrusiveness of advertisements: Scale development and validation. *Journal of Advertising, 31*, 37–47.

Lian, J. W., & Lin, T. M. (2008). Effects of consumer characteristics on their acceptance of online shopping: Comparisons among different product types. *Computers in Human Behavior, 24*, 48–65. doi:10.1016/j.chb.2007.01.002

Linstone, H., & Turloff, M. (1975). *The Delphi method: Techniques and applications*. London, UK: Addison-Wesley.

Locander, W. B., & Hermann, P. W. (1979). The effect of self-confidence and anxiety on information seeking in consumer risk reduction. *JMR, Journal of Marketing Research, 16*, 268–274. doi:10.2307/3150690

Luarn, P., & Lin, H. H. (2005). Toward an understanding of the behavioral intention to use mobile banking. *Computers in Human Behavior, 21*, 873–891. doi:10.1016/j.chb.2004.03.003

Mahatanankoon, P. (2007). The effects of personality traits and optimum stimulation level on text-messaging activities and m-commerce intention. *International Journal of Electronic Commerce, 12*, 7–30. doi:10.2753/JEC1086-4415120101

Malhotra, N. K., Kim, S. S., & Agarwal, J. (2004). Internet users' information privacy concerns (iuipc): The construct, the scale, and a causal model. *Information Systems Research, 15*, 336–355. doi:10.1287/isre.1040.0032

Marez, L., Vyncke, P., Berte, K., Schurman, D., & Moor, K. (2007). Adopter segments, adoption determinants and mobile marketing. *Journal of Targeting. Measurement and Analysis for Marketing, 16*, 78–96. doi:10.1057/palgrave.jt.5750057

McCrae, R. R., & Costa, P. T. (2003). *Personality in adulthood, a five-factor theory perspective*. New York, NY: Guilford Press. doi:10.4324/9780203428412

McKnight, D. H., Cummings, L. L., & Chervany, N. L. (1998). Initial trust formation in new organizational relationships. *Academy of Management Review, 23*, 473–490.

Mehrabian, A., & Russell, J. A. (1974). *An approach to environmental psychology*. Cambridge, MA: MIT Press.

Menon, S., & Kahn, B. (1995). The impact of context on variety seeking in product choices. *The Journal of Consumer Research, 22*, 285–295. doi:10.1086/209450

Midgley, D. F., & Dowling, G. R. (1978). Innovativeness: The concept and its measurement. *The Journal of Consumer Research, 4*, 229–241. doi:10.1086/208701

Mitchell, V., & Boustani, P. (1993). Market development using new products and new customers: A role for perceived risk. *European Journal of Marketing, 27*, 17–32. doi:10.1108/03090569310026385

Mooradian, T. A. (1996). Personality and ad-evoked feelings: The case for extraversion and neuroticism. *Journal of the Academy of Marketing Science, 24*, 99–110. doi:10.1177/0092070396242001

Mooradian, T. A., & Olver, J. M. (1997). I can't get no satisfaction: The impact of personality and emotion on post purchase processes. *Psychology and Marketing, 14*, 379–393. doi:10.1002/(SICI)1520-6793(199707)14:4<379::AID-MAR5>3.0.CO;2-6

Mort, G. S., & Drennan, J. (2005). Marketing m-services: Establishing a usage benefit typology related to mobile user characteristics. *Journal of Database Marketing and Customer Strategy Management, 12*, 327–342. doi:10.1057/palgrave.dbm.3240269

Mowen, J. (2000). *The 3M model of motivation and personality.* Norwell, MA: Kluwer Academic.

Nysveen, H., Pedersen, P. E., & Thorbjørnsen, H. (2005). Intentions to use mobile services: Antecedents and cross-service comparisons. *Journal of the Academy of Marketing Science, 33*, 330–347. doi:10.1177/0092070305276149

Okazaki, S. (2004). How do Japanese consumers perceive wireless ads? A multivariate analysis. *International Journal of Advertising, 23*, 429–454.

Okazaki, S. (2009). Social influence model and electronic word of mouth PC versus mobile internet. *International Journal of Advertising, 28*, 439–472. doi:10.2501/S0265048709200692

Parasuraman, A. (2000). Technology readiness index (tri): A multiple-item scale to measure readiness to embrace new technologies. *Journal of Service Research, 2*, 307–320. doi:10.1177/109467050024001

Park, C. (2006). Hedonic and utilitarian values of mobile internet in Korea. *International Journal of Mobile Communications, 4*, 497–508.

Pervin, L. A. (1994). A critical analysis of current trait theory. *Psychological Inquiry, 5*, 103–113. doi:10.1207/s15327965pli0502_1

Pihlström, M. (2007). Committed to content provider or mobile channel? Determinants of continuous mobile multimedia service use. *Journal of Information Technology Theory and Application, 9*, 1–24.

Raju, P. S. (1980). Optimum stimulation level: Its relationship to personality, demographics and exploratory behavior. *The Journal of Consumer Research, 7*, 272–282. doi:10.1086/208815

Rettie, R., Grandcolas, U., & Deakins, B. (2005). Text message advertising: Response rates and branding effects. *Journal of Targeting. Measurement and Analysis for Marketing, 13*, 304–313. doi:10.1057/palgrave.jt.5740158

Richins, M. (1983). An analysis of consumer interaction styles in the marketplace. *The Journal of Consumer Research, 10*, 73–82. doi:10.1086/208946

Riding, R. (2001). *Cognitive style analysis – research administration.* Birmingham, AL: Learning and Training Technology.

Riesman, D. (1950). *The lonely crowd.* New Haven, CT: Yale University Press.

Roberton, T. S., & Myers, J. H. (1969). Personality correlates of opinion leadership and innovative buying behavior. *JMR, Journal of Marketing Research, 6*, 164–168. doi:10.2307/3149667

Rodin, J. (1990). Control by any other name: Definitions, concepts, and processes. In Rodin, J., Schooler, C., & Schaie, K. W. (Eds.), *Self-directedness: Causes and effects throughout the life course.* Mahwah, NJ: Lawrence Erlbaum.

Rook, D. W., & Fisher, R. J. (1995). Normative influences on impulsive buying behavior. *The Journal of Consumer Research, 22*, 305–313. doi:10.1086/209452

Ross, C., Orr, E. S., Sisic, M., Arseneault, J. M., Simmering, M. G., & Orr, R. R. (2009). Personality and motivations associated with Facebook use. *Computers in Human Behavior*, *25*, 578–586. doi:10.1016/j.chb.2008.12.024

Rowe, G., & Wright, G. (1999). The Delphi technique as a forecasting tool: Issues and analysis. *International Journal of Forecasting*, *15*, 353–375. doi:10.1016/S0169-2070(99)00018-7

Scheier, M. F., & Carver, C. S. (1985). Optimism, coping, and health: Assessment and implications of generalized outcome expectancies. *Health Psychology*, *4*, 219–247. doi:10.1037/0278-6133.4.3.219

Schiffman, L. G., & Kanuk, L. L. (2007). *Consumer behavior*. Upper Saddle River, NJ: Prentice Hall.

Shin, D. H. (2009). Towards an understanding of the consumer acceptance of mobile wallet. *Computers in Human Behavior*, *25*, 1343–1354. doi:10.1016/j.chb.2009.06.001

Shugan, S. M. (2004). The impact of advancing technology on marketing and academic research. *Marketing Science*, *23*, 469–476. doi:10.1287/mksc.1040.0096

Silvera, D. H., Lavack, A. M., & Kropp, F. (2008). Impulse buying: The role of affect, social influence, and subjective wellbeing. *Journal of Consumer Marketing*, *25*, 23–33. doi:10.1108/07363760810845381

Simon, H. A. (1960). *The new science of management decision*. New York, NY: Harper and Row.

Skinner, E. A. (1996). A guide to constructs of control. *Journal of Personality and Social Psychology*, *71*, 549–570. doi:10.1037/0022-3514.71.3.549

Sultan, F., & Rohm, A. J. (2008). How to market to generation m(obile). *MIT Sloan Management Review*, *49*, 35–41.

Sweeney, J. C., Soutar, G. N., & Johnson, L. W. (1999). The role of perceived risk in the quality-value relationship: A study in a retail environment. *Journal of Retailing*, *75*, 77–105. doi:10.1016/S0022-4359(99)80005-0

Varnali, K., & Toker, A. (2010). Mobile marketing: The-state-of-the-art. *International Journal of Information Management*, *30*, 144–151. doi:10.1016/j.ijinfomgt.2009.08.009

Wais, J. S., & Clemons, E. K. (2008). Understanding and implementing mobile social advertising. *International Journal of Mobile Marketing*, *3*, 12–18.

Watson, R. T., Pitt, L. F., Berthon, P., & Zinkhan, G. M. (2002). U-commerce: Expanding the universe of marketing. *Journal of the Academy of Marketing Science*, *30*, 333–348. doi:10.1177/009207002236909

Webster, J., & Martocchio, J. (1992). Microcomputer playfulness: Development of a measure with workplace implication. *Management Information Systems Quarterly*, *16*, 201–225. doi:10.2307/249576

Wind, Y., & Mahajan, V. (2002). Convergence marketing. *Journal of Interactive Marketing*, *16*, 64–79. doi:10.1002/dir.10009

Wood, R., & Bandura, A. (1989). Impact of conceptions of ability on self-regulatory mechanisms and complex decision making. *Journal of Personality and Social Psychology*, *56*, 407–415. doi:10.1037/0022-3514.56.3.407

Yager, S. E., Kappelman, L. A., Maples, G. A., & Prybutok, V. R. (1997). Microcomputer playfulness: Stable or dynamic trait? *The Data Base for Advances in Information Systems*, *28*, 43–52.

Yang, K. C. (2005). Exploring factors affecting the adoption of mobile commerce in Singapore. *Telematics and Informatics*, *22*, 257–277. doi:10.1016/j.tele.2004.11.003

Yang, K. C. (2007). Exploring factors affecting consumer intention to use mobile advertising in Taiwan. *Journal of International Consumer Marketing, 20*, 33–49. doi:10.1300/J046v20n01_04

Zeithaml, V. A. (1988). Consumer perceptions of price, quality, and value: A means-end model and synthesis of evidence. *Journal of Marketing, 52*, 2–22. doi:10.2307/1251446

Zhang, J., & Mao, E. (2008). Understanding the acceptance of mobile SMS advertising among young Chinese consumers. *Psychology and Marketing, 25*, 787–805. doi:10.1002/mar.20239

This work was previously published in the International Journal of E-Services and Mobile Applications, Volume 3, Issue 4, edited by Ada Scupola, pp. 1-20, copyright 2011 by IGI Publishing (an imprint of IGI Global).

Chapter 2
The Role of Consumer Innovativeness and Trust for Adopting Internet Phone Services

JungKun Park
University of Houston, USA

Te-Lin Chung
Purdue University, USA

Won-Moo Hur
Pukyung National University, South Korea

ABSTRACT

The Internet phone, which was recently introduced to the communication market, is characterized by its lower cost and increased compatibility. However, its popularity and diffusion is relatively low in the North American market compared with Asian or European markets. This study examines the adoption of Internet phone service in Korea to develop a better understanding of consumers' acceptance of the Internet phone service application. The Unified Theory of Acceptance and Use of Technology model (UTAUT) is used as theoretical background with two additional constructs (consumer innovativeness and perceived trust) in the proposed model. Using a mail survey with 437 responses collected in Korea, the results indicate that consumers' trust is the major factor affecting adoption of new phone service. The results also indicate that consumers' innovativeness would influence the effects of trust and facilitating conditions toward intention to use. The performance expectancy shows a dominant effect on trust toward phone services. Although effort expectancy and social influence also significantly contribute to consumers' trust, the effects are relatively minor. The effect of performance expectancy would be stronger on consumers with lower income.

DOI: 10.4018/978-1-4666-2654-6.ch002

INTRODUCTION

During the past few years, Internet and Internet Protocol (IP) based applications, such as Voice over Internet Protocol (VOIP), have made the Internet a ubiquitous means of communication (Rao, Angelov, & Nov, 2006). VOIP is the IP-based application that allows telecommunications to take place through the Internet; for example, Skype and Vonage, which are major VOIP service providers in the USA. Compared with the traditional Public Switched Telephone Network (PSTN), VOIP is more affordable for both private consumers and businesses (De Biji & Peitz, 2005). The development of social networking sites and the lowered costs of international calls have made the VOIP service a very important innovation. The VOIP service market is forecasted to grow about ten-fold in terms of revenue from 2004 to 2010, and the Asian market is expected to develop quicker and faster than the North American market (The Yankee Group, 2007).

As for any other new technology application, there are some factors that might affect consumer adoption of VOIP. Based on the empirical data from South African consumers, some factors that affect the adoption of VOIP have already been addressed: high bandwidth costs, confusion in the market, security issues, Quality of Service (QoS) issues, and regulation concerns (Tobin & Bidoli, 2005). Other than competing with other telecommunication services, such as cell phones and PSTN, the VOIP service providers are also facing serious competition among themselves. Based on the different types of providers, there are currently three types of available service: Self-provided, e.g. Skype, with which users can make free calls between PCs; IP telephony, e.g., Vonage, and Yahoo!BB, which allow outgoing and incoming calls through virtual numbers; and Corporate LAN/WAN, which works as a replacement of private branch exchanges (PBX) and allows any incoming and outgoing calls between telephones and personal computers. Due to the competition,

the service providers need to educate consumers about Internet protocol based applications, and also need to provide more valuable services than their competitors. Thus, a systematic examination of the factors that affect promoting the new technology is important for service providers.

Since VOIP works over the Internet, it transmits data with digital packets rather than analog signals as traditional telecommunication does. In order to maintain good quality of service, the Internet that VOIP utilizes should be capable of handling high downloading and uploading speeds. Korea developed an advanced wireless technology, "wireless broadband" (Wibro), which allows up to 18 megabytes per second of download speed (Cherry, 2005). Wibro has the ability to carry video data and voice data, specifically VOIP packets. The popularity of the wireless Internet has made Korea a great model for other countries to study in terms of the adoption process of wireless Internet applications, especially VOIP. Based on a review of the current VOIP industry, the main research question is: What factors would affect consumers' decisions to adopt the Internet phone services? To further understand the previous question, it must be determined if trust plays a role in consumers' decision for adopting a new technology such as VOIP. It is also necessary to discover what situational conditions would affect the effects of those factors.

This study attempts to examine the factors that would affect consumers' adoption of VOIP in a larger consumer context than Korea. To accomplish this goal, the Unified Theory of Acceptance and Use of Technology model (UTAUT) was applied as the conceptual framework (Venkatesh, Morris, Davis, & Davis, 2003). Also, consumers' trust toward VOIP services is included in this study. Trust was considered to have a key role in many economic activities (Fukuyama, 1995) as well as to be a governance mechanism in dependent relationships (Brandach & Eccles, 1989), such as a contract relationship between VOIP service providers and subscribers. Following Hoffman, Novack, and Peralta (1998)'s study of online shopping adoption, trust is considered to be essential; online companies must establish trust-based strategies to build positive relationships with their customers in order to increase their market share and profits. Lack of trust is one of the most frequently cited reasons to explain why consumers do not purchase from online sources (Slyke, Belanger, & Comunale, 2004). Furthermore, the concept of innovativeness has been denoted as an important factor in explaining consumers' acceptance of new products from the aspect of consumers' personal characteristics (Rogers & Shoemaker, 1971). In order to take the individual differences of sensitivity to innovations into consideration, innovativeness was proposed as another moderator besides the demographic characteristics proposed in the original UTAUT model (Venkatesh et al., 2003). This study attempts to illuminate the importance of trust and consumer innovativeness in consumers' acceptance of VOIP technology utilizing modified UTAUT for the Korean market.

CONCEPTUAL DEVELOPMENT AND REVIEW OF LITERATURE

UTAUT and Applications

Studies of technology adoption and acceptance have developed many models including the Theory of Reasoned Action (TRA) (Fishbein & Ajzen, 1975), the Technology Acceptance Model (TAM) (Davis, 1989), the Theory of Planned Behavior (TPB) (Ajzen, 1991), and the Innovation Diffusion Theory (IDT) (Rogers, 1995). Venkatesh et al. (2003) empirically reviewed the above models and identified the constructs with strongest explanatory power. Based on these constructs, the authors proposed an integrated model, UTAUT, which provides a greater explanation power compared with other adoption models.

The UTAUT model suggests that consumers' intention to use information technology is subjected to performance expectancy, effort expectancy, and social influences. Consumers' actual use of the information technology is contingent on their intention to use and their perceived facilitating condition. Explicitly, performance expectancy refers to the degree that the consumers believe that using the system is helpful in achieving their goals. Performance expectancy was considered to have more explanation power over other constructs in previous studies (Venkatesh et al., 2003). Effort expectancy explains the degree of ease that consumers believe is associated with using the technology. Social influence, derived from subjective norms in TPB and TRA, is defined as the degree to which consumers perceived that others expect them to use the new technology. The effect of social influence on consumers' use intention depends on the mandatory or voluntary situations in previous models (Venkatesh et al., 2003). In mandatory settings, social influence is driven by consumers' intention to comply with pressure. On the other hand, in voluntary settings, the social influence is driven by consumers' internalization or identification with other consumers' opinions and expectations of using the technology (Venkatesh & Davis, 2000). Thus, the importance of social influence is subjected to contingent situations. The facilitating condition refers to the degree to which users believe there is some support for using the technology when needed, for example, the resources and the knowledge to use it, and knowing which people to contact for help (Venkatesh et al., 2003). However, facilitating conditions were not expected to influence consumers' intention to use, only to influence their actual use directly.

Due to situational factors that influence the explanation power of the UTAUT model, age, gender, experience, and voluntariness were proposed as moderators in the original UTAUT model (Venkatesh et al., 2003). The original UTAUT study concluded that the effect of performance expectancy on behavioral intention is stronger on younger workers and men. On the other hand, the effect of effort expectancy is stronger for women, older workers, and those with limited experience. As for the effect of social influence on behavioral intention, the effect is stronger for women, older workers, those with limited experience, and those who use the product under mandatory situations. The effect of facilitating conditions on actual use of the technology is stronger for older workers and more experienced workers.

Earlier studies have explained the role of age in work performance and indicate that younger workers may value extrinsic rewards more than older workers do (Hall & Mansfield, 1995). Thus, performance expectancy has greater effect on younger workers, and effort expectancy, which concerns the ease of usage, doesn't have as much effect on younger workers. Moreover, because the needs of affiliation increases with age, older people would be more influenced by social influence (Rhodes, 1983).

Gender plays a moderating role based on people's gender cognition (Lynott & McCandless, 2000). Men are expected to pay more attention to their performance compared to women. Meanwhile, because women are more sensitive to others' expectations, they are more likely to be affected by social influences (Venkatesh, Morris, & Ackerman, 2000). The moderating effect of voluntariness and experience of using the technology, however, is derived from the empirical examinations of the models at different time points (Venkatesh et al., 2003). Within the given time points, the relationship among effort expectancy, social influence, and behavior intention were found to be less significant for consumers who had some experience with the technology (Venkatesh et al., 2003). Also, the effects of social influences were found to be significant mostly while the usage is mandatory.

The UTAUT model has been validated and applied to many different settings. In education, UTAUT has been tested on faculty members' adoption of tablet PCs in higher education (Anderson,

Schwager, & Kerns, 2006) and the results suggest that performance expectancy and voluntariness are the most salient drivers for faculty's actual use of tablet PCs. Similarly, UTAUT was used in a pilot study to establish a cyber-infrastructure for the Smithsonian Center for Education and Museum Studies (SCEMS) and the Smithsonian Council of Education Directors (SCED) (Pappas & Volk, 2007). In retailing, UTAUT was adapted for Chinese mobile commerce settings), suggesting that users' education level and cultural background should be taken into consideration when applying UTAUT (Park et al., 2007).

With earlier studies, it is worthwhile to note that although UTAUT could be adapted for most technology adoptions, the explanation power of the constructs depends on situational influences (Ventakesh et al., 2003) and cultural background (Park et al., 2007). Moreover, the construct "use behavior," which indicates consumers' actual usage, is usually not accessible when consumers are still evaluating the construct "intention to use" in the pilot study stages. Thus, it is necessary to examine the constructs in the model with a more integrated perspective while trying to apply the model in VOIP services, which is relatively new and not yet understood by consumers or service providers.

The Role of Perceived Trust

Traditionally, trust played an important role in consumers' decision making processes. Business-to-consumer commerce cannot be realized unless consumers feel comfortable in the new medium with unfamiliar vendors (Gefen & Straub, 2003). Trust also plays a crucial role in many economic activities that can result in undesirable behavior (Fukuyama, 1995). Trust was considered as a governance mechanism in exchange relationships involved with uncertainty, vulnerability, and dependence (Brandach & Eccles, 1989). Trust is built on the information that enables consumers to understand the social environment in which they are interacting and to believe that things are controllable (Gundlach & Murphy, 1993). Trust is considered an essential factor in relational exchange, being "the cornerstone of the strategic partnership" between the seller and the buyer (Spekman, 1988). Moreover, trust can provide buyers with a high level of expectation to satisfy exchange relationships (Hawes, Mast, & Swan, 1989). For a new Internet application, such as VOIP, consumers have not yet acquired enough information and experience to feel comfortable enough to adopt it, and thus trust is more necessary in order for consumers to make the adoption decision.

Previous studies had shown the connection between trust and the UTAUT model. For example, from the point of view of social psychology, trust encompasses the expectations in one transaction and the risks associated with the interaction with the trusted party (Lee & Turban, 2001). While trust can be a kind of economical analysis, such as evaluating costs and benefits (Doney, Cannon, & Mullen, 1998), in this study the expectations are considered as the expected benefits and the costs brought by using VOIP. Similarly, the ability of the trusted party,, refers to the skills and competencies that enable users to have influence within some special domain (Mayer, Davis, & Schoorman, 1995). In other words, trust is partially contributed from the users' expectation of the possible profits and the expectation of costs, such as the effort of using the technology. For example, consumers may consider the convenience and high quality of telecommunication as benefits, and consider the privacy concerns and the effort of skills acquisition as costs of using Internet phones. Therefore the following hypothesis was developed:

H1: Increased performance expectancy is likely to increase consumer's perceived trust toward VOIP services.

H2: Increased effort expectancy is likely to increase consumer's perceived trust toward VOIP services.

From the point of view of personality, in initial stages, trust toward a new object is driven by the consumer's personality (Mayer et al., 1995) and socialized disposition. A reduction in feelings of uncertainty might result from familiarity or from others' opinions and social influences (Gefen, 2000), such as peer pressure and the intention to identify with others. When applied to VOIP adoption settings, the social influences that encourage consumers to use the service may be related to consumers' tendency to trus, thus:

H3: Increased social influence is likely to increase consumer's perceived trust toward VOIP service.

There are studies that indicate the consequence of trust in technology adoption scenarios. For example, trust was directly related with consumers' purchase decisions in e-commerce settings (Gefen, 2000). Also, the concepts of trust were integrated in previous studies and applied to the TRA model in e-commerce (McKinght, Choudhury, & Kacmar, 2002). The results showed that trust is an antecedent of use behavior. Also, encompassed within the TAM model, trust is considered as an antecedent for intention to use (Gefen, Karahanna, & Staub, 2003).

H4: Increased perceived trust toward VOIP service is likely to increase consumer intention to use VOIP service.

On the other hand, there are rare supports for the relation between trust and facilitating condition, and there is also some controversy over the influence of facilitating condition on behavioral intention. Facilitating condition, the technological and/or organizational supports used to remove barriers to using new technology applications, is directly related with the actual use of the technology in the UTAUT model. The influence of facilitating condition on actual behavior is salient in both mandatory and voluntary situations in initial stages (Venkatesh et al., 2003). In the original UTAUT study, facilitating condition did not show a significant influence on behavioral intention because the effect is overshadowed by the presence of effort expectancy and performance expectancy. However, facilitating condition significantly contributes to the behavioral intentions in TPB (Venkatesh et al., 2003). Along with previous hypotheses, effort expectancy and performance expectancy are not directly related with the use intention in this study, so the effect of facilitating control on use intention is not supposed to be overshadowed by the two constructs. Thus, the facilitating condition is hypothesized to have influence on consumers' intention to use VOIP services while the available resources could enable consumers to have more access to the new technology.

H5: Increased facilitating condition is likely to increase consumer intention to use VOIP service.

Moderating Effects in Modified UTAUT for VOIP

While the application of UTAUT depends on variations of the situational influences, moderating effects should also be reexamined. Gender difference was considered as one of the moderators in the UTAUT model based on gender scheme theory, which states that females pay more attention to the norms, so social influences would have more salient effects on females than on males. Nevertheless, recent studies of Internet technology applications demonstrated that the gap in gender has been diminished (Zhou, Dai, & Zhang, 2007). For example, gender difference did not significantly contribute to different mobile commerce shopping behaviors in Spain (Bigne, Ruiz, & Sanz, 2005). However, since VOIP technology is still in the introductory stages, the possible moderating effect of gender should still be taken into consideration.

Age is expected to have a positive moderating effect on the influence of effort expectancy and social influences, and it is also expected to have a negative moderating effect on the influence of performance expectancy, because older workers may be more socialized and younger workers may value work performance more (Venkatesh et al., 2003). Lascu, Bearden, and Rose (1995) also support the moderating effect of age, but in a contrary way. Their results suggested that greater social influence would occur in younger consumers.

A longitudinal study by Venkatesh, et al. (2000) showed that the individual technology acceptance decision making process would not be changed across income levels. However, VOIP service is still in its introductory stages compared to traditional telecommuting competitors, and it might not be able to provide an ideal price due to its current manufacturing scale, so price could be influential. Therefore, income should be included as a moderating factor. Considering the possible moderating effects of demographics, this study proposes the following hypotheses:

H6: The relationship between performance expectancy and trust will be moderated by gender, age, and income.
H7: The relationship between effort expectancy and trust will be moderated by gender, age, and income.
H8: The relationship between social influence and trust will be moderated by gender, age, and income.

The most important moderator that has been neglected in the UTAUT model and its applications is consumer innovativeness. Innovativeness is a well-researched topic in both organizational (e.g., Hurley & Hult, 1998) and consumer research (Blythe, 1999; Goldsmith, d'Hauteville, & Flynn, 1998). Consumer innovativeness is a personality construct that is employed to predict consumers' tendencies to adopt a wide range of technological products (Wood & Swait, 2002). Innate innovativeness was considered as "the degree to which an individual is receptive to new ideas and makes innovation decisions independently of the communicated experience of others" (Midgley & Dowling, 1978, p. 236). In the context of business and marketing, consumers' innovativeness is closely related to the adoption of products (Blythe, 1999; Rogers, 1995) and this influences the speed with which the adoption takes place after a product enters the market (Goldsmith & Flynn, 1992). As in the study by Rogers and Shoemaker (1971), the definition of innovativeness is "the degree to which an individual is relatively earlier in adopting an innovation than other members of his system" (Rogers & Shoemaker, 1971). Also, personal innovativeness serves as a key moderator in the relationships among the information source for the new technology, consumers' perception of the new technology, and their intention to adopt the new technology (Agarwal & Prasad, 1998). Considering the importance of consumer innovativeness in adopting innovative technology, the moderating effect is examined in the relationships among facilitating condition, trust, and intention of using VOIP. Considering the possible moderating effects of consumer innovativeness, this study proposes the following hypotheses:

H9: The relationship between trust and intention to use VOIP services will be moderated by gender, age, incomes, and consumer innovativeness.

H10: The relationship between facilitating condition and intention to use VOIP services will be moderated by gender, age, income and consumer innovativeness.

Together with the moderators, the proposed model is shown in Figure 1.

METHODOLOGY

Data Collection and Demographic Characteristics

Data for this study were collected using a self-administered mail survey. Participants were panels from a professional marketing survey company; 1,000 randomly selected respondents in Korea were contacted and asked to participate in this study via postal mail with 437 respondents completing the survey (r=43.7%). Their participation made them eligible for five $100 cash prize drawings. The demographic information of the sample is shown in Table 1.

Figure 1. Conceptual model

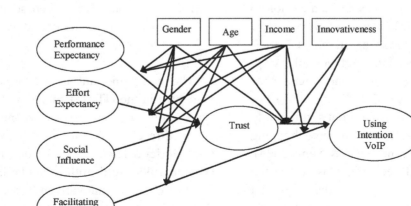

Table 1. Sample characteristics

Age	20-24	18.8%	Income	below $20,000	9.2%
	25-29	23.8%		$20,001~$30,000	15.8%
	30-34	26.8%		$30,001~$40,000	25.2%
	35-39	31.1%		$40,001~$50,000	16.7%
Gender	Male	50.1%		$50,001~$60,000	17.6%
	Female	49.9%		$60,001 or over	15.6%

Measurements

To develop the questionnaire, existing scales for all seven constructs were examined. Twenty-three survey items were adopted and modified to the VOIP context by researchers to measure variables in the model using seven-point Likert scales (See Appendix). With regards to the consistency between the English version and the Korean version, questionnaires in the original English were first translated into Korean and back translated into English by two multilingual interpreters. The questionnaire was finally validated by two other faculty members who are experts in consumer research. Correlations, reliabilities, means, and standard deviations of the constructs are provided in Table 2 and Table 3.

RESULTS

The proposed model was tested by structural equation modeling. The results indicate a good fit of the model and all the proposed paths are supported (Table 4). The $\chi2$ to degree of freedom ratio is between 2.0 and 3.0, as suggested (Tate, 1998), which indicates that the data fits the model well, $\chi2=312.98$, and degree of freedom= 124. Model fit indices also show a good fit of the model, RMSEA=.06, CFI=.96, and GFI=.93. Consumers' performance expectancy has a strong influence on consumers' perceived trust, $\beta= .71$, t=6.84, and P<.01. Thus, hypothesis 1 is supported. As for Hypothesis 2, the effect of effort expectancy on trust is significant but small, $\beta=.12$, t=1.67, and P< .1. So hypothesis 2 is not supported. Hypothesis 3 is supported. The effect of social influence on trust is significant too, $\beta=.19$, t=3.18, and P< .01. Hypothesis 4 is also supported. Consumers' perceived trust of VOIP has a large effect on their behavior intention of using VOIP services, $\beta=.67$, t=9.96, and P< .01. As to hypothesis 5, the effect of facilitating conditions has a significant effect on consumers' behavior intention to use VOIP services as well, but the effect is small, $\beta=.17$, t=2.95, and P < .01.

The results for moderating effects are mixed (Table 5). To examine the moderating effect, the moderating constructs were divided into high and low levels by their means, excepting the gender construct. For hypothesis 6, only income has a

Table 2. Reliability and validity of measurement

Variables*	# of items	Factor loadings	t-vales	Reliability	AVE
Performance Expectancy	3	0.71-0.79	16.36-18.94	0.80	0.58
Effort Expectancy	3	0.74-0.81	16.98-19.30	0.82	0.60
Social Influence	3	0.78-0.81	18.29-19.22	0.83	0.63
Facilitating Conditions	3	0.56-0.76	9.87-16.73	0.74	0.52
Behavior Intention	3	0.87-0.90	22.31-23.87	0.92	0.79
Trust	3	0.69-0.76	15.57-17.47	0.77	0.53

*Goodness-of-fit: $\chi2120 = 230.84$, p=0.00; CFI=0.97; NFI=0.95; GFI=0.94; AGFI=0.92; RMSEA=0.05

Table 3. Construct means, standard deviations, and correlations

	PF	EF	SI	FC	BI	TR
F	1.00					
EF	0.77***	1.00				
SI	0.68***	0.45***	1.00			
FC	0.51***	0.73***	0.53***	1.00		
BI	0.72***	0.56***	0.76***	0.54***	1.00	
TR	0.81***	0.76***	0.62***	0.57***	0.70***	1.00
Means	2.89	3.12	2.26	3.01	2.61	2.95
S.D	0.75	0.74	0.79	0.78	0.88	0.65

*p<0.1, **p<0.05, ***p<0.01

significant moderating effect on the path from performance expectation to trust, and the result showed that performance expectancy will have larger influence on trust for lower income consumers than for higher income consumers, $\Delta \chi 2 = 5.22$, P<.01.

Hypothesis 7, the path from effort expectancy to trust, is significantly moderated by gender and income. The effect of effort expectancy on trust is larger for female consumers than for male consumers, $\Delta \chi 2 = 8.55$, P<.05. The effect is also larger for higher income consumers than lower income consumers, $\Delta \chi 2 = 5.20$, P<.01.

In hypothesis 8, the moderating effect of age and income are significant. For consumers who are under 29, social influence will have a larger influence on their trust of VOIP services ($\Delta \chi 2 = 9.40$, P<.05). For consumers who have higher income level, social influences will be more likely to influence consumers' trust building in new technology such as VOIP ($\Delta \chi 2 = 11.26$, P<.01).

Hypothesis 9, the path from trust to behavior intention, is moderated by gender, income, and innovativeness. For female consumers, trust of VOIP will have a larger influence on their intention to use the VOIP services than for male consumers, $\Delta \chi 2 = 3.54$, P<.1. Also, trust will have a larger influence on intention to use for consumers who have higher income than for those who have lower income, $\Delta \chi 2 = 6.27$, P<.01. The effect is also larger for consumers who have lower innovativeness than those who have higher innovativeness ($\Delta \chi 2 = 3.67$, P<.1).

Hypothesis 10, the path from facilitating condition to behavior intention, is significantly moderated by income and innovativeness. The

Table 4. Structural models results

Structural Path	Coefficient	t-Value
Performance Expectancy → Trust	0.71	6.84***
Effort Expectancy → Trust	0.12	1.67*
Social Influence → Trust	0.19	3.18***
Trust → Behavior Intention	0.67	9.96***
Facilitating Conditions → Behavior Intention	0.17	2.95***
Goodness-of-fit: ÷2124 = 312.98, p=0.00; CFI=0.96; NFI=0.93; GFI=0.93; AGFI=0.90; RMSEA=0.06		

*p<0.1, **p<0.05, ***p<0.01

Table 5. The moderating effects

Structural Path	Gender			Age			Income			Innovativeness		
	Male	Female	χ2	Under 29	Over 30	χ 2	Lower Gr.	Higher Gr.	χ 2	Lower Gr.	Higher Gr.	χ 2
	(n=219)	(n=218)	df=1	(n=184)	(n=253)	df=1	(n=184)	(n=253)	df=1	(n=223)	(n=214)	df=1
Performance Expectancy → Trust	0.73 (5.04)	0.52 (2.59)	0.88	0.62 (1.90)	0.58 (2.62)	1.06	0.86 (2.91)	0.52 (2.24)	5.22***	--	--	--
Effort Expectancy → Trust	0 (-.05)	0.36 (2.34)	8.55**	0.12 (0.93)	0.20 (1.71)	-0.03	0.04 (0.30)	0.20 (1.61)	5.20***	--	--	--
Social Influence → Trust	0.29 (3.17)	0.23 (2.37)	-0.52	0.37 (1.91)	0.29 (2.37)	9.40**	0.14 (1.25)	0.36 (2.31)	11.26***	--	--	--
Trust → Behavior Intention	0.71 (7.71)	0.79 (3.42)	3.54*	0.73 (2.02)	0.79 (3.12)	-0.06	0.60 (3.10)	0.78 (2.58)	6.27***	0.91 (2.51)	0.66 (2.22)	3.67*
Facilitating Conditions→ Behavior Intention	0.17 (2.41)	-.01 (.11)	0.83	0.17 (1.87)	-0.01 (-.12)	0.04	0.19 (1.97)	0.05 (0.60)	5.67***	-0.09 (-.93)	0.12 (1.22)	5.22**

*p<0.1, **p<0.05, ***p<0.01

result shows that facilitating conditions will have a higher influence on consumers' intention to use VOIP services for consumers with lower income than higher income, (Δ χ2= 5.67, P<.05). For consumers have a higher degree of innovativeness, facilitating conditions will have a higher influence on consumers' intention to use VOIP services than for consumers who have lower innovativeness (Δ χ2= 5.22, P<.01).

DISCUSSIONS

This study attempts to illuminate the importance of trust and consumer innovativeness in consumers' acceptance of VOIP technology in the Korean market. As expected from the literature review, the data concerning Korean consumers' attitude toward the VOIP service fit the adjusted UTAUT model well. Consumers' trust toward the VOIP services is the major contribution to consumers' intention to use the services, while the contribu-tion of facilitating condition is also significant, but lower. The results support the importance of trust and conform to the study by Gefen et al. (2003), which demonstrated the influence of trust in the TAM model. The results of this study also imply that the effect of trust on intention to use VOIP would be stronger on female consumers with a higher income and lower innovativeness. The role of innovativeness in this relationship should be reasonable. Consumers who are less innovative would need more emotional support to make the decision to adopt the new technology, and trust would allow the consumers to adapt and acquire the ability for new things (Erikson, 1968); thus, the influence of trust would be more influential for less innovative consumers.

On the other hand, the effect of facilitating condition on intention to use VOIP services would be stronger on consumers with lower income and higher innovativeness. Contrary to trust, facilitat-ing control enables consumers to adopt VOIP services without worrying that they do not have enough knowledge or support to use the service. It is more important for consumers with lower income and higher innovativeness because these consumers may adopt the new technology before they could acquire enough information from other consumers and before building up their trust of the new technology. From the findings of the relationships among trust, facilitating condition, intention to use, and innovativeness, there are some managerial implications. While consumer innovativeness cannot be changed by VOIP service providers, increasing consumer trust and facilitating control becomes important means for service providers to stimulate consumers' intention to use the service depending on consumer's level of innovativeness.

CONCLUSION

The performance expectancy shows a dominant effect on trust toward VOIP services. Although effort expectancy and social influence also significantly contribute to consumers' trust, the effects are relatively minor. The results conform to previous studies that denoted trust as the economic evaluation of the costs and benefits of the trusted party (Doney et al., 1998). In terms of managerial application, the salience of performance expectancy could support the idea that the marketing strategy of Korean VOIP service providers is on the right track. Korean service providers mainly promote the functions of VOIP services, such as modifiability and personalization, which are reasonable from a technical perspective (Seong, 2008) and are attractive for consumers who value the applications' performance. The effect of performance expectancy would be stronger on consumers with lower income. An explanation of this result is that consumers with lower income are more sensitive to value returned for dollar spent. A service bundle which combines Internet phone service and other services could attract the price-sensitive groups more.

The influence of effort expectancy shows that consumers would trust the VOIP if they believe that the technology is easy to use. The influence indicates the importance of adequate information about actual use while the VOIP service is in its introductory stages. Also, the influence of effort expectancy is stronger for female consumers with higher income.

Social influence is also found to be a factor that affects consumers' trust toward VOIP services. Consumers' opinions could mitigate their uncertainty, thus influencing their trust of the new technology (Gefen, 2000). In practice, social influence could be more effective in East Asian cultures, where non-conformity is more likely to be taken as deviance (Kim & Markus, 1999). The differences between cultures suggest that a different outcome might occur in the North American market. The effect of social influence on trust was stronger for consumers who are younger and with higher income. The moderating role of age here departs from the original UTAUT model, which suggested that older workers are more likely to be affected by social influence (Venkatesh et al., 2003). However, it conforms to the study by Lascu et al. (1995), in which the authors indicated larger effect of social influence on younger people.

This study highlights the importance of trust of the VOIP services and the moderating role of innovativeness in consumers' acceptance of VOIP. The results provide researchers with an adapted model of UTAUT, which is validated in the Korean VOIP market. Trust is supported to be an important indicator of consumers' intention to use a new technology. The results of moderating effect also provide the VOIP service providers in Korea with strategies for different market segments based on age, gender, income, and consumers' innovativeness; it could also be a good predictor for the VOIP service providers in other global markets.

LIMITATIONS AND FUTURE RESEARCH

Some limitations of this study should be noted. First, this research uniquely utilized trust as an attitudinal variable instead of directly using attitudes toward Internet phone service between antecedents and behavioral intention. Although trust may illustrate consumers' evaluation of adopting a new service, this study might have ignored other attitudinal factors. Future research should accommodate other attitudinal variables inclusively. Second, consumers' previous experience or voluntariness of usage was not measured. The inclusion of these two variables with consumer innovativeness might show separate effects on usage intention. For future research, this theoretical framework can be replicated for other advanced technologies or countries with different socioeconomic variables.

REFERENCES

Agarwal, R., & Prasad, J. (1998). A conceptual and operational definition of personal innovativeness in the domain of information technology. *Information Systems Research, 9*(2), 204–215. doi:10.1287/isre.9.2.204

Ajzen, I. (1991). The theory of planned behavior. *Organizational Behavior and Human Decision Processes, 50*(2), 179–211. doi:10.1016/0749-5978(91)90020-T

Anderson, J. E., Schwager, P. H., & Kerns, R. L. (2006). The drivers for acceptance of Tablet PCs by faculty in a college of business. *Journal of Information Systems Education, 17*(4), 429–440.

Bhattacherjee, A. (2000). Acceptance of E-commerce services: The case of electronic brokerages. *IEEE Transactions on Systems Man and Cybernetics and Humans, 30*(4), 411–420. doi:10.1109/3468.852435

Bigne, E., Ruiz, C., & Sanz, S. (2005). The impact of Internet user shopping patterns and demographics on consumer mobile buying behavior. *Journal of Electronic Commerce Research, 6*(3), 193–209.

Blythe, J. (1999). Innovativeness and newness in high-tech consumer durables. *Journal of Product and Brand Management, 8*(5), 415–429. doi:10.1108/10610429910296028

Brandach, J. L., & Eccles, R. G. (1989). Market versus hierarchies: From ideal types to plural forms. *Annual Review of Sociology, 15*(1), 97–118. doi:10.1146/annurev.so.15.080189.000525

Cherry, S. (2005). South Korea pushes mobile broadband- The WiBro scheme advances. *IEEE Spectrum*, 14–16. doi:10.1109/MSPEC.2005.1502522

Davis, F. D. (1989). Perceived usefulness, perceived ease of use, and user acceptance of information technology. *Management Information Systems Quarterly, 13*(3), 319–339. doi:10.2307/249008

De Biji, P. W. J., & Peitz, M. (2005). Local loop unbundling in Europe: Experience, prospects, and policy challenges. *Communications and Strategies, 57*(1), 33–57.

Doney, P. M., Cannon, J. P., & Mullen, M. R. (1998). Understanding the influence of national culture on the development of trust. *Academy of Management Review, 23*(3), 601–620. doi:10.2307/259297

Erikson, E. H. (1968). *Identity: Youth and Crisis.* New York: Norton.

Fishbein, M., & Ajzen, I. (1975). *Belief, Attitude, Intention and Behavior: An Introduction to Theory and Research.* Reading, MA: Addison-Wesley.

Fukuyama, F. (1995). *Trust: The Social Virtues and the Creation of Prosperity.* New York: Free Press.

Gefen, D. (2000). E-commerce: the role of familiarity and trust. *Omega, 28*, 725–777. doi:10.1016/S0305-0483(00)00021-9

Gefen, D., Karahanna, E., & Straub, D. W. (2003). Trust and TAM in online shopping: An integrated model. *Management Information Systems Quarterly, 27*(1), 51–89.

Gefen, D., & Straub, D. W. (2003). Managing user trust in B2C e-services. *E-Service Journal, 2*(2), 7–24. doi:10.2979/ESJ.2003.2.2.7

Goldsmith, R. E., d'Hauteville, F., & Flynn, L. R. (1998). Theory and measurement of consumer innovativenss: a transactional evaluation. *European Journal of Marketing, 32*(3/4), 340–353. doi:10.1108/03090569810204634

Goldsmith, R. E., & Flynn, L. R. (1992). Identifying innovators in consumer product markets. *European Journal of Marketing, 26*(2), 42–55. doi:10.1108/03090569210022498

Gundlach, G. T., & Murphy, P. E. (1993). Ethical and legal foundations of relational marketing exchanges. *Journal of Marketing, 57*(4), 35–46. doi:10.2307/1252217

Hall, D., & Mansfield, R. (1995). Relationships of age and seniority with career variables of engineers and scientists. *The Journal of Applied Psychology, 60*(2), 201–210. doi:10.1037/h0076549

Hawes, J. M., Mast, K. E., & Swan, J. E. (1989). Trust earning perceptions of sellers and buyers. *Journal of Personal Selling & Sales Management, 9*(1), 1–8.

Hoffman, D. L., Novack, T. P., & Peralta, M. (1998). Building consumer trust online. *Communications of the ACM, 42*(4), 80–85. doi:10.1145/299157.299175

Hurley, R. F., & Hult, G. T. M. (1998). Innovation, market orientation, and organizational integration and empirical examination. *Journal of Marketing, 62*(3), 42–54. doi:10.2307/1251742

Joseph, B., & Vyas, S. J. (1984). Concurrent validity of a measure of innovative cognitive style. *Journal of the Academy of Marketing Science, 12*(2), 159–175. doi:10.1007/BF02729494

Kim, H., & Markus, H. R. (1999). Deviance or uniqueness, harmony or conformity? A ultural analysis. *Journal of Personality and Social Psychology, 77*(4), 785–800. doi:10.1037/0022-3514.77.4.785

Lascu, D. N., Bearden, W. O., & Rose, R. L. (1995). Norm extremity and interpersonal influences on consumer conformity. *Journal of Business Research, 32*(3), 201–212. doi:10.1016/0148-2963(94)00046-H

Lee, M. K. O., & Turban, E. (2001). A trust model for consumer Internet shopping. *International Journal of Electronic Commerce, 6*(1), 75–91.

Lynott, P. P., & McCandless, N. J. (2000). The impact of age vs. life experiences on gender role attitudes of women in different cohorts. *Journal of Women & Aging, 12*(2), 5–21. doi:10.1300/J074v12n01_02

Mayer, R. C., Davis, J. H., & Schoorman, F. D. (1995). An integration model of organizational trust. *Academy of Management Review, 20*(3), 709–734. doi:10.2307/258792

McKnight, D. H., Choudhury, V., & Kacmar, C. (2002). Developing and validating trust measures for e-commerce: An integrated typology. *Information Systems Research, 13*(3), 334–359. doi:10.1287/isre.13.3.334.81

Midgley, D. F., & Dowling, G. R. (1978). Innovativeness: The concept and its measurement. *The Journal of Consumer Research, 4*(4), 229–242. doi:10.1086/208701

Pappas, F. C., & Volk, F. (2007). Audience counts and reporting system: Establishing a cyber-infrastructure for museum educators. *Journal of Computer-Mediated Communication, 12*(2), 752–768. doi:10.1111/j.1083-6101.2007.00348.x

Park, J., Yang, S., & Lehto, X. (2007). Adoption of mobile technologies for Chinese consumers. *Journal of Electronic Commerce Research, 8*(3), 196–206.

Rao, B., Angelov, B., & Nov, O. (2006). Fusion of disruptive technologies: Lessons from the Skype case. *European Management Journal, 24*(2-3), 174–188. doi:10.1016/j.emj.2006.03.007

Rhodes, S. R. (1983). Age-related differences in work attitudes and behavior: A review and conceptual analysis. *Psychological Bulletin, 93*(2), 328–367. doi:10.1037/0033-2909.93.2.328

Rogers, E. (1995). *Diffusion of Innovations*. New York: Free Press.

Rogers, E. M., & Shoemaker, F. F. (1971). *Communication of Innovations: A Cross-Cultural Approach*. New York: Free Press.

Seong, S. (2008). *VOIP in Japan and Korea*. London: Ovum.

Slyke, C. V., Belanger, F., & Comunale, C. L. (2004). Factors influencing the adoption of web-based shopping: The impact of trust. *Databases for Advances in Information Systems, 35*(2), 32–49.

Spekman, R. E. (1988). Strategic supplier selection: Understanding ling-term buyer relationships. *Business Horizons, 31*(4), 75–81. doi:10.1016/0007-6813(88)90072-9

Tate, R. (1998). *An introduction to modeling outcomes in the behavioral and social science* (2nd ed.). Minneapolis, MN: Burgess.

The Yankee Group Report. (2007). *How open are the new VOIP market? A global perspective.* Retrieved from http://www.yankeegroup.com

Tobin, P. K. J., & Bidoli, M. (2005). Factors affecting the adoption of Voice over Internet Protocol (VoIP) and other converged IP services in South Africa. *South African Journal of Business Management, 37*(1), 31–39.

Venkatesh, V., & Davis, F. D. (2000). A theoretical extension of the technology acceptance model: Four longitudinal field studies. *Management Science, 45*(2), 186–204. doi:10.1287/mnsc.46.2.186.11926

Venkatesh, V., Morris, M. G., & Ackerman, P. L. (2000). A longitudinal field investigation of gender differences in individual technology adoption decision making processes. *Organizational Behavior and Human Decision Processes, 83*(1), 33–60. doi:10.1006/obhd.2000.2896

Venkatesh, V., Morris, M. G., Davis, G. B., & Davis, F. D. (2003). User acceptance of information technology: Toward a unified view. *Management Information Systems Quarterly, 27*(3), 425–478.

Wood, S. L., & Swait, J. (2002). Psychological indicator of innovation adoption: Cross-classification based on need for cognition and need for change. *Journal of Consumer Psychology, 12*(1), 1–13. doi:10.1207/S15327663JCP1201_01

Zhou, L., Dai, L., & Zhang, D. (2007). Online shopping acceptance model- A critical survey of consumer factors in online shopping. *Journal of Electronic Commerce Research, 8*(1), 41–62.

APPENDIX

Table 6.

Measure	Reference	Items
Performance Expectancy	Venkatesh et al., 2003	Using an internet phone service would enable me to accomplish tasks more quickly. Using an internet phone service could increase my efficiency. If I used an internet phone service, my phone calls would be more productive.
Effort Expectancy	Venkatesh et al., 2003	My use of an internet phone service would be clear and understandable. It would be easy for me to become skilled at using an internet phone service. Learning to operate an internet phone service would be easy for me.
Social Influence	Venkatesh et al., 2003	People who are important to me think that I should use an internet phone service. My family says that I should use an internet phone service. My friends tell me that using internet phone service could be good for me.
Facilitating Conditions	Venkatesh et al., 2003	I have the resources necessary to use an internet phone service. I have the knowledge necessary to use an internet phone service. If I adopt internet phone service, I know that a specific person or group will be available for assistance with any technical problem I may encounter.
Trust on VOIP	Bhattacherjee, 2000	The technology of VOIP will be trustworthy. I think this technology of VOIP will work well. This technology of VOIP will meet my expectations.
Customer Innovativeness	Joseph & Vyas, 1984	I like to try new and different things. I often try new brands before my friends and neighbors do. When I see a new brand on the shelf, I often buy it to see what it is like. I like to wait until something has been proven before I try it.(r) I feel apprehensive about trying out new things.
Using Intention	Venkatesh et al., 2003	I intend to use an Internet phone service in the next 12 months. I predict I will use an Internet phone service in the next 12 months. I plan to use an Internet phone service in the next 12 months.

This work was previously published in the International Journal of E-Services and Mobile Applications, Volume 3, Issue 1, edited by Ada Scupola, pp. 1-16, copyright 2011 by IGI Publishing (an imprint of IGI Global).

Chapter 3

Examining Behavioral Intention Toward Mobile Services:
An Empirical Investigation in Greece

Theodora Zarmpou
University of Macedonia, Greece

Vaggelis Saprikis
University of Macedonia, Greece

Maro Vlachopoulou
University of Macedonia, Greece

ABSTRACT

This study presents a conceptual model that combines perceived ease of use, perceived usefulness, innovativeness, trust, demographic characteristics and relationship drivers in order to examine their influence on the mobile services' adoption intention. The proposed model is empirically tested using data collected from a survey with questionnaires conducted in Greece. The results are analyzed through factor analysis, stepwise regression analysis, and ANOVAs. The findings show that individuals' innovativeness, their educational level, and the relationship ties between the users and the mobile services are key factors to encourage m-services' adoption. The results provide interesting insights and useful hints to practitioners and researchers.

INTRODUCTION

The transition from the cabled internet and the electronic services to the wireless internet and the mobile services is a fact. Thanks to the progress of the wireless communication technologies and devices (smartphones, PDAs, Palmtops, etc.),

there is an increasing interest from both the industry and the public sector in exploring the expanding possibilities of their businesses or the betterment in the fulfillment of the individuals' every day needs.

The mobile data services mainly refer to the communication services (e-mails, SMS, MMS,

DOI: 10.4018/978-1-4666-2654-6.ch003

etc.), web information services (weather information, sports, banking information, news, etc.), database services (telephone directories, map guides, etc.), entertainment (ringtones, videos, games, etc.) and commercial transactions through the mobile devices (buying products, making reservations, banking, stock trading, etc.) (Lu, Yao, & Yu, 2005).

A basic research question is whether these services are worth being used by the wide part of the population or not. A first impression would be that they are quite popular since the statistics show that the mobile phones in Greece have a high penetration degree (146%) (Athens University of Economics and Business & ICAP GROUP, 2008). Surveys show, however, that in spite of the high penetration rate, 4 out of 5 Greeks have never used any of the aforementioned services, whereas the majority of the users seldom utilize such services (Information Systems Technologies Lab [IST LAB], 2007). So, it is a big challenge to find those specific attributes of the mobile services that still keep them rather unpopular in Greece given the wide adoption of mobile devices. It is an even bigger challenge, though, to find possible solutions and make suggestions regarding the set of factors that affect their adoption. The aim of the paper is to find out users' reaction towards different parameters that would influence the individual's intention to use the mobile services in the current Greek reality.

There are several behavioral intention theories. The most popular and widely used ones are following. The *Diffusion Of Innovations* (DOI) perspective is introduced by Rogers (1995) investigating a variety of factors which are considered to be determinants for the actual adoption and usage of Information Systems. According to DOI, potential adopters evaluate an innovation based on innovation attributes (relative advantage, compatibility, complexity, trialability and observability) (Rogers, 1995). The *Theory of Reasoned Action* (TRA) was first proposed by Fishbein and Ajzen (1975) and it supported that users' intention to adopt a technology is determined by two factors: personal in nature (attitude) and social influence (social or subjective norm). TRA was later evolved to the *Theory of Planned Behavior* (TPB) by adding perceived behavioral control to the initial determinants (Ajzen, 1991). The TPB was also enriched with stable, decomposed beliefs structures for the TPB model and proposed the Decomposed Theory of Planned Behavior (Taylor & Todd, 1995). Finally, the *Technology Acceptance Model* (TAM) indicates that perceived ease of use and perceived usefulness are the two main beliefs that determine one's intention to use technology (Davis, 1989). The most recent one, however, is the *Unified Theory of Acceptance and Use of Technology* (UTAUT). Venkatesh, Morris, Davis, and Davis (2003) combined eight models (the above theories' models plus a few of their extensions) in a unified technology acceptance model, which contains five determinants (performance expectancy, effort expectancy, social influence, facilitating conditions, and behavioral intention). It is used so far for a number of technology types such as e-government (Alawadhi & Morris, 2008), wireless LAN (Anderson, & Schwager, 2003) and m-commerce (Pedersen, Methlie, & Thorbjornsen, 2002).

Although TAM is negatively criticized by a team of researchers, it compares favorably to TRA and TPB (El-Kasheir, Ashour, & Yacout, 2009). When deeper explanation of user adoption intention is desired, it allows other factors to be incorporated easily into its basic model (Hong, Thong, & Tam, 2006). Hence, in the current study, we keep the basic variables of TAM- perceived ease of use, perceived usefulness and behavioral intention- and through literature research we contribute with new variables- trust, innovativeness, relationship drivers and demographics- which are expected to have influence on the mobile services adoption intention, especially in the Greek area. All these constructs in this paper, are defined in a strict mobile context (trust, for instance, is used regarding the different mobile data services dimensions). The objective of this study is to construct an instrument in order to provide an explanation of the determinants of mobile data services acceptance. The study suggests a conceptual model that shows how its different constructs influence the Greeks' adoption towards the above services. Such a study in the mobile domain has not yet been investigated for this culture.

Consequently, the contribution of this article is threefold: First, the article contributes to the development of a general theoretical and conceptual model for examining the effects of a number of variables on mobile services usage intention in an unexplored for this field's related literature country. It builds on the growing literature at the technology acceptance models. Second, exploring the factors that impact on the users' decision to adopt a new service, which is essential for gaining a holistic view of the consumers' reaction on the new technology. Researchers and practitioners get empirical insights for predicting the acceptance in the market, and especially in Greece, where scarce research has been conducted around the topic. Third, the article uses perspectives from the marketing area for a technology acceptance issue by including the relationship drivers construct. It combines ideas from the relationship marketing with the unique characteristics of the mobile technologies. The idea of correlating the adoption intention of the m-services with the relationship ties that are possible to be created by their own use is quite innovative and very specialized in the mobile features of the considered technology.

The paper is organized as follows: There is an explanation of the hypotheses formed based on literature review and the description of the constructs that are included in the theoretical model. It is followed by a description of the applied methodology and presents the survey's results. The final sections conclude with a discussion commenting on the data gathered and recommend some ideas for future research directions.

CONCEPTUAL MODEL AND RESEARCH HYPOTHESES

A survey research is conducted using a questionnaire to examine the factors that affect the users' behavioral intention to adopt mobile services. Based on the literature review, a conceptual model is formulated (Figure 1) with the initial hypotheses. The model includes the following variables (Table 1): M-Services' Adoption Intention, Perceived Usefulness, Perceived Ease of Use, Demographics, Innovativeness, Trust and Relationship Drivers. In this section, the variables are explained, as well as, the related hypotheses.

M-Services Adoption Intention

In most of the well-established models of behavioral intention theories, such as the Technology Adoption Model (TAM) and the Theory of Reasoned Action (TRA), there has been an attempt to examine the factors that affect the consumers' decision on using a technology studied (Wu & Wang, 2005). Fishbein and Ajzen (1975) first defined the term "Behavioral Intention" to depict "a person's subjective probability that he will perform some behavior". Davis, also, follows up with this idea to give shape to TAM (Davis, 1989), which finally concludes to the "Actual System Use". Based on these concepts, in the paper herein, there is a construct included in the proposed model entitled "m-services adoption intention" to describe a person's subjective probability that he or she will perform mobile data services.

Perceived Usefulness

"Perceived usefulness" has been an instrumental construct in many of the technology adoption models that have been proposed since 1989, when Davis first used this term. It is defined as "the degree to which a person believes that using a particular system would enhance his or her job performance" (Davis, 1989). Perceived usefulness

Figure 1. Initial conceptual model

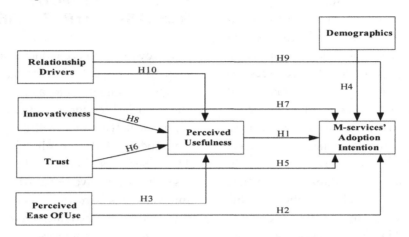

Table 1. The conceptual and operational definitions of the research variables

Research Variables	Conceptual Definition	Operational Definition
M-services' Adoption Intention	"A person's subjective probability that he or she will perform some m-servicess" (Fishbein & Ajzen, 1975).	AI1: I intend to use/continue using m-services in the near future AI2: I believe my interest towards m-services will increase in the future AI3: I intend to use m-services as much as possible AI4: I recommend others to use m-services
Perceived Usefulness	"The degree to which a person believes that using m-services would enhance his or her job performance" (Davis, 1989).	PU1: I think using m-services makes it easier for me to conduct transactions PU2.:I think using m-services makes it easier for me to follow up my transactions PU3: I think using m-services increases my productivity PU4: I think using m-services increases my effectiveness PU5: I think using m-services increases my efficiency
Perceived Ease of Use	"The degree to which a person believes that using m-services would be free of effort" (Davis, 1989).	PEU1: I think using m-services is easy PEU2: I think learning to use m-services is easy PEU3: I think finding what I want via m-services is easy PEU4: I think becoming skilful at using m-services is easy
Trust	The security in mobile payments, the confidentiality of personal data, the trust-worthiness in the results of the m-services and the integrity of the terms of use of the m-services.	TR1: I think using m-services in monetary transactions is safe TR2: I think my personal data are in confidence while using m-services TR3: I think the terms of use are strictly followed while using m-services TR4: I think using m-services for my transactions is trustworthy
Innovativeness	"The willingness of an individual to try out any new information technology" (Midgley & Dowling, 1978).	INN1: I am usually among the first to try m-services INN2: I am eager to learn about new technologies INN3: I am eager to try new technologies INN4: My friends and neighbours often come to me for advice about new technologies and innovation
Relationship Drivers	The time and location personalization of the m-services, their adaptation to the consumers' profile, the consumers' dynamic permission option and the consumers' reward by the use of the m-services.	RLDR1: I think using m-services should reward me with sales, coupons, etc. RLDR2: I think m-services are customized to my profile RLDR3: I think m-services are customized to the location and time I am, when I use them RLDR4: I think using m-services gives me the opportunity to control the start, continuation and end of my transactions

has been included as a construct in a number of surveys for different types of technologies and systems, such as mobile payments (Chen, 2008), mobile commerce (Min, Ji, & Qu, 2008), mobile data services in China (Qi, Li, Li, & Shu, 2009), application frameworks (Polančič, Heričko, & Rozman, 2010), and generally, technology adoption models (Im, Kim, & Han, 2008; Sun, Wang, & Cao, 2009; Wang, Lo, & Fang, 2008; Wu & Wang, 2005). In all the above studies, the perceived usefulness of the technology concerned influences positively individuals' adoption intention of this technology. Thus, it is reasonably expected that the same relationship between the model's constructs could also be applied to the present study. Therefore:

H1: Perceived Usefulness will have a positive effect on m-services adoption intention.

Perceived Ease of Use

Additionally to perceived usefulness, "Perceived Ease of Use" has been a vital concept in many of technology adoption models, too. It is defined as "the degree to which a person believes that using a particular system would be free of effort" (Davis, 1989). It has been included as a construct in a number of studies for different types of technologies and systems, such as mobile data services in China (Qi et al., 2009), application frameworks (Polančič et al., 2010), and generally new technologies (Im et al., 2008; Sun et al., 2009; Wang et al., 2008; Wu & Wang, 2005). In all the above studies, the perceived ease of use of the technology concerned has a direct positive effect on the behavioral intention to use each technology studied every time. Thus, regarding the behavioral intention to use the mobile services, it is hypothesized that:

H2: Perceived Ease of Use will have a positive effect on m-services adoption intention.

In the literature studied, there is a relationship between the perceived ease of use and perceived usefulness as influential factors on consumers' behavioral intention. Researchers proposing adoption intention models verify the positive influence of perceived ease of use on perceived usefulness regarding a variety of technology topics: mobile commerce (Aldas-Manzano, Ruiz-Mafe, & Sanz-Blas, 2009; Sun et al., 2009; Wu & Wang, 2005), mobile data services (Qi et al., 2009) or technology acceptance models in general (Im et al., 2008; Polančič et al., 2010; Schepers & Wetzels, 2007; Walczuch, Lemmink, & Streukens, 2007; Wang et al., 2008; Yi, Jackson, Park, & Probst, 2006). Hence, it is reasonable to hypothesize that the same relationship is valid for our research regarding the mobile services:

H3: Perceived Ease of Use will have a positive effect on perceived usefulness.

Demographic Characteristics

Based on the available literature, there is a noticeable relationship between the consumers' demographics and the technology studied acceptance. Li, Glass, and Records (2008) studied the differences between males' and females' attitude towards mobile commerce, as well as Min et al. (2008) verified the user's demographics relationship with m-commerce's acceptance in Japan. There are, also, findings which prove that individual differences – age, gender, educational background, occupation and income- do affect the adoption of mobile banking services (Crabbe, Standing, & Standing, 2009; Sulaiman, Jaafar, & Mohezar, 2007).

Regarding m-services, which is our area of interest, there are studies that examine the influence of the consumers' demographics on adoption intention (Mort & Drennan, 2005). Mylonakis (2004) published his research conclusions around the m-commerce services in Greece. The results showed that Greek men have a higher percentage of contact with the internet, compared to women. Thus, we hypothesize that demographics in the current study have a significant impact on the research goal:

H4: Demographics will have a significant effect on m-services adoption intention.

Trust

In previous studies, trust has been a significant factor in influencing consumers' behavior towards a specific technology, especially when it comes to cases of uncertain environments, such

As e-commerce (Chong, Ooi, & Arumugam, 2009; Gefen & Straub, 2003; Holsapple & Sasidharan, 2005; Pavlou, 2003; Wei, Marthandan, Chong, Ooi, & Arumugam, 2009). It is strongly recommended that trust should also be examined as a driving factor in the area of mobile commerce (Min et al., 2008). Mobile commerce is exposed to greater danger of insecurity than e-commerce and therefore the importance of trust is relatively higher in m-commerce (Wei et al., 2009).

In order to define and measure trust, there have been many suggestions in literature attributing it to meanings like privacy protection permitting a user to choose how his or her personal information is used (Bhattacherjee, 2002), or perceived credibility showing that one partner believes that the other partner has the required expertise to perform a job effectively and reliably (Cho, Kwon, & Lee, 2007; Crabbe et al., 2009). Pavlou (2003) stated that "trust in e-commerce is the belief that allows consumers to willingly become vulnerable to the online retailers after having considering the retailers' characteristics" including goodwill trust (benevolence) and credibility (honesty, reliability, and integrity). Min et al. (2008) divided the entity of trust in two sub-entities: trust in technology and trust in service providers. Trust in technology redirects to technical protocols, transaction standards, regulating policies, and payment systems (Min et al., 2008), whereas, according to

Bhattacherjee's analysis for e-commerce services (2002), trust in service providers refers to ability- the user's perception of provider's competencies and knowledge salient to the expected behavior, integrity- the user's perception that the service providers will adhere to a set of principles or rules of exchange acceptable to the users during and after the exchange, and benevolence- the service provider is believed to intend doing good to the users, beyond its own profit motive.

In this study, by trust we refer to the security in mobile payments when needed, to confidentiality of personal data (such as sending credit card details while using mobile services), to trustworthiness in the results after a mobile service is conducted and to the integrity of the terms of use of the mobile services.

Min et al. (2008) studying mobile commerce, Pavlou (2003) examining consumers' acceptance of electronic commerce, Wei et al. (2009) analysing the m-commerce adoption in Malaysia, Suh, and Han (2002) contributing to e-banking, and Gefen and Straub (2003) talking about B2C e-Services detected a positive influence of trust on consumers' behavioral intention. In specific, Pavlou (2003) said that: "trust reduces behavioral uncertainty related to the actions of the Web retailer, giving a consumer a perception of some control over a potentially uncertain transaction. On the other hand, Ha and Stoel (2009) proved that there is an influence of trust on intention to e-shop through its influence on usefulness. This sense of overall control over their on-line transactions positively influences consumers' purchase intentions." Thus, it is reasonable to assume that there is, also, a positive relationship between trust and behavioral intention when it comes to the adoption of mobile commerce services in Greece. So the following hypotheses can be stated:

H5: Trust will have a positive effect on consumers' m-services adoption intention.
H6: Trust will have a positive effect on perceived usefulness.

Innovativeness

Innovativeness in Information Technology is the "willingness of an individual to try out any new information technology" (Flynn & Goldsmith, 1993; Midgley & Dowling, 1978). In free interpretation in the field of technology acceptance, innovation refers to the degree of interest in trying a new thing, new concept, or innovative product or service (Rogers, 1995).

Innovativeness as a personality trait has been correlated with technology adoption in previous studies as an integrated factor along with optimism, discomfort and insecurity in the framework of the Technology Readiness Index (TRI) theory (Walczuch et al., 2007). Individuals, who are respected by their peers for their first-hand knowledge of an innovation and are considered as competent technically, consider the complexity of technology less troublesome suggesting a direct positive effect on perceived ease of use (Yi et al., 2006). It is also concluded by Walczuch et al. (2007) that innovativeness has a positive impact on perceived ease of use and negative impact on perceived usefulness regarding the adoption process of IT from service employees. Kuo and Yen (2009) also examined the relationship of innovativeness with perceived ease of use and perceived usefulness. Their study showed that innovativeness influences positively the perceived ease of use of 3G mobile value-added services, whereas the influence on perceived usefulness of 3G mobile value-added services is insignificant. Lu et al. (2005) studied the influence of personal innovativeness and social influence on wireless internet services and found that they do not have a direct significant effect on user intention to adopt the wireless internet services via their mobile devices. Innovativeness has also been examined as a factor influencing the use of Internet (Lam, Chiang, & Parasuraman, 2008). Chen and Tong (2003) discussed the importance of innovation for the mobile telecom industry, as well as, Sulaiman et al. (2007) showed that

personal innovativeness reflects on the adoption intention of mobile banking in Malaysia.

The positive results of studies around the direct effect of innovativeness on mobile commerce in Singapore (Yang, 2005) and on mobile shopping (Aldas-Manzano et al., 2009) have aroused the curiosity of the writers to examine the effect that innovativeness has on the adoption and perceived usefulness of mobile services in the Greek market. Based on the above literature, the following hypotheses can be stated:

H7: Innovativeness will have a positive effect on consumers' m-services adoption intention.
H8: Innovativeness will have a positive effect on perceived usefulness.

Relationship Drivers

Modern marketing indicates that relationship building between the customers and the brand is recommended to earn customers' loyalty and hence, increase the purchase interest for this brand's products. Striving towards this direction, one of the rules for a firm to build successful relationships with its customers is to have distinctive competencies (Morgan, 2000). Among other potential sources of such distinctive competencies, Lacey (2007) referred to customized services, which can reflect superior value or psychological benefits to the customers. More precisely, his research work concluded to the fact that preferential treatment, as a variable of resource drivers, has a direct positive influence on customer's commitment to the firm, which finally leads to increased purchase intentions. Since 1988, Edvardsson studying the service quality has pointed out the importance of adapted services to the individuals' special requirements. Until today, though, individualization is a determining user requirement (Büyüközkan, 2009).

Experts in the marketing field have recognized some unique features of the mobile technology, which can help exclusively with the relationship building between the customers and the brand. Mobile services, mainly due to the functions of SMS and MMS exchange, have the exclusiveness of ubiquitous and universal information accessibility, of information personalization and information dissemination (Nysveen, Pedersen, Thorbjornsen, & Berthon, 2005). Mobile and wireless technology do not set information access restrictions regarding time and space (Balasubramanian, Peterson, & Jarvenpaa, 2002). Consumers can reach the information and use the applications from everywhere (Nysveen et al., 2005; Siau et al., 2001) and anytime (Chang & Whinston, 2001; Kannan, Yunos, Gao, & Shim, 2003; Nysveen et al., 2005). This gives a strong advantage for time-sensitive and location-based content and services to be received by the consumers at the point accustomed to their personal identity (Doyle, 2001; Kannan et al., 2001; Nysveen et al., 2005; Siau et al., 2001; Watson, Pitt, Berthon, & Zinkhan, 2002). Thus, personalized information and/or services are able to be achieved.

Because of these features, the marketing experts seized the chance to exploit the mobile services as an adding channel to the existing line of channels in order to promote specific products, give new buying medium opportunities and, finally, increase consumers' intention to use these products (Nysveen et al., 2005). In this study, however, we change the roles and view the mobile services as the final product. Borrowing the aforementioned marketing ideas, we make the hypothesis that the characteristics of personalization and preferential treatment that the mobile services can offer might increase their own adoption, as they do with any kind of products.

Scharl, Dickinger, and Murphy (2005) examining the success factors of mobile marketing referred again to personalization as an influential factor. They gave examples, such as SMS advertising campaigns that will be sent preferably to phone numbers kept into categorized databases according to the customers' previous actions; examples of such actions are past leisure activi-

ties, music interests or occupation (preferential treatment). Taking advantage of mobile services' location and time personalization people can ask for directions to the nearest gas station open at the time of request or can receive alerts and flight delay notices through their mobile devices.

Deeper study dwelling on parameters, which lead to successful mobile services implementation, includes another significant characteristic of m-services: individuals need to decide when to respond to a mobile transaction, if at all (Scharl et al., 2005). Consumers need to have the access control (Geser, 2004) and freedom to give permission for their participation in mobile marketing activities (Barwise & Strong, 2001). And the permission needs to be dynamic; it does not only refer to the consumers' "opt-in" agreement, but also to the opportunity to change their preferences or stop their participation in the mobile transaction (Barnes & Scornavacca, 2004) whenever they feel to do so.

Among other factors that contribute to successful mobile advertising, Scharl et al. (2005) concluded to the fact that some common mobile applications urge the recipients to act on the spot. An example of such an application is the mobile couponing. Mobile coupons are stored in the mobile phone's memory (they are hard to get lost or forgotten) and so, they are easy to redeem. If consumers are subject to get "rewarded" with mobile coupons, discount prices, prizes, newsletters, free call time, etc. when they participate in a mobile transaction, they would be motivated to use the mobile services more frequently (Androulidakis, & Androulidakis, 2005).

We organize all the above mobile services usage indicators into one factor entitled Relationship Drivers. To sum up, relationship drivers is a term used in our study to declare the time and location personalization of the m-services, their adaptation to the consumers' profile, the consumers' dynamic permission option and the consumers' reward by the use of the m-services. They are teamed up in one construct, since we hypothesize that these dimensions have a common characteristic; they create a relationship between the consumers and the m-services. This relationship, finally, influences the consumers' mobile services adoption intention. Thus in our model:

H9: Relationship Drivers will have a positive effect on the Adoption Intention.

After a relationship is evolved between the consumers and the mobile services, it is reasonable to assume that the individuals might view the mobile services as an integrated part in their lives as they do with the mobile phones (IST Lab, 2007). They are expected to be emotionally attached to the mobile services and, hence increase their actual use. After the frequent use, though, people are getting used to fulfill their transaction needs via the mobile services and finally find them more useful. So, we hypothesize that:

H10: Relationship Drivers will have a positive effect on the Usefulness

METHODOLOGY AND RESULTS

Data were collected through a questionnaire both in a hard-copy form and in an electronic version of it. The electronic questionnaire was uploaded on a website for one month - January 15th to February 15th 2010. Additionally, an e-mail was sent to members of various lists (students and non-students) asking to respond to the questions and it was, also, posted on two popular social networks, Facebook and Twitter. Regarding applied questions, they are based on prior surveys approved for their validity and reliability. The questionnaire was pretested before being widely distributed, whereas a pilot study using a sample of thirty responses helped to identify possible problems in terms of clarity and accuracy. Thus, comments and feedback from respondents improved the final presentation of the items. Fifty-

seven participants gave incomplete answers and their results were dropped from the study. Finally, a total of 445 consumers from Greece provided data for the study.

In order to test the conceptual model, a data analysis was conducted in three stages. The first step employed factor analysis using principal component analysis (PCA) and orthogonal rotation (VARIMAX) in order to test the data validity and reliability, followed by two separate stepwise regression analyses and ANOVAs in order to examine the ten hypotheses. According to Hair, Black, Babin, Anderson, and Tatham (2006), the results of factors analysis are appropriate variables for subsequent application to other statistical techniques, that is stepwise regression analysis. These methodology steps have been previously applied in relevant scientific researches (Al-Qirim, 2005; Claycomb, Iyer, & Germain, 2005; Crabbe et al., 2009; El-Kasheir et al., 2009; Gottschalk & Abrahamsen, 2002; Grewal, Comer, & Mehta, 2001; Le, Rao, & Truong, 2004; McCloskey, 2006; Rao, Truong, Senecal, & Le, 2007; Wei et al., 2009) in order to give the final, valid constructs of the suggested model.

Demographic Characteristics and Descriptive Statistics

The demographic profile of respondents presented in Table 2 indicates that 51.5% are (229) male and 48.5% (216) are female. The vast majority of them (88.1%) are between 18-34 years old, whereas only 2.9% are above the age of 44. In terms of their educational background, 80.4% have received higher education studies, whereas with respect to their occupation, about one out of four (40.1%) are students, followed by private employees (18.9%) and freelancers (17.6%). Finally, regarding respondents' monthly income, 67.1% of them get paid up to 1,500€, whereas about one out of five (19.1%) preferred to avoid revealing his/her wages. The comprehensive

Table 2. Demographic characteristics of the respondents

Demographics	Frequency	Percent (%)
Gender	229	51.5%
Male	216	48.5%
Female		
Age	157	35.3%
18-24	235	52.8%
25-34	40	9.0%
35-44	13	2.9%
>44		
Education	4	0.9%
Elementary school	83	18.7%
High school	213	47.9%
University/Tech. Col.	145	32.5%
Master/PhD		
Occupation	178	40.1%
Student	84	18.9%
Private Employee	48	10.8%
Public Servant	78	17.6%
Freelancer	24	5.4%
Unemployed	33	7.2%
Other		
Monthly Income (€)	102	22.8%
<600	80	18.0%
601-900	55	12.4%
901-1200	62	13.9%
1201-1500	23	5.2%
1501-1800	20	4.5%
1801-2400	18	4.1%
>2400	85	19.1%
I don't answer		

demographic characteristics of the sample are presented as follows.

Operationalization of the Variables

Operational definitions of the study instruments are shown in Table 1. For each variable, a multiple-item scale was developed where each item was measured based on a 5-point Likert scale, ranging from 1-"Completely Disagree" to 5-"Completely Agree". In specific, four items were used to measure perceived ease of use, trust, innovativeness, relationship drivers and m-services' adoption intention, whereas five items were used to measure perceived usefulness.

Data Validity and Reliability

Factor analysis was applied to test the validity of the variables, classify and reduce questions into sub-variables when possible, and calculate factor loadings. Specifically, the principal component analysis (PCA) using orthogonal rotation (VARIMAX) was firstly performed to assess the underlying structure of the data. The PCA method is particularly suited to summarize the most of the original information (variance) in a minimun number of factors for prediction purposes (Hair et al., 2006). Orthogonal extraction, using VARIMAX rotation, suited for research goals and the need to reduce a large number of variables to a smaller set of uncorrelated variables. Additionally, VARIMAX rotation attempts to minimise the number of variables that have high loadings on a factor; hence enhancing the interpretability of the factors (Hair et al., 2006).

Nevertheless, in order to test the appropriateness of the data for factor analysis, several measures were applied to the entire population matrix. Specifically, Bartlett's test of sphericity ($p = 0.000$) confirmed the statistical probability that the correlation matrix has significant correlations among the variables, whereas the result of Kaiser-Meyer-Olkin (KMO) measure of sampling adequacy was 0.898, which is meritorious. Additionally, the measures of sampling adequacy (MSA) values all exceed 0.50 for both the overall test and each individual variable (Hair et al., 2006). All the aforementioned measures indicated the suitabily of factor analysis.

By applying the Kaiser eigenvalues criterion, six factors extracted that collectively explained (with eigenvalues 8.160, 2.117, 2.080, 1.747, 1.681 and 1.239 respectively) 68.099% of the variance in all items. Regarding construct validity, which testifies how well the results obtained from the use of the measure fit the theories around which the test is designed (Crabbe et al., 2009), it was tested by the use of two broadly applied tests, convergent and discriminant validity. In specific, "convergent validity is demonstrated if the items load strongly (>0.50) on their associated factors, whereas discriminant validity is achieved if each item loads stronger on its associated factor than on any other factor" (Hair et al., 2006, p. 137). Table 3 shows that all items have loading greater than 0.50 and load stronger on their associated factors than on other factors. Thus, convergent and discriminant validity are demonstrated. The six factors (perceived usefulness, trust, innovativeness, relationship drivers, perceived ease of use and m-services' adoption intention) proved to be relatively easy to interpret, owing to the strong variable loadings. Finally, construct reliability was assessed using Cronbach's alpha. Table 3 also shows that values ranged from 0.650 to 0.913. According to Hair et al. (2006, p. 137), "the generally agreed upon lower limit for Cronbach's alpha is 0.70, although it may decrease to 0.60 in exploratory research".

Hypotheses' Testing

Factors affecting m-services' adoption intention:

An inter-item correlation analysis (Table 4) revealed moderate positive correlation, but significant relationships between m-services' adoption intention and all of the predictors at the 0.01 level. Specifically, innovativeness has the highest correlation coefficient ($r = 0.526$), followed by perceived usefulness ($r = 0.511$), trust ($r = 0.363$), relationship drivers ($r = 0.325$) and perceived ease of use ($r = 0.312$) in correspondence. The results substantiates hypotheses H2, H5, H1, H7 and H9 that 99 out of 100 times there will be significant positive relationship between m-services' adoption intention and these variables.

A stepwise regression analysis was conducted to assess the best predictors among the independent variables believed to impact on m-services' adoption intention. Additionally, it revealed which of the five hypotheses were supported. According to Hair et al. (2006), stepwise regression is considered as the most popular sequential approach to variable selection. The results presented in

Table 5 indicate that 41% of the variance in m-services' adoption intention is explained by three predictors in the model. Innovativeness has the highest explanatory value of 27.7% ($b = 0.355$, $t = 8.715$, $p = 0.000$), followed by perceived usefulness 10.0% ($b = 0.317$, $t = 7.749$, $p = 0.000$) and relationship drivers 3.3% ($b = 0.189$, $t = 4.999$, $p = 0.00$). From the five hypotheses only two, namely H2 and H5 (the impact of perceived ease of use and trust on m-services' adoption intention), were not supported.

Factors affecting perceived usefulness:

The correlation analysis (Table 4) indicated that perceived usefulness significantly correlated positively with all variables ($p = 0.01$). Thus, the results substantiates hypotheses H3, H6, H8 and H10 that 99 out of 100 times there will be significant positive relationship between perceived usefulness and relationship drivers, innovativeness, ease of use and trust.

A second stepwise regression analysis was conducted in order to assess the best predictors among the independent variables believed to impact on perceived usefulness, as well as, to reveal which of the four hypotheses were supported. The results presented in Table 6 indicate that 34.2% of the variance in perceived usefulness is explained by three predictors in the model. Trust has the highest explanatory value of 22.4% ($b = 0.323$, $t = 7.633$, $p = 0.000$), followed by innovativeness 7.4% ($b = 0.245$, $t = 5.759$, $p = 0.000$) and perceived ease of use 4.4% ($b = 0.224$, $t = 5.465$, $p = 0.00$). From the four hypotheses only one, namely H10 (the impact of relationship drivers on perceived usefulness), was not supported.

Influences of the demographic factors on m-services' adoption intention:

ANOVA was conducted to examine the influence of demographics on adoption intention. In specific, age, educational level, occupation and monthly income were tested. Of particular interest was to find whether significant differences in adoption intention towards m- services can be attributable to demographic groupings, as well

Table 3. Rotated component matrix

Items	Factors					
	Perceived Usefulness	M-services' Adoption Intention	Trust	Innovativeness	Perceived Ease of Use	Relationship Drivers
PEOU1	0.109	0.070	0.047	0.103	**0.791**	0.116
PEOU2	0.030	0.000	0.040	0.054	**0.815**	0.064
PEOU3	0.167	0.207	0.112	-0.044	**0.664**	0.141
PEOU4	0.199	0.091	0.104	0.243	**0.674**	0.051
PU1	**0.653**	0.275	0.138	0.121	0.311	0.113
PU2	**0.797**	0.169	0.184	0.190	0.108	0.026
PU3	**0.852**	0.147	0.216	0.141	0.131	0.053
PU4	**0.824**	0.205	0.155	0.064	0.087	0.060
PU5	**0.841**	0.148	0.183	0.162	0.075	0.079
TR1	0.211	0.202	**0.777**	0.112	0.065	-0.002
TR2	0.103	0.058	**0.835**	0.076	-0.016	0.109
TR3	0.161	0.057	**0.791**	0.113	0.144	0.103
TR4	0.285	0.143	**0.685**	0.119	0.147	0.060
INN1	0.204	0.326	0.306	**0.587**	0.020	0.080
INN2	0.166	0.172	0.119	**0.827**	0.143	0..033
INN3	0.112	0.181	0.108	**0.856**	0.121	0.081
INN4	0.131	0.163	0.038	**0.794**	0.062	0.019
RLDR1	-0.003	0.276	-0.067	0.007	-0.148	**0.583**
RLDR2	0.091	0.180	0.117	0.094	0.118	**0.734**
RLDR3	0.059	0.057	0.095	-0.010	0.184	**0.781**
RLDR4	0.068	-0.072	0.089	0.067	0.175	**0.617**
BI1	0.194	**0.787**	0.161	0.199	0.123	0.112
BI2	0.215	**0.797**	0.124	0.187	0.077	0.072
BI3	0.215	**0.841**	0.119	0.185	0.123	0.112
BI4	0.208	**0.748**	0.097	0.228	0.100	0.154
Cronbach's alpha	**a = 0.913**	**a = 0.894**	**a = 0.836**	**a = 0.838**	**a = 0.771**	**a = 0.650**

Table 4. Results of correlations analysis for all variables

Factors	BI	PU	TR	INN	PEOU	RLDR
BI	1					
PU	0.511**	1				
TR	0.363**	0.473**	1			
INN	0.526**	0.428**	0.370**	1		
PEOU	0.312**	0.379**	0.267**	0.284**	1	
RLDR	0.325**	0.219**	.214**	0.189**	0.275**	1

** $p = 0.001$

Table 5. Results of the stepwise regression analysis for adoption intention towards m-services

Model	R	R Square	Adjusted R Square	Std. Error of the Estimate	Change Statistics				
					R Square Change	F Change	df1	df2	Sig. F Change
1	0.526	0.277	0.275	0.72875	0.277	169.710	1	443	0.000
2	0.614	0.377	0.374	0.67740	0.100	70.706	1	442	0.000
3	0.640[a]	0.410	0.406	0.65973	0.033	24.994	1	441	0.000

[a] Predictors: (constant), innovativeness, perceived usefulness, relationship drivers

as, to establish whether sample's variances differ from each other. Regarding age, it was tested by the use of independent-samples t-test in order to statistically compare male and female respondents. As shown in Table 7 above, there are no significant differences in m-services' adoption intention as a result of age, occupation, monthly income and gender, whereas educational level reveals significant differences. Thus, with the exception of educational level, all the demographic hypotheses were not supported.

DISCUSSION

The studied model consists of seven constructs: m-services' adoption intention, the perceived usefulness of m-services, the perceived ease of their use, the demographics of the survey's participants, innovativeness, trust, and relationship drivers. Based on published literature, possible relationships between all the constructs and the m-services' adoption intention are examined, as well as the influence of trust, innovativeness, perceived ease of use and relationship drivers on

perceived usefulness. The initial hypotheses form the model in Figure 1. After testing these hypotheses through stepwise regression analyses and ANOVAs we conclude to the conceptual model, which is designed in Figure 2.

As it is obvious from the final figure, not all of the initial hypotheses are verified. So, the literature regarding other types of technology, which we followed as an example from pre-studied acceptance models, does not always apply the occasion of mobile data services adoption in Greece. More specifically, the analysis of the studied relationships gives the following remarks:

- The impact of perceived usefulness, innovativeness and relationship drivers on m-services' adoption intention is verified. Among these approved hypotheses, innovativeness has the strongest effect on adoption intention. Perceived usefulness and relationship drivers, in response, are following. The mobile data services are expected to be successful, if they provide new technology or new ideas; trivial services might not be intriguing to adopt. In

Table 6. Results of the stepwise regression analysis for perceived usefulness

Model	R	R Square	Adjusted R Square	Std. Error of the Estimate	Change Statistics				
					R Square Change	F Change	df1	df2	Sig. F Change
1	0.473	0.224	0.222	0.65581	0.224	127.708	1	443	0.000
2	0.546	0.298	0.295	0.63398	0.074	46.591	1	442	0.000
3	0.585[b]	0.342	0.338	0.61424	0.045	29.863	1	441	0.000

[b] Predictors: (constant), trust, innovativeness, perceived ease of use

Table 7. Results of ANOVA for demographic variables

Demographic variables	Categories	Mean (SD)	Demographic variables	Categories	Mean (SD)
Age	18-24 25-34 35-44 >44	3.46 (0.82) 3.67 (0.88) 3.61 (0.84) 3.73 (0.79)	Occupation	Student Private Employee Public Servant Freelancer Unemployed Other	3.55 (0.83) 3.61 (0.87) 3.68 (0.76) 3.65 (0.93) 3.43 (0.76) 3.59 (1.02)
	F value (p)	1.545 (0.188)		F value (p)	0.441 (0.82)
Gender	Male Female	3.60 (0.83) 3.58 (0.89)	Monthly income (€)	<600 601-900 901-1200 1201-1500 1501-1800 1801-2400 >2400 I don't answer	3.48 (0.84) 3.69 (0.88) 3.71 (0.82) 3.57 (0.86) 3.58 (1.07) 3.73 (0.77) 3.95 (1.13) 3.45 (0.78)
	t value (p)	0.251 (0.268)			
Educational level	Elementary school High school University/Tech. Col. Master/PhD	3.06 (0.52) 3.38 (0.81) 3.58 (0.87) 3.76 (0.86)		F value (p)	1.579 (0.129)
	F value (p)	4.163 (0.006)			

addition, they are expected to succeed, if they provide something that makes people's routine transactions easier and increase their productivity, effectiveness and efficiency. If mobile services are personalized, provide preferential treatment and their use gives discount or coupons in return are likely to be adopted.

Figure 2. Final model

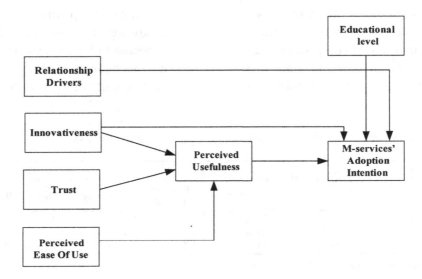

- It is noteworthy that trust and perceived ease of use do not affect directly the adoption intention. Both of them, however, have a significant indirect effect on adoption intention through perceived usefulness. Perceived ease of use and trust were expected to influence positively the users' adoption intention. This initial belief, however, comes in contrast to the findings. It seems that Greeks are not skeptic about non-secure m-payments or losing confidential personal data, and do not doubt about the trustworthiness of the m-services' results or the integrity of their terms of use. At least, these are not what they think about, when it comes to the usage of a mobile service.
- The impact of perceived ease of use, innovativeness and trust on perceived usefulness is verified. Trust has the strongest effect, followed by innovativeness and perceived ease of use.
- Relationship drivers does not seem to have a significant effect on perceived usefulness.
- With the exception of educational level, demographic characteristics cannot be judged as an important influential factor towards the m-services' adoption intention. The statistics showed that the more educated the people are, the more they intend to use m-services.

CONCLUSION

Prior studies show an intense interest in exploring the factors that influence people's behavior towards the adoption of a variety of new technologies and systems. In order to accomplish this goal, there are multiple behavioral theories introduced suggesting conceptual models and recommending different influential factors. This research contributes with new dimensions to these theories adding a marketing perspective

and concluding to the variables that form people's perception about mobile services usage in Greece. Consequently, it might be clearer when it comes to the Greek's culture and their intention to use mobile data services.

The survey's results reflecting the Greek reality could be a strong tool at the hands of companies and organizations involved with the mobile services development, mobile commerce investment and mobile marketing. We could mention some real-life ideas, which based on the above discussion, the m-services are likely to succeed. First of all, mobile services firms or governmental organizations that provide mobile functions should take advantage of the relationship ties that m-services are able to create with their users. A pregnant woman, for example, would trust in buying a mobile subscription to a baby center, which could notify her with SMS of new products and birth consultancies in the different stages of her pregnancy giving her the chance to withdraw from the service whenever she likes; it is a useful service, personalized to each woman offering the opt-in/out decision. In addition, airline companies trying to avoid the long check-in queues at the airports could motivate the travelers to use mobile check-in by giving them in return m-coupons for a coffee/food at the airports; this useful service includes users' reward.

Despite the fact that results provide meaningful implications, the research has three limitations. First, it is restricted to the examination of six factors. Second, it is limited to the Greek territory. Third, the hypotheses refer only to the impact of the factors on the adoption intention and perceived usefulness; more correlations between the factors are worth being examined. For future research, more influential factors can be added to the conceptual model, so that we gain a more holistic view of the incentives to adopt mobile services. Examples of such factors are cost- especially defined for the mobile expenses (equipment, subscription, transaction costs etc.) and mobile functionality in Greece (enough access points, net speed, etc.). The results are going to be compared with non-Greek similar studies. Such a comparison and taking into consideration the different cultural notions could be beneficial for the mobile industry targeting to multicultural services and global utilities. Finally, another data analysis method, such as Structural Equation Modeling (SEM), can be applied, in order to develop an even more concrete model including the relationships of the factors with each other.

REFERENCES

Ajzen, I. (1991). Theory of planned behavior. *Organizational Behavior and Human Decision Processes*, *50*(2), 179–211. doi:10.1016/0749-5978(91)90020-T

Al-Qirim, N. (2005). An empirical investigation of an e-commerce adoption-capability model in small businesses in New Zealand. *Electronic Markets*, *15*(4), 418–437. doi:10.1080/10196780500303136

AlAwadhi, S., & Morris, A. (2008). The use of the UTAUT model in the adoption of e-government services in Kuwait. In *Proceedings of the 41st Hawaii International Conference on System Sciences* (pp. 1-5).

Aldas-Manzano, J., Ruiz-Mafe, C., & Sanz-Blas, S. (2009). Exploring individual personality factors as drivers of M-shopping acceptance. *Industrial Management & Data Systems*, *109*(6), 739–757. doi:10.1108/02635570910968018

Anderson, J., & Schwager, P. (2003). SME adoption of wireless LAN technology: Applying the UTAUT model. In *Proceedings of the 7th Annual Conference of the Southern Association for Information Systems*, Savannah, GA (pp. 39-43).

Androulidakis, N., & Androulidakis, I. (2005). Perspectives of mobile advertising in Greece. In *Proceedings of the International Conference on Mobile Business*, Sydney, Australia (pp. 441-444).

Athens University of Economics and Business & ICAP GROUP. (2008). *Social-financial overview of the mobile phone industry in Greece.* Retrieved from http://www.sepe.gr/files/pdf/Executive%20 Summary.pdf

Balasubramanian, S., Peterson, R. A., & Jarvenpaa, S. L. (2002). Exploring the implications of M-commerce for markets and marketing. *Journal of the Academy of Marketing Science*, *30*(4), 348–361. doi:10.1177/009207002236910

Barnes, S. J., & Scornavacca, E. (2004). Mobile marketing: The role of permission and acceptance. *International Journal of Mobile Communications*, *2*(2), 128–138. doi:10.1504/IJMC.2004.004663

Barwise, P., & Strong, C. (2001). Permission-based mobile advertising. *Journal of Interactive Marketing*, *16*(1), 14–24. doi:10.1002/dir.10000

Bhattacherjee, A. (2002). Individual trust in online firms: Scale development and initial test. *Journal of Management Information Systems*, *19*(1), 211–241.

Büyüközkan, G. (2009). Determining the mobile commerce user requirements using an analytic approach. *Computer Standards & Interfaces*, *31*, 144–152. doi:10.1016/j.csi.2007.11.006

Chen, J., & Tong, L. (2003). Analysis of mobile phone's innovative will and leading customers. *Science Management*, *24*(3), 25–31.

Chen, L. (2008). A model of consumer acceptance of mobile payment. *International Journal of Mobile Communications*, *6*(1), 32–52. doi:10.1504/ IJMC.2008.015997

Cho, D. Y., Kwon, H. J., & Lee, H. Y. (2007). Analysis of trust in internet and mobile commerce adoption. In *Proceedings of the 40th Hawaii International Conference on System Science* (p. 50).

Claycomb, C., Iyer, K., & Germain, R. (2005). Predicting the level of B2B e-commerce in industrial organizations. *Industrial Marketing Management, 34*(3), 221–234. doi:10.1016/j.indmarman.2004.01.009

Crabbe, M., Standing, C., & Standing, S. (2009). An adoption model for mobile banking in Ghana. *International Journal of Mobile Communications, 7*(5), 515–543. doi:10.1504/IJMC.2009.024391

Davis, D. (1989). Perceived usefulness, perceived ease of use and user acceptance of information technology. *Management Information Systems Quarterly, 13*, 319–340. doi:10.2307/249008

Doyle, S. (2001). Software review: Using short message services as a marketing tool. *Journal of Database Marketing, 8*(3), 273–277. doi:10.1057/palgrave.jdm.3240043

Edvardsson, B. (1988). Service quality in customer relationships: A study of critical incidents in mechanical engineering companies. *The Service Industries Journal,* [8(4), 427-445.

El-Kasheir, D., Ashour, A., & Yacout, O. (2009). Factors affecting continued usage of internet banking among Egyptian customers. *Communications of the IBIMA, 9*, 252–263.

Fishbein, M., & Ajzen, I. (1975). *Belief, attitude, intention and behavior: An introduction to theory and research.* Reading, MA: Addison-Wesley.

Flynn, L., & Goldsmith, R. (1993). A validation of the Goldsmith and Hofacker innovativeness scale. *Educational and Psychological Measurement, 53*, 1105–1116. doi:10.1177/0013164493053004023

Gefen, D., & Straub, D. W. (2003). Managing user trust in B2C e-services. *E-Service Journal, 2*(2), 7–24. doi:10.2979/ESJ.2003.2.2.7

Geser, H. (2004). Towards a sociological theory of the mobile phone. In H. Geser (Ed.), *Sociology in Switzerland: Sociology of the mobile phone.* Zurich, Switzerland: University of Zurich. Retrieved from http://socio.ch/mobile/t_geser1.htm

Gottschalk, P., & Abrahamsen, A. F. (2002). Plans to utilize electronic marketplaces: The case of B2B procurement markets in Norway. *Industrial Management & Data Systems, 102*(6), 325–331. doi:10.1108/02635570210432028

Grewal, R., Comer, J. M., & Mehta, R. (2001). An investigation into the antecedents of organizational participation in business-to-business electronic markets. *Journal of Marketing, 65*(3), 17–33. doi:10.1509/jmkg.65.3.17.18331

Ha, S., & Stoel, L. (2009). Consumer e-shopping acceptance: Antecedents in a technology acceptance model. *Journal of Business Research, 62*, 565–571. doi:10.1016/j.jbusres.2008.06.016

Hair, J., Black, W., Babin, B., Anderson, R., & Tatham, R. (2006). *Multivariate data analysis.* Upper Saddle River, NJ: Prentice Hall.

Holsapple, C. W., & Sasidharan, S. (2005). The dynamics of trust in online B2C e-commerce: A research model and agenda. *Information Systems and E-business Management, 3*(4), 377–403. doi:10.1007/s10257-005-0022-5

Hong, S., Thong, J., & Tam, K. (2006). Understanding continued information technology usage behavior: A comparison of three models in the context of Mobile Internet. *Decision Support Systems, 42*, 1819–1834. doi:10.1016/j.dss.2006.03.009

Im, I., Kim, Y., & Han, H. (2008). The effects of perceived risk and technology type on users' acceptance of technologies. *Information & Management, 45*(1), 1–9.

Information Systems Technologies Laboratory (IST Lab). (2007). *Research about the tendency in the use of mobile data services in Greece (Comparison study 2006-2007)*. Athens, Greece: Athens University of Economics Wireless Research Center.

Kannan, P. K., Chang, A., & Whinston, A. B. (2001). Wireless commerce: Marketing issues and possibilities. In *Proceedings of the 34th Hawaii International Conference on System Sciences* (pp. 1-6).

Kuo, Y., & Yen, S. (2009). Towards an understanding of the behavioral intention to use 3G mobile value-added services. *Computers in Human Behavior*, *25*(1), 103–110. doi:10.1016/j.chb.2008.07.007

Lacey, R. (2007). Relationship drivers of customer commitment. *Journal of Marketing Theory and Practice*, *15*(4), 315–333. doi:10.2753/MTP1069-6679150403

Lam, S., Chiang, J., & Parasuraman, A. (2008). The effects of the dimensions of technology readiness on technology acceptance: An empirical analysis. *Journal of Interactive Marketing*, *22*(4), 19–39. doi:10.1002/dir.20119

Le, T., Rao, S., & Truong, D. (2004). Industry-sponsored marketplaces: A platform for supply chain integration or a vehicle for market aggregation? *Electronic Markets*, *14*(4), 295–307. doi:10.1080/10196780412331311748

Li, S., Glass, R., & Records, H. (2008). The influence of gender on new technology adoption and use- mobile commerce. *Journal of Internet Commerce*, *7*(2), 270–289. doi:10.1080/15332860802067748

Lu, J., Yao, J., & Yu, C. (2005). Personal innovativeness, social influences and adoption of wireless internet services via mobile technology. *The Journal of Strategic Information Systems*, *14*(3), 245–268. doi:10.1016/j.jsis.2005.07.003

McCloskey, D. W. (2006). The importance of ease of use, usefulness, and trust to online consumers: An examination of the technology acceptance model with older consumers. *Journal of Organizational and End User Computing*, *18*(4), 47–65. doi:10.4018/joeuc.2006070103

Midgley, D., & Dowling, G. (1978). Innovativeness: The concept and its measurement. *The Journal of Consumer Research*, *4*(4), 229–242. doi:10.1086/208701

Min, Q., Ji, S., & Qu, G. (2008). Mobile commerce user acceptance study in China: A revised UTAUT model. *Tsinghua Science and Technology*, *13*(3), 257–264. doi:10.1016/S1007-0214(08)70042-7

Morgan, R. M. (2000). Relationship marketing and marketing strategy: The evolution of relationship marketing within the organization. In Sheth, J. N., & Parvatiyar, A. (Eds.), *Handbook of relationship marketing* (pp. 481–505). Thousand Oaks, CA: Sage.

Mort, G., & Drennan, J. (2005). Marketing m-services: Establishing a usage benefit typology related to mobile user characteristics. *Database Marketing & Customer Strategy Management*, *12*(4), 327–341. doi:10.1057/palgrave.dbm.3240269

Mylonakis, J. (2004). Can mobile services facilitate commerce? Findings from the Greek telecommunication market. *International Journal of Mobile Communications*, *2*(2), 188–198. doi:10.1504/IJMC.2004.004667

Nysveen, H., Pedersen, P., Thorbjornsen, H., & Berthon, P. (2005). Mobilizing the brand: The effects of mobile services on brand relationships and main channel use. *Journal of Service Research*, *7*(3), 257–276. doi:10.1177/1094670504271151

Pavlou, P. (2003). Consumer acceptance of electronic commerce: Integrating trust and risk with the technology acceptance model. *International Journal of Electronic Commerce*, *7*(3), 101–134.

Pedersen, P., Methlie, L., & Thorbjornsen, H. (2002). Understanding mobile commerce end-user adoption: A triangulation perspective and suggestions for an exploratory service evaluation framework. In *Proceedings of the 35th Hawaii International Conference on System Sciences* (p. 8).

Polančič, G., Heričko, M., & Rozman, I. (2010). An empirical examination of application frameworks success based on technology acceptance model. *Journal of Systems and Software, 83*(4), 574–584. doi:10.1016/j.jss.2009.10.036

Qi, J., Li, L., Li, Y., & Shu, H. (2009). An extension of technology acceptance model: Analysis of the adoption of mobile data services in China. *Systems Research and Behavioral Science, 26*(3), 391–407. doi:10.1002/sres.964

Rao, S., Truong, D., Senecal, S., & Le, T. (2007). How buyers' expected benefits, perceived risks, and e-business readiness influence their e-marketplaces usage. *Industrial Marketing Management, 36*, 1035–1045. doi:10.1016/j.indmarman.2006.08.001

Rogers, M. (1995). *Diffusion of innovations*. New York, NY: Free Press.

Scharl, A., Dickinger, A., & Murphy, J. (2005). Diffusion and success factors of mobile marketing. *Electronic Commerce Research and Applications, 4*, 159–173. doi:10.1016/j.elerap.2004.10.006

Suh, B., & Han, I. (2002). Effect of trust on customer acceptance of Internet banking. *Electronic Commerce Research and Applications, 1*(3-4), 247–263. doi:10.1016/S1567-4223(02)00017-0

Sulaiman, A., Jaafar, N. I., & Mohezar, S. (2007). An overview of mobile banking adoption among the urban community. *International Journal of Mobile Communications, 5*(2), 157–168. doi:10.1504/IJMC.2007.011814

Sun, Q., Wang, C., & Cao, H. (2009). An extended TAM for analyzing adoption behavior of mobile commerce. In *Proceedings of the Eighth International Conference on Mobile Business* (pp. 52-56).

Taylor, S., & Todd, P. A. (1995). Understanding information technology usage: A test of competing models. *Information Systems Research, 6*(2), 144–176. doi:10.1287/isre.6.2.144

Venkatesh, V., Morris, M., Davis, G., & Davis, F. (2003). User acceptance of information technology: Toward a unified View. *Management Information Systems Quarterly, 27*, 425–478.

Walczuch, R., Lemmink, J., & Streukens, S. (2007). The effect of service employees' technology readiness on technology acceptance. *Information & Management, 44*(2), 206–215. doi:10.1016/j.im.2006.12.005

Wang, C., Lo, S., & Fang, W. (2008). Extending the technology acceptance model to mobile telecommunication innovation: The existence of network externalities. *Journal of Consumer Behaviour, 7*(2), 101–110. doi:10.1002/cb.240

Watson, R., Pitt, F., Berthon, P., & Zinkhan, G. (2002). U-Commerce: Expanding the universe of marketing. *Journal of the Academy of Marketing Science, 30*(4), 333–347. doi:10.1177/009207002236909

Wei, T., Marthandan, G., Chong, A., Ooi, K., & Arumugam, S. (2009). What drives Malaysian m-commerce adoption? An empirical analysis. *Industrial Management & Data Systems, 109*(3), 370–388. doi:10.1108/02635570910939399

Wu, J., & Wang, S. (2005). What drives mobile commerce? An empirical evaluation of the revised technology acceptance model. *Information & Management, 42*(5), 719–729. doi:10.1016/j.im.2004.07.001

Yang, K. (2005). Exploring factors affecting the adoption of mobile commerce in Singapore. *Telematics and Informatics*, *22*, 257–277. doi:10.1016/j.tele.2004.11.003

Yi, M., Jackson, J., Park, J., & Probst, J. (2006). Understanding information technology acceptance by individual professionals: Toward an integrative view. *Information & Management*, *43*(3), 350–363. doi:10.1016/j.im.2005.08.006

Yunos, H. M., Gao, J. Z., & Shim, S. (2003). Wireless advertising's challenges and opportunities. *IEEE Computer*, *36*(5), 30–37.

This work was previously published in the International Journal of E-Services and Mobile Applications, Volume 3, Issue 2, edited by Ada Scupola, pp. 1-19, copyright 2011 by IGI Publishing (an imprint of IGI Global).

Chapter 4
Colored vs. Black Screens or How Color Can Favor Green E-Commerce

Jean-Eric Pelet
University of Nantes, France

Panagiota Papadopoulou
University of Athens, Greece

ABSTRACT

This paper presents the results of an exploratory qualitative study conducted with 26 consumers about their use of computer screen savers. The results show how the use of screen savers remains almost nonexistent. Unknown or taking too long to apply, this feature is not attractive to persons interviewed who do not use it for sustainable development purposes. The paper presents the results of this qualitative study, offering an interpretive analysis of the reasons and factors explaining this type of computer user's behavior. The paper also discusses the potential of using screensaver functionality in e-commerce websites, particularly in the Mediterranean region. In this direction it looks into how this could be provided by the establishment of two elements - a browser and a website extension, which will be tested in a future online experiment.

INTRODUCTION

In the European Commission, Information and Communication Technology (ICT) contribute to 2% of global greenhouse gas emissions (Gartner, 2007). Their environmental impact is therefore a concern that gradually receives research attention. A principal approach to address the environmental impact of Information Technology is to adopt a Green use — 'reducing the energy consumption of computers and other information systems as well as using them in an environmentally sound man-

DOI: 10.4018/978-1-4666-2654-6.ch004

ner' (Murugesan, 2008). In this vein, in terms of management, information systems and marketing could potentially contribute to the effort to decrease these effects through electronic commerce. In particular, this could hold true with regards to the management of the screen saver for computers aiming to reduce their energy consumption and darken the color appearance of e-commerce websites. The display of a predominantly white interface is in fact more likely to "fatigue" than if the screen is black, or dark. But most of the e-commerce websites visited by French display dominant white colors (Figure 1).

The aim of this paper is to understand whether consumers are likely to become more responsible in playing for sustainable development in the context of electronic commerce, through action on the screen saver of the computer or an equivalent system. A simple feature to implement such as screen saver systems of existing computers can be a first line of research. An exploratory approach seems necessary in the attempt to answer this question. Indeed, despite the influence of the trend "sustainable development" on consumer

behavior, there is not to our knowledge a model unified and useful for its study. As a result, the subject remains unexplored, both in terms of information systems or marketing.

By looking at consumer responsibility, we seek to grasp the opportunity for e-merchants, e-learning websites and the actors of information systems with a human machine interface in general, to contribute to the effort for decreasing greenhouse gas emissions. The user action on the light emitted by the screen seems indeed likely to reduce unnecessary power consumption as easy as clicking a button. This could be articulated as the research question: "the ability to click a button present on the browser or on an e-commerce website, changing the appearance of the interface to pollute less while guaranteeing the same content readability through a "intelligent curtain screen saver" after a certain period of inactivity, would it allow e-commerce players to promote reduction of greenhouse gases?"

Our work shows that by improving the appearance of colorful interfaces of e-commerce websites, the proper choice of contrasts between

Figure 1. Presentation of the 10 most visited e-commerce sites in France (01/2010): all have a dominant color white or very clear (source: FEVAD 2010)

the dominant colors – screen background – and dynamic colors – text, buttons, images, tables, search engines – would allow to reduce the energy consumption of screens when the curtain screen saver is used. This would appear after a period of inactivity of the user cursor corresponding to the standby screen.

In this research, we see that 1) screen savers are not used to save energy, regardless of the computer, public or private, 2) the majority of the most visited e-commerce websites in France in 2009 - 2010 have a dominant white color corresponding to the hue which is the largest consumer of energy for a screen. Hence the choice of a black screen saver allows for saving as much energy as possible. This paper presents the interest to use it in a sustainable development perspective.

To determine the perception and use of consumers about the effects of a screen saver, the results of an exploratory study conducted among 26 persons are presented. They follow a review of the literature on the energy consumption of different types of computer screens in organizations and among individuals, and a comprehensive definition of responsible consumer. The paper presents and analyzes the results of the preliminary study and offers a discussion of screen saver use in e-commerce for environment protection before reaching its conclusion describing future research.

CONCEPTUAL RESEARCH FRAMEWORK

The Screens and Footprint of Greenhouse Gas Emissions

This part initially presents the consumption of websites of white colors, which are more energy consuming than those displaying more black. A review of the literature on energy consumption depending on the type of monitor screen follows.

The current proposed solutions to display more black than white (for example the Blackle website[1] like a *"black google"*) do not have interesting results in terms of satisfaction since a negative contrast (white on black) has bad results in terms of readability and memorization (Pelet & Papadopoulou, 2010a). The contrast caused by the dynamic color (gray) on the dominant color (black) is likely to cause effects opposite to those expected, as it doesn't help users, on the contrary, it makes the content more difficult to read. Indeed, it has been shown that negative contrast (light text on dark background) caused by a dark dominant color and a light dynamic color cause negative moods that can lead to leave the site because it is more tiring than a positive contrast (Pelet & Papadopoulou, 2009).

The ecological footprint defined by the WWF (World Wide Foundation) as *"a measure of the pressure exerted by man over nature"* shows that a user can help to reduce it by a different use of its computer screen. The footprint represents a tool allowing to assess the area necessary for a population to respond to its consumption of resources in order to absorb its waste.

At company level, the contribution of ICT in the emission of greenhouse gases until 2007 was about 2%. Some predictions state that the global footprint of ICT could note a growth of 6% for 2020 (Infos-Industrielles, 2007). Faced with this alarming situation and within the perspective to promote and encourage better use of energy efficiency, several proactive measures have emerged, including: the European Programmes, the *"Energy Efficiency"* Programme of the general direction of the "Information and Media Society" and the U.S. program "EnergyStar" (Gartner, 2007). The latter is the origin of EC/106/2008 Regulation on labeling of office equipment. The label is a certification guaranteeing the energy efficiency of office equipment at both economical and ecological level (Breuil et al., 2008). The EnergyStar label offers with the proposed office equipment, the opportunity for companies to become more successful in terms of efficiency and energy return. The label proposes to companies a range of products, enabling them to optimize and use energy more efficiently, for example by promoting the use of the laptop that uses 50% to 80% less energy than a desktop computer. EnergyStar also offers the opportunity for companies to have supporting data allowing to compare the energy consumption of LCD screens, plasma and CRT (Figure 2).

We present these three technologies, LCD, Plasma and CRT, used in computer monitors in Appendix A.

Studies have shown that the production of a CRT screen computer accounted alone for an emission of 680kg of CO_2 and 1 250kg for a flat screen computer. A basic workstation represents, for use of 5 years, about 2000kg of CO_2 per year. A server is itself 536kg of CO_2 per year (Breuil et al., 2008). Meanwhile, the replacement of CRTs with flat screen requires 10 times more CO_2 in terms of energy consumption. The energy saved by better management of PCs and monitors in the offices would decrease from 23 TWh / year (Tera Watt hours / year) to 17 TWh / year if the screen saver was activated according to Kawamoto et al. (2001), about CRTs and flat screens.

Thus, a 15" LCD consumes 30% less energy than a CRT 15". Meanwhile, an LCD 17" consumes about 50% less than a CRT 17". This difference tends to decrease as the size of the screen increases. The results given by EnergyStar show that the rate of 8 hours per day, for two computers

Figure 2. Format of LCD computer screens - Plasma – CRT

LCD - PLASMA   CRT

(LCD and CRT) of similar size, choosing an LCD monitor offers a saving of energy reaching over 100kWh/year (EnergyStar, 2009).

For its part, the European Union has set a major goal of reducing its energy costs by 20% by 2020 (Melquiot, 2009). The forecast for the achievement of these objectives are optimistic to the extent where 50% of the reduction of energy use comes from just ICT. They would thus save 1 to 4 times their own emissions on the rest of the economy. This is justified by example with firms that generalize the teleconference, allowing no movement.

At the individual level, energy consumption appears as an area where efforts can be further enhanced. The British for example represent alone one third of the total energy consumption (Melquiot, 2009). Among the proposed improvements, a visual display system regarding the use of electrical appliances active, standby or off is strongly recommended. This alternative aims to educate "people who simply do not make a direct link between the energy consumed by an appliance in standby and the pollution it causes," according to the Directorate General for Information Society and Media of the European Commission. French consumers have tended to replace their CRT screens with flat screens, with CRT televisions threatened of extinction. But the difference in energy between the two screens is remarkable as we have seen.

Screen Savers

The screen saver, originally developed to extend the life of the battery of laptops, is now commonly used to automatically reduce the energy used by a computer switched on, when it is idle or is inactive (Webber et al., 2001). However, it is rather advisable to activate the "energy consumption management" as the screen saver does not save energy, on the contrary, unless it is black. The screens operating in standby mode permanently or display colors in the form of texture rather than solid colors in fact require more system resources and therefore energy, as evidenced by this small collection of screen savers offered by default by the Windows operating system (Figure 3):

In addition, the screen saver presents an inconvenience for some users: when an activity has been started on the computer for example, the sudden release of the screen saver disturbs the user most of the time. It can be difficult for some users to get rid of it, since settings are not so obvious to understand and then modify. Setting the screen saver can then become time consuming because of the skills that are necessary for some users to configure it. These characteristics are evidenced in the exploratory analysis that we present later. Here is an example of processes allowing the configuring of a screen saver on most computer workstations equipped with Windows (French version) (Table 1, Appendix C).

With so many steps, it may seem simpler and ecological to disable the "standby" feature that most modern electrical appliances, such as computers and monitors, have, unless a system really protective of the environment exists. The "standby" functionality of screens is strongly energy consuming since it contributes to a power consumption varying from 40 to 120 kWh / year. In this perspective, it is more advisable to use the latest version of the management of energy consumption, namely "SpeedStep", which is an advanced feature of the computer processor.

Public organizations are called to adopt in turn a policy to green procurement to support innovation in the field of energy efficiency. The action of the user deciding to modify its ecological footprint to reduce this consumption becomes that of a responsible consumer. By choosing when visiting an e-commerce website, to trigger a screen saver appropriate for it, the Internet user can help the reduction of releases of greenhouse gases.

Towards a Responsible Consumer

Research conducted by Autio, Heinonen, and Heiskanen (2009) examines how young consumers construct their images of green consumption based on the narrative of ecological discourse. The participants recognize the differences in these green discourses; students speak in their dissertations to "save the planet Earth", "do things responsibly" and "differentiate yourself from all to make a difference". However, collective action to promote sustainable consumption

Figure 3. Screen savers - energy consumers of Windows

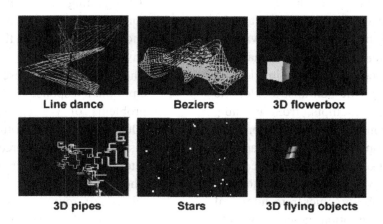

does not appear in their research. According to the latter, green consumers do not complain, do not advocate or do not organize consumer action. The green consumption seems to be an action only on an individual basis where the activity is conducted in parallel but not as collective action among consumers. This brings us to the responsible consumer, more willing when it is only to adjust the screen saver on his computer or the computer he uses at work, than when being part of a group using the same computer.

The concept of « socially responsible consumption » has received considerable attention from the second half of the twentieth century. Researchers have suggested several names to describe such a profile such as "socially conscious consumer" (Anderson & Cinningham, 1972; Webster, 1975; Brooker, 1976; Engel & Blackwell, 1982), "socially concerned consumer" or "concerned" (Belch, 1979, 1982).

Mohr et al. (2001) defined the socially responsible consumer as "a person that bases its acquisition, use and possession of products on the desire to minimize or eliminate adverse effects and the desire to maximize long-term beneficial impact on society." Note from this definition that social responsibility stems from an awareness of consequences of purchase and use on the community and the belief that the power a consumer holds can change entirely a way of consumption. Thus, the socially responsible consumer sacrifices his personal welfare at the expense of that of the community. Here, a person who would visit an e-commerce website, would it see it changing appearance after a certain period, in view of saving screen energy: the administrator of the e-commerce website attempts, from its side, to protect the environment by such action. The important thing becomes the public welfare and interests of the environment in accordance with the norms of society (Velasquez & Rostankowski, 1985). Social responsibility comes from individual accountability of consumption choices in order to achieve collective well-being (Bisaillon, 2005).

Socially Responsible Consumption: A Social Concern and Concern for the Environment

Several research studies show that socially responsible consumption includes both a social and an environmental dimension (Roberts, 1995; Antil, 1984; François-Lecompte & Florence, 2004). Consuming in a socially responsible way means, on the one hand, meeting the collective needs of the society in which the consumer represents an active member, and on the other hand, preserving the physical environment that surrounds us. At this level, attention must be paid to ecological concerns defined by Maloney and Ward (1973)[2] as "the set of specific knowledge and emotions, the level of susceptibility and the extent of behavior to respond to environmental problems and pollution.

This definition, as well as several other works on ecological concern (Balderjahn, 1988; Tucker et al., 1981; Dembkowsky & Hammer-Lloyd, 1994; Anderson & Cunningham, 1972; Webster, 1975; Singhapakdi & Tower, 1991; Zaiem, 2005) show that this concept has been studied following the three-dimensionality of the attitude:

- The cognitive dimension referring to "the subjective knowledge on environmental issues (Zaiem, 2005) reflects all the information and beliefs available to the individual regarding the screens and their energy consumption, as well as the standby mode and its direct link with the protection of the environment;
- The affect which represents in ecology "all emotional responses related to perceived problems of the environment" (Zaiem, 2005) reflects the attitude adopted by a visitor of an e-commerce website using an intelligent screen saver system;
- The conative dimension of environmental concern represents a tendency to act or even the actual behavior of an individual wishing to reduce energy consumption,

this act being realized by the permanent and systematic activation and configuration of the screen standby energy saver.

In the next section we describe the exploratory study we conducted and then present the principal results that will provide answers to these three dimensions (cognition, affect and behavior) of consumer behavior.

RESEARCH METHODOLOGY: THE EXPLORATORY STUDY LEADING TO CONTENT ANALYSIS

To position our research with regard to literature and the perspective to verify and validate the experience we want to put in place, we are interested in the perception that consumers have about the screen saver on their computers. The criterion of data saturation being retained (Mucchielli, 1991), 26 people were interviewed individually in order to better understand the variables suggest an influence consumer behavior. We interviewed these individuals in order to gather information on their Internet habits and their perception of sustainable development in this regard. The interview guide used is presented in Appendix B and the description of our sample is shown in the table in Appendix C.

We have adopted a neutral attitude towards the respondents so as not to influence them in the way they respond. After each interview was transcribed, with an average duration ranging from 4 to 17 minutes, we got a 93-page verbatim. The interviews allowed us to notice that the screen saver appeared as a subject of a very limited interest to consumers. The qualitative data have been analyzed with a table summarizing all the results of our respondents, where each construct referred to a "1" weight as explained in the following example (Figure 4):

We present the results we obtained in the form of a synthesis of themes, constructs and modes evoked in this exploratory qualitative analysis in the following section.

RESULTS

Four trends emerge from the exploratory analysis that was conducted. Ignorance of what a screen-saver is; the fact that it is used for purposes other than setting computer in standby mode; skepticism about its usefulness and the restrictive nature of its setting and finally the almost total lack of feelings experienced by users on the future of the screen saver to protect the environment. These results will be categorized according to the three components of the classical attitude model, namely the cognitive (information, beliefs), the affective (attitude, emotions) and conative or behavioral dimension (the tendency of behavior or behavior itself). We develop these points in the following lines.

Cognitive Responses of the Respondents: Ignorance of What a Screen Saver is and Lack of Information

Setting a screen saver black is not a priority; however, the screen saver described by the interviewees often does not correspond to what may be a economic screen saver, that is black. For the respondents, a screen saver represents a way to save the battery of the computer (12/26) and a way to have fun when we no longer use the screen (7/26) which means that the communication made on the energy consumption of peripheral devices such as screens seems that it could be largely increased. None of the respondents lacked a precise idea about the impact of a black screen saver on the protection of the environment. Its use and especially the "sleep" function of the computer is not a widespread practice. In most cases, the persons interviewed did not know how to use or activate it with the computer settings. Some even do not know what a screen saver is, or how to set it. The screen saver is mostly perceived as an animated screen with images representing a landscape or past holidays. Indeed, for some people, the screen saver should be aesthetically pleasing, pretty to look at like when it is animated by a slide show (9/26).

Figure 4. Part of the table used to concatenate the results of the 26 interviews

Each respondent [n°1 in the picture] has his own table file in which all the questions extracted from the interview guide were written. If a label [n°4] linked to a *cognitive, affective* or *behavioral* dimension [n°6] appeared when reading the transcript of a respondent's interview, a figure"1" [n°3] was marked in a *result* column. Then, we grouped the "result" columns of all our respondents in the final table in order to understand precisely what the answers meant, no matter who the respondents were. This yielded a total of 26 columns, with "1" or empty. Rows in this table indicated the different themes [n°5] issued from the interview guide. Topics and words related to a precise field appeared in a "result" column [n°2], showing the weight of each construct. This finally allowed us to formulate assumptions linking the answers with the respondent's profiles. Thus, each construct had a weight in a first table, made of rows presenting or not figures "1". Another table served to concatenate the 26 files in a summarizing table, enabling us to use the columns of the"1" to add rows and finally write the results obtained by adding all the figures "1"

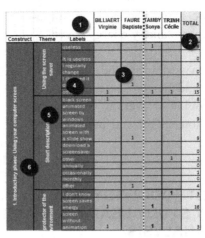

Affective Responses of the Respondents: Indifference and Skepticism

The interviewees in general attach no importance to the shape or color of the screen saver. They do not spontaneously bring value and give it only very little importance as they only set it very rarely or not at all. However, many recognize a direct link between the screensaver and the protection of the environment, constituting potential users of a black screen saver.

On the other hand, some respondents seem skeptical about the usefulness of a screen saver sometimes thinking that their computer does not have one (9/26). Its only use would be according to them of economic order: a reason that has nothing to do with the protection of the environment. Setting a screen saver is also increasingly perceived as a constraint (14/26), so simpler settings would make more sense. Some people have installed a screen saver on their computer, chosen among others because they liked the animation and the colors (9/26). Other people interviewed had no particular interest for their screen saver, but are still susceptible to the "friendly" aspect that it could generate. The screen saver is recognized as practical and economic, though not necessarily an

essential feature. According to some respondents, it would be necessary to explain the use and usefulness of a screensaver. It seems that users, in their majority, do not seem against setting their screen saver if the topic on energy conservation is introduced before. This represents above all something fun and many of those interviewed did not see the screen saver as a tool so indispensable for sustainable development. The aspect of "sustainable development" and "energy conservation" is not a subject well known to computer users in connection with the screen saver. Even if they think that a screen for sustainable development is a screen that is activated quickly, they do not necessarily set it so as not to be annoyed when watching movies for example. The description made by some of them shows it is not used for its primary function of battery saver or protector of the environment, but more like a slide show or a system for amusement (18/26).

Behavioural Responses of the Respondents Use the Screen Saver for Other Reasons / Binding Settings

According to the survey, the screen saver is not used (15/26). Of those questioned, some use very little of their screen saver by having changed it only once (5/26). Others have a black one that "can be configured at the time of purchase" as they mention (8/26). This screen is convenient for them. They use it in the perspective of saving energy so that their battery is discharged more slowly.

On the other hand, some respondents use the default settings offered by the system: black screen (8/26) or slide show (9/26). The majority mentions that they even "force" sleep mode by closing the computer. Ideally, a screen that would consume less energy would quickly resume its functionality when you use it again. Respondents generally did not really remember anything related to the settings of the screen saver. Having mostly not met any problems in setting it, some admit however that for a novice, the configuration is not very intuitive. It stems from this survey that the computers are considered as personal items to which very few people have access. For the interviewee to be interested in a black screen, its installation should be easy and fast to be able to be then configured itself. Hardly anyone configures their computer to automatically switch into sleep mode as the clicks are many and compelling. Thus, for some respondents, the screen saver is too complicated to configure in order for users to be sufficiently aware of.

ANALYSIS OF RESULTS

It is therefore clear from the interviews that if the energy saving which represents an intelligent screen saver setting turns out to be an important factor, and *"if the screen saver is nice, easy to implement and above all initiated by means of some information or an external person, one's sensitivity to "energy saver" screen savers grows"*. Some respondents feel no particular involvement with regard to the protection of the environment having little or never changed the screen settings (11/26). No particular emotion seems to be felt

with the idea of protecting the environment through action on it. This group of respondents is not any more against the idea of acting for its protection because they see some usefulness. Indeed, some respondents may be more interested in a more "interactive" and "attractive" screen saver without explicitly specifying what they want. The expression "useful screen saver" is not so familiar to all respondents. Communicating this new concept to the general public and vendors is therefore essential. The same holds with the definition and operation of a screen saver.

Interviewees appear in all sensitive to the fact that a screen saver could be a good way to save energy. The feelings experienced in response to possible settings indicate that consumers do not yet seem ready to spend time to set the computer screen so that it participates in the protection of the environment. The answers collected allow us to estimate that this is related more to ignorance of the existence of the screen saver, or a lack of information by which the screen saver is capable of protecting the environment, rather than to a deliberate act of malevolence. Indeed, the projection made during the exploratory analysis (see interview guide Topic 3, Appendix B), suggests the idea that awareness would change the situation in favor of the reduction of energy consumption, beneficial for the planet. 16 out of 26 respondents interviewed said that from the moment they learned that a gesture so little as setting the screen saver allowed to contribute in protecting the environment, they would make these changes in parameters in their computer.

This exploratory study was conducted to understand some phenomena related to computer use and sustainable development. The study we have presented aims to serve as a first step to extract the constructs that we wish to measure in a questionnaire taking part in a future quantitative study. This study shows the relative low interest or lack of interest of computer users in sustainable development. Regarding the power consumption of their computer, they don't seem aware about the possibility to protect the environment by switching their screen off when not using it. The exploratory study attests of this unfortunate state.

With these findings which are in no way representative of a possible action at a global level, we gained an initial understanding needed to allow us consider the subsequent confirmatory analysis. We wonder if an e-commerce website should, in an eco-citizen approach, change its appearance when the user does not use his mouse for example within one minute. We wish to examine whether a company that adopts a button to put a curtain of sleep helps position itself as an ethical company, oriented towards sustainable development. This interruption of the use being apprehended by the lack of movement of the mouse does not imply that the user is not reading the information displayed. If so, we can investigate on a company's interest to:

- Choose to dress its website with colors that consume less energy for screens that display them. This would facilitate the emergence of a transparent black screen, optimized for reading to remain enjoyable
- Allow continue reading content displayed on the screen even when it goes into sleep mode. Reading the text portions remains pleasant when the "curtain" falls through the use of a contrast selected not to affect the reading quality. It thus takes into account:
- A contrast less tiring for the eyes of people whose eyesight is reduced,
- A contrast allowing everyone to read the contents of the screen, there including people with deficiencies in color vision, such as color blindness,
- Functionality less laborious to set up than the system configuration of Windows screen saver.

DISCUSSION

Internet users now have to take into account an increasingly sophisticated Internet with technical aspects to learn in order to surf easily and in parallel, ethical consideration for those who caution the sustainable development as a priority. Indeed, the prominence of advertisements, widgets or transparent flash used as pop up or inner windows on the Internet can easily be interpreted as an indicator for the increasing growth of the web as a collection of dynamic sites providing e-commerce, news, entertainment and social network facilities that have become the most popular destinations, whereas ten years ago web usage was focused on mostly static information and content delivery. These changes of the web also induce changes in user behavior. The growing number of dynamic web pages and web applications suggests that interaction with the web client is changing from single-window hypertext navigation to a new mode where several interaction paths are followed in parallel. Users are then faced with a new cognitive overhead when they browse the Web (Weinreich et al., 2006). This overhead is a factor inducing consumers to choose easy to remember paths to attain their target on the website, in order to simplify their foreseeable visit. This implies that they prefer to choose websites that are "intelligent" and on top of it, interested in the protection of the environment. As an example, simple websites are often smartly designed, with simplicity and with only essential features. They work without harnessing the computer too much and are not too hungry of energy. In this sense, the overhead resulting from the current web arena contributes to the adoption of technologies in favor of the protection of the environment.

On top of it, the ecological footprint becomes increasingly important in nowadays, making users more responsible than ever. This is true for commerce in general but also for e-commerce. It is now common to buy an item and deliver a percentage of the sale for the recycling process of the product called "Eco-out"3.

With this paper, we try to go a little further and help organizations to learn how to adopt these new solutions and participate in the protection of the environment. A simple action on the screen can apparently change a lot of aspects linked to the protection of the environment. Customers in their majority are ready to act in favor of the environment when using their computer, this is why it would seem normal to log on a website and then activate an energy saver curtain.

As the protection of the environment becomes a key issue for the socially responsible consumer, it affects his behavior, including the selection of vendors to purchase from. Within the context of electronic commerce, responsible consumers will seek to conduct online purchases and transactions in general on websites of vendors that actively care and contribute to energy saving. Companies that show their involvement by providing systems that are environment friendly or 'green' should be more appealing to customers, new or existing ones, and would be preferred over others that do not. In this perspective, the interest and the actions of an online vendor towards reducing energy consumption and environment protection will be another, new factor for customer attraction and retention. Websites enabling a 'green' use of e-commerce could be a good step towards this direction. In particular, the deployment of an intelligent curtain screen saver in e-commerce websites can be an easy and simple solution, showing effectively the ecological concern of an online vendor.

Projecting an interest for the environment and contributing to its protection, such features can create or enhance consumer trust in an online vendor. Following the typology of McKnight et al. (2002) and McKnight and Chervany (2001, 2002), trust in an online vendor can be defined in terms of consumer beliefs in the benevolence, competence, integrity and predictability of the vendor. These trusting beliefs in an online vendor are defined as follows:

- Trusting Belief - Benevolence is the belief that the online vendor cares about the customer and is motivated to act in the customer's interest and not opportunistically.
- Trusting Belief - Competence is the belief that the online vendor has the ability or power to do for the customer what the customer needs done.
- Trusting Belief - Integrity is the belief that the online vendor makes agreements in good faith, communicates honesty and fulfills promises.
- Trusting Belief - Predictability is the belief that the online vendor's actions (good or bad) are consistent enough that the customer can forecast them in a given situation.

Thus, by offering functionality that contributes to energy saving on their websites, online vendors can invoke consumer trust as they convey benevolence. The provision of such a feature would be perceived as a signal of a benevolent online vendor; a vendor concerned for the environment and for the customers who are also eco-citizens and who, therefore, acts in their interest contributing to the protection of the eco-citizen customers and the environment. In this way, an online vendor can build trust, as it creates customers trusting belief in the online vendor benevolence.

At the same time, the actual implementation and availability of website functionality enabling reduction of energy consumption conveys the vendor's competence. When customers visit e-commerce websites with such ecological, energy saving functionality, they will perceive that the online vendor is not only benevolent and cares for the consumer, but it also has the ability and the means to actually manifest this interest in practice and do what the consumer needs. In this way, online vendors can further succeed in building trust, as consumers will form a trusting belief in the online vendor competence.

The potential value of energy saving features in e-commerce websites, such as the 'curtain' screen saver can be even higher in the Mediterranean region, where one need to take into account the sun and the difficulty to read on screens when the brightness is high. Nowadays, more and more people use their computer outside the house or the office, in order to take advantage of available Wi-Fi connections and enjoy being outside. The Mediterranean weather facilitates this kind of computer use. However, in the Mediterranean region, the sun shines strongly, making laptops screens hard to read when you work outside. The readability of content on a screen when the luminosity of the environment is very high can be problematic. Indeed, the low contrast occurred by the sun and the screen won't help the visitors of e-commerce websites at all. Previous research has shown that the contrast created between the dynamic colors (text) and the dominant one (background) could slow the reading of contents (Hall & Hanna, 2003) and damage the user's eyes that make efforts to read the content. This can lead to negative mood feelings towards an online vendor and, eventually to a degradation of consumer trust (Pelet & Papadopoulou, 2010a, 2010b). A 'curtain' screen saver, which provides a dark appearance of an e-commerce website by increasing the contrast between screen colors and the physical light, could be a useful solution to this issue. Its potential can be viewed as two-fold. First, it will facilitate the readability of the content of the e-store and result in a positive mood of consumer which will ultimately increase trust in the online vendor, both in terms of benevolence and competence. Second, it will meet the requirements for energy saving, further enhancing consumer trust in the online vendor.

Thus it could be interesting to consider what the OLPC (One Laptop Per Child) project managed to do, with the "XO" project: "A small machine with a big mission" (OLPC, 2010). The XO is a potent learning tool designed and built especially for children in developing countries, living in some of the most remote environments. It's about the

size of a small textbook. It has built-in wireless and a unique screen that is readable under direct sunlight for children who go to school outdoors. It's extremely durable, brilliantly functional, energy-efficient, and fun. The idea of a screen enabling readability under sunlight could possibly be adopted in the context of e-commerce, through the provision of a 'curtain' screen saver which offers a similar effect.

This approach towards energy saving and environment protection is not restricted to e-commerce but can also be extended to mobile commerce. In the mobile industry, m-commerce players could be even more interested in sustainable development since mobile constructors have been trying to build batteries for mobile devices that enhance their duration. By making 'curtain' screen savers appear after a short period of inactivity when users don't touch the screen or press any button, mobile devices preserve battery energy and thus reduce the greenhouse gas emissions. In this way, such energy saving features on m-commerce websites should potentially be more valuable and welcome in the mobile industry. Thus, it can be posited that the 'green' use of IT and online commerce is a topic where m-commerce could is in advance and more fruitful in comparison to e-commerce.

LIMITATIONS AND FUTURE RESEARCH

We are aware that this study is not without limits. An exploratory analysis is not sufficient at all to offer results on which to build a strategy. This study needs to be followed by a confirmatory analysis showing that people don't use or don't know how to use or don't know the existence of their screensavers. Some figures are not recent enough to be sure that they represent the actual reality of the screen market. In some developed countries, flat screen (LCD and Plasma) are becoming more and more used in households as well as in universities, schools and companies, but in others, CRT screens remain the most popular. Thus a study focused on a precise area might be more precise.

The responsible consumer online appears ready to use a system to rest the screen, if there is such a command that enables that. A system extension for the Firefox browser, displaying on the inactive screen a screen saver, like a curtain of sleep, resting the screen by consuming less energy while ensuring optimal readability, is being designed. It is the same with regard to the button for Websites, currently being finalized. Aiming to confirm the content analysis conducted during the exploratory phase in a forthcoming study on the Internet, this in vivo experiment was designed to verify the interest of the e-commerce websites to rethink the colors they display when the screen becomes inactive. The experiment allows visiting an e-commerce website, which is under construction, featuring two buttons triggering the screen saver. The experiment will measure the use made of these buttons, and the perception of users about them. A button, designed with a programmer of extensions[4] for Firefox, will serve to activate the screen saver, enabling customers to participate in sustainable development from that browser. Another button provides the same functions as the extension mentioned above, being adapted to an e-commerce website. Once activated, this button allows the emergence of a "curtain of sleep" that falls over the website and provides a dark version of its appearance. This will be offered as a Grüner Punkt type button, symbol of sustainable development on a website (Figure 5).

As part of the Firefox browser, the button can exist provided that the proposed extension is accepted by the Mozilla community. As part of an e-commerce website, companies decided to opt for a policy of "sustainable electronic commerce" could offer this type of button on their interface.

Figure 5. Logo representing the "green dot" (die Grüne Punkt), symbol of sustainable development

The simple "grüne punkt" button provided on the website could thus be seen as a tool that affords to save the earth. The ability to test this button that makes the curtain screen saver falling after a short period of time when the mouse is not used could be interesting. This is part of our future research endeavours.

REFERENCES

Anderson, W. T. Jr, & Cunningham, W. L. (1972). The socially conscious consumer. *Journal of Marketing*, 36(3), 23. doi:10.2307/1251036

Antil, J. (1984). Socially responsible consumers: Profile and implications for public policy. *Journal of Macromarketing*, 4, 19–32. doi:10.1177/027614678400400203

Autio, M., Heiskanen, E., & Heinonen, V. (2009). Narratives of green consumers – the antihero, the environmental hero and the anarchist. *Journal of Consumer Behaviour*, 8(1), 40–53. doi:10.1002/cb.272

Balderjahn, I. (1988). Personality variables and environmental attitudes as predictors of ecologically responsible consumption patterns. *Journal of Business Research*, 17(1), 51–56. doi:10.1016/0148-2963(88)90022-7

Belch, M. A. (1979). Identifying the socially and ecologically concerned segment through lifestyle research: Initial findings. In Henion, K. E. II, & Kinnear, T. C. (Eds.), *The conserver society* (pp. 69–81). Chicago, IL: American Marketing Association.

Belch, M. A. (1982). A segmentation strategy for the 1980's: Profiling the socially-concerned market through life-style analysis. *Journal of the Academy of Marketing Science, 10*(4), 345–358. doi:10.1007/BF02729340

Bisaillon, V. (2005). Le consumérisme politique comme nouveau mouvement social économique, in Consumérisme politique I: Du boycott au boycott. Chaire de responsabilité sociale et de développement durable ESG-UQAM, Receuil de textes CEH/RT-30-2005. In *Proceedings of the 8ème Séminaire de la Série Annuelle sur les Nouveaux Mouvements Sociaux Économiques* (pp. 6-17).

Breuil, H., Burette, D., & Flüry-Hérard, B. (2008, Décembre). *TIC et Développement durable, (Rapport), Ministère de l'Ecologie, de l'Energie, du Développement Durable et de l'Aménagement du Territoire, Conseil général de l'environnement et du développement durable, N° 005815-01.* Retrieved from http://www.telecom.gouv.fr/fonds_documentaire/rapports/09/090311rapport-ticdd.pdf

Brooker, G. (1976). The self-actualizing socially conscious consumer. *The Journal of Consumer Research, 3*(2), 107–112. doi:10.1086/208658

Dembkowsky, S., & Hammer-Lloyd, S. (1994). The environmental value-attitude system model: A framework to guide the understanding of environmentally-conscious consumer behavior. *Journal of Marketing Management, 10*(4), 593–603. doi:10.1080/0267257X.1994.9964307

Engel, J. F., & Blackwell, R. D. (1982). *Consumer behaviour.* New York, NY: Oxford University Press.

FEVAD. (2010). *Fédération du e-commerce et de la vente à distance, Chiffres Clés disponibles sur le site web de cet organism.* Retrieved from http://www.fevad.com

Gartner. (2007, October 7-12). *Green IT: The new industry shockwave.* Paper presented at the ITXPO Symposium, Orlando, FL.

Hall, R. H., & Hanna, P. (2003). The impact of web page text-background color combinations on readability, retention, aesthetics, and behavioral intention. *Behaviour & Information Technology, 23*(3), 183–195. doi:10.1080/01449290410001669932

Hollands, J. G., Parker, H. A., McFadden, S., & Boothby, R. (2002). LCD versus CRT displays: A comparison of visual search performance for colored symbols. *Human Factors, 44*, 210. doi:10.1518/0018720024497862

Infos-Industrielles. (2007). *Les TIC au service de l'efficacité énergétique.* Retrieved from http://www.infos-industrielles.com/dossiers/1156.asp

Jiun-Haw, L., Liu, D. N., & Wu, S.-T. (2008). *Introduction to flat panel displays.* New York, NY: John Wiley & Sons.

Kawamoto, K., Koomey, J. G., Nordman, B., Brown, R. E., Piette, M. A., Ting, M., et al. (2001). Electricity used by office equipment and network equipment in the US (Tech. Rep. No. LBNL-45917) Berkeley, CA: National Laboratory.

Luder, E. (1997). Active matrix addressing of LCDs: Merits and shortcomings. In MacDonald, L. W., & Lowe, A. C. (Eds.), *Display systems* (pp. 157–172). Chichester, UK: John Wiley & Sons.

Maloney, M. P., & Ward, M. P. (1973). Ecology, let's hear it from the people. *The American Psychologist, 28*, 583–586. doi:10.1037/h0034936

McKnight, D. H., & Chervany, N. L. (2001-2002). What trust means in e-commerce customer relationships: An interdisciplinary conceptual typology. *International Journal of Electronic Commerce, 6*(2), 35–59.

McKnight, D. H., Choudhury, V., & Kacmar, C. (2002). The impact of initial consumer trust on intentions to transact with a web site: A trust building model. *The Journal of Strategic Information Systems, 11*, 297–323. doi:10.1016/S0963-8687(02)00020-3

Melquiot, P. (2009). *Technologies de l'information et de la communication (TIC), impacts sur l'environnement et le climat.* Retrieved from http://www.actualites-news-environnement. com/19871-Technologies-information-communication-TIC-environnement-climat.html

Menozzi, M., Napflin, U., & Krueger, H. (1999). CRT versus LCD: A pilot study on visual performance and suitability of two display technologies for use in office work. *Displays, 20,* 3–10. doi:10.1016/S0141-9382(98)00051-1

Mucchielli, A. (1991). *Les Méthodes de Contenus, Que sais-je?* Paris, France: Presses Universitaires de France.

Murugesan, S. (2008). Harnessing green IT: Principles and practices. *IEEE IT Professional,* 24-33.

OLPC. (2010). *One laptop per child.* Retrieved from http://laptop.org/en/

Pelet, J.-É., & Papadopoulou, P. (2009, September 25-27). The effects of colors of e-commerce websites on mood, memorization and buying intention. Paper presented at the 4th Mediterranean Conference on Information Systems, Greece.

Pelet, J.-E., & Papadopoulou, P. (2010a). Consumer responses to colors of e-commerce websites: An empirical investigation. In Kang, K. (Ed.), *E-Commerce.* Rijeka, Croatia: In-Tech. doi:10.5772/8897

Pelet, J.-E., & Papadopoulou, P. (2010b). The effect of e-commerce websites colors on consumer trust. *International Journal of E-Business Research.*

Roberts, J. A. (1995). Profiling levels of socially responsible consumer behaviour: A cluster analytic approach and its implications for marketing. *Journal of Marketing Theory and Practice,* 97-117.

Singhapakdi, A., & La Tour, M. S. (1991). The link between social responsibility orientation, motive appeals, and voting intention: A case of an anti-littering campaign. *Journal of Public Policy & Marketing, 10*(2), 118–129.

Sustainable Energy Europe Campaign. (2008). *Directorate-general for energy and transport.* Retrieved from http://www.sustenergy.org/

Tucker, L. R., Dolich, I. J., & Wilson, D. T. (1981). Profiling environmentally responsible consumer-citizens. *Journal of the Academy of Marketing Science, 9*(4), 454–478. doi:10.1007/BF02729884

UE ENERGY STAR. (2009). *Introduction au Programme européen ENERGY STAR.* Retrieved from http://www.eu-energystar.org/fr/index.html

Velasquez, M. G., & Rostankowski, C. (1985). *Ethics: Theory and practice.* Upper Saddle River, NJ: Prentice-Hall.

Webb, D. J., Mohr, L. A., & Harris, K. E. (2008). A re-examination of socially responsible consumption and its measurement. *Journal of Business Research, 68,* 91–98. doi:10.1016/j.jbusres.2007.05.007

Webber, C. A., Roberson, J. A., McWhinney, M. C., Brown, R. E., Pinckard, M. J., & Busch, J. F. (2006). After-hours power status of office equipment in the USA. *Energy, 31,* 2823–2838. doi:10.1016/j.energy.2005.11.007

Webster, F. E. (1975). Determining the characteristics of the socially conscious consumer. *The Journal of Consumer Research, 2*(3), 188–196. doi:10.1086/208631

Wright, S. L., Bailey, I. L., Tuan, K.-M., & Wacker, R. T. (1999). Resolution and legibility: A comparison of TFT-LCDs and CRTs. *Journal of the Society for Information Display, 7,* 253–256. doi:10.1889/1.1985290

Zaiem, I. (2005). Le Comportement Ecologique du Consommateur: Modélisation des Relations et Déterminants. *La Revue des Sciences de Gestion: Direction et Gestion*, *40*, 75–88. doi:10.1051/larsg:2005032

ENDNOTES

[1] http://www.blackle.com/

[2] Quoted by Ling-yee, 1997

[3] Since 2006 all electrical and electronic products sold online or in store are subject to an Eco-participation in addition to the normal price. It is visible and transparent information which concerns the price of each product. In accordance with European Directive 2002/96-CE and decree of July 20, 2005, Eco-systems provide a mission of general interest: setting up throughout the French territory. This is a national scheme for collect, clean up and recycle waste electrical and electronic equipment at end of life equipment.

[4] Extensions (or add-ons) of Firefox are programs designed to provide functionality to the browser.

APPENDIX A: PRESENTATION OF THE THREE TECHNOLOGIES LCD, PLASMA AND CRT, USED IN COMPUTER MONITORS

- LCD Technology (Liquid Crystal Display) is based on a screen composed of two transparent grooved parallel plates, oriented at 90°, between which there is a thin layer of liquid containing molecules (liquid crystals) which have the property to move when subjected to electric current. LCDs are composed of many tiny liquid crystals arranged in rows and columns. Liquid crystal molecules can be reoriented by an electric field. LCDs function by twisting the axis of polarization of the light as it passes through the liquid crystal such that when the light reaches the front polarizer, it is oriented correctly to pass through, allowing it to be seen by an observer (Hollands, 2002). When an electric field is applied, the structure is untwisted and no light is emitted (Jiun-Haw, Liu & Wu, 2008). The LCDs used for computer displays are most commonly active matrix, with an electronic switch at each pixel location (Luder, 1997);

- Plasma Technology (PDP Plasma Display Panel) is based on an emission of light through the excitation of a gas. We distinguish between screens called "matrix" passive technology TN (Twisted Nematic), the pixels of which are controlled by row and column, and screens called "active matrix" technology TFT (Thin Film Transistor), in which each pixel is controlled individually;

- The cathode ray tube (CRT Cathode Ray Tube) is composed of a heated filament, cathodes and anodes in the form of pierced lenses subjected to a potential difference which create an electric field accelerating electrons. They just hit the screen on which a fluorescent layer is put reacting to the shock of electrons by creating a spot of light. Conventional CRT displays rely on an evacuated glass tube with a display screen at one end and electron guns at the other. The guns emit electron beams that are deflected to various screen locations by use of magnetic fields generated by a deflection yoke. The beams strike phosphors near the screen, converting the electrons to observable light energy. A particular pixel is illuminated or not by coordinating the timing of the gun output and the magnetic fields controlling the deflection yoke.

Given the vastly different technologies underlying each display type, it is no wonder that the images rendered with LCD and CRT displays differ. Among other differences, the pixel definition with LCDs is much sharper than for CRTs (Wright, Bailey, Tuan, & Wacker, 1999). CRTs tend to produce a blurred Gaussian distribution of light at each pixel, whereas LCDs produce a sharp edge to each pixel (Menozzi, Napflin, & Krueger, 1999). This sharp edge has clear advantages, but the edge may be problematic when rendering curves (leading to aliasing problems) and may introduce a high spatial frequency noise component into the display.

APPENDIX B: INTERVIEW GUIDE USED IN THE CONTEXT OF THE EXPLORATORY ANALYSIS

Interview Guide

The interview can be conducted only if the respondent has a computer, or can modify the settings on the computer screen saver if it is for professional use. He/she should be aware of this information before starting.

Terms of Service
Respondent No:
  ○ Date: ... ... / ... ... / ... ..
  ○ Location (home, work or other please specify): ... ...
  ○ Top of the interview: ... .. h ... . Min
  ○ End of interview: ... .. h ... . Min
  ○ Interview duration (in minutes): min
  ○ The computer used by the respondent is:
Personal Computer: □ Computer located in the workplace: □ Computer loaned by the company: □
• Screen Type: Flat Screen: □ CRT Monitor: □

Start of interview

Thank you for accepting to devote to me a little of your time to help me achieve this study. The topic of the interview is computer screens. I invite you to speak spontaneously and freely, all information is likely to be interesting to me.

1. Introductory phase: Using your computer screen
    1. Could you talk about the use you make of the screen saver on your computer?
    2. If you have a screen saver configured on your screen, can you briefly to describe it?
    3. What is a screen protector of the environment for you?
2. Focusing on subject phase

   Now I would like you to remember the last time you set the screen saver on your computer.

4. Can you remember your setting of the screen saver?
5. What do you think of the ease that you have to set the screen saver on your computer?
    3. Phase of deepening
       Theme 1: the constituent steps of setting the screen saver
6. What does the screen saver on your computer represent for you?
7. What is a useful computer screen saver for you?
8. What settings of the screen saver allow you to have a positive impression on the computer?
   Theme 2: The emotions and feelings felt after setting the screen saver on your computer
9. Have you ever felt something deep in setting your screen saver?

10. Could you explain in detail?
    Theme 3: The feelings and reactions vis-à-vis a computer screen saver
    - Thank you for reading the following text:

Xavier is sensitive to sustainable development. Since he knows that a screen saver configured to turn on after a few minutes of inactivity of the computer mouse can save the environment, he is interested in the question. When invited to dine with friends who appreciate good food and good wine, he spent a little time with Tonio on his computer, and found that the screen saver never appeared, after they had left the computer during dinner. A fine gourmet, Xavier can appreciate the quality of foods. The atmosphere is friendly. Suddenly, he is urged to get up to go set the screen saver, and he starts explaining why this action can save the planet ...

Reread several times if necessary

1. Can you describe *the first impressions* of Tonio when he sees Xavier getting up hastily to go set the screen saver?
2. In your opinion, what does he feel?
3. And to you, what does this gesture inspire?
4. What do you think will be the reaction of Tonio?
5. And you, what would you do?

Data on respondent

- Sex
- Age: ()

Put a cross in the specified location:

- less than 18 years old, (....)
  between 18 and 30 years old (....)
  between 30 and 40 years old (....)
  more than 40 years old (....)
- Education: ... ...
- Occupation: ... ... Nationality: ... ...

Your actions and your statements were observed and recorded: and you have the option to hear the part of the recording in question and, if desired, to have its removal or destruction.

APPENDIX C

Table 1. Description of the sample used in the exploratory analysis

	Sample Description
Nature	Consumers who have French nationality and live in France
Sex	- 65% women - 35% men
Age	- 88.5% $\in$ [15-25] - 11.5% $\in$ [26-35]
Number	26 (semantic saturation point reached)
Study Type	Exploratory Study
Selection	Diverse population of respondents in terms of gender, ages and professions
Recruitment	Respondents were recruited following the selection criteria listed above

This work was previously published in the International Journal of E-Services and Mobile Applications, Volume 3, Issue 2, edited by Ada Scupola, pp. 20-38, copyright 2011 by IGI Publishing (an imprint of IGI Global).

Chapter 5
Identifying the Direct Effect of Experience and the Moderating Effect of Satisfaction in the Greek Online Market

Michail N. Giannakos
Ionian University, Greece

Adamantia G. Pateli
Ionian University, Greece

Ilias O. Pappas
Ionian University, Greece

ABSTRACT

The scope of this paper is to examine the perceptions which induce the Greek customers to purchase over the Internet, testing the direct effect of experience and the moderating effect of satisfaction. A review of research conducted in the Greek online market demonstrates that satisfaction, self-efficacy, and trust keep a prominent role in the Greek customers' shopping behavior. To increase understanding of this behavior, two parameters of the UTAUT model, performance expectancy and effort expectancy, are incorporated. The findings demonstrate that customers' perceptions about all of the parameters do not remain constant, as the experience acquired from past purchases increases. Moreover, the relationship of experience with self-efficacy and intention to repurchase changes, as satisfaction gained from previous purchases increases. The implications of this study are interesting not only for the Greek but also for the Meditterranean researchers and e-retailers, since the Mediterranean ebusiness market shares several cultural similarities with the Greek market.

INTRODUCTION

Retaining online customers is of great importance for all firms as it makes them capable of gaining advantage over their competitors. Customers who spend more money, buy more often, refuse to re-spond to competitors' promotions and use positive word-of-mouth constitute the loyal customers of a firm. Those customers are the most important ones (Dick & Basu, 1994; Bolton, 1998; Rust & Donthu, 1995). Studies have shown that by increasing customer retention there are increased

DOI: 10.4018/978-1-4666-2654-6.ch005

profits for companies that compete in mature and highly competitive markets, especially service industries, such as banking, hotels and airlines (Fornell & Wernerfelt, 1987; Reichheld & Sasser, 1990). For instance, Reichheld and Schefter (2000) found that increasing customer retention by just 5 percent could increase firms' profitability by 25 percent to 95 percent.

The economic growth, the technological infrastructure, the regulatory framework, the living standards and the weather conditions are just some of the factors affecting the digital profile of a country (Observatory of the Greek Information Society, 2010). Specifically, compared to the rest European countries, the Mediterranean countries are considered as laggards of e-commerce adoption (Vehovar, 2003). According to Turk et al. (2008), the character, the culture and the lifestyle of the Mediterranean inhabitants set the intention to repurchase as a matter of high priority. This study aims at increasing understanding of the special features of online Mediterranean customers and their behavior towards repurchasing.

Greece, as a country with low levels of Internet and ecommerce adoption (Papazafeiropoulou et al., 2001; Buhalis & Deimezi, 2003; European Commission, 2009), as well as a member of the European Union, geographically located in the Mediterranean area, comprises a very interesting case study. According to a recent study of the Observatory of the Greek Information Society (2010), consumer awareness of Greeks is underdeveloped and this seems to be a matter of idiosyncrasy. Although Greece is rated fifth among European counties in complaining about online retailers' services, Greeks (72%) are first among Europeans on not doing anything to dispense justice.

Recent research in the area of online customer behavior has revealed that customers' previous online experience is likely to have an effect on their future intention to re-use an online application (Chiu et al., 2009). Saprikis et al. (2010) point out differences between the perceptions of adopters and non-adopters (non experienced users) in the Greek market. Since the Greek electronic market is currently in growth (European Commission, 2009), young Greeks are becoming more willing to adopt e-commerce practices (Angeli & Kyriakoullis, 2006). Nevertheless, the Greek market comprises a mixed market, consisting of users with diverse levels of experience and great differences in behavior and perceptions.

The overall purpose of the study is to examine the key factors of the B2C e-commerce in the Greek market. The detailed objectives include: a) identifying the key factors affecting the online shopping behavior of the Greek consumers, b) empirically validating and testing the constructs corresponding to these factors using data gathered from a sample of online consumers in Greece, c) investigating the direct effect of online shopping experience on the key factors identified, and d) investigating the moderating effect of satisfaction on the key factors identified.

More specifically, in this paper, we investigate the relationship between the Greek users' online experience and the following key factors affecting their online shopping behavior; self-efficacy (SEF), effort expectancy (EE), performance expectancy (PE), trust (TR) and intention to repurchase (IR). These key factors have stemmed from review of studies on the special features of the Greek online market as well as from dominant theoretical research in the area of technology acceptance. We argue that they are positively associated with users' number of purchases and they are also indirectly affected by satisfaction that derives from previous experience. In order to support our arguments, we are based on a set of models and theories, such as Unified Theory of Acceptance and Use of Technology (UTAUT) (Venkatesh et. al., 2003), Expectation Confirmation Theory (ECT) (Oliver & DeSarbo, 1988) and Social Cognitive Theory (SCT) (Bandura, 1986).

The paper is organized as follows. In the next section, we discuss existing status of research concerning the Greek online shopping market, as well as theories underlying the key factors discussed in the paper. Next we present the methodology as well as the measures adopted for collecting data on the online shopping behavior of the Greek market. Then we present the empirical results derived and discuss these results. The last section of the paper raises the key theoretical and practical implications of this research and discusses several ideas on further research in the area.

LITERATURE REVIEW AND RESEARCH HYPOTHESES

Existing Research on the Greek Online Shopping Market

There are several recent studies investigating the special features of the Greek e-commerce market (Table 1).

According to Barbonis and Laspita (2005), Greeks are highly affected by remarks made from friends and relatives. No matter whether they are positive or negative, they influence their satisfaction in a significant way. Gounaris et al. (2010) argue that the word-of-mouth communication influences the Greek consumers' satisfaction and their attitude towards online shopping. Moreover, Xanthidis and Nicholas (2007) support that the Greek customers see the buying process more as a type of interaction rather than just as a procedure. Most Greek consumers seem to prefer to talk to salesmen before completing a purchase (Buhalis & Deimezi, 2003), while some others prefer to touch and feel the products they intend to buy (Xanthidis & Nicholas, 2007).

Previous research has indicated that it is interesting to study self-efficacy in cultures with high levels of uncertainty avoidance (Hernandez et al., 2009). Greeks tend to avoid uncertainty (Barbonis & Laspita, 2005). Self-efficacy affects users' intention to adopt online shopping (Luarn

& Lin, 2005) and is related with experience, as the latter leads to higher self-efficacy (Bandura, 1986). Several research studies in the area have also revealed that security and privacy affect in a significant way the Greek Internet users' feeling of trust towards online shopping (Buhalis & Deimezi, 2003; Harkiolakis & Halkias, 2007; Zorotheos & Kafeza, 2009; Saprikis et al., 2010). This feeling is different for Greeks and British, as Greeks feel more anxious about using ecommerce, which makes it more difficult for them to trust online retailers (Angeli & Kyriakoullis, 2006). Moreover, Greeks present the highest levels of distrust among Europeans for independent and public authorities that are supposed to protect them. Only 30% of them feel safe from the measures taken for online protection (Observatory of the Greek Information Society, 2010).

Review of all these studies has resulted in identifying the following key parameters defining the Greek online shopping behavior; satisfaction, self-efficacy, and trust.

Background Theories

Throughout the information systems literature, several models and theories have been proposed that routinely explain over 40 percent of the variance in individuals' intention to adopt (Venkatesh & Morris, 2000; Davis et al., 1989). More recent research has resulted in proposing a Unified Theory of Acceptance and Use of Technology (UTAUT), which can explain as much as 70 percent of the variance in intention to adopt (Venkatesh et al., 2003). Nevertheless, there are several complementary theories that explain intention to adopt, such as the theory of reasoned action (TRA), the motivational model (MM), the theory of planned behavior (TPB), the expectation confirmation theory (ECT) and the social cognitive theory (SCT). In our attempt to examine the Greek online market, we are based on a synthesis of UTAUT (Venkatesh et. al., 2003) with SCT (Bandura,

Table 1. Current research concerning the Greek online shopping market

Research Study	Research Subject	Results
Papazafeiropoulou et al. (2001)	Examines the case of Greece as an example of a country that currently presents low levels of electronic commerce adoption but has the potential to grow fast.	The membership of Greece in the European Union is positive since it helps policy makers in the country to follow directives adopted at an international level.
Buhalis & Deimezi (2003)	Demonstrates a number of indicators that synthesize the level of e-commerce penetration.	23% of the sample does not trust Internet transactions, 21% of the sample does not have the required knowledge/skills. Mediterranean countries tend to approach their purchasing process as a form of social exchange and interaction.
Barbonis & Laspita (2005)	Explores the Greek consumers' attitude towards e-commerce, given their cultural identity.	58% of the sample believes that positive comments from their friends and relatives could relieve their anxiety in buying through the Internet.
Harkiolakis & Halkias (2007)	Investigates the perceptions of online buyers in Greece and studies their online behavior when participating in e-communities.	No bias regarding sex and profession. Primary motivation was the convenience offered. Security was considered the most important issue.
Angeli & Kyriakoullis (2006)	Addresses issues of cross-cultural validity of 'trust attributes' by comparing two European nations (UK and Cyprus) which are characterized by different cultural values.	Greeks were more anxious about using e-commerce than British. Young Greeks are becoming more individualist and willing to undertake risk-taking activities in e-commerce. Greek people living in a collectivist society prefer to follow the habits of other people like them. The study showed an important evolution of cultural values within the young generation.
Zorotheos & Kafeza (2009)	Examines whether the Greek internet users' privacy concerns and perceived privacy control affect their willingness to use internet web places in order to transact.	Both privacy concerns and perceived privacy control have direct impact not only to users' trust toward the web site as expected but also to the willingness to transact through the Internet.
Xanthidis & Nicholas (2007)	Identifies the reasons why Greek digital consumers do not commit to eCommerce transactions, explains whether this phenomenon relates only to local businesses or, if it is a general phenomenon, suggests possible "incentives" by the businesses that could trigger positive reactions on the part of digital consumers.	Greek consumers do not trust plastic money, which is essential part of every electronic transaction. Not quite comfortable with the process of making a transaction and prefer instead to touch and feel the products they intend to buy. They tend to trust well- known international brands and firms rather than the relevant Greek ones.
Theodoridis & Chatzipanagiotou (2009)	Extends the test of the functional relationship between store image attributes and customer satisfaction in the market environment of Greece.	Pricing and Products-related attributes were equally significant among all types of customers that occurred from the study.
Gounaris et al. (2010)	Examines the effects of service quality and satisfaction on three consumer behavioral intentions, namely word-of-mouth, site revisit, and purchase intentions in the context of internet shopping.	E-service quality has a positive effect on e-satisfaction behavioral intentions, namely site revisit, word-of-mouth communication and repeat purchase.
Maditinos et al. (2010)	Develops a model to predict and explain consumers' intentions to transact with an internet based B2C e-commerce system, based on TAM.	Suggests a "narrower view" of the overall level of perceived risk dividing it into two basic sub-factors: the transactional security and the product delivery and services.
Saprikis et al. (2010)	Examines the perceptions of Greek university students' adopters and non-adopters of online shopping in terms of demographic profile, expectations of online stores.	Greatly significant difference was identified between adopters and non-adopters. 55% of the sample reports security and privacy reasons for not buying online. 19% of the sample mentions unawareness of the buying procedure through the Internet.

1986), providing arguments for including self-efficacy, and ECT (Oliver & DeSarbo, 1988), discussing the key role of satisfaction.

Unified Theory of Acceptance and Use of Technology (UTAUT)

Technology Acceptance Model (TAM) constitutes the most commonly used framework in IS contexts, designed to predict information technology acceptance and usage on the job (Davis, 1989). TAM proposes the belief, attitude, intention and behavior causal relationship to explain and predict technology acceptance within a group of users (Davis, 1989; Davis et al., 1989). This model proposes that perceived usefulness and perceived ease of use influence a person's attitude toward using a new technology, which in turn influences his intention to use it (Shih, 2004). In 2003, Venkatesh et al. propose the Unified Theory of Acceptance and Use of Technology (UTAUT), which combines a great number of previous TAM-related studies. In the UTAUT model, performance expectancy and effort expectancy were used to resemble the traditional constructs of 'perceived usefulness' and 'perceived ease of use', respectively, from the original TAM study.

In the context of online shopping in Greek market, this study measures performance expectancy and effort expectancy using constructs adapted from previous studies (Devaraj et al., 2002 ; Pavlou, 2003). Since effort expectancy and performance expectancy are widely studied for predicting acceptance and use of information systems, our research for the Greek market would be incomplete without these two key constructs.

Social Cognitive Theory (SCT)

The Social Cognitive Theory (SCT) (Bandura, 1986) posits that cognition exerts a considerable influence on the construction of one's reality, as it selectively encodes information, and imposes structure on actions (Jones, 1989). This theory is used to explain how people acquire and maintain certain behavioral patterns, while it also provides the basis for intervention strategies (Bandura, 1986). Environment, people and behavior are three factors that affect the evaluation of behavioral change. People learn through observing others' behavior, attitudes, and outcomes. Compeau and Higgins (1995) applied and extended SCT to the context of computer utilization. Nevertheless, the nature of the model and the underlying theory allows it to be extended to acceptance and use of information technology in general (Venkatesh et al., 2003). Within the social cognitive theory, self–efficacy is addressed as a form of self-evaluation that influences the decision and effort to undertake a certain behavior. Self-efficacy reflects an individual's belief in his/her capability to perform a task and, thus, it promotes the sharing of knowledge (Gravill & Compeau, 2008).

In the Greek market, the lack of knowledge and skills (awareness) about online shopping procedures is widely accepted, even from the customers themselves (Buhalis & Deimezi, 2003). Hence, an interesting issue raised is how self efficacy influences the Greek consumer's online shopping behavior.

Expectation Confirmation Theory (ECT)

The concept of Information Systems (IS) continuance has been examined a lot and in different ways. Existing studies agree that continuance behavior assumes committing IS use as a part of a normal activity in progress. Hence, continued use of IS can be evident of continuance behavior (Lee & Kwon, 2009). The IS continuance model originates from the Expectation Confirmation Theory (ECT). The ECT has been used to investigate consumers' repeat decision in the consumer behavior literature (Oliver, 1993). ECT considers satisfaction as a key variable for customers' continuance intention. Satisfaction depends on beliefs, experiences, relationships, and other psychological factors (Giese & Cote, 2000).

Based on arguments provided by previous research on the Greek online shopping market (Angeli & Kyriakoullis, 2006; Gounaris et al., 2010), we herein identify satisfaction as a critical factor moderating the effect of previous experience on other parameters of the Greek consumers' behavior.

Research Hypotheses

Based on the aforementioned theoretical and empirical research, a set of hypotheses have been formulated and examined in this paper.

Self-Efficacy (SEF)

Self-efficacy beliefs determine how people feel, think, motivate themselves and behave (Bandura, 1994). Taylor and Todd (1995) argue that the capability to perform a task is greater for experienced users. Therefore, the increase of online shopping experience is likely to enforce consumer's self efficacy. Our hypothesis proposes that:

H1: Users' previous experience has a positive and significant effect on their self-efficacy in online shopping.

Effort Expectancy (EE)

Effort expectancy refers to consumers' perspective that online shopping is free of effort. It is likely that improvements in ease of use may also depend on the increased experience, especially in the Greek market, where the online shopping adoption is quite low. Therefore, the following hypothesis is formulated:

H2: Users' previous experience has a positive and significant effect on their effort expectancy from online shopping.

Performance Expectancy (PE)

Performance expectancy refers to the degree to which consumers believe that online shopping improves their transaction experience. Since an online shop serves as an improved communication interface between retailers and customers, it is expected that online shopping experience will be positively and significantly related to perceived usefulness (Ajzen & Fishbein, 1980). Hence, we hypothesize that:

H3: Users' previous experience has a positive and significant effect on their performance expectancy from online shopping.

Trust (TR)

In the online shopping context, trust is defined as the buyer's belief that the e-vendor is behaving ethically (Pavlou & Fygenson, 2006). It is clear that, although trust may not be essential for a customer to visit online shops, it is indispensable if the customer has to engage in a transaction or in any other kind of ongoing relationship. In order to develop long-term relationships with their customers, it is important for e-retailers to both develop and nurture consumer trust (Palvia, 2009). Specifically, they need to provide their customers with a continuance sense of security, privacy and reliability. Previous positive experience is likely to have an impact on the customers' sense of trust (Chiu et al., 2009). Therefore, we argue that:

H4: Users' previous experience has a positive and significant effect on their trust in online shopping.

Intention to Repurchase (IR)

Intention to repurchase (continuance intention) is herein defined as an intention to continue using online shops for making their purchases. Previous research has proven that current shopping experience is positively associated with customers' loyalty (Chiu et al., 2009). Here, we wish to examine if the effect of the Greek consumers' online shopping experience on their intention to repurchase is significant. Therefore, we hypothesize:

H5: Users' previous experience has a positive and significant effect on their intention to repurchase.

Satisfaction (STF)

Customer satisfaction is a measure of subjective evaluation of any outcome or experience associated with the purchase of a product/service (Westbrook, 1980). For e-retailers, customer satisfaction often leads to favorable results, such as improved customer retention, positive word of mouth and increased profits (Zeithaml, 2000). Studies have suggested that customer perceptions of service quality and satisfaction influence in a positive way their purchasing intentions (Lee & Lin, 2005).

In our research, we assume an even more important role of satisfaction, that of moderating the effects of experience on the five aforementioned factors; self-efficacy, effort expectancy, performance expectancy, trust and intention to repurchase. Specifically, we argue that, if previous experience is positively evaluated, and hence incurs customers' satisfaction, then it has an even higher impact on customers' shopping behavior (beliefs on capabilities, perceptions on ease to use and usefulness, trust and intention to repurchase).

Hence, the following five hypotheses are formulated:

H6a: Satisfaction from previous use of online shopping moderates the influence of experience on users' self-efficacy.

H6b: Satisfaction from previous use of online shopping moderates the influence of experience on users' effort expectancy.

H6c: Satisfaction from previous use of online shopping moderates the influence of experience on users' performance expectancy.

H6d: Satisfaction from previous use of online shopping moderates the influence of experience on users' trust.

H6e: Satisfaction from previous use of online shopping moderates the influence of experience on users' intention to repurchase.

RESEARCH METHODOLOGY

Sampling

Our research methodology included a survey conducted through the delivery and collection of individual questionnaires. A number of different methods were recruited for attracting respondents; questionnaires distributed in various places (universities, public areas) and e-mails were sent to different mailing lists. The survey was open during the last two weeks of April 2010. We aimed at about 800 Greek users of online shopping, 282 of which finally responded.

As Table 2 shows, the sample of respondents was composed of more men (68,2%) than women (31,8%). In terms of age, the majority of the respondents (49.5%) were between 25 and 34, while the second more frequent age group (31,8%) involved people between 19 and 24. Finally, the great majority of the respondents (84,4%) included graduates or post-graduate students. In order to participate in the survey, respondents should have made at least one online purchase within the last year.

From the total of 282 respondents, 136 (48, 23%) had made at least five online purchases in the past six months (high experience users), whereas 146 (51, 77%) had a limited number (no more than four) of online purchases within the past six months (low experience users). Table 3 presents some critical demographic features of the two groups of respondents.

Table 2. Users' demographic profile

Demographic Profile	No	%
Gender		
Male	193	68,4%
Female	89	31,6%
Marital Status		
Single	236	83,7%
Married	41	14,5%
Divorced	5	1,8%
Age		
0-18	12	4,3%
19-24	90	31,9%
25-34	140	49,6%
35-44	35	12,4%
45+	5	1,8%
Education		
Primary School	1	0,4%
Gymnasium	2	0,7%
High School	40	14,2%
University	152	53,9%
Post Graduate	87	30,9%

Measures

The questionnaire was divided into two parts. The first part included questions on the demographics of the sample (age, gender, education). The second part included measures of the various constructs identified in the literature review section. Table 4 lists the questionnaire items used to measure each construct. In almost all cases, with an exception standing for the 'online shopping experience' variable, 7-point Likert scales were used to measure the model's variables.

RESEARCH FINDINGS

First, we carried out an analysis of reliability and dimensionality to check the validity of the scales used in the questionnaire. Regarding the reliability of the scales, we were based on Cronbach alpha indicators and the inter-item correlations for the items of each variable. As we can see in Table 5, the results of the tests showed acceptable indices of internal consistency in the five scales considered.

In the next stage, we proceeded to evaluate the uni-dimensionality of the scales developed by carrying out a principal components analysis. The existence of uni-dimensionality is very important, since it allows calculating the average of the indicators that compose each construct. Consequently, it is possible to use a solely factor for representing each theoretical construct. Factorial analysis, with principal components and varimax rotation, was carried out to test uni-dimensionality of our five scales. As Table 5 presents, all items exhibited factor loadings that were higher than 0, 5.

The factor analysis identified five distinct factors; 1) self-efficacy, 2) effort expectancy, 3) performance expectancy, 4) trust, and 5) intention to repurchase/ satisfaction (Table 5). An interesting observation concerns the loading of the satisfaction's items under the sample factor with the items of the intention to repurchase. This might arise from the strong influence of satisfaction on intention to repurchase. Furthermore, this is an

Table 3. High and low experience users' demographic profiles

Demographic Profile	High Experience Users		Low Experience Users	
	No	%	No	%
Gender				
Male	109	80,1	84	57,5%
Female	27	19,9	62	42,5%
Marital Status				
Single	110	80,9%	126	86%
Married	24	17,6%	17	11,6%
Divorced	2	1,5%	3	2,1%
Age				
0-18	7	5,1%	5	3,4%
19-24	37	27,2%	53	36,3%
25-34	67	49,3%	73	50%
35-44	22	16,2%	13	8,9%
45+	3	2,2%	2	1,4%
Education				
Primary School				
Gymnasium	2	1,5%	1	0,7%
High School	21	15,4%	19	13,0%
University	68	50%	84	57,5%
Post Graduate	45	33,1%	42	28,8%

Table 4. Key factors' constructs and items

Constructs	Items	Sources
Self-Efficacy (SEF)	I feel capable of using the Internet for purchasing products. (SEF1)	(Luarn & Lin, 2005; Hernandez et al.,2009)
	I feel capable of locating shopping sites on the Internet. (SEF2)	
	I feel comfortable searching for information about a product on the Internet. (SEF3)	
Effort Expectancy (EE)	It is easy to become skilful at using online shop. (EE1)	(Devaraj et al., 2002; Pavlou, 2003; Chiu et al., 2009)
	Learning to operate online shop is easy. (EE2)	
	Online shops are flexible to interact with. (EE3)	
	My interaction with online shop is clear and understandable. (EE4)	
	Online shops are easy to use. (EE5)	
Performance Expectancy (PE)	Online shopping enables me to search and buy goods faster. (PE1)	(Devaraj et al., 2002; Pavlou, 2003; Chiu et al., 2009)
	Online shopping enhances my effectiveness in goods searching and buying. (PE2)	
	Online shopping makes it easier to search for and purchase goods. (PE3)	
	Online shopping increases my productivity in searching and purchasing goods. (PE4)	
	Online shopping is useful for searching and buying goods. (PE5)	
Trust (TR)	Based on my online shopping experience, I know that online shops are honest. (TR1)	(Pavlou, 2003; Pavlou & Chai, 2002; Wu & Chen, 2005; Chiu et al., 2009)
	Based on my experience with online shop, I know they are not opportunistic. (TR2)	
	Based on my experience with online shops, I know they keep their promises to customers. (TR3)	
	Based on my experience with online shops, I know they are trustworthy. (TR4)	
Satisfaction (STF)	I am satisfied with the online shopping experience. (STF1)	(Lin et al., 2005; Hsu et al., 2006)
	I am pleased with the online shopping experience. (STF2)	
	My decision to use online shopping was a wise one. (STF3)	
	My feeling with using online shops was good. (STF4)	
Intention to Repurchase (IR)	I intend to continue online shopping in the future. (IR1)	(Bhattacherjee, 2001; Lin et al., 2005; Hsu et al., 2006)
	I will continue online shopping in the future. (IR2)	
	I will regularly use online shops in the future. (IR3)	
Online Shopping Experience (EXP)	How many times (approximately) have you purchased products from an online store in the past six months?	(Chiu et al., 2009)

indication that satisfaction plays an important role on customers' intention in the Greek online market.

To examine the hypotheses H1-H5, a multivariate analysis of variance (MANOVA) including the five dependent variables and the one independent variable was executed. As we can see in Table 6, the experience exhibits a highly significant impact on users' self-efficacy (F=29,072; p<0,001), supporting H1. In addition, we found that online shopping experience has a significant effect on performance expectancy (F=13,411; p<0,001) as well as on intention to repurchase (F=29,918;

p<0,001). These results provide strong support for hypotheses H3 and H5. Regarding the impact of experience on users' effort expectancy, the findings showed that users with higher experience of online shopping feel more capable of using the Internet to make their purchase. While this effect is less significant than the effect of experience on users' performance expectancy, it is still strong enough (F=8,220; p<0,005) to support H2. Finally, as Table 6 shows, the effect of experience on trust is marginally acceptable, thus providing weak support for H4.

Table 5. Summary of measurement scales

Construct/Items	Mean	S.D	CR	Factor 1	Factor 2	Factor 3	Factor 4	Factor 5
Self Efficacy (SEF)			0,806					
SEF1	6,15	1,22		,474	,210	,154	,215	**,629**
SEF2	6,00	1,24		,236	,228	,153	,154	**,808**
SEF3	6,51	0,91		,243	,342	,054	,181	**,670**
Effort Expectancy (EE)			0,853					
EE1	5,58	1,21		,133	,113	,166	**,784**	,130
EE2	5,26	1,35		,080	,066	,178	**,834**	,048
EE3	5,02	1,22		,111	,280	,106	**,689**	,135
EE4	5,65	1,15		,272	,285	,208	**,520**	,390
EE5	5,46	1,22		,256	,257	,194	**,700**	,137
Performance Expectancy (PE)			0,898					
PE1	5,79	1,38		,147	**,754**	,130	,174	,209
PE2	5,77	1,28		,214	**,757**	,134	,159	,317
PE3	5,97	1,18		,280	**,762**	,171	,234	,191
PE4	5,64	1,36		,297	**,746**	,219	,138	,078
PE5	6,05	1,17		,318	**,662**	,194	,238	,136
Trust (TR)			0,877					
TR1	5,09	1,28		,312	,218	**,763**	,151	,039
TR2	4,56	1,37		,121	,163	**,752**	,131	,089
TR3	5,13	1,21		,276	,131	**,807**	,196	,132
TR4	5,23	1,22		,255	,147	**,782**	,249	,108
Satisfaction (STF)			0,911					
STF1	5,84	1,05		**,710**	,176	,382	,210	,169
STF2	5,78	1,11		**,629**	,226	,405	,256	,202
STF3	5,87	1,20		**,696**	,342	,305	,098	,165
STF4	5,75	1,21		**,669**	,349	,313	,208	,054
Intention to Repurchase (IR)								
IR1	6,26	1,21	0,897	**,822**	,237	,156	,132	,210
IR2	6,32	1,14		**,799**	,199	,178	,183	,270
IR3	5,70	1,53		**,791**	,177	,106	,073	,171

Table 6. Hypotheses testing using MANOVA

Dependent Variable	Mean (SD)				
	Low Experience	High Experience	Mean Diff.	F	Sign.
Self-efficacy	5,93 (1,13)	6,53 (0,61)	0,60	29,072	0,000****
Effort Expectancy	5,24 (1,05)	5,57 (0,87)	0,33	8,220	0,004***
Performance Expectancy	5,62 (1,22)	6,08 (0,84)	0,46	13,411	0,000****
Trust	4,88 (1,18)	5,14 (0,96)	0,26	3,954	0,048*
Intention to Repurchase	5,74 (1,41)	6,47 (0,73)	0,73	28,918	0,000****

**** $p < 0,001$ *** $p < 0,005$ ** $p < 0,01$ * $p < 0,05$

In order to test the hypotheses referring to the moderator effects of satisfaction, we performed a multivariate analysis of covariance (MANCOVA) test. To do so, we classified the participants into three groups: a) low satisfied, b) medium satisfied, and c) highly satisfied. The threshold for medium satisfied was defined as the mean value of satisfaction being smaller than 5 and greater than 3. Accordingly, the two extreme categories of low and highly satisfied users were defined. The results of the test provided evidence for the significant effect of satisfaction on the five dependent variables (Table 7). Nevertheless, the test did not provide support for all the hypothesized moderating effects of satisfaction. Specifically, as Table 7 presents, when satisfaction is handled as moderator, then the effects of experience on effort expectancy, performance expectancy and trust turn to be non-significant. Instead, the moderating effects of satisfaction prove to be highly significant in the relationship of experience with self-efficacy (F=13,780; p<0,000) and intention to repurchase (F=11,603, p<0,001). Hence, our empirical results provide support for the moderating relationships expressed through hypotheses H6a and H6e. Instead, hypotheses H6b-d are rejected.

DISCUSSION

The results of our research allow us to argue that online shopping experience affects in a highly significant way the three out of the five variables proposed for investigating the online shopping Greek market: self-efficacy, performance expectancy and intention to repurchase. Specifically, compared to users having low experience of online shopping, high experienced users exhibit higher belief in their capabilities to carry out an online purchase, higher expectancy for the usefulness of online shopping and higher intention to reuse the online shopping channel. After that, the degree of users' previous experience turns into a key feature to continue using Internet for shopping. This research has also identified, although in a lower degree of significance, the effects of experience on effort expectancy and trust. Hence, we can argue that low experienced users are more suspicious of repurchasing through Internet. This is due to their belief that it requires significant effort as well as their lack of trust in either the medium or the e-retailers.

These results are concurrent with previous findings in the literature (Bandura, 1986) stating that experience is the strongest generator of self-efficacy (H1). Self-efficacy is more important for customers with online shopping experience because they feel confident of themselves to make online purchases (Hernandez et al., 2009). The more confident they feel, the more purchases they make, and vice versa. Also, users with experience in online shopping find it more useful than the rest shopping channels, since they can search for a wide variety of products, many of which are provided in lower prices (Hernandez et al., 2009; Saprikis et al., 2010). This is concurrent with our findings under which experience has a significant effect on performance expectancy (H3). Moreover, according to Shim et al. (2001), the number of a buyer's online transactions affects his online shopping intention, which consorts with our results under which experience has a significant effect on intention to repurchase (H5). Users with none experience are much worried and have greater trust issues from users with high experience (Saprikis et al., 2010). Our empirical research demonstrated that previous experience affects the customers' trust on online shopping (H4). Moreover, according to Venkatesh et al. (2003), effort expectancy is considered more significant for users of low technological experience. Our empirical research confirms the above relationship in the case of the Greek market by demonstrating that the degree of experience a user has influences significantly his effort expectancy (H2).

Regarding the moderating effects, we found that satisfaction moderated the impact of two of

Table 7. Hypotheses Testing using MANCOVA

Source	Dependent Variable	Type III Sum of Squares	df	Mean Square	F	Sig.
Corrected Model	SEF	88,926[a]	2	44,463	71,996	,000
	EE	43,312[b]	2	21,656	26,823	,000
	PE	98,133[c]	2	49,066	60,184	,000
	TR	90,727[d]	2	45,364	52,433	,000
	IR	233,046[e]	2	116,523	200,166	,000
Intercept	SEF	49,621	1	49,621	80,348	,000
	EE	50,213	1	50,213	62,194	,000
	PE	25,386	1	25,386	31,138	,000
	TR	8,192	1	8,192	9,469	,002
	IR	,688	1	,688	1,182	,278
Satisfaction	SEF	64,333	1	64,333	104,169	,000
	EE	35,649	1	35,649	44,155	,000
	PE	83,238	1	83,238	102,098	,000
	TR	86,098	1	86,098	99,515	,000
	IR	195,960	1	195,960	336,626	,000
Experience	SEF	8,510	1	8,510	13,780	,000
	EE	1,624	1	1,624	2,012	,157
	PE	2,520	1	2,520	3,091	,080
	TR	,012	1	,012	,014	,907
	IR	6,755	1	6,755	11,603	,001
Error	SEF	171,687	278	,618		
	EE	224,446	278	,807		
	PE	226,645	278	,815		
	TR	240,519	278	,865		
	IR	161,832	278	,582		
Total	SEF	11126,000	281			
	EE	8453,080	281			
	PE	9910,320	281			
	TR	7366,250	281			
	IR	10821,222	281			
Corrected Total	SEF	260,613	280			
	EE	267,758	280			
	PE	324,778	280			
	TR	331,246	280			
	IR	394,878	280			
a. R Squared =, 341 (Adjusted R Squared =, 336)						
b. R Squared =, 162 (Adjusted R Squared =, 156)						
c. R Squared =, 302 (Adjusted R Squared =, 297)						
d. R Squared =, 274 (Adjusted R Squared =, 269)						
e. R Squared =, 590 (Adjusted R Squared =, 587)						

the five relationships proposed. Specifically, the more satisfied the users were, the more intense the positive effect of experience on users' self-efficacy became (H6a, Figure 1). This could be explained by the fact that satisfaction gained from previous experience counts a lot on people's motivation to re-purchase via Internet. Also, the more satisfied the users were, the more intense the positive effect of experience on users' intention to re-purchase was (H6e, Figure 2). As explained earlier, the Greek consumers have rated satisfaction highly as an antecedent of their intention to continue online shopping. Thus, previous positive experience of online purchases is expected to affect their intention to repeat their online purchases.

On the contrary, the moderating effects of satisfaction on performance expectancy, effort expectancy and trust were not significant at all. This could be better explained using the corresponding plots of these moderator relationships. While in Figure 1 and 2 the lines depicting the different satisfaction levels tend to intersect, in Figures 3, Figure 4 and Figure 5, the satisfaction lines appear to be almost completely parallel. This means that satisfaction can not affect neither in positive nor in negative direction the effect of experience on users' performance expectancy, effort expectancy and trust. Hence, we can argue that no matter whether they have been satisfied or not, highly experienced consumers keep having high perceptions for the effectiveness of online medium in goods' searching and buying (performance expectancy) as well as for the degree of ease associated with online shopping (effort expectancy). Moreover, consistent with previous studies in the marketing literature, having identified trust as antecedent of customer satisfaction (Chiu et al., 2009; Lin & Wang, 2006), our research shows that any argument on the reverse effect of satisfaction on trust cannot be efficiently supported.

CONCLUSION AND FURTHER WORK

While several Greek studies examining e-commerce adoption have been conducted (Angeli & Kyriakoullis, 2006; Harkiolakis & Halkias, 2007), few of them have taken into consideration the Greek consumers' cultural behavior and beliefs (Gounaris et al., 2010; Barbonis & Laspita, 2005). Building on previous studies' outcomes regarding the Greek consumers' habits (Buhalis & Deimezi, 2003), this study formulates a set of hypotheses describing their online shopping behavior. Specifically, we investigate the key role of experience gained and satisfaction raised from it, and thus test their direct and moderating effects, respectively, on key parameters of the Greek online shopping behavior (beliefs on capabilities, perceptions of ease to use and usefulness, trust, and intention to repurchase).

By investigating the Greek market, we wish to provide useful insight for the Mediterranean online shopping market, which holds several similarities with the Greek one, due to their culture and their level of e-commerce adoption. Reviewing previous research on the Greek consumers' specificities, we have deduced that trust, self-efficacy and satisfaction constitute crucial factors affecting the Greek consumers' online shopping behavior. Furthermore, several socio-cultural characteristics of the Greek people, such as beliefs (lack of trust, unawareness), relationships (strong relation with friends and relatives) and experiences (complaints not heeded) have been identified. According to Giese and Cote (2000), these characteristics have a direct effect on the consumers' sense of satisfaction.

Our empirical research has demonstrated that the three theoretical models TAM-UTAUT, SCT and ECT, and their constructs exhibit a high degree of reliability and credibility in the Greek market. Furthermore, we have made several important observations regarding the effects of online shopping experience and satisfaction in the Greek consumers' behavior. First, we provided evidence for the

Figure 1. Moderator Effect of Satisfaction on Self-Efficacy (SEF)

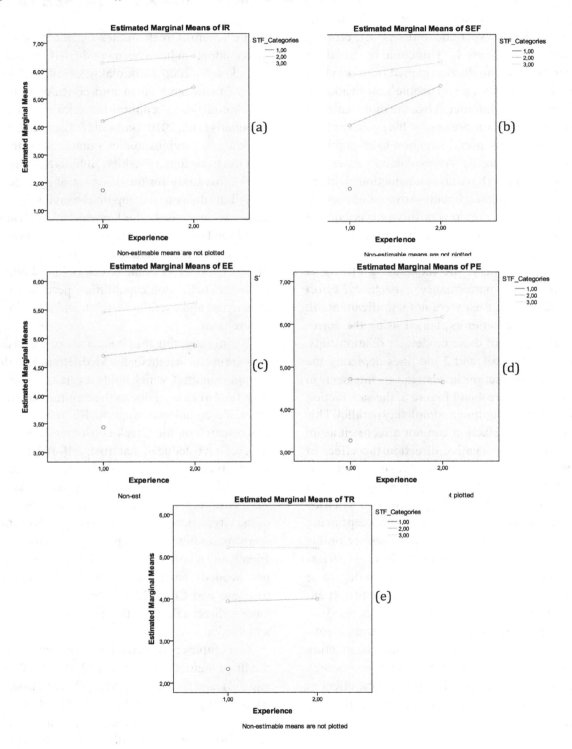

Figure 2. Moderator Effect of Satisfaction on Intention to Repurchase (IR)

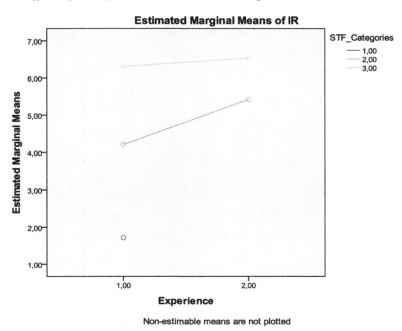

Figure 3. Moderator Effect of Satisfaction on Effort Expectancy (EE)

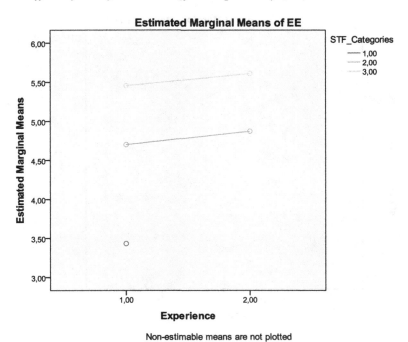

Figure 4. Moderator Effect of Satisfaction on Performance Expectancy (PE)

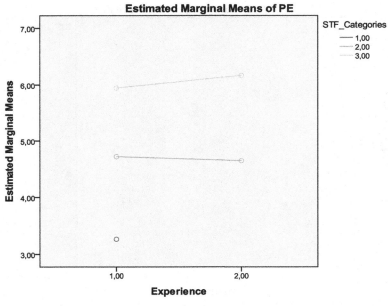

Figure 5. Moderator Effect of Satisfaction on Trust (TR)

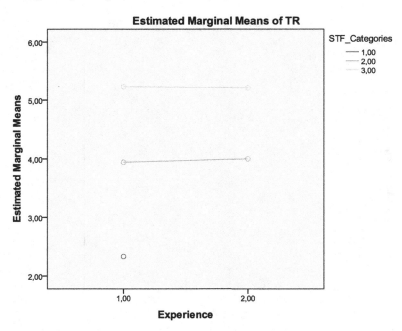

significant impact of users' previous experience on self-efficacy, performance expectancy, effort expectancy, trust and eventually intention to re-purchase. Moreover, considering satisfaction as a moderating variable, our research demonstrated that it can reinforce the existing positive relationship of experience on self-efficacy and intention to re-purchase. Following, satisfaction was regarded as a fundamental factor affecting the ultimate intention of the customer. This was illustrated by the factor analysis results, based on which, satisfaction and intention to repurchase are identified under the same factor. Moreover, we found that satisfaction moderates the influence of experience on both intention to repurchase and self-efficacy. This reinforces the importance of satisfaction in the Greek market, as experience is a powerful factor that affects, either more or less multiple aspects of the Greek consumers' online shopping behavior (trust, satisfaction, effort expectancy, performance expectancy, intention to repurchase).

Our empirical research has addressed several shortcomings of previous studies in the area. Specifically, Saprikis et al. (2010) investigate the Greek users' perceptions by dividing their sample into adopters and non-adopters. Hence, their results do not examine users with low experience. Likewise, Hernandez et al. (2009) do not include low experienced customers in their sample. Moreover, Xanthidis and Nicholas (2007) take into account the number of users who have completed online purchases and investigate how their experience affects their trust towards online retailers. Nevertheless, they do not investigate the satisfaction gained and how that moderates the effect of experience on the relationship developed with the retailers.

Our study is one of the few studies, which includes satisfaction as a variable moderating the relationship of experience with self-efficacy and intention to repurchase. Researchers should distinguish customers into highly satisfied, medium satisfied and low satisfied in order to establish

behavior patterns for each group. Furthermore, a deeper research should be made to recognize the key features that increase the customers' satisfaction in online markets. Herein, we derived the significance of the sense of satisfaction in the Greek online market. Further research is encouraged towards investigating the effect of satisfaction on the online shopping behavior of the rest Mediterranean markets.

Our results identify the Greek socio-cultural differences having an effect on customers' satisfaction. These results could help the Greek, and by extension the Mediterranean, markets to enforce their online shops with social and cultural characteristics for increasing customers' loyalty. Online shops should provide customers with more precise information about the products, their policies and their after-sales support in order to increase customers' trust. Moreover, the Web site designers must develop a friendly environment to reduce customers' unawareness. Also, as social interaction affects the Greek consumers' shopping behavior, some social capabilities should be added in web sites to increase customers' satisfaction. Finally, e-retailers should provide mechanisms for profiling their Greek customers based on their experience and satisfaction, in order to offer them adapted services. This could increase online shopping penetration and loyal customers' percentage.

Although our findings provide meaningful implications for online shopping, our study has several limitations. First, the convenient (random) sampling process may limit the validity of our findings to other contexts of online shopping (e.g. using a high percentage - about 50% - of online customers in the age of 25-34). Second, our study was carried out in Greece. As such, its results may not be adequately generalized to other Mediterranean countries. Finally, the current model tested in this research has not taken into consideration the interrelationships existing between the five key factors affecting the online behavior of the Greek consumers, such as the effect of effort expectancy on performance expectancy and the effect of performance expectancy on satisfaction. Testing such interrelationships requires developing a new model that prescribes a complex system of factors affecting the customers' intention to repurchase. However, in order to do so, a larger sample size than that used in this research, is required.

In the next part of this ongoing research we plan to develop such a model and apply a Structural Equation Modeling (SEM) method in order to examine relationships developed among the five factors identified in this paper. In that model, the intention to repurchase is going to be handled as dependent variable, while trust, self-efficacy, performance expectancy and effort expectancy will be handled as independent variables. Based on this study's results, our model will also include examination of the satisfaction's moderating role, and will discriminate results into those concerning low versus high experienced users. Future research might draw from a wider sample of online shopping users to ensure the sample is even more representative of the typical e-commerce user in Greece. Last, but not least, testing model with users from the rest Mediterranean countries could reveal the online shopping profile of the Mediterranean customers.

REFERENCES

Ajzen, I., & Fishbein, M. (1980). *Understanding attitudes and predicting social behavior*. Upper Saddle River, NJ: Prentice- Hall.

Angeli, A. D., & Kyriakoullis, L. (2006). Globalization vs. localization in e-commerce: Cultural-aware interaction design. In *Proceedings of the Working Conference on Advanced Visual Interfaces*.

Bandura, A. (1986). *Social foundations of thought and action: a social cognitive theory*. Upper Saddle River, NJ: Prentice-Hall.

Bandura, A. (1994). Self-efficacy. In Ramachaudran, V. S. (Ed.), *Encyclopedia of human behavior* (*Vol. 4*, pp. 71–81). New York, NY: Academic Press.

Barbonis, P. A., & Laspita, S. (2005). Some factors influencing adoption of e-commerce in Greece. In. *Proceedings of the IEEE International Conference on Engineering Management, 1*, 31–35. doi:10.1109/IEMC.2005.1559082

Bhattacherjee, A. (2001). Understanding information systems continuance: An expectation-confirmation model. *Management Information Systems Quarterly, 25*(3), 351–370. doi:10.2307/3250921

Bolton, R. N. (1998). A dynamic model of the duration of the customer's relationships with a continuous service provider: the role of satisfaction. *Marketing Science, 17*(1), 45–65. doi:10.1287/mksc.17.1.45

Brown, I., & Jayakody, R. (2008). B2C e-commerce success: A test and validation of a revised conceptual model. *The Electronic Journal Information Systems Evaluation, 11*(3), 167–184.

Buhalis, D., & Deimezi, O. (2003). Information technology penetration and ecommerce developments in Greece, with a focus on small to medium-sized enterprises. *Tourism Research, 13*(4), 309–324.

Chiu, C. M., Lin, H. Y., Sun, S. Y., & Hsu, M. H. (2009). Understanding customers' loyalty intentions towards online shopping: An integration of technology acceptance model and fairness theory. *Behaviour & Information Technology, 28*(4), 347–360. doi:10.1080/01449290801892492

Compeau, D. R., & Higgins, C. A. (1995). Computer self-efficacy: Development of a measure and initial test. *Management Information Systems Quarterly, 19*(2), 189–211. doi:10.2307/249688

Davis, F. D. (1989). Perceived usefulness, perceived ease of use, and user acceptance of information technology. *Management Information Systems Quarterly*, *13*(3), 319–340. doi:10.2307/249008

Davis, F. D., Bagozzi, R., & Warsaw, P. (1989). User acceptance of computer technology: A comparison of two theoretical models. *Management Science*, *35*(8), 982–1003. doi:10.1287/mnsc.35.8.982

Devaraj, S., Fan, M., & Kohli, R. (2002). Antecedents of B2C channel satisfaction and preference: Validating e-commerce metrics. *Information Systems Research*, *13*(3), 316–333. doi:10.1287/isre.13.3.316.77

Dick, A., & Basu, K. (1994). Customer loyalty: Toward an integrated conceptual framework. *Journal of the Academy of Marketing Science*, *22*(2), 99–113. doi:10.1177/0092070394222001

European Commission. (2009). *Report on cross-border e-commerce in the EU*. Brussels, Belgium: Commission Staff Working Document.

Fornell, C., & Wernerfelt, B. (1987). Defensive marketing strategy by customer complaint management. *JMR, Journal of Marketing Research*, *24*(4), 337–346. doi:10.2307/3151381

Giese, J. L., & Cote, J. A. (2000). Defining consumer satisfaction. *Academy of Marketing Science Review*. Retrieved from http://www.amsreview.org/articles/giese01-2000.pdf

Gounaris, S., Dimitriadis, S., & Stathakopoulos, V. (2010). An examination of the effects of service quality and satisfaction on customers' behavioral intentions in e-shopping. *Journal of Services Marketing*, *24*(2), 142–156. doi:10.1108/08876041011031118

Gravill, J., & Compeau, D. (2008). Self-regulated learning strategies and software training. *Information & Management*, *45*(5), 288–296. doi:10.1016/j.im.2008.03.001

Harkiolakis, N., & Halkias, D. (2007). Online buyer behaviour and perceptions in Greece. *International Journal of Applied Systemic Studies*, *1*(3), 317–328. doi:10.1504/IJASS.2007.017714

Hernández, B., Jiménez, J., & Martín, M. J. (2009). Customer behavior in electronic commerce: The moderating effect of e-purchasing experience. *Journal of Business Research*, *63*(9-10), 964–971. doi:10.1016/j.jbusres.2009.01.019

Hsu, M., Yen, C., Chiu, C., & Chang, C. (2006). A longitudinal investigation of continued online shopping behavior: An extension of the theory of planned behavior. *International Journal of Human-Computer Studies*, *64*(9), 889–904. doi:10.1016/j.ijhcs.2006.04.004

Jones, J. W. (1989). Personality and epistemology: Cognitive social learning theory as a philosophy of science. *Zygon*, *24*(1), 23–38. doi:10.1111/j.1467-9744.1989.tb00974.x

Lee, G. G., & Lin, H. F. (2005). Customer perceptions of e-service quality in online shopping. *International Journal of Retail and Distribution Management*, *33*(2), 161–175. doi:10.1108/09590550510581485

Lee, Y., & Kwon, O. (2009). Can affective factors contribute to explain continuance intention of web-based services? In *Proceedings of the 11th International Conference on Electronic Commerce* (pp. 302-310).

Lin, C. S., Wu, S., & Tsai, R. J. (2005). Integrated perceived playfulness into expectation–confirmation model for web portal context. *Information & Management*, *42*(5), 683–693. doi:10.1016/j.im.2004.04.003

Lin, H. H., & Wang, Y. S. (2006). An examination of the determinants of customer loyalty in mobile commerce contexts. *Information & Management*, *43*(3), 271–282. doi:10.1016/j.im.2005.08.001

Luarn, P., & Lin, H. H. (2005). Toward an understanding of the behavioral intention to use mobile banking. *Computers in Human Behavior, 21*(6), 873–891. doi:10.1016/j.chb.2004.03.003

Maditinos, D., Sarigiannidis, L., & Dimitriadis, E. (2010). The role of perceived risk on Greek internet users' purchasing intention: An extended TAM approach. *International Journal of Trade and Global Markets, 3*(1), 99–114. doi:10.1504/IJTGM.2010.030411

Observatory of the Greek Information Society. (2010). *The attitude of Greeks against online shopping, indicators of consumer's behavior.* Retrieved from http://www.observatory.gr/page/default.asp?la=1&id=2101&pk=439&return=183

Oliver, R. L. (1993). Cognitive, affective, and attribute bases of the satisfaction response. *The Journal of Consumer Research, 20*(3), 418–430. doi:10.1086/209358

Oliver, R. L., & DeSarbo, W. S. (1988). Response determinants in satisfaction judgments. *The Journal of Consumer Research, 14*(4), 495–507. doi:10.1086/209131

Palvia, P. (2009). The role of trust in e-commerce relational exchange: A unified model. *Information & Management, 46*(4), 213–220. doi:10.1016/j.im.2009.02.003

Papazafeiropoulou, A., Pouloudi, A., & Doukidis, G. (2001). *Electronic commerce policy making in Greece.* Melbourne, Australia: Center for Strategic Information Systems.

Pavlou, P. A. (2003). Consumer acceptance of electronic commerce: integrating trust and risk with the technology acceptance model. *International Journal of Electronic Commerce, 7*(3), 101–134.

Pavlou, P. A., & Chai, L. (2002). What drives electronic commerce across cultures? A cross-cultural empirical investigation of the theory of planned behavior. *Journal of Electronic Commerce Research, 3*(4), 240–253.

Pavlou, P. A., & Fygenson, M. (2006). Understanding and predicting electronic commerce adoption: an extension of the theory of planned behavior. *Management Information Systems Quarterly, 30,* 115–143.

Reichheld, F. F., & Sasser, E. W. (1990). Zero defections: Quality comes to services. *Harvard Business Review, 68*(5), 105–111.

Reichheld, F. F., & Schefter, P. (2000). E-loyalty: Your secret weapon on the web. *Harvard Business Review, 78*(4), 105–113.

Rust, R. T., & Donthu, N. (1995). Capturing geographically localized misspecification error in retail store choice models. *JMR, Journal of Marketing Research, 32*(1), 103–110. doi:10.2307/3152115

Saprikis, V., Chouliara, A., & Vlachopoulou, M. (2010). Perceptions towards online shopping: Analyzing the Greek University students' attitude. *Communications of the IBIMA,* 1-13.

Shih, H. (2004). Extended technology acceptance model of internet utilization behavior. *Information & Management, 41*(6), 719–729. doi:10.1016/j.im.2003.08.009

Shim, S., Eastlick, M. A., Lotz, S. L., & Warrington, P. (2001). An online prepurchase intentions model: The role of intention to search. *Journal of Retailing, 77*(3), 397–416. doi:10.1016/S0022-4359(01)00051-3

Taylor, S., & Todd, P. A. (1995). Assessing IT usage: The role of prior experience. *Management Information Systems Quarterly, 19*(2), 561–570. doi:10.2307/249633

Theodoridis, P., & Chatzipanagiotou, K. (2009). Store image attributes and customer satisfaction across different customer profiles within the supermarket sector in Greece. *European Journal of Marketing, 43*(5-6), 708–734. doi:10.1108/03090560910947016

Turk, T., Blazic, B., & Trkman, P. (2008). Factors and sustainable strategies fostering the adoption of broadband communications in an enlarged European Union. *Technological Forecasting and Social Change*, *75*(7), 933–951. doi:10.1016/j.techfore.2007.08.004

Vehovar, V. (2003*). Security concern and on-line shopping: International study of the credibility of consumer information on the Internet.* Retrieved from http://www.consumerwebwatch.org/pdfs/Slovenia.pdf

Venkatesh, V., & Morris, M. G. (2000). Why don't men ever stop to ask for directions? Gender, social influence, and their role in technology acceptance and usage behavior. *Management Information Systems Quarterly*, *24*(1), 115–139. doi:10.2307/3250981

Venkatesh, V., Morris, M. G., Davis, G. B., & Davis, F. D. (2003). User acceptance of information technology: toward a unified view. *Management Information Systems Quarterly*, *27*(3), 425–478.

Westbrook, R. A. (1980). A rating scale for measuring product/service satisfaction. *Journal of Marketing*, *44*, 68–72. doi:10.2307/1251232

Wu, I. L., & Chen, J. L. (2005). An extension of trust and TAM model with TPB in the initial adoption of on-line tax: An empirical study. *International Journal of Human-Computer Studies*, *62*(6), 784–808. doi:10.1016/j.ijhcs.2005.03.003

Xanthidis, D., & Nicholas, D. (2007). Consumer preferences and attitudes towards eCommerce activities. Case study: Greece. In *Proceedings of the 6th WSEAS International Conference on E-ACTIVITIES* (pp. 134-139).

Zeithaml, V. A. (2000). Service quality, profitability and the economic worth of customers: What we know and what we need to learn. *Journal of the Academy of Marketing Science*, *28*(1), 67–85. doi:10.1177/0092070300281007

Zorotheos, A., & Kafeza, E. (2009). Users' perceptions on privacy and their intention to transact online: a study on Greek Internet users. *Direct Marketing: An International Journal*, *3*(2), 139–153. doi:10.1108/17505930910964795

This work was previously published in the International Journal of E-Services and Mobile Applications, Volume 3, Issue 2, edited by Ada Scupola, pp. 39-58, copyright 2011 by IGI Publishing (an imprint of IGI Global).

Chapter 6
Behavioral Intention Towards Mobile Banking in India:
The Case of State Bank of India (SBI)

N. Thamarai Selvan
National Institute of Technology, India

B. Senthil Arasu
National Institute of Technology, India

M. Sivagnanasundaram
Kirloskar Institute of Advanced Management Studies, India

ABSTRACT

The rapid growth of mobile technologies and devices makes it possible for the customers of banking services to conduct banking at any place and at any time. Today, most of the banks in the world provide mobile access to its customers for banking as mobile banking systems improve their efficiency and reduce transaction costs. Banks invested heavily in the mobile banking system hoping that its customers would embrace it with open arms. Contrary to the expectation, the lukewarm patronage to mobile banking makes it crucial to understand the factors that contribute to users' intention to use mobile banking. This study extends the applicability of technology acceptance model (TAM) to the mobile banking context. Based on the review of literature, few additional constructs were added to the TAM. Structural equation modeling (SEM) was used to test the casual relationships proposed. Findings of the study support the proposed model's ability of explaining the users' intention to adopt mobile banking.

INTRODUCTION

In the last two decades, service industry has witnessed tremendous changes in the way business is conducted while comparing to the previous era. Convergence of technologies has made the distribution of services more convenient than ever before. Automatic Teller Machines, bill payment kiosks, internet based services and phone based services (both voice and text), automated hotel check out, automated check-in for flights, automated food ordering system in restaurants, vending

DOI: 10.4018/978-1-4666-2654-6.ch006

machines, Interactive voice response systems are examples of technology based service delivery channels. In case of retail banking, banks have traditionally delivered services through face-to-face interactions with consumers at branch offices (Lee, 2002). But traditional delivery channels are being challenged and complemented by new electronic channels (Meuter et al., 2000; Morrison et al., 1998). The new electronic channels are Automatic Teller Machines (ATM), Internet. The most recent addition to the existing electronic channels is Mobile banking. M-banking is defined as provision and availment of banking services with the help of mobile telecommunication devices such as mobile phones (Mallat et al., 2004).

Recent statistics indicate that, the number of mobile phone users in India is 584.32 Million as of March 31, 2010 (Telecom Regulatory Authority of India, 2010) and expected to grow further at a brisk rate. Rapid proliferation of mobile phones and recent advancement in wireless technology could definitely be a catalyst for mobile banking. However, research shows a poor linkage between growth in sales of mobile phones and advanced mobile services (Blechar et al., 2006). Mobile banking services in India are still in the nascent stage, promising to be medium to reach the unbanked rural mass which stands at 41%. The transaction volumes with mobile banking are very low (Chakrabarty, 2010). As technology started to occupy the center stage of banking services delivery today, it becomes imperative to understand users' acceptance of mobile banking and to identify the factors affecting their intentions to use mobile banking. This information can assist developers in the building of mobile banking systems that consumers want to use, or help them discover why potential users avoid using the existing system. While a growing body of literature exists, limited empirical evidence indicates how people perceive mobile banking in India, what factors influence Indian user's adoption of mobile banking.

The objective of this paper is to investigate factors that influence SBI customers' adoption

of mobile banking. For the purpose, the foundational and classical Technology Acceptance Model (TAM) (Davis, 1989) has been chosen as the base model. The TAM used for this study is modified to incorporate the factors related to the mobile banking context because in the original TAM model, perceived usefulness and perceived ease of use are addressed as the most important constructs in predicting information system (IS) acceptance in the work environment.

The next section provides the over view of State Bank of India (SBI) and its mobile banking services. We then present the research model and hypotheses. Then analytical results are reported in the ensuing section. The final sections present the conclusions and discuss the implications of the findings of this research.

Overview of State Bank of India (SBI) and its Mobile Banking Service

The State Bank of India is the largest commercial bank in India in terms of profits, Assets, deposits, branches and employees. The evolution of State Bank of India can be traced back to the first decade of the 19th century. After few amalgamations and re designations it has got its name and establishment by the act of the parliament of India. The corporate center of SBI is located in Mumbai. In order to cater to different functions, there are several other establishments in and outside Mumbai, apart from the corporate center, and the bank boasts of having as many as 14 local head offices and 57 Zonal Offices, located at major cities throughout India. It is recorded that SBI has about 10000 branches, well networked to cater to its customers throughout India. In addition to banking, through its various subsidiaries, it also provides a whole range of financial services, which include life insurance, merchant banking, mutual funds, security trading, pension fund management and primary dealership in the money market. It operates in four business segments: Treasury, Corporate/Wholesale Banking, Retail Banking and Other Banking Business. The technology deployment initiative of State bank of India started with its need for core banking solution by the year 2000. But it has not stopped there. As the market grows increasingly competitive, the dependence on technology also increases making SBI to deploy latest developments in the technological front. It is proposed to grow their network to more than 30,000 ATMs during 2010.

The State Bank Mobile Banking Service, "State Bank freedoM", enables customers to move funds, check balance, make payment of bills through mobile phones without visiting any branch for Registration for the Service. Although mobile banking is a free service offered by the bank, the customers will have to bear the charges imposed by the telecom operators for the SMS/ GPRS. Apart from this, mobile money transfer and mobile wallet are the two new initiatives that were planned to be undertaken through mobile banking.

Review of Literature

As mobile banking is still at the nascent stage, there exist only a few studies on adoption of mobile banking by consumers. Like the studies on internet banking, some of the studies on mobile banking explored the reasons behind consumer adoption/non-adoption of mobile banking. The results indicated that factors associated with adoption are belief that MB helps to fulfill personal banking needs, location free convenience, and cost effectiveness. The factors associated with resistance are system configuration and safety concerns, functional issues, basic fee for mobile banking web connections (Matilla, 2003; Yang, 2005). Whilst the prior researches on adoption of mobile banking which intended to develop causal models were based on TAM (e.g., Luarn & Lin, 2005; Gu, Lee, & Suh, 2009).

The classical Technology Acceptance Model (TAM) has been a foundational model in understanding an individual's decision to use technology rooted in the Theory of Reasoned Action (TRA)

(Ajzen & Fishbein, 1980). TAM postulates that user acceptance of a new system is determined by the users' intention to use the system, which is influenced by the users' beliefs about the system's perceived usefulness and perceived ease of use. Perceived usefulness is defined as the extent to which a person believes that using a particular system will enhance his or her performance, and perceived ease of use refers to the extent to which a person believes that using a particular system will be free of effort. TAM has been empirically replicated or extended to explain various behaviors with adopting technology (e.g., Jackson et al., 1997; Gefen & Straub, 1997; Vankatesh & Davis, 2000; Legris et al., 2003; Gefen, 2003). Most of these studies have aimed at relatively simple IT, such as personal computer, e-mail system, and word processing and spreadsheet software. Many researchers have used Technology Acceptance Model (TAM) to explain an individual's acceptance of new Information Technology (IT) and verified that the perceived usefulness and the perceived ease-of-use are key constructs of individual acceptance (Davis, 1989; Davis, Bagozzi, & Warshaw, 1989; Mathieson, 1991; Segars & Grover, 1993; Adams, Nelson, & Todd, 1992; Hendrickson, Massey, & Cronan, 1993; Jackson et al., 1997; Gefen & Straub, 1997; Doll, Hendrickson, & Deng, 1998; Vankatesh & Davis, 2000; Gefen, 2003; Agarwal & Karahanna, 2000; Gefen, Karahanna & Straub, 2003; Lu, Yu, Liu, & Yao, 2003; Pavlou, 2003; Wang, Wang, Lin, & Tang, 2003; Legris, Ingham, & Collerette, 2003).

In the mobile banking context, the antecedent beliefs of TAM: Perceived usefulness, Perceived ease of use was found to have an impact on adoption of mobile banking (Luarn & Lin, 2005; Amin, Rizal, Hamid, Lada, & Anis, 2008). However, the TAM's fundamental constructs do not fully reflect the specific influences of technological and usage-context factors that may alter the users' acceptance (Moon & Kim, 2001). Hence, caution needs to be taken when applying the findings developed for the earlier generations of IT to the new virtual environment (Chen et al., 2000). Besides, the user group of mobile banking is very complex in composition, having more diversified education and socioeconomic background than those of other information systems. This complex user group composition makes it necessary to examine the acceptance of new technologies with different user populations in different contexts. As a wireless technology, the usage context of mobile banking is quite different from that of the stand-alone application software in the IS context. Therefore, the traditional TAM variables (i.e., perceived ease of use and perceived usefulness) may not fully reflect the users' intention to adopt mobile banking, necessitating a search for additional factors that better predict the acceptance of mobile banking.

The additional constructs added thus have to better reflect newly emerging technologies. The other factors found to have an influence on Mobile banking adoption along with the original TAM variables are Perceived credibility (Luarn & Lin, 2005; Amin et al., 2008), Normative pressure (Amin et al., 2008), Perceived financial cost, Perceived self efficacy (Luarn & Lin, 2005), Perceived risk (Lee, Lee, & Kim, 2007) and Trust (Lee, Lee, & Kim, 2007; Gu, Lee, & Suh, 2009).

In this study, normative influence and trust with the bank were considered in addition to the constructs proposed by Luarn and Lin (2005). The additional constructs considered are expected to enhance the understanding of customer acceptance of mobile banking further to the general constructs used in base TAM.

RESEARCH MODEL AND HYPOTHESIS

The research model for our study is shown in Figure 1. In the model, we considered the key determinants of technology adoption lay down by the TAM and augmented them with factors based on empirical findings reported in the related literature. In the proposed model, the "attitudes" construct has been removed for simplification.

This was done in line with other studies of the TAM (e.g., Lu & Gustafson, 1994; Chau, 1996; Hong et al., 2001; Featherman & Pavlou, 2003; Luarn & Lin, 2005; Guh, Lee, & Suh, 2009). The proposed constructs and hypotheses are supported by prior researches.

Perceived Usefulness (PU)

Perceived usefulness can be defined as the degree to which a person believes that using a particular system would enhance his or her job performance (Davis et al., 1989). Perceived usefulness is strongly associated with productivity. It suggests that using computer in the workplace would increase user's productivity, improve job performance, enhance job effectiveness and be useful in the job. The reason behind why people use mobile banking is that they find it to be very useful. People use mobile banking as it enhances their effectiveness. There is also extensive research support in the area of IS adoption that there exists a significant effect of perceived usefulness on usage intention (Agarwal & Prasad, 1999; Davis et al., 1989; Hu et al., 1999; Jackson et al., 1997; Venkatesh, 1999, 2000; Venkatesh & Davis, 1996, 2000; Venkatesh & Morris, 2000; Luarn & Lin, 2005; Cheong & Park, 2005; Chiu et al., 2005; Wang, et al., 2003). Further, Luarn and Lin (2005) as well as Guh, Lee and Suh (2009) have established that perceived usefulness has significant impact in the development of initial willingness to use mobile banking. Hence we propose that

H1: Perceived usefulness will have a positive effect on the behavioral intention to use mobile banking

Perceived Ease of Use (PEOU)

Perceived ease of use refers to the degree to which a person believes that using a particular system would be free of effort (Davis, 1989). Channels that are perceived to be easier to use have a higher likelihood of being chosen by consumers

Figure 1. Research model

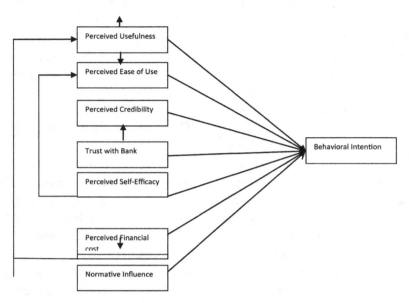

(Pavlou et al., 2006). Prior research have always indicated that there is a positive relationship between perceived ease of use and usage intention (Agarwal & Prasad, 1999; Davis et al., 1989; Hu, Chau, Sheng, and Tam, 1999; Jackson, Chow, & Leitch, 1997; Venkatesh, 1999, 2000; Venkatesh & Davis, 1996, 2000; Venkatesh & Morris, 2000; Luarn & Lin, 2005; Cheong & Park, 2005; Chiu et al., 2005; Wang et al., 2003). Further, Luarn and Lin (2005) and Guh, Lee, and Suh (2009) have established that perceived ease of use has significant impact in the development of initial willingness to use mobile banking.

In the mobile services context, perceived ease of use can be described as the extent to which individuals perceive freedom of difficulty with the use of mobile technology and services in day to day usage (Knutsen, Constantiou, & Damsgaard, 2005). Customers may find using mobile technology and services less complex or tedious to use. For example, browsing Internet on mobile devices has been perceived as tedious (Teo & Pok, 2003). At the same time, mobile services which are perceived to be easier to use than others are more likely to be accepted by users (Pikkarainen et al., 2004). The usage experience of mobile

services could be negatively affected by the small screen size and the miniature key of the mobile phones. Further, user friendly and usable intuitive man-machine interfaces, including clear and visible steps, suitable content and graphical layouts, help functions, clear commands, symbols and meaningful error messages are likely to influence adoption (Condos et al., 2002). In the internet banking context, Wang et al. (2003) showed that perceived ease of use is an antecedent predictor of perceived credibility. Hence we propose that

H2: Perceived ease of use will have a positive effect on the behavioral intention to use mobile banking

H2a: Perceived ease of use will have a positive effect on perceived credibility

H2b: Perceived ease of use will have a positive effect on perceived usefulness.

Perceived Credibility (PC)

Another important factor that can affect the adoption of mobile banking is perceived credibility. Perceived credibility is concerned with security and privacy of the system with which personal information given and monetary transaction is carried out. Perceived credibility in the mobile banking context refers to the extent to which a person believes that the use of mobile banking will have no security or privacy threats. It has been established that perceived credibility has significant influence on user willingness to engage in online exchanges of money and personal sensitive information (e.g., Hoffman et al., 1999; Friedman et al., 2000). In general, the perceived credibility that people have in the system, to securely conclude their transactions and maintain the privacy of their personal information, affects their voluntary acceptance of mobile banking. Wang et al. (2003) examined the impact of perceived credibility on usage intention, and found that perceived credibility had a significant effect on intention to use online banking systems. As mobile banking is similar to online banking it has been proposed that

H3: Perceived credibility will have a positive effect on behavioral intention to use mobile banking

Trust with the Bank (TB)

Trust refers to the belief one has that the promise of another can be relied upon and that, in unpredictable circumstances, the other will act in a spirit of goodwill and in a caring way towards the trustor (Grazioli & Jarvenpaa, 2000). Trust has always been considered as a facilitator in many buyer–seller transactions that can provide consumers with high expectations of satisfying exchange relationships (Hawes et al., 1989). Trust is a very important factor for new technologies adoption (Fukuyama, 1995). Dimitriadis and Kyrezis (2008) suggests that trust in the bank already known to the customer could be hypothesized as a variable influencing the trust in the new channel of the bank as one could expect a transfer of trust from the bank towards its new channel. In the mobile

banking context consumers' behavioral intention to adopt mobile banking is expected to be enhanced by higher trust belief towards the bank (Luo et al., 2010). When users trust banks, they will perceive mobile banking to be useful and they are willing to use it (Gu, Lee, & Suh, 2009). We thus hypothesize

H4: Trust with the bank will have a positive effect on behavioral intention to use mobile banking
H4a: Trust with the bank will have a positive effect on perceived credibility of the channel

Perceived Self-Efficacy (SE)

Self-efficacy is a person's belief in his or her ability to do a behavior (Bandura, 1982, 1997). Self efficacy can be defined as "people's judgments of their capabilities to organize and execute courses of action required to attain designated types of performance. It is concerned not with the skills one has but with judgments of what one can do with whatever skills one possesses" (Bandura, 1986). Thus technology self efficacy means an individual's perception of his or her ability to technology in the accomplishment of task rather than reflecting simple component skills (Compeau & Higgins, 1995). It has been observed that self-efficacy is an important factor to consider for adopting new technologies (Compeau, Higgins, & Huff, 1999). Even if the prospects are convinced about the benefits a new technology may offer, if they are not confident in their own ability to use them, they may not want to use them (Compeau, Higgins, & Huff, 1999). Hence the consumers behavioral intention to use mobile banking is influenced by their technology self efficacy. Further there exists a strong empirical evidence that self efficacy influences the Perceived ease of use (e.g., Agarwal & Karahanna, 2000; Igbaria & Iivari, 1995; Venkatesh, 2000; Venkatesh & Davis, 1996; Ong et al., 2004; Wang, Wang, Lin,

& Tang, 2003). Hence we propose the following hypotheses

H5: Perceived self efficacy will have a positive effect on behavioral intention to use mobile banking

H5a: Perceive self efficacy will have a positive effect on perceived ease of use

Perceived Financial Cost (PFC)

Economic motivations and outcomes are most often the focus of IS acceptance studies (Mathieson et al., 2001). According to Constantinides (2002) a customer switching from Electronic commerce to Mobile Commerce involves three additional expenses viz., Equipment costs, access cost, and transaction fees. This makes mobile commerce costlier to the customer. As mobile banking is a type of mobile commerce (Eastin, 2002); it can be implied that mobile banking involves additional costs for the customer. Perceived cost was found to be a significant antecedent of the behavioral intention to use mobile commerce (Pagani, 2004; Okazaki, 2005; Kim et al., 2007). Luarn and Lin (2005) found in their study on individual adoption of mobile banking that perceived financial cost has significantly negative effects on users' behavioral intention. Wu and Wang (2005) also found support for the negative effect of perceived cost on user adoption of mobile commerce. Hung et al. (2003) studied the relationship between the cost of value-added services and consumer attitudes, and found that the cost of value-added services negatively influences consumer attitudes toward the use of WAP services. Based on the above premises we propose that

H6: Perceived financial cost will have a negative effect on behavioral intention to use mobile banking

H6a: Perceived financial cost will have a negative effect on normative influence

Normative Influence (NI)

Normative influence is defined as the tendency to comply with the positive expectations of others (Bachmann et al., 1993; Bearden et al., 1989; Grimm et al., 1999). Normative influence occurs when a person conforms to the expectations of others to obtain a reward or avoid a punishment (Deutsch & Gerard, 1995). The influence of social factor over user behavior is a well established effect through various researches. Several theories suggest that social influence is a critical factor that formulates the user behavior. According to the Theory of Reasoned Action (TRA), a person's behavioral intentions are influenced by subjective norms besides attitude. Innovation diffusion theory (IDT) suggests that user adoption decisions are influenced by a social system beyond an individual's decision style and the characteristics of the IT. Shi et al. (2008) studied adoption of Internet banking from consumers perspectives and found that normative and coercive pressure significantly influence the attitude and intention to adopt of Internet banking. Further, Venkatesh and Davis (2000), Rose and Fogarty (2006) and Gu et al. (2009) found that subjective norms have a significant influence on perceived usefulness and behavioral intentions. Hence it is proposed that

H7: Normative Influence will have a positive effect on behavioral intention to use mobile banking

H7a: Normative Influence will have a positive effect on Perceived Usefulness.

METHODOLOGY

Instrument Development

In order to ensure content validity, the items used for the constructs in this study were mostly adapted from prior studies. Table 1 shows the constructs

Table 1. Constructs and their source

Construct	Source
Perceived ease of use	Luarn and Lin (2005) and Curran and Meuter (2005)
Perceived usefulness	Gu et al. (2009) and Luarn and Lin (2005)
Perceived self-efficacy	Compeau and Higgins (1995)
Perceived financial cost	Wu and Wang (2005)
Trust with the bank	Gu et al. (2009) and Dimitriadis and Kyrezis (2008)
Perceived credibility	Wang et al. (2003)
Normative influence	Gu et al. (2009)
Behavioral intention	Venkatesh and Davis (1996, 2000)

and the source. Seven point Likert scales (1–7), with anchors ranging from "strongly disagree" to "strongly agree" were used for all Items. The items used in this study are given in the Appendix.

Sample and Procedure

Data used in this study to evaluate the proposed research model were obtained from the employees working in the ancillary units located in an Industrial estate, having savings bank account with a specific branch of SBI in that locality. The Industrial estate is located near to the Institute where the authors are affiliated. These ancillary units are involved in fabrication of heavy metal components. Almost all the employees have salary account with that specific branch of SBI. The data was obtained from the respondents by visiting them at their work premises. After ensuring that they have the mobile handsets through which mobile banking can be carried out, they were explained about mobile banking and the necessary software was installed in their hand sets. Following this those prospective mobile banking customers were asked to respond to the self-administered questionnaire. The total number of responses obtained thus was 303. Ninety four percent of the respondents were male as quite a few female employees were

employed in heavy metal manufacturing units. The age of the respondents ranged from 18 to 41 years. Fifty six percent of the respondents were diploma holders while forty eight percent were graduates; a further 6% were post-graduates.

Data Analysis and Results

The Structural Equation Modeling (SEM) approach was used to validate the research model. SEM approach was selected because of its ability to test causal relationships between the constructs with multiple measurement items. Further SEM approach is capable of testing the measurement characteristics of constructs. Amos 16.0 was used to carry out the analysis.

We followed the two-stage procedure suggested by Anderson and Gerbing (1988) for analyzing the collected data. First, we examined the measurement model to measure reliability, convergent and discriminant validity. Then, we examined the structural model to investigate the strength and direction of the relationships among the theoretical constructs.

Common Method Variance (CMV)

Since this study has obtained the responses for both independent and dependent variable from the same source, it becomes necessary to check for the method bias. The extent of spurious covariance shared among variables as a result of application of common method used to collect the data is termed as common method variance (Buckley et al., 1990). Harman's single factor test with CFA approach was employed to assess the CMV. In this method all of the manifested items are modeled as the indicators of a single factor that represents method effects. As the fitness indices of the CFA model showed unsatisfactory model fit with *chi-square/df* ratio of 10.717; CFI =0.431; RMSEA =0.179, indicating that CMV is not the major source of the variations in the observed items.

Analysis of Measurement Model

A confirmatory measurement model that specifies the posited relations of the observed variables to the underlying constructs, with the constructs allowed to intercorrelate freely, was validated.

The model fit indices were within the acceptable limits with chi-square/df = 2.482, CFI = 0.923, and RMSEA = 0.70, hence the measurement model was fit; we proceeded to evaluate for reliability and validity. The adequacy of the individual items and the composites were assessed by measures of reliability and validity.

Reliability for all items of a construct should be evaluated jointly by investigating composite reliability (CR) and the average variance extracted (AVE). The composite reliability estimates is calculated by the formula provided by Fornell and Larcker (1981). CR is given by (square of the summation of the factor loadings) / {(square of the summation of the factor loadings) + (summation of error variance)}. The CR and AVE calculated for all the eight constructs of this study were given in Table 1. For a construct to possess good reliability, CR should be at least 0.60 and the AVE should be at least 0.5 (Hair et al., 1995).

Two types of validity measures, convergent and discriminant were examined. Convergent validity is the degree to which multiple attempts to measure the same concept are in agreement.

The convergent validity is evaluated by investigating the value of standardized factor loadings. Items should load at least 0.60 on their respective hypothesized component and all loadings need to be significant. The loading for each item on their respective construct was significant (p < 0.001) and loadings were above 0.60. The results shown in Table 2 confirmed the convergent validity of the instrument.

Discriminant validity is the degree to which the measures of different concepts are distinct. A comparison of the squared multiple correlations between each of the constructs to the average variance extracted estimates for those constructs will provide proof of discriminant validity between the constructs (Curran et al., 2003).

The average variance extracted for each construct was calculated using standardized factor loadings and the indicator measurement errors (Hair et al., 1998) and ranged from 0.525 to 0.770 which is shown as the diagonal elements in the Table 3. If the average variance extracted for each construct is greater than the square of the correlation between the constructs, then discriminant validity is demonstrated (Chaudhuri & Holbrook, 2001; Fornell & Larcke,r 1981). In each case in the measurement model, this is true, by which the discriminant validity is verified. In summary, the measurement model demonstrated adequate reliability, convergent validity and discriminant validity.

Analysis of Structural Model

After assessing the reliability and validity with CFA, we tested the overall fit of the path model. The overall model fit evaluates the correspondence of the actual or observed input matrix with that predicted from the proposed model. Analysis of the proposed model given in Figure 1, shows that this model offers best explanation of the data. The chi-square value is 643.22 and the degrees of freedom are 234 resulting in a *chi-square/df* ratio of 2.749, (should be <.3.00). The RMSEA for the model is 0.076 (close to the moderate fit limit of .80). Similarly the other fit indices are well in the limits or close to it. The CFI is 0.905(should be >.9) indicating that the model has a good fit. The overall fit indices of our model found to be fit by satisfying the recommended levels in previous research (Hu & Bentler, 1999; Shumacker & Lomax, 2004). We were assured therefore, that our research model is an adequate representation of the entire set of causal relationships.

Having assessed the fit indices for the measurement model and overall model, we examined the estimated coefficients of the causal relationships between constructs, which would validate the hy-

Table 2. Descriptive statistics, composite reliability, average variance extracted

Construct	Item	Mean	Std. Deviation	Factor loading	Composite Reliability	Average Variance Extracted
Perceived Ease of Use	PE1	5.50	1.276	0.785	0.862	0.677
	PE2	5.52	1.299	0.819		
	PE3	5.46	1.418	0.862		
Perceived Use-fulness	PU1	4.90	1.229	0.840	0.879	0.707
	PU2	4.90	1.226	0.847		
	PU3	4.86	1.184	0.836		
Self Efficacy	SE1	5.34	1.229	0.736	0.767	0.525
	SE2	5.29	1.263	0.645		
	SE3	5.47	1.153	0.786		
Perceived Fi-nancial Cost	PFC1	2.69	1.060	0.755	0.836	0.632
	PFC2	2.57	1.122	0.885		
	PFC3	2.61	1.104	0.736		
Normative Influence	NI1	4.28	1.637	0.863	0.909	0.770
	NI2	4.22	1.531	0.877		
	NI3	4.29	1.532	0.892		
Trust with the Bank	TB1	5.22	1.137	0.847	0.881	0.712
	TB2	5.19	1.243	0.848		
	TB3	5.18	1.147	0.837		
Perceived Cred-ibility	PC1	4.73	1.307	0.787	0.868	0.687
	PC2	4.67	1.240	0.919		
	PC3	4.68	1.329	0.773		
Behavioral Intention	BI1	4.80	1.135	0.750	0.792	0.559
	BI2	4.81	1.097	0.766		
	BI3	4.80	1.163	0.726		

Table 3. Average variance extracted (AVE), shared variances

Construct	PE	PU	SE	PFC	NI	TB	PC	BI
PE	0.677							
PU	0.043	0.707						
SE	0.294	0.181	0.525					
PFC	0.197	0.067	0.256	0.632				
NI	0.019	0.008	0.012	0.014	0.770			
TB	0.126	0.116	0.342	0.137	0.020	0.712		
PC	0.105	0.159	0.203	0.007	0.008	0.070	0.687	
BI	0.514	0.309	0.430	0.511	0.033	0.411	0.243	0.559

Note: The diagonal elements are AVE of the specific constructs. Off diagonal elements are shared variances between the constructs.

Figure 2. Model with results (***significant p <0.001, ** significant p<0.01,*significant p<0.05, ns – not significant p>0.05)

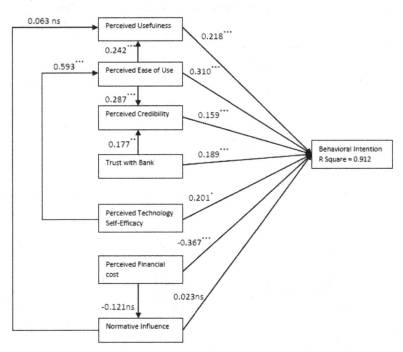

pothesized effects. Properties of the causal paths, including standardized path coefficients, p values and explanation of variance for each equation in the hypothesized model are presented in Figure 2. The behavioral intention to use mobile banking was predicted by Perceived usefulness (β=0.218,p <0.001), Perceived ease of use(β=0.218,p <0.001), Perceived credibility(β=0.218,p <0.001), Trust with the bank (β=0.218,p <0.001), Perceived self efficacy (β=0.218,p <0.001) and Perceived financial cost (β=0.218,p <0.001). Altogether these constructs accounted for 91% variation in behavioral intention towards mobile banking. As a result hypotheses H1, H2, H3, H4, H5 and H6 were supported. Normative Influence (β=0.023, p >0.05) did not affect.

Behavioral intention and perceived usefulness as well (β=0.063, p >0.05) hence hypotheses H7, H7a were not supported. Hypothesis H6a was also found to be not supported as perceived financial cost (β= -0.121, p >0.05) did not affect

normative influence. Perceived ease of use was found to have have a significant impact on perceived usefulness (β=0.242, p <0.001) and perceived credibility (β=0.287, p <0.001). Further, trust with the bank (β=0.177, p <0.001) was found to have a significant effect on perceived credibility and perceived self efficacy (β=0.593, p <0.001) was found to be a strong predictor of perceived ease of use. As a result hypotheses H2a, H2b, H4a and H5a were all supported. The results of the hypothesis tests are summarized in the Table 4.

DISCUSSION

Ever since TAM was proposed by Davis (1989), it has been replicated and extended to explain the adoption of IS. TAM has also been used successfully to elucidate the adoption of mobile banking (e.g., Luarn & Lin, 2005; Gu et al., 2009).We extended the model proposed by Luarn and Lin

Table 4. Summary of hypothesis testing

Hypo-thesis	Relationship	Expected direction	Observed direction	β	p-value	Result
1	Perceived usefulness → Behavioral intention	+	+	0.218	0.000	Accepted
2	Perceived ease of use → Behavioral intention	+	+	0.310	0.000	Accepted
2a	Perceived ease of use → Perceived usefulness	+	+	0.242	0.000	Accepted
2b	Perceived ease of use → Perceived credibility	+	+	0.287	0.000	Accepted
3	Perceived credibility → Behavioral intention	+	+	0.159	0.000	Rejected
4	Trust with the bank → Behavioral intention	+	+	0.189	0.000	Accepted
4a	Trust with the bank → Perceived credibility	+	+	0.177	0.008	Accepted
5	Perceived self efficacy → Behavioral intention	+	+	0.201	0.015	Accepted
5a	Perceived self efficacy → Perceived ease of use	+	+	0.593	0.000	Accepted
6	Perceived financial cost → Behavioral intention	-	-	-0.367	0.000	Accepted
6a	Perceived financial cost → Normative influence	-	-	-0.121	0.058	Rejected
7	Normative influence → Behavioral intention	+	+	0.023	0.550	Rejected
7a	Normative influence → Perceived usefulness	+	+	0.063	0.331	Rejected

(2005) by incorporating certain additional variables, so that the explaining power of the model is improved. Thus two variables were added; social norm and trust with the bank. Social norm or normative influence was added to the model as Luran and Lin (2005) recommended it. As we felt that trust in the bank is the most important factor that helps the customer to try new channels we have incorporated trust with the bank in the model. The same has been supported by the studies of Dimitriadis and Kyrezis (2008) and Gu et al. (2009).

Our findings strongly support the appropriateness of using this extended TAM to understand the intentions of people towards the use of mobile banking. It also reinforces the results of the previous research on adoption of mobile banking using TAM (e.g., Luarn & Lin, 2005; Gu et al., 2009). Perceived usefulness, ease of use, credibility, self-efficacy, financial costs and trust with the bank were found to have a significant impact on behavioral intention. Perceived financial cost was found to have stronger influence on behavioral intention towards mobile banking as the customer has to bear the brunt for carrying out transactions

through mobile phones. Perceived ease of use followed by perceived usefulness was observed to be the significant factors.

Normative influence did not have a bearing on both behavioral intention and perceived usefulness. This finding is in tandem with previous research, which showed that social influence has no influence on perceived usefulness and behavioral intention of financial services in voluntary context (Venkatesh & Davis, 2000; Venkatesh et al., 2003). As the use of mobile banking is voluntary, users are not influenced by normative pressure to concur with the behavior of others, rather customers resort to mobile banking based on their need.

Consistent with prior research (Agarwal & Karahanna, 2000; Wang, Wang, Lin, & Tang, 2003; Ong, Lai, & Wang, 2004; Luarn & Lin, 2005; Reid & Levy, 2008; Gu et al., 2009), perceived self efficacy is found to be a strong predictor of perceived ease of use. Perceived ease of use in turn affects the perceived usefulness and perceived credibility. This is also supported by previous research that also found that perceived ease of use affects perceived usefulness (Wang et al., 2003;

Ong, Lai, & Wang, 2004; Luarn & Lin, 2005; Gu et al., 2009) and perceived credibility (Wang et al., 2003; Ong, Lai, & Wang, 2004; Luarn & Lin, 2005). As perceived self efficacy directly or indirectly influences the behavioral intention, the banks have to focus on developing the self efficacy of the consumers by way of imparting training to the customers on the usage of mobile banking.

In contrast to the study of Lu et al. (2010), the linkage between the trust in the bank and behavioural intention towards mobile banking was supported in this study. Comparing to the previous studies on adoption of mobile banking, the outcome of our study strongly advocate that our model is better in terms of explaining the variation in behavioral intention. The R-square of our model was 91.2% while it was 82% in the study of Luarn and Lin (2005) and 72.2% in the study of Gu et al. (2009).

Implications for Research and Practice

Ever since TAM was proposed by Davis (1989), it has been replicated and extended to explain the adoption of IS. TAM has also been used successfully to elucidate the adoption of mobile banking (e.g., Luarn & Lin, 2005; Gu et al., 2009) though mobile banking is very different from the systems examined in prior studies that use TAM. This study, examining the factors influencing the behavioural intention towards mobile banking, extended the model proposed by Luarn and Lin (2005) by incorporating certain additional variables, so that the explaining power of the model is improved. Thus two variables were added; social norm and trust with the bank. Social norm or normative influence was added to the model as Luran and Lin (2005) recommended it. As trust in the bank is the most important factor that helps the customer to try new channels; trust with the bank was incorporated in the model. The same has been supported by the studies of Dimitriadis and Kyrezis (2008) and Gu et al. (2009). The findings of this study

strongly support the appropriateness of using this extended TAM to understand the intentions of people towards the use of mobile banking. It also reinforces the results of the previous research on adoption of mobile banking using TAM (e.g., Luarn & Lin., 2005; Gu et al., 2009).

As suggested by the extended TAM, perceived self-efficacy influences behavioral intention, either directly or indirectly, through its effect on perceived ease of use. Thus, management attention might be more fruitfully focused on the development of perceived self-efficacy. Mobile banking practitioners can increase their customers' intent to use the system through perceived self-efficacy and the mediating TAM variables (i.e., perceived ease of use, perceived usefulness and perceived credibility). In order to promote a customer's perception of self-efficacy in mobile banking, managers could organize training courses in various mobile commerce applications, thus increasing customers' familiarity with mobile technologies.

The usage of mobile banking is completely based on the individual's willingness. The user group in India has diversified background in terms of culture and literacy level. The service providers charge higher prices for the SMS related to mobile banking. In order to attract more users towards mobile banking, service providers must reduce the users' perceived financial cost through strategic alliances with mobile service providers. Mobile banking should be user friendly. Like ATMs the mobile banking service should be accessible through vernacular language as it may aid in easy comprehension which impacts the perceived ease of use. The trustworthiness of the system in protecting the security and privacy of the users also need to be addressed.

CONCLUSION

Considering the extended TAM proposed by Luarn and Lin (2005) as a base model, normative influence and trust with the bank has been added as two new additional factors. The two additional factors were added based on the literature regarding TAM and TPB. These two factors were incorporated in order to develop a unified model on adoption of mobile banking. This model is empirically demonstrated by using the actual data from the customers of State bank of India (SBI). First, this study finds that explanatory power of the suggested model is much higher than others and the validity of the model was strongly supported. Secondly, this study verified the effect of perceived usefulness, perceived ease-of-use, perceived credibility, perceived self efficacy, trust in the bank, and perceived financial cost on behavioral intention in mobile banking. The findings of the study would be useful for the bankers in devising the strategies for enhancing the acceptance of mobile banking.

Although the findings are encouraging and useful, the present study has certain limitations and needs further research. First, the study was confined to the customers of one specific bank and one specific branch. The responses are based on the respondents' perceptions and experiences with regard to the specific bank they are exposed. The small size and homogeneity of the sample means that the findings are only representative of the specific bank and the particular branch that was considered. Consequently, they do not represent views held by the various segments of the population in India. Moreover, most of the respondents in this study are educated and have exposure to new technologies. As such, they are not representative of Indian society. Future research on less educated and experienced consumers will offer additional validation of the model proposed in this study. Second the data presented is cross-sectional. Longitudinal data will be needed in the future to assess what factors will influence consumers' intentions to use mobile banking. Third, the usage of mobile banking is currently voluntary. The findings may not be generalized to the mandatory environment, where the customer is not having an option to no to use it. Thus, further research

is necessary to verify the differences between the voluntary and mandatory environments. Finally a comparative study with customers of different bank will yield a better insight for the banks to know their position in terms of customers' adoption of mobile banking. Similarly validating the model with users and non-users of mobile banking will throw light on the impediments that affect the adoption of mobile banking.

REFERENCES

Adams, D. A., Nelson, R. R., & Todd, P. A. (1992). Perceived usefulness, ease of use, and usage of information technology: A replication. *Management Information Systems Quarterly*, *16*(2), 227–247. doi:10.2307/249577

Agarwal, R., & Karahanna, E. (2000). Time flies when you're having fun: Cognitive absorption and beliefs about information technology usage. *Management Information Systems Quarterly*, *24*(4), 665–694. doi:10.2307/3250951

Agarwal, R., & Prasad, J. (1999). Are individual differences germane to the acceptance of new information technologies. *Decision Sciences*, *30*(2), 361–391. doi:10.1111/j.1540-5915.1999. tb01614.x

Ajzen, I., & Fishbein, M. (1980). *Understanding attitudes and predicting social behavior*. Upper Saddle River, NJ: Prentice Hall.

Amin, H., Rizal, M., Hamid, A., Lada, S., & Anis, Z. (2008). The adoption of mobile banking in Malaysia: The case of Bank Islam Malaysia Berhad (BIMB). *International Journal of Business and Society*, *9*(2), 43–53.

Anderson, J. C., & Gerbing, D. W. (1988). Structural equation modeling in practice: A review and recommended two-step approach. *Psychological Bulletin*, *103*(3), 411–423. doi:10.1037/0033-2909.103.3.411

Bachmann, G. R., John, D. R., & Rao, A. R. (1993). Children's susceptibility to peer group purchase influence: An exploratory investigation. *Advances in Consumer Research. Association for Consumer Research (U. S.), 20,* 463–468.

Bandura, A. (1982). Self-efficacy mechanism in human agency. *The American Psychologist, 37,* 122–147. doi:10.1037/0003-066X.37.2.122

Bandura, A. (1986). *Social foundations of thought and action.* Upper Saddle River, NJ: Prentice Hall.

Bandura, A. (1997). *Self-efficacy: The exercise of control.* New York, NY: W.H. Freeman.

Bearden, W. O., Netemeyer, R. G., & Teel, R. J. (1989). Measurement of consumer susceptibility to interpersonal influence. *The Journal of Consumer Research, 15*(4), 473–481. doi:10.1086/209186

Blechar, J., Constantiou, I. D., & Damsgaard, J. (2006). Exploring the influence of reference situations and reference pricing on mobile service user behaviour. *European Journal of IS, 15*(3), 285–291.

Buckley, M. R., Cote, J. A., & Comstock, S. M. (1990). Measurement errors in behavioral sciences: The case of personality/attitude research. *Educational and Psychological Measurement, 50*(3), 447–474. doi:10.1177/0013164490503001

Chakrabarty, K. C. (2010). Mobile commerce, mobile banking: The emerging paradigm reserve bank of India. *Monthly Bulletin,* 23-30.

Chau, P. Y. K. (1996). An empirical assessment of a modified technology acceptance model. *Journal of Management Information Systems, 13*(2), 185–204.

Chaudhuri, A., & Holbrook, M. B. (2001). The chain of effects from brand trust and brand affect to brand performance: The role of brand loyalty. *Journal of Marketing, 65*(2), 81–93. doi:10.1509/jmkg.65.2.81.18255

Chen, C., Czerwinski, M., & Macredie, R. (2000). Individual differences in virtual environments - Introduction and overview. *Journal of the American Society for Information Science American Society for Information Science, 51*(6), 499–507. doi:10.1002/(SICI)1097-4571(2000)51:6<499::AID-ASI2>3.0.CO;2-K

Cheong, J. H., & Park, M. C. (2005). Mobile internet acceptance in Korea. *Internet Research, 15,* 125–140. doi:10.1108/10662240510590324

Chiu, Y. B., Lin, C. P., & Tang, L. L. (2005). Gender differs: Assessing a model of online purchase intentions in e-tail service. *International Journal of Service Industry Management, 16,* 416–435. doi:10.1108/09564230510625741

Compeau, D. R., & Higgins, C. A. (1995). Computer self-efficacy: Development of a measure and initial test. *Management Information Systems Quarterly, 19,* 189–211. doi:10.2307/249688

Compeau, D. R., Higgins, C. A., & Huff, S. (1999). Social cognitive theory and individual reaction to computing technology: A longitudinal study. *Management Information Systems Quarterly, 23*(2), 145–158. doi:10.2307/249749

Condos, C., James, A., Every, P., & Simpson, T. (2002). Ten usability principles for the development of effective WAP and m-commerce services. *Aslib Proceedings, 54*(6), 345–355. doi:10.1108/00012530210452546

Constantinides, E. (2002). The 4S web-marketing mix model. *Electronic Commerce Research and Applications, 1*(1), 57–76. doi:10.1016/S1567-4223(02)00006-6

Curran, J. M., & Meuter, M. L. (2005). Self-service technology adoption: Comparing three technologies. *Journal of Services Marketing, 19*(2), 103–113. doi:10.1108/08876040510591411

Curran, J. M., Meuter, M. L., & Surprenant, C. F. (2003). Intentions to use self-service technologies: A confluence of multiple attitudes. *Journal of Service Research*, *5*(3), 209–224. doi:10.1177/1094670502238916

Davis, F. D. (1989). Perceived usefulness, perceived ease of use and user acceptance of information technology. *Management Information Systems Quarterly*, 319–340. doi:10.2307/249008

Davis, F. D., Bagozzi, R. P., & Warshaw, P. R. (1989). User acceptance of computer technology: A comparison of two theoretical models. *Management Science*, *35*(8), 982–1003. doi:10.1287/mnsc.35.8.982

Deutsch, M., & Gerard, H. (1995). A study of normative and informational social influences upon individual judgment. *Journal of Abnormal and Social Psychology*, *51*, 624–636.

Dimitriadis, S., & Kyrezis, N. (2008). Does trust in the bank build trust in its technology based channels? *Journal of Financial Services Marketing*, *23*(1), 28–38. doi:10.1057/fsm.2008.3

Doll, W. J., Hendrickson, A., & Deng, X. (1998). Using Davis's perceived usefulness and ease of use instruments for decision making: A confirmatory and multi group invariance analysis. *Decision Sciences*, *29*(4), 839–870. doi:10.1111/j.1540-5915.1998.tb00879.x

Eastin, M. S. (2002). Diffusion of e-commerce: An analysis of the adoption of four e-commerce activities. *Telematics and Informatics*, *19*(3), 251–267. doi:10.1016/S0736-5853(01)00005-3

Featherman, M. S., & Pavlou, P. A. (2003). Predicting e-services adoption: A perceived risk facets perspective. *International Journal of Human-Computer Studies*, *59*, 451–474. doi:10.1016/S1071-5819(03)00111-3

Fornell, C., & Larcker, D. (1981). Structural equation models with unobservable variables and measurement error. *JMR, Journal of Marketing Research*, *18*(1), 39–50. doi:10.2307/3151312

Friedman, B., Kahn, P. H. Jr, & Howe, D. C. (2000). Trust online. *Communications of the ACM*, *43*(12), 34–40. doi:10.1145/355112.355120

Fukuyama, E. (1995). *Trust: The social virtues & the creation of prosperity*. New York, NY: Free Press.

Gefen, D. (2003). TAM or just plain habit: A look at experienced online shoppers. *Journal of End User Computing*, *15*, 1–13. doi:10.4018/joeuc.2003070101

Gefen, D., Karahanna, E., & Straub, D. W. (2003). Trust and TAM in online shopping: An integrated model. *Management Information Systems Quarterly*, *27*, 51–90.

Gefen, D., & Straub, D. W. (1997). Gender differences in the perception and use of e-mail: An extension to the technology acceptance model. *Management Information Systems Quarterly*, *21*, 389–400. doi:10.2307/249720

Grazioli, S., & Jarvenpaa, S. L. (2000). Perils of Internet fraud: An empirical investigation of deception and trust with experienced Internet. *IEEE Transactions on Systems, Man, and Cybernetics. Part A, Systems and Humans*, *30*(4), 395–410. doi:10.1109/3468.852434

Grimm, P. E., Agrawal, J., & Richardson, P. S. (1999). Product conspicuousness and buying motives as determinants of reference group influences. *European Advances in Consumer Research*, *4*, 97–103.

Gu, J. C., Lee, S. C., & Suh, Y. H. (2009). Determinants of behavioral intention to mobile banking. *Expert Systems with Applications*, *36*, 11605–11616. doi:10.1016/j.eswa.2009.03.024

Hair, J. Jr, Anderson, J., Norman, J., & Black, W. (1995). *Multivariate data analysis with readings* (4th ed.). Upper Saddle River, NJ: Prentice Hall.

Hair, J. F., Tatham, R. L., Anderson, R. E., & Black, W. (1998). *Multivariate data analysis*. Upper Saddle River, NJ: Prentice Hall.

Hawes, J. M., Kenneth, E. M., & Swan, J. E. (1989). Trust earning perceptions of sellers and buyers. *Journal of Personal Selling & Sales Management*, *9*, 1–8.

Hendrickson, A. R., Massey, P. D., & Cronan, T. P. (1993). On the test–retest reliability of perceived ease of use scales. *Management Information Systems Quarterly*, *17*(2), 227–230. doi:10.2307/249803

Hoffman, D. L., Novak, T. P., & Peralta, M. (1999). Building consumer trust online. *Communications of the ACM*, *42*(4), 80–85. doi:10.1145/299157.299175

Hong, W., Thong, J. Y. L., Wong, W. M., & Tam, K. Y. (2001). Determinants of user acceptance of digital libraries: An empirical examination of individual differences and system characteristics. *Journal of Management Information Systems*, *18*(3), 97–124.

Hu, L. T., & Bentler, P. M. (1999). Cutoff criteria for fit indexes in covariance structure analysis: Conventional criteria versus new alternatives. *Structural Equation Modeling*, *6*, 1–55. doi:10.1080/10705519909540118

Hu, P. J., Chau, P. Y. K., Liu Sheng, O. R., & Yan Tam, K. (1999). Examining the technology acceptance model using physician acceptance of telemedicine technology. *Journal of Management Information Systems*, *16*(2), 91–112.

Hung, S. Y., Ku, C. Y., & Chang, C. M. (2003). Critical factors of WAP services adoption: An empirical study. *Electronic Commerce Research and Applications*, *2*(1), 46–60. doi:10.1016/S1567-4223(03)00008-5

Igbaria, M., & Iivari, J. (1995). The effects of self-efficacy on computer usage. *Omega*, *23*(6), 587–605. doi:10.1016/0305-0483(95)00035-6

Jackson, C. M., Chow, S., & Leitch, R. A. (1997). Towards an understanding of the behavioral intention to use an information system. *Decision Sciences*, *28*(2), 357–389. doi:10.1111/j.1540-5915.1997.tb01315.x

Kim, H., Chan, H. C., & Gupta, S. (2007). Value-based adoption of mobile Internet: An empirical investigation. *Decision Support Systems*, *43*, 111–126. doi:10.1016/j.dss.2005.05.009

Knutsen, L., Constantiou, I. D., & Damsgaard, J. (2005). Acceptance and perceptions of advanced mobile services: Alterations during a field study. In *Proceedings of the International Conference on Mobile Business*, Sydney, Australia (pp. 326-331).

Lee, J. (2002). A key to marketing financial services: The right mix of products, services, channels and customers. *Journal of Services Marketing*, *16*(3), 238–258. doi:10.1108/08876040210427227

Lee, K. S., Lee, H. S., & Kim, S. Y. (2007). Factors influencing the adoption behavior of mobile banking: A South Korean perspective. *Journal of Internet Banking and Commerce*, *12*(2).

Legris, P., Ingham, J., & Collerette, P. (2003). Why do people use information technology? A critical review of the technology acceptance model. *Information & Management*, *40*(2), 191–204. doi:10.1016/S0378-7206(01)00143-4

Lu, H. P., & Gustafson, D. H. (1994). An empirical study of perceived usefulness and perceived ease of use on computerized support system use over time. *International Journal of Information Management*, *14*(5), 317–329. doi:10.1016/0268-4012(94)90070-1

Lu, J., Yu, C., Liu, C., & Yao, J. E. (2003). Technology acceptance model for wireless Internet. *Internet Research: Electronic Networking Applications and Policy, 13,* 206–222. doi:10.1108/10662240310478222

Luarn, P., & Lin, H. H. (2005). Toward an understanding of the behavioural intention to use mobile banking. *Computers in Human Behavior, 21,* 873–891. doi:10.1016/j.chb.2004.03.003

Luo, X., Li, H., Zhang, J., & Shim, J. P. (2010). Examining multi-dimensional trust and multi-faceted risk in initial acceptance of emerging technologies: An empirical study of mobile banking services. *Decision Support Systems, 49,* 222–234. doi:10.1016/j.dss.2010.02.008

Mallat, N., Rossi, M., & Tuunainen, V. K. (2004). Mobile banking services. *Communications of the ACM, 47*(5), 42–46. doi:10.1145/986213.986236

Mathieson, K. (1991). Predicting user intentions: Comparing the technology acceptance model with the theory of planned behavior. *Information Systems Research, 84*(1), 123–136.

Mathieson, K., Peacock, E., & Chin, W. W. (2001). Extending the technology acceptance model: The influence of perceived user resources. *The Data Base for Advances in Information Systems, 32*(3), 86–112.

Mattila, M. (2003). Factors affecting the adoption of mobile banking services. *Journal of Internet Banking and Commerce, 8*(1).

Meuter, M. L., Ostrom, A. L., Roundtree, R. I., & Bitner, M. J. (2000). Self-service technologies: Understanding customer satisfaction with technology-based service. *Journal of Marketing, 64*(3), 50–64. doi:10.1509/jmkg.64.3.50.18024

Moon, J.-W., & Kim, Y.-G. (2001). Extending the TAM for the World-Wide-Web context. *Information & Management, 38,* 217–230. doi:10.1016/S0378-7206(00)00061-6

Morrison, P. D., & Roberts, J. H. (1998). Matching electronic distribution channels to product characteristics: Role of congruence in consideration set formation. *Journal of Business Research, 41*(3), 223–229. doi:10.1016/S0148-2963(97)00065-9

Okazaki, S. (2005). Mobile advertising adoption by multinationals: Senior executives initial responses. *Internet Research, 15*(2), 160–180. doi:10.1108/10662240510590342

Ong, C. S., Laia, J. Y., & Wang, Y. S. (2004). Factors affecting engineers' acceptance of asynchronous e-learning systems in high-tech companies. *Information & Management, 41*(6), 795–804. doi:10.1016/j.im.2003.08.012

Pagani, M. (2004). Determinants of adoption of third generation mobile multimedia services. *Journal of Interactive Marketing, 18*(3), 46–59. doi:10.1002/dir.20011

Pavlou, P., & Fygenson, M. (2006). Understanding and predicting electronic commerce adoption: An extension of the theory of planned behavior. *Management Information Systems Quarterly, 30*(1), 44.

Pikkarainen, T., Pikkarainen, K., Karjaluoto, H., & Pahnila, S. (2004). Consumer acceptance of online banking: An extension of the technology acceptance model. *Internet Research, 14*(3), 224–235. doi:10.1108/10662240410542652

Reid, M., & Levy, Y. (2008). Integrating trust and computer self-efficacy with TAM: An empirical assessment of customers' acceptance of banking information systems (BIS) in Jamaica. *Journal of Internet Banking and Commerce, 12*(3).

Rose, J., & Fogarty, G. (2006). Determinants of perceived usefulness and perceived ease of use in the technology acceptance model: Senior consumers' adoption of self-service banking technologies. In *Proceedings of the 2nd Biennial Academy of World Business, Marketing & Management Development Conference* (Vol. 2, pp. 122-129).

Schumacker, R. E., & Lomax, R. G. (2004). *A beginner's guide to structural equation modeling.* Mahwah, NJ: Lawrence Erlbaum.

Segars, A. H., & Grover, V. (1993). Re-examining perceived ease of use and usefulness: A confirmatory factors analysis. *Management Information Systems Quarterly, 17*(4), 517–526. doi:10.2307/249590

Shi, W., Shambare, N., & Wang, J. (2008). The adoption of internet banking: An institutional theory perspective. *Journal of Financial Services Marketing, 12*(4), 272–286. doi:10.1057/palgrave.fsm.4760081

Teo, T. S. H., & Pok, S. H. (2003). Adoption of WAP-enabled mobile phones among Internet users. *Omega. International Journal of Management Science, 31*(6), 483–498.

Venkatesh, V. (1999). Creation of favourable user perceptions: Exploring the role of intrinsic motivation. *Management Information Systems Quarterly, 23*(2), 239–260. doi:10.2307/249753

Venkatesh, V. (2000). Determinants of perceived ease of use: Integrating control, intrinsic motivation, and emotion into the technology acceptance model. *Information Systems Research, 11*(4), 342–365. doi:10.1287/isre.11.4.342.11872

Venkatesh, V., & Davis, F. D. (1996). A model of the antecedents of perceived ease of use: Development and test. *Decision Sciences, 27*(3), 451–481. doi:10.1111/j.1540-5915.1996.tb01822.x

Venkatesh, V., & Davis, F. D. (2000). A theoretical extension of the technology acceptance model: Four longitudinal field studies. *Management Science, 46*(2), 186–204. doi:10.1287/mnsc.46.2.186.11926

Venkatesh, V., & Morris, M. G. (2000). Why don't men ever stop to ask for directions? Gender, social influence, and their role in technology acceptance and usage behavior. *Management Information Systems Quarterly, 24*(1), 115–139. doi:10.2307/3250981

Venkatesh, V., Morris, M. G., Davis, G. B., & Davis, F. D. (2003). User acceptance of information technology: Toward a unified view. *Management Information Systems Quarterly, 27*(3), 425–478.

Wang, Y. S., Wang, Y. M., Lin, Y. M., & Tang, T. I. (2003). Determinants of user acceptance of internet banking: An empirical study. *International Journal of Service Industry Management, 14*(5), 501–519. doi:10.1108/09564230310500192

Wu, J. H., & Wang, S. C. (2005). what drives mobile commerce? An empirical evaluation of the revised technology acceptance model. *Information & Management, 42*, 719–729. doi:10.1016/j.im.2004.07.001

Yang, K. (2005). Exploring factors affecting the adoption of mobile commerce in Singapore. *Telematics and Informatics, 22*, 257–277. doi:10.1016/j.tele.2004.11.003

APPENDIX

Perceived Usefulness

Using Mobile banking would improve my performance in conducting banking transactions.
Using Mobile banking would make it easier for me to conduct banking transactions.
I would find Mobile banking useful in conducting my banking transactions.

Perceived Ease of Use

Learning to use Mobile banking is easy for me.
It would be easy for me to become skillful at using Mobile banking.
I would find Mobile banking easy to use.

Perceived Credibility

Using Mobile banking would not divulge my personal information.
I would find Mobile banking secure in conducting my banking transactions.
Mobile phone is a reliable enough medium for carrying out banking activities.

Trust with the Bank

I believe my bank is trustworthy.
I believe my bank keeps its promises and commitments.
My bank is remarkably expert and specialized in its field.

Perceived Self-Efficacy

I could use the Mobile banking systems, if I had just the built-in help facility for assistance.
I could use the Mobile banking systems, if I had seen someone else using it before trying it myself.
I could use the Mobile banking systems, if someone showed me how to do it first.

Perceived Financial Cost

I think the equipment cost is expensive for using Mobile banking.
I think the access cost is expensive for using Mobile banking.
I think the transaction fee is expensive for using Mobile banking.

Behavioral Intention

I use Mobile banking because people think I should use Mobile banking.
I use Mobile banking because it is very famous.
I use Mobile banking because many people use Mobile banking.

Normative Influence

Assuming I have access to the Mobile banking, I intend to use it.

Given that I have access to the Mobile banking, I predict that I would use it.

If I have access to the Mobile banking, I want to use it as much as possible.

This work was previously published in the International Journal of E-Services and Mobile Applications, Volume 3, Issue 4, edited by Ada Scupola, pp. 37-56, copyright 2011 by IGI Publishing (an imprint of IGI Global).

Section 2
Organizational and Inter-Organizational Issues in the Online Environment

Chapter 7

Defining, Applying and Customizing Store Atmosphere in Virtual Reality Commerce:
Back to Basics?

Ioannis G. Krasonikolakis
Athens University of Economics and Business, Greece

Adam P. Vrechopoulos
Athens University of Economics and Business, Greece

Athanasia Pouloudi
Athens University of Economics and Business, Greece

ABSTRACT

This paper studies the concept of Store Atmosphere in Virtual Commerce (V-Commerce) through the Web in order to empirically define its determinants and investigate their applicability and customization capabilities. A series of in depth interviews with field experts (study #1) along with an online question-naire survey (study #2) served as the data collection mechanisms of the study. The empirical findings suggest that while the social aspect dimension of V-Commerce limits customization capabilities, it provides several innovative options for manipulating Store Atmosphere. Additionally, the results indicate that Store Atmosphere attributes can be grouped in three factors with high average scores concerning the importance users attach to them. Specifically, storefront, store theatrics, colors, music and graphics are grouped in Factor #1 and reflect the "Store's Appeal". Crowding, product display techniques and innovative store atmosphere services are grouped in Factor #2 labeled "Innovative Atmosphere", while store layout constitutes the only attribute included in Factor #3. The paper outlines the theoretical and managerial implications of these research results.

DOI: 10.4018/978-1-4666-2654-6.ch007

INTRODUCTION

Virtual worlds (VWs) are three-dimensional environments where users engage in numerous activities through their in-world representatives, the so-called "avatars". Several VWs adopted characteristics and applications of social networks, but also exploit the integration of VWs and other Web 2.0 applications as well as e-commerce (Messinger et al., 2009; Spence, 2008). In this paper we focus on these latter types of VWs, excluding purely game-oriented environments such as World of Warcraft from the scope of this research.

In VWs, users, through their avatars, can talk with their friends or make new friends (socialization), play in-world electronic games (entertainment), build houses (interior and exterior decoration), buy and sell both virtual and real products/services (v-commerce) and numerous other activities (Krasonikolakis & Vrechopoulos, 2009). Consequently, social presence in a VW store does not merely imply the co-existence of many avatars within the same store at the same time but also many other options that are also applicable in conventional retailing (i.e. in-world communication with other people). For example, these options may include the product rating, suggest-to-friend, shopping with friends, etc.

A number of business reports demonstrate the growing significance of VWs as a market. Indicatively, according to eMarketer (2010), users of the Second Life virtual world spent $567 million on user-to-user transactions (i.e. user-generated virtual items for avatars) in 2009 (65% annual increase), while it is predicted that more than $6 billion will be spent worldwide on virtual items by 2013. Also, virtual goods sales in this virtual world reached the amount of $6.1 million in 2009. In Second Life alone, the number of users increased by 15% reaching 769 million worldwide in 2009 (eMarketer, 2010). At the same time, the increased interaction for real economic

purposes within Virtual Retailing Environments (VREs) in recent years have been noted (O'Reilly, 2006). Saiman (2009), a virtual business owner, claims that the virtual "lifecycle of business" has many similarities with the real one. Studying the consumption of virtual goods in the context of virtual environments and commodities, Martin (2008) claimed that people consume virtual products to meet exchange- and symbolic-value than use-value.

In the context of VWs, several studies address the need to understand how VWs influence user behaviour (Messinger, Ge, Stroulia, Lyons, Smirnov, & Bone, 2008). Becerra and Stutts (2008), employing sociometer theory indicated that the willingness to become different from real life is one of the driving forces of the use of VWs. Landay (2008) stated that the avatar appearance affects the owner's social behaviour and this, in turn, influences real world behaviour. One step further, Messinger et al. (2008), through both qualitative and quantitative research, found that behaviour in virtual worlds influences real world behaviour and vice versa. They also designated that users' avatars are often similar to themselves but better-looking. On the contrary, Vicdan and Ulusoy (2008), through a netnographic approach, found that the "body concept" is regarded differently in the virtual context, compared to the real environment.

The need to express their identity and find a specific group to belong, urges users to the consumption of virtual goods (Boostrom, 2008). Dechow (2008) suggests that data from past user behaviour combined with the development of artificially intelligent avatars would help entrepreneurs to influence user behaviour in various stages of the shopping process (i.e. direct change of the prices of the products, direct offer of value added services).

This dynamic role of VWs as an alternative e-shopping channel has also drawn research attention to Virtual Reality Commerce through the Web (V-Commerce) (Haven, Bernoff, Glass, & Feffer, 2007; Hendaoui, Limayem, & Thompson, 2008). For example, Frost, Chance, Norton, and Ariely (2008) note that virtual reality features enhance the value offered to customers, while in a similar vein Kim and Forsythe (2008) report that virtual reality applications through the web enhance the entertainment value of the shopping experience. Highly vivid interfaces such as 3D virtual stores provide motives, emotions, meanings and communication which are represented objectively (Mazursky & Vinitzky, 2005). Back in 1996, Burke stated that 3D effectiveness in e-commerce applications lies in their ability to generate a virtual environment for the end-user in which his/her experiences will affect shopping in the physical environment. While retailing activity in the VRE context is active, research on designing the atmosphere of these stores is generally deficient (Krasonikolakis & Vrechopoulos, 2009).

In order to address this research gap, the present paper aims to provide an initial understanding of the nature of store atmosphere in this fast evolving e-shopping landscape. This is achieved by building on earlier theoretical work defining this concept in the context of e-business and then empirically defining the determinants and exploring the applicability and customization capabilities of Virtual Store Atmosphere in V-Commerce.

The paper is structured as follows. First we define Virtual Store Atmosphere determinants (based on a review of the relevant literature and expert interviews) by comparing the traditional, web and virtual reality retail environments. The resulting theoretical framework is used to define the determinants of V-Commerce Store Atmosphere through factor analysis. The implications of the empirical findings are discussed next. Finally, we summarize the conclusions and limitations of the study and presents managerial implications and directions for further research.

EXPLORING THE APPLICABILITY AND CUSTOMIZATION CAPABILITIES OF VIRTUAL STORE ATMOSPHERE

In order to investigate store atmosphere in the context of VWs, this paper employs earlier work on store atmosphere (Lewison, 1994), web atmospherics (Dailey, 1999), as well as the concept of "Virtual Store Atmosphere" (Vrechopoulos, O' Keefe, & Doukidis, 2000). Specifically, the paper compares store atmosphere determinants in three alternative retailing channels: the traditional (physical) environment, the web environment and the virtual retailing environment.

Store atmosphere was defined by Lewison (1994) as the overall emotional and aesthetic effect which is created by a store's physical features. According to Lewison's framework, in conventional retailing, the store atmosphere is determined by three major factors: store image, store atmospherics and store theatrics. Store image includes external (storefront) and internal impressions (e.g. layout, product display techniques). Store atmospherics refer to the five human senses (scent, touch, smell, taste and sight), while store theatrics includes décor themes and store events. These factors have been found relevant in the web retailing environment as well in the work of Vrechopoulos et al. (2000). In their work, Lewison's (1994) framework was applied to the context of Web retailing indicating that all its components could be applied online, except for the touch and taste dimensions of store atmospherics (these are applicable today as discussed below).

As Kotler (1973, 1974) first stated, retailers design their stores in ways that produce specific emotional effects to buyers, which influence their behavior. Similarly, online retailers can create an atmosphere via their website which can affect the shoppers' image and experience with the online store (Eroglu, Machleit, & Davis, 2000). Indeed, "Virtual Store Atmosphere" of a web retail store has been defined as an element of the virtual retail mix (Siomkos & Vrechopoulos, 2002).

In order to enrich Lewison's (1994) framework in the context of V-Commerce, eight in-depth interviews with experts were conducted (Study #1). These experts are active researchers in the field of e-commerce. They were selected through a convenience sampling approach (i.e. those that are accessible) in the context of a Business School. The decision for selecting researchers as experts was based both on budget and accessibility constraints but mainly on the innovative character of the investigated topic. Therefore it was quite difficult to find experienced users as well as a large pool of experts to select from. Experts were asked to propose store atmosphere dimensions that could be potentially manipulated in a virtual reality environment. Then, we tried to match their answers to Lewison's framework (as adapted by Vrechopoulos et al., 2000). Some of their answers were directly related to the existing attributes, while new dimensions lead us to the development of new attributes. Specifically, the experts confirmed the relevance of store image, store atmospherics and store theatrics but also identified crowding and innovative store atmosphere services (e.g. flying within the store) as important consumer behavior influencing factors in the case of V-Commerce. The final set of attributes (i.e. those derived through theory and confirmed through the personal interviews as well as the new attributes identified through the personal interviews) are displayed in the left column of Table 1.

In order to compare these determinants across three alternative retailing channels, namely the traditional (physical) environment, the web environment and the virtual reality environment, we explore and compare their applicability and customization potential in these environments in the following paragraphs. Customization, according to Strauss and Frost (2009), refers to the process of tailoring marketing/retailing mixes to meet consumer needs even at the individual

level. Similarly, applicability denotes whether the attributes can be offered/applied in the respective shopping channel. Indicatively, sound applicability online is low due to the option that users have to turn off the music while music applicability in conventional retailing is high. We employed a scale (i.e., low, medium, high) which indicates the degree of applicability and customization offered to retailers for each determinant/attribute in the sense that these attributes are manipulated by them. The results of this comparison process are presented in the following paragraphs and subsequently summarized in Table 1.

Store Image

Store image is the store's 'personality' as perceived by consumers, and it consists of external and internal impressions (Lewison, 1994). In conventional stores, changing elements of a building or even the layout of departments is difficult, time consuming and costly. Similarly, it is not possible to customize internal (e.g. layout) and external (e.g. building) variables except in some cases where the conventional retailer employs technology to customize the appearance of the store (e.g. personalized product displays through video walls).

On the Web and VREs, in contrast, changes in store layout are easier in terms of cost, time and effort. Similarly, an advantage of VREs is that they offer high capabilities in terms of product display techniques (e.g. 3D). Flavian, Gurrea, and Orus (2009) report that the visual aspects of a web site constitute key factors for achieving a successfully e-service website. However, in virtual worlds the store image customization capabilities are typically low due to the social presence dimension (e.g., customization of store layout when more than one avatar is present within the store at the same time). Such an evolution constitutes a paradigm shift as far as the customization and personalization capabilities of electronic commerce are concerned. In other words, while in Web 1.0 commerce customization and personalization are technology-enabled and applicable, it is clear that in VREs this situation is different at least for some Store Atmosphere variables (e.g., layout, product display techniques) due to the social presence dimension. Therefore, regarding the applicability and customization capabilities of store image, the VRE looks more like a conventional store rather than a Web one that we could describe as a 'back to the basics' trend.

At the same time, it should be acknowledged that technology enables customization across these environments in new ways. For example, this technological impact may be witnessed in some conventional retail stores (e.g., electronic shopping carts enable personalized product recommendation through a computer screen placed on the cart; mobile phones enable customized interaction of the consumer with the shelf; RFID technologies enable personalized advertising; etc.). However, as far as the layout is concerned, customization seems not to be applicable in VWs in cases where more than one avatar is in the same store at the same time (e.g. communicate while navigating through the stores' aisles). Similarly, customization is not applicable to product display techniques when two avatars are standing in front of the same virtual shelf discussing about a specific product placed on this shelf (as would be the case in a conventional store). Certainly, several other store atmosphere variables, such as music or colors, can be customized. However, even for these variables the retailer needs to take in to account the social dimension aspect. For example, two avatars may want to listen to the same music when shopping, in order to share the same experience (as would be the case in conventional stores). Similarly, as far as store decoration is concerned, avatars may want to see the same colors, signs, etc. in order to be able to discuss them and, therefore, enjoy a similar shopping experience. In sum, while the majority of store atmosphere features can be customized within the context of a VW retail store, the social aspect dimension may moderate this option in the sense that a considerable num-

ber of customers may wish to enjoy and share a common shopping experience in a one-to-many store environment similarly to the traditional way of shopping.

Store Atmospherics

The social presence aspect dominating in VREs affects store atmospherics as well. As far as sight appeal is concerned, a virtual reality retailer can smartly guide the customers' eyes via the store through the appropriate lighting manipulations. Similarly, special products can be highlighted in a prominent place in the store and with 3Ds' capabilities one can present a virtual environment as very similar to brick-and-mortar. However, while in the traditional Web's Graphical User Interface (GUI) sight customization is applicable; in the VRE such a capability is limited, due to the presence of more than one customer at the same time in the store. The same stands for music. Specifically, while web site visitors can easily turn off music (i.e. low applicability) customization is potentially high (i.e. when customers turn on the sound mechanism). However, in a Virtual store, sound applicability is high, despite the fact that avatars can turn off the sound as well. This is explained by the social aspect dimension, in the sense that avatars visiting a retail store usually wish to enjoy the same services (including music) that other avatars enjoy within the same store at the same time and, thus, do not turn off the sound. As a result, virtual reality retailers have low customization capabilities concerning music since they design and offer one shopping environment for more than one customer that visit the same store simultaneously. Finally, as far as scent, touch and taste are concerned, technology enables web sites and VREs to offer such capabilities to their customers (e.g. intelligent gloves, electronic tongue). However, similarly to the earlier discussion, customization capabilities in VRE are limited due to social presence.

Store Theatrics

VREs provide a high interactive channel and can virtually support all the décor themes and the events that can take place in a conventional store and make shopping a more entertaining experience. The same stands for the web but with comparatively fewer animation capabilities mainly due to the absence of 3D features. However, store theatrics' customization on the web is higher than the corresponding one in VREs, again due to the social aspect dimension.

The owner of a virtual reality store can organize for his customers, virtual live events and happenings such as concerts, or movies so as to enhance the value of products or brand name and strengthen the relationship with consumers. Along these lines, consumers could be engaged in the production phase of products (experience marketing) through contests (prize money or products to winners) organized by the owners of virtual stores. These activities generate favorable emotional and aesthetical responses to consumers since they consider themselves part of the store's 'family'. These activities can also take place in a traditional environment, but customization capabilities are restricted due to the presence of more than one consumer in the same place (the same stands for VREs).

Crowding

In a social virtual environment the impact of crowding on consumers can be twofold. First, avatars may have to face system lag when there are more avatars in a specific place than the system can support. Also, there may be avatars that do not like shopping in a crowded place. Yet, in a virtual reality setting the owner has the possibility to retain several similar or not virtual stores in order to administer crowding. Traffic can be manipulated through the teleporting capability which is available in virtual environments such as Second Life. On the web, the technology capabili-

ties we have witnessed in recent years diminish system failures while in traditional environments retailers can't directly control crowding as far as space constraints are concerned. In other words, crowding in traditional retailing is controlled to a great extent by customers themselves.

Innovative Store Atmosphere Services

Store image, store atmospherics and store theatrics constitute challenging areas where innovative store atmosphere services can be applied. A virtual reality environment can adopt services both from traditional and web environment and also provide new services, not applicable in other retailing channels. Griffith and Chen (2004) indicated that 3-D advertisements have a greater influence on consumers compared to 2-D online advertisements. Kim, Fioreb, and Lee (2007) stated that presentations of 3D virtual objects offer higher interactivity instead of 2D and could be used as a promising promotional tool (Nikolaou, Bettany, & Larsen, 2010). Also, interactivity creates positive reactions to consumers, thus offering many benefits (Fiore, Jin, & Kim, 2005). Interactive kiosks that are placed in or outside a store and the existence of virtual employees are some examples of the new services offered in virtual environments that were not available in web stores. Virtual employees welcome the visitors entering the store and are willing to help and advise consumers about their choices, as in a traditional store.

DEFINING THE DETERMINANTS OF V-COMMERCE STORE ATMOSPHERE RESEARCH SETTING

In order to empirically define the determinants of V-Commerce store atmosphere, we build on earlier research work in conventional and web retailing. We used 9 store atmosphere variables that according to the literature (Lewison, 1994;

Table 1. Store atmosphere determinants' applicability and customization capabilities in alternative retailing channels

Store Atmosphere Determinants		Applicability; Customization Capabilities of Store Atmosphere in 3 alternative retailing channels:		
		Traditional environment	*Web environment*	*VW retail environment*
Store image		High; Low	High; High	High; Low
Store atmospherics	*Sight*	High; Low	High; High	High; Low
	Sound	High; Low	Low; High	Medium; Low
	Scent	High; Low	Low; High	Low; Low
	Touch	High; Low	N/A	Low; High
	Taste	High; Low	N/A	N/A
Store theatrics		High; Low	Medium; High	High; Low
Crowding		Low; Low	Medium; Medium	High; High
Store Atmosphere Services		Medium; Low	Medium; High	High; High

Vrechopoulos et al., 2000), in-depth interviews and the personal judgment of the researchers are the most common ones in the context of a virtual reality retailing.

The variables included in Table 1 served as the attributes that were factor analyzed towards providing a list of underlying factors. Specifically, we did not use the predefined categories but rather tested whether these variables are grouped in a similar fashion in the context of virtual reality retailing. Scent, touch and taste were excluded from this analysis due to the fact that most users may not be aware of their applicability and existence (e.g. e-tongue, e-gloves, etc.) and, therefore, may not provide reliable answers. For the sound option we referred to music while for the sight one we used colors and graphics since they are the most relevant and common sight attributes of a graphical user interface. An electronic questionnaire was developed and served as the data collection instrument of this study (Study #2).

Consumers were asked to indicate the importance they attach to each of these variables (1-5 Likert scale) when they select a retail store in the context of V-Commerce. Pre-tests were conducted in order to test the questionnaire's reliability and

to modify any unclear questions. Due to the explorative nature of the research, a convenience sample was adopted. The population from which the sample was drawn included Internet users that visit VWs available through the web and, therefore, are suitable research subjects. The sampling frame of the present study was Facebook and Second Life. Specifically, 400 invitations to complete the questionnaire were sent to users that were likely to have visited a virtual environment (these were identified through groups on Facebook that are funs of Virtual Worlds). Also, virtual questionnaire kiosks were placed in two Greek regions in the virtual world "Second Life" where avatars crossing by were able to take part in the research by filling in the questionnaire and getting an award of 20 Linden Dollars (approximately 6 cents of euro) as a participation motive.

It should be noted that since VWs constitute a novel retail channel, we selected an exploratory instead of a conclusive research design that was executed through a consumer quantitative survey, in order to obtain an initial understanding of the nature of the basic marketing phenomena in this emerging shopping channel (Malhotra, & Birks, 2000). We employed a quantitative rather than qualitative research design in order to be able to ensure a large sample of respondents and, therefore, be able to generalize results more safely. We believe that the results of this quantitative research could motivate an elaborate qualitative research project to study in depth some findings of the present study that were unanticipated or counterintuitive.

RESULTS

A total of 104 users, 61 by the invitation through the groups of Facebook and 43 by the questionnaires' kiosks in-world, took part in the study. The percentage of females was 53.8%. The vast majority (81.8%) of the sample was below 36 years old; approximately 40% were aged between 18 to 25 years old. This finding is probably explained by the fact that in their early stages, most virtual reality worlds were primarily game-oriented. So it is reasonable they attract younger users.

A minimum of five subjects per variable is required for factor analysis (Malhotra & Birks, 2000). This requirement is fully met in the case of this research that involves 9 variables and 104 subjects. Tests of normality (Kolmogorov-Smirnov and Shapiro-Wilk) and linearity support the appropriateness of the factor analytical model. Furthermore, the several sizable correlations resulted from the correlation matrix, imply that the matrix is appropriate for factor analysis (Hair, Black, Babin, Anderson, & Tatham, 2005). Also, multicollinearity and singularity were conducted to check if any of the squared multiple correlations are near or equal to one. Finally, Bartlett's test of sphericity (Approx. Chi- Square 138.716, df 36.000, Sig 0.000) and Kaiser-Mayer-Olkin measure (0.643) were conducted in order to prove the appropriateness of the model (Coakes, Steed, & Ong, 2009).

Table 2 displays the three factors that were extracted. Storefront, store theatrics, colors, music

and graphics were grouped in one factor (Factor #1). We label this factor *Store Appeal* because all these attributes are related to the "artistic" part of a store (e.g., the store as a theater), the way the aesthetics of the store are perceived by customers. Crowding, product display techniques and innovative store atmosphere services were grouped in a second factor (Factor #2). We label this factor *Innovative Atmosphere*; these elements are directly related to the innovative aspect offered by VREs in the sense that 3D technology provides such capabilities for displaying products, providing services and manipulating crowding that are new to the world of retailing. Also, innovative product display techniques (this is actually a core retail service) guide avatars' navigational behavior within the store and, therefore, affect the crowding dimension. Finally, *Store Layout* constitutes the only attribute included in Factor #3. This finding highlights the importance of this graphical user interface dimension as a major consumer influencing factor in V-Commerce, in the sense that consumers perceive it as a selection criterion that is not related to others. Therefore, this factor should be investigated on its own, similarly to the

Table 2. Rotated factor matrix

Store Environment/ Atmosphere Determinants (Variables)	Factors		
	Store Appeal (1)	Innovative Atmosphere (2)	Store layout (3)
Storefront	,690		
StoreTheatrics	,565		
Colors	,551		
Music	,488		-,334
Graphics	,380		
Crowding		,627	
Product DisplayTechniques		,493	
Innovative Store Atmosphere Services		,466	,407
StoreLayout			,638

*extraction method: Principal Axis Factoring
**rotation method: Varimax with Kaiser Normalization

relevant research practice. This finding confirms the available knowledge on the topic of online store layout effects on consumer behaviour in the context of multichannel retailing (Baker, Grewal, & Parasuraman, 1994; Burke, 2002; Grewal & Baker, 1994; Griffith, 2005; Lohse & Spiller, 1999; Merrilees & Miller, 2001; Simonson, 1999).

Furthermore, it should be noted that all factor scores indicate that consumers attach significant importance to them when they select a V-Commerce store to conduct purchases (Average scores: Factor 1: 3,42, Factor 2: 3,88, Factor 3: 3,84). This finding is consistent with an earlier study in Web retailing by Vrechopoulos, Siomkos, and Doukidis (2001). Specifically, that study found that consumers attach high importance to store atmosphere variables when they select a Web based retail store to conduct their purchases. It also reported that the score consumers attached to importance of store selection criteria is higher for potential shoppers compared to the current ones. This finding was attributed to the various concerns (e.g. security, effectiveness, etc.) that a shopper has when he/she uses a new retail channel to conduct purchases. Similarly, since the percentage of current V-Commerce shoppers is lower that the potential ones it is expected to obtain such high average scores for the store selection criteria. In other words, consumers that plan to adopt a new shopping channel, compared to the current ones, usually attach higher importance to the majority of the potential criteria in order to select a particular store (Vrechopoulos et al., 2001).

Finally, it should be underlined that the resulted factors' content (i.e. variables) is not in line with the available knowledge from both conventional and traditional web retailing, implying that VWs' visitors perceive them differently. Thus, factor analysis results do not confirm established knowledge; this finding, along with the implications of all findings of this empirical research is discussed extensively in the next session.

DISCUSSION OF RESULTS

Several studies have demonstrated the social aspect of Virtual Worlds (Jung & Kang, 2010; Lin, 2008; Ridings & Gefen, 2004). Indicatively, Jung and Kang (2010) report that social relations are one of the major goals of people visiting VWs. Similarly, eMarketer (2010) predicts that virtual worlds such as Second Life would fall behind in the virtual-goods economy while social network players like Facebook become more involved in this virtual space.

Regardless of the predictions about which business model will survive, it is clear that the social aspect dimension dominating Virtual Worlds re-shapes consumer behaviour and, correspondingly, revolutionizes the way in which online research should be conducted. The findings of the present study indicate that the potential simultaneous presence of more than one online customer (i.e. avatar) at the same virtual reality store (as is the case in conventional retailing where many customers may be co-located at the same time) sets several restrictions as far as customization capabilities are concerned (e.g. layout customization) but at the same time provides several challenging business and research opportunities (e.g. social communication effects, store layout effects, etc.).

Similarly, the factor analysis showed that layout is perceived by VWs users as an important store selection criterion that is considered separately (i.e. not grouped with other variables) during the decision making process regarding store choice. This finding is also explained by the fact that the participants in this survey are experienced VWs users and, therefore, they are aware of the advanced and innovative navigational capabilities (i.e. flying) that could potentially be offered to them through a virtual reality environment on the Web.

It should be underlined, however, that 3D shopping environments through the Web could be also offered as innovative features by retail web sites that do not belong in Virtual Worlds like Second Life. In other words, "traditional" web sites could just add some 3D features either in their existing online stores or as alternative versions of their online presence. Current business practice (e.g. fashion industry) indicates that several web sites already offer such services. In that case, the social aspect dimension is not present, at least not in the same sense as it operates in a virtual reality environment such as Second Life. This implies that online users visiting these web sites do not use avatars and in general do not interact with other online shoppers as is the case in virtual worlds like Second Life. Thus, customization in this case is applicable in a similar way as in conventional web stores. For example, the virtual layout of the 3D store can be customized at the individual level since one customer visits the store each time. However, such 3D Web sites could offer several social communication services (e.g. forums, blogs, etc.), enabling customers to communicate through the Web site. In this case, social communication would be just an e-service (as is already the case on many web sites) and not the core differentiating characteristic of the shopping environment. Along these lines, a common e-service offered by many web sites today is the information offered to customers about the current online users as a specific Web site. Apparently, however, this service does not create obstacles in the GUI customization process but simply provides a social flavour that may affect consumer behavior (i.e. knowing how many customers are online in a retail web site may affect the way customers perceive this web site – e.g. effects on store image, consumer trust, etc.).

In sum, Virtual Reality Retailing could take two forms: through a "traditional" web site with embedded 3D features or through a VW like Second Life. Regardless of the Virtual Reality Retailing form, however, it is evident that the simultaneous presence of more than two customers (i.e. avatars) on a 3D retail store creates several challenging business opportunities and research questions. In other words, the available knowledge from Web 1.0 as far as GUI interface and customization effects on consumer behavior are concerned should be reconsidered.

CONCLUSIONS, IMPLICATIONS AND FURTHER RESEARCH

The social aspect dominating VREs affects the way that store atmosphere determinants could be applied and manipulated towards supporting and influencing consumer behavior. While technology enables customization, in the case of V-Commerce such an option is not a panacea. In contrast to conventional web retailing, V-retailers cannot easily customize their online stores to the unique preferences and wishes of their customers, simply because they face a one-to-many instead of a one-to-one situation. Obviously, this has been the case in the physical retailing world as well. It seems, therefore, that the established knowledge in conventional retailing may be more appropriate as a starting point for formulating and testing research hypotheses through experimental conclusive designs, instead of the extant knowledge on web retailing.

Managers should realize the importance of their stores' atmosphere as a major consumer behavior influencing factor, thus, preventing the business' failures that several web retail stores faced in the previous decade. Specifically, they should be both aware of the applicability and customization options of the available store atmosphere determinants (Table 1) and advised on how consumers perceive and group in their minds store atmosphere variables (Table 2). For example, in a number of VWs such as "Second life", avatars enjoy flying or being teleported. However, the ability of flying throughout a shop to watch all, or most of the available products or services might bring about opportunities for changes in the store layout (Prasolova, 2008).

One major limitation of the study is the relatively small sample size (i.e., 104 users). Also, the number store atmosphere variables used for creating the underlying factors could be larger and more detailed. Future research could elaborate more on that. Also, another limitation of the study is reflected to the fact that shopping in Virtual worlds seems to be a novel consumer behavior, which is currently rather fluid and subject to changes. Therefore, the elaborate quantitative study presented here might become quickly outdated, as the medium and the behaviors that are afforded by it are transformed. However, we consider that it is worth looking at behavior in virtual worlds as it evolves. First, this contributes to our understanding of what is currently happening in this environment. Through our quantitative investigation we have the opportunity to explore the applicability of different metrics and to provide insights for further research in this field. This second contribution is particularly important, in our opinion, as it sets the scene for exploring how this new environment will evolve. Finally, it might be the case that the online virtual word consumer behavior is exactly the same like the one offline, but until there is a critical mass of evidence on this, the practical significance of this research is only limited to early adopters of the respective medium. However, according to the Diffusion of Innovation Theory (Rogers, 1983) a new to the world product or service is firstly adopted by innovators and then by early adopters. Thus, the present study investigates the behavior of these users-consumers that have adopted the innovation of VWs first. Besides, this is the only available sample that could answer to a questionnaire like the one used by the present study, simply because the remaining Internet users have no experience in interacting with VWs and, therefore, they are not able to answer the corresponding questions. In sum, it is common research practice to conduct quantitative consumers surveys targeting innovators and early adopters of any given innovation (e.g., Internet shoppers were considered innovators in the previous decade (Vrechopoulos et al., 2001).

The research results of this paper provide a basis for several interesting further research directions. Indicatively, researchers could investigate how crowding and "innovation atmosphere" affect consumer behavior in Virtual Worlds and provide design guidelines to practitioners for creating suitable stores similarly to the traditional physical store's research and business practice.

Conversely, there is a need to explore in more depth characteristics such as telepresense, vividness, interaction and virtual object touch. In this respect, there is a considerable challenge in this area for researchers to investigate store layout effects on consumer buying behavior in order to meet the need of designing suitable layout types for this emerging retailing channel. Specifically, researchers could attempt to classify the available types of VWs retail store layouts and investigate through experimental designs whether there are significant differences between them as far as consumer behaviour dependent variables are concerned. Alternatively, researchers could transform and adapt the available knowledge from conventional and Web 1.0 retailing to V-Commerce and proceed to empirically test through experiments whether this theory is confirmed.

Other innovations that should also be considered in the context of VW stores include product highlights, product search guidance (e.g. arrows pointing to the product self location), virtual product trying-on (in case of clothes and shoes), 3D glasses or advanced 3D input methods, such as gestures, etc. Future research should elaborate on how these services affect and support consumer behavior. To that end, researchers could develop use case scenarios employing innovative consumer services that could be potentially offered within the context of a VW retail store and evaluate their attractiveness through empirical consumer surveys (e.g. paper-and-pencil, lab or field experiments, focus groups, surveys, etc.). These scenarios could be derived through the combined use of qualitative research initiatives (e.g. in-depth interviews with consumers and experts) and literature.

Finally, it should be underlined that potential future research attempts on this topic should consider that the GUI design of virtual reality retail stores should target the "average" user (i.e. no customization) as far as several manipulated variables are concerned (e.g. layout). In other words, in Web 1.0 and also in "traditional" Web sites that just adopt 3D features but not belong to virtual worlds (e.g. Second Life) managers and researchers could potentially create "segments of one customer" (i.e. the essence of customization). This situation is completely different for virtual worlds in the sense that more than one customer could be co-located at the same time within the same virtual reality retail store similarly to what happens in conventional retailing (i.e. "back to basics"). This implies that customization of several store atmosphere variables (e.g. layout) is not applicable in this online environment. Such an evolution may form the basis of a paradigm shift in store atmosphere studies in VWs compared to the established knowledge and available research practice of store atmosphere studies in the context of the "traditional" Web environment.

ACKNOWLEDGMENT

We would like to thank the Editors and the two anonymous reviewers for their useful comments that helped us to revise the present paper. This research has been partially funded by the Program Supporting Basic Research (PEVE II) of Athens University of Economics and Business.

REFERENCES

Baker, J., Grewal, D., & Parasuraman, A. (1994). The influence of store environment on quality inferences and store image. *Journal of the Academy of Marketing Science*, *22*, 328–339. doi:10.1177/0092070394224002

Becerra, E., & Stutts, M. (2008). Ugly duckling by day, super model by night: The influence of body image on the use of virtual worlds. *Journal of Virtual Worlds Research*, *1*(2), 1–19.

Boostrom, R. (2008). The social construction of virtual reality and the stigmatized identity of the newbie. *Journal of Virtual Worlds Research*, *1*(2), 1–19.

Burke, R. R. (1996). Virtual shopping: Breakthrough in marketing research. *Harvard Business Review*, *74*(2), 120–131.

Burke, R. R. (2002). Technology and the customer interface: What consumers want in the physical and virtual store. *Journal of the Academy of Marketing Science*, *30*(4), 411–432. doi:10.1177/009207002236914

Coakes, S., Steed, L., & Ong, C. (2009). SPSS: *Vol. 16. Analysis without anguish*. Chichester, UK: John Wiley & Sons.

Dailey, L. (1999). Designing the world we surf in: A conceptual model of web atmospherics. In *Proceedings of the AMA Summer Educator's Conference*, Chicago, IL.

Dechow, D. (2008). Surveillance, consumers, and virtual worlds. *Journal of Virtual Worlds Research*, *1*(2), 1–4.

eMarketer. (2010). *Boom time in second life: No recession for virtual economy*. Retrieved from http://www.emarketer.com/Article.aspx?R=1007482

Eroglu, S. A., Machleit, K. A., & Davis, L. M. (2000). Online retail atmospherics: Empirical test of a cue typology. In *Retailing 2000: Launching the new millennium: Proceedings of the Sixth Triennial National Retailing Conference* (pp. 144-150).

Fiore, A. M., Jin, H. J., & Kim, J. (2005). For fun and profit: Hedonic value from image interactivity and responses toward an online store. *Psychology and Marketing*, *22*(8), 669–694. doi:10.1002/mar.20079

Flavian, C., Gurrea, R., & Orus, C. (2009). The impact of online product presentation on consumers' perceptions: an experimental analysis. *International Journal of E-Services and Mobile Applications*, *1*(3), 17–37. doi:10.4018/jesma.2009070102

Frost, J. H., Chance, Z., Norton, M. I., & Ariely, D. (2008). People are experience goods: Improving online dating with virtual dates. *Journal of Interactive Marketing, 22*(1), 51–61. doi:10.1002/dir.20107

Grewal, D., & Baker, J. (1994). Do retail store environment cues affect consumer price perceptions? An empirical examination. *International Journal of Research in Marketing, 11*(2), 107–115. doi:10.1016/0167-8116(94)90022-1

Griffith, D. A. (2005). An examination of the influences of store layout in online retailing. *Journal of Business Research, 58*(10), 1391–1396. doi:10.1016/j.jbusres.2002.08.001

Griffith, D. A., & Chen, Q. (2004). The influence of virtual direct experience (vde) on on-line ad message effectiveness. *Journal of Advertising, 33*(1), 55–68.

Hair, J., Black, W., Babin, B., Anderson, R., & Tatham, R. (2005). *Multivariate data analysis* (6th ed.). Upper Saddle River, NJ: Prentice Hall.

Haven, B., Bernoff, J., Glass, S., & Feffer, K. A. (2007). *A second life for marketers?* Cambridge, MA: Forrester Research.

Hendaoui, A., Limayem, A., & Thompson, C. W. (2008). 3D social virtual world: Research issues and challenges. *IEEE Internet Computing,* 88–92. doi:10.1109/MIC.2008.1

Jung, Y., & Kang, H. (2010). User goals in social virtual worlds: A means-end chain approach. *Computers in Human Behavior, 26*(2), 218–225. doi:10.1016/j.chb.2009.10.002

Kim, J., Fiore, M. A., & Lee, H. H. (2007). Influences of online store perception, shopping enjoyment, and shopping involvement on consumer patronage behavior towards an online retailer. *Journal of Retailing and Consumer Services, 14,* 95–107. doi:10.1016/j.jretconser.2006.05.001

Kim, J., & Forsythe, S. (2008). Adoption of virtual try-on technology for online apparel shopping. *Journal of Interactive Marketing, 22*(2), 45–59. doi:10.1002/dir.20113

Kotler, P. (1973-4). Atmospherics as a marketing tool. *Journal of Retailing, 49,* 48–63.

Krasonikolakis, I., & Vrechopoulos, A. (2009, September 25-27). Setting the research agenda for store atmosphere studies in virtual reality retailing. In *Proceedings of the 4th Mediterranean Conference on Information Systems*, Athens, Greece.

Landay, L. (2008). Having but not holding: Consumerism & commodification in second life. *Journal of Virtual Worlds Research, 1*(2), 1–5.

Lewison, M. (1994). *Retailing* (5th ed.). New York, NY: Macmillan.

Lin, H.-F. (2008). Determinants of successful virtual communities: Contributions from system characteristics and social factors. *Information & Management, 45*(8), 522–527. doi:10.1016/j.im.2008.08.002

Lohse, L. G., & Spiller, P. (1999). Internet retail store design: How the user interface influences traffic and sales. *Journal of Computer-Mediated Communication, 5*(2).

Malhotra, N. K., & Birks, D. F. (2000). *Marketing research: An applied approach*. London, UK: Pearson.

Martin, J. (2008). Consuming code: Use-value, exchange-value, and the role of virtual goods in second life. *Journal of Virtual Worlds Research, 1*(2), 1–21.

Mazursky, D., & Vinitzky, G. (2005). Modifying consumer search processes in enhanced on-line interfaces. *Journal of Business Research, 58,* 1299–1309. doi:10.1016/j.jbusres.2005.01.003

Merrilees, B., & Miller, D. (2001). Superstore interactivity: A new self-service paradigm of retail service. *International Journal of Retail & Distribution Management, 29*(8), 379–389. doi:10.1108/09590550110396953

Messinger, P. R., Ge, X., Stroulia, E., Lyons, K., Smirnov, K., & Bone, M. (2008). On the relationship between my avatar and myself. *Journal of Virtual Worlds Research, 1*(2), 1–17.

Messinger, P. R., Stroulia, E., Lyons, K., Bone, M., Niu, H., & Smirnov, K. (2009). Virtual worlds — past, present, and future: New directions in social computing. *Decision Support Systems, 47*, 204–228. doi:10.1016/j.dss.2009.02.014

Nikolaou, I., Bettany, S., & Larsen, G. (2010). Brands and consumption in virtual worlds. *Journal of Virtual Worlds Research, 2*(5), 1–15.

O'Reilly, T. (2006). *What is Web 2.0: Design patterns and business models for the next generation of software.* Retrieved from http://www.oreillynet.com/pub/a/oreilly/tim/news/2005/09/30/what-is-Web-20.html

Prasolova-Forland, E. (2008). Analyzing place metaphors in 3D educational collaborative virtual environments. *Computers in Human Behavior, 24*(2), 185–204. doi:10.1016/j.chb.2007.01.009

Ridings, C. M., & Gefen, D. (2004). Virtual community attraction: Why people hang out online. *Journal of Computer-Mediated Communication, 10*, 4.

Rogers, E. M. (1983). *Diffusion of innovations* (3rd ed.). New York, NY: Free Press.

Saiman, A. (2009). Barriers to efficient virtual business transactions. *Journal of Virtual Worlds Research, 2*(3), 1–14.

Simonson, I. (1999). The effect of product assortment on buyer preferences. *Journal of Retailing, 75*(3), 347–370. doi:10.1016/S0022-4359(99)00012-3

Siomkos, G., & Vrechopoulos, A. (2002). Strategic marketing planning for competitive advantage in electronic commerce. *International Journal of Services Technology and Management, 3*(1), 22–38. doi:10.1504/IJSTM.2002.001614

Spence, J. (2008). Demographics of virtual worlds. *Journal of Virtual Worlds Research, 1*(2), 1–45.

Strauss, J., & Frost, R. (2009). *E-marketing* (5th ed.). Upper Saddle River, NJ: Prentice-Hall.

Vicdan, H., & Ulusoy, E. (2008). Symbolic and experiential consumption of body in virtual worlds: From (Dis)embodiment to symembodiment. *Journal of Virtual Worlds Research, 1*(2), 1–22.

Vrechopoulos, P. A., O'Keefe, M. R., & Doukidis, I. G. (2000, June 19-21). Virtual store atmosphere in internet retailing. In *Proceedings of the 13th International Conference on Bled Electronic Commerce*, Slovenia.

Vrechopoulos, P. A., Siomkos, G., & Doukidis, G. (2001). Internet shopping adoption by Greek consumers. [f]. *European Journal of Innovation Management, 4*(3), 142–152. doi:10.1108/14601060110399306

This work was previously published in the International Journal of E-Services and Mobile Applications, Volume 3, Issue 2, edited by Ada Scupola, pp. 59-72, copyright 2011 by IGI Publishing (an imprint of IGI Global).

Chapter 8

Information Technology and Supply Chain Management Coordination:
The Role of Third Party Logistics Providers

Pier Paolo Carrus
University of Cagliari, Italy

Roberta Pinna
University of Cagliari, Italy

ABSTRACT

Logistics Service Providers (3PL) have become important players in supply chain management. In a highly competitive context characterized by "time compression", a successful strategy depends increasingly on the performance of Logistics Service Providers as they play a key integrative role linking different supply chain elements more effectively. However, the role of the information technology capability of these 3PL has not drawn much attention. The research question is: can IT be viewed as a fundamental supply chain management coordination mechanism? If so, does IT capability of third party logistics providers to improve performance in the supply chain and become a bigger factor in a strategic buyer-3PL relationship? By drawing on earlier research on the supply chain management coordination mechanism, the IT capability of third party logistics providers, a case study is conducted.

INTRODUCTION

In this paper we defined supply chain as a set of three or more entities (organizations or individuals) directly involved in the upstream and downstream flows of products, services, finances, and information from a source to a customer (cf. Mentzer et al., 2001). In other words, the philosophy of supply chain management extends the concept of partnerships into a multifirm effort to manage the total flow of goods from the supplier to the ultimate customer to achieve greater

DOI: 10.4018/978-1-4666-2654-6.ch008

benefits. A supply chain management involves three distinct interrelated flows: product/service, information and financial flow. Successful supply chain management requires planning, managing and controlling these three flows through the integration of key processes, from original suppliers through manufacturers, retailers to the end-users, which produce values to the ultimate consumers (Lambert et al.,1998; Bowersox et al., 1996).

Encompassed within this definition, Mentzer et al. (2001) identifies three degrees of supply chain complexity: a "direct supply chain," an "extended supply chain," and an "ultimate supply chain." A direct supply chain consists of a company, a supplier, and a customer involved in the upstream and/or downstream flows of products, services, finances, and/or information (Figure 1a). An extended supply chain includes suppliers of the immediate supplier and customers of the immediate customer, all involved in the upstream and/or downstream flows of products, services, finances, and/or information (Figure 1b). An ultimate supply chain includes all the organizations involved in all the upstream and downstream flows of products, services, finances, and information from the ultimate supplier to the ultimate customer (Figure 1c). Figure 1 illustrates the complexity that ultimate supply chains can reach. In this example, a third party logistics (3PL) provider may be defined as an external supplier that performs all or part of a company's logistics function. With the increasing globalization of markets, companies began to view logistics as more than simply a source of cost savings and recognize it as a source of enhancing product or service offerings as part of the broader supply chain process to create competitive advantage (Novack, Langley, & Rinehart 1995; McDuffie, West, Welsh, & Baker 2001). However, because logistics users often lack the competence to operate logistics activities internally, they tend to outsource to third-party logistics (3PL). In this way, firms can better focus on their core competencies, such as manufacturing and retailing, while allowing third-

party specialists to take care of functions such as transportation, distribution, and warehousing to satisfy the ultimate needs of their customers. The outsourcing of logistics activities requires creating synergetic relationship between the partners with the objective of maximizing customer value and providing a profit to each supply chain member.

All the companies involved in the network are important in establishing a desired level of customer service in the supply chain and satisfying their customers' requirements. These companies are interdependent in such a way that an individual company's performance affects the performance of other members of the supply chain. If there is a problem in one company, the problem consequently causes other problems in other areas and weakens the effectiveness of the whole supply chain. Forrester's (1958) seminal study of industrial dynamics in a four channel supply chain illustrates how rational decision-makers acting independently can cause customer demand information to distort and amplify while moving upstream in the supply chain, resulting in inaccurate forecasts, inefficient asset utilization and poor customer service. In the 1990s, this phenomenon was re-introduced by Lee, Padmanabhan, and Whang (1997), when they coined the term "bullwhip effect" in supply chains to refer to the sub-optimization phenomenon. It is so-called because small order variability at the customer level amplifies the orders for upstream players, such as wholesalers and manufacturers, as the order moves up along a supply chain.

Coordination within a supply chain is a strategic response to the challenges that arise from these dependencies. Benefits from coordination of supply chain activities are well-documented in the literature. There is a growing body of academic research, in a variety of disciplines, on coordination in supply chains, particularly addressing the potential coordination mechanisms available to eliminate sub-optimization within supply chains. Similarly, there is a growing interest in industry to better understand supply chain coordination and

Figure 1. Types of supply chain relationships (Mentzer et al., 2001)

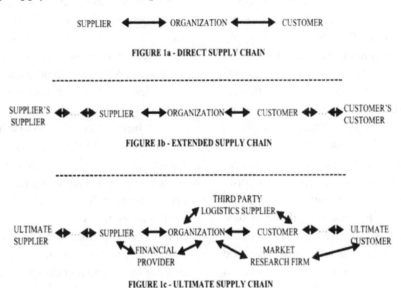

the coordination mechanisms that are available to assist the supply chain manager.

Sahin and Robinson (2002) surveyed the literature on supply chain integration and proposed information sharing and coordination among supply chain members as the primary drivers of supply chain performance. The evolution of technologies used by logistics service providers is essentially driven by the development of warehouse management systems, product follow-up techniques and automatic identification. IT allows supply chain partners to communicate directly over data-rich, easy to use information channels, which reduces coordination costs faster than in house production costs and promotes the trend toward outsourcing.

In the present paper, the role played by IT in supply chain coordination is investigated. It is hypothesized that information technology capability of third party logistics providers improves performance in the chain and becomes a bigger factor in a strategic buyer-3PL relationship.

The paper is structured as follows. The introduction presents the background and research question. The second session presents theoretical background of the paper. The following session

introduces the hypotheses and theoretical model. The last two sessions present the analysis and results as well as discussion and conclusion.

Theoretical Background

Coordination and IT

The importance of coordination in the SCM has been the subject of SCM theory over the last decade or so, as highlighted by some of the definitions of SCM. Benefits from coordination of supply chain activities are well documented in the literature (Porter, 1985; Sahin & Robinson, 2005; Cachon, 2004; Min, 2001; Christiaanse & Zaccour, 2003). Central to collaboration is the mutual sharing of large information among supply chain members (Cooper et al., 1997; Novack, Langley, & Rinehart, 1995). The Global Logistics Research Team at Michigan State University (1995) defines information sharing as the willingness to make strategic and tactical data available to other members of the supply chain. Numerous researchers have found that when buyers and logistics services providers communicate and share information they are more likely to improve the quality of their products or services; cycle time reduction; reduce the costs of protecting against opportunistic behaviour and improve cost savings through greater product design and operational efficiencies (Carr & Pearson, 1999; Kotabe et al., 2003; Prahinksi & Benton, 2004; Giunipero et al., 2006). There is a growing body of academic research on coordination in supply chain, particularly addressing the potential coordination mechanism available to eliminate sub-optimization within supply chains. Similarly, there is a growing interest in industry to better understand supply chain coordination and the coordination mechanisms that are available to assist the supply chain manager.

Some approaches to coordination include identifying coordination types as structured or unstructured and formal or informal, depending on the extent to which firms use formal design of roles and mechanisms to synchronize activities and flows within the supply chain (Lusch & Brown 1996). Earlier research efforts define norms as mutual understanding of behaviour that is approved and required of supply chain members and unwritten mechanisms that are critical in achieving supply chain coordination (Dahlstrom, McNeilly, & Speh, 1996; Fisher, Maltz, & Jaworski, 1997). Gundlach, Achrol, and Mentzer (1995, p. 81) advocating the use of norms as a coordination mechanism. Flexibility, solidarity, mutuality, harmonization of conflict, restraint in the use of power, concern for reputation, and information sharing are examples of some of the norms often discussed in the coordination literature (Achrol & Gundlach, 1999; Gundlach, Achrol, & Mentzer, 1995; Maloni & Benton, 2000). Fugate, Sahin, and Mentzer (2006) proposing price, non-price, and flow coordination as major categories of coordination mechanisms. Flow coordination mechanisms are designed to manage product and information flows in supply chains. Sahin and Robinson (2002) provide an extensive literature review on product flow coordination and information sharing in supply chains, classifying the literature based on the degree of information sharing and

coordination. Vendor Managed Inventory (VMI), Quick Response (QR), Collaborative Planning, Forecasting and Replenishment (CPFR), Efficient Consumer Response (ECR), and postponement are among some of the initiatives used for product and information flows.

3PLs and SCM Optimization

As the practice of logistics outsourcing and SCM have become increasingly important as a source of competitive advantage, researchers have started to explore various research topics related to logistics outsourcing and the role of 3PLs in SCM (Cheng & Grimm, 2006; Maloni & Carter, 2006; Selviaridis & Spring, 2006; Lai et al., 2008). In a highly competitive context, effective leadership of logistic service providers lies in their capacity to innovate in the area of information flow management. In today's SCM practices, a successful strategy depends on the performance of 3PLs as they play a key integrative role linking the different supply chain elements more effectively. Information and communication technology (ICT) has become an important element of 3PLs competitive capability as it enables higher levels of supply chain integration. Specifically, the type of IT used, largely determines the nature and quality of interactions the company has with customers, suppliers, and trading partners. A high level of IT capability has been shown to provide a clear competitive advantage and can be a differentiating factor in terms of company performance (Kathuria et al., 1999). Sauvage (2003) noted that in a highly competitive business characterised by time compression, technological effort becomes a critical variable and a significant tool for differentiation of logistics services. Van Hoek (2002) assigned a specific role to ICT for 3PLs aiming to perform customised operations for service users. The author pointed out that the use of specific technological capabilities may leverage transport and logistics services and facilitate more effective integration across companies in the supply chain. For 3PLs, ICT capabilities can assure the rapid customisation of products and maintain competitive lead-times. The result is that competitive advantage in the 3PL industry will be based increasingly on creating value for customers as many value-added activities are directly or indirectly dependent on ICT applications (Crowley, 1998). Nevertheless, the use of ICT in the 3PL sector is unevenly distributed between large and small-medium sized logistics service providers. Large firms have heavily invested in ICT and have actively developed information systems (ISs). Furthermore, they have been using in-house ISs to support their operations for a long time. Small logistics service providers, on the other hand, have more difficulties in setting up ICT applications due to reluctance to change and insufficient human and financial resources. Although ICT development has strongly affected the international logistics service industry in recent years, the adoption of new technologies in the Italian logistics service market appears relatively low. A number of surveys confirm this situation. Merlino and Testa (1998) analysed the level of computerisation and ICT investment by 3PLs in Northern Italy. The study, carried out on a sample of 197 firms, revealed that these companies are only at the initial stage of adopting ICT. The survey highlighted that the dissemination of new technologies is proceeding at an intermittent and non-homogeneous pace. Investments in new technology are still motivated by a tactical rather than a strategic logic. The authors attribute this to the history of the firm and its entrepreneurial culture. Another survey, aimed at assessing the relationship between company culture and the usage of ICT, was conducted on a sample of 48 shipping agents and freight forwarders located in Southern Italy, specifically in the Campania region (Minguzzi & Morvillo, 1999). The results showed that investments in computer hardware and software are mainly associated with entrepreneurial culture rather than with economic and business matters. Other recent surveys report a number of interesting issues. KPMG (2003) pointed out that

in comparison to other industries; ICT investment in the Italian 3PL industry is limited. Furthermore, the level of outsourcing of ICT and e-business applications is very low. Finally, a recent survey shows that the most important reasons for not investing in ICT are related to financial factors. The size of investment and the implementation costs, together with running costs, are considered the most influential factors inhibiting ICT investment (Evangelista & Sweeney, 2006).

Conceptual Model and Hypotheses

This study adopts the RBV to conceptually frame our model because the RBV is an appropriate theory for supply chain and logistics management research (Cheng & Grimm, 2006; Hunt & Davis, 2008). The RBV has been increasingly adopted to examine logistics-related capability and performance. For example, Richey, Daughert,y and Roath (2007) adopted the RBV to investigate how technological readiness influences logistics service quality and organizational performance. Autry, Griffis, Goldsby, and Bobbin (2005) proposed and tested a resource-based model to examine how resource commitment influences warehouse management capabilities and organizational performance. Daugherty, Autry, and Ellinger (2001) examined the impact of resource commitment on reverse logistics performance. Furthermore, the RBV has also been used to examine the relationship development between logistics users and service providers. For example, Sinkovics and Roath (2004) examined the development and impact of the capabilities of logistics users (i.e., operational flexibility and collaboration) on their outsourcing relationships with 3PL providers. The RBV suggests that the possession and development of a set of heterogeneous resources leads a firm to establish a competitive advantage over its competitors in the marketplace (Barney, 1991). According to Barney (1991), firms can achieve a competitive advantage based on resources that are firm-specific, valuable, rare, imperfectly imitable and not strategically substitutable by other resources. Resources such as capital equipment, employee skills and patents are key inputs into a firm's business processes, but a competitive advantage cannot be achieved based only on the possession of these resources. Without the distinctive competence or capacity to manage and make better use of these resources, a firm cannot achieve a competitive advantage in the short term or a sustained competitive advantage in the long run. Such competence or capacity is called "capability" (Mahoney & Pandain, 1992). According to Barney (1991), capability means that a firm needs to be so managed and organized that it can exploit the full potential of its resources. IT capability can therefore be defined as the "ability to mobilize and deploy IT-based resources in combination or co-present with other resources and capabilities" (Bharadwaj, 2000, p. 171), or, to be more specific, "a firm's IT infrastructure, its human IT skills, and its ability to leverage IT for intangible benefits serve as firm-specific resources, which in combination create a firm-wide IT capability" (Bharadwaj, 2000, p. 176). Similarly, the logistics literature suggests that IT capability is "the application of hardware, software and networks to enhance information flow and facilitate decisions" (Closs, Goldsby, & Clinton, 1997, p. 6). These definitions suggest that the ability of 3PL firms to utilize and combine IT-based resources with other resources may determine their business performance.

In this study, we switch the focus on IT capability of 3PL providers. The conceptual model of the study is depicted in Figure 2. In the model, we propose that the link between strategic buyer – 3PL relationship, performance and satisfaction is not direct and linear. Indeed, there are different elements that have an impact on each one. Competitive priorities and Logistic Service Provider capabilities are two antecedents of strategic buyer-3PL relationship and that IT integration and social mechanism influence the buyer/3PL competitive advantage.

Competitive Priorities

The model differentiates five unique competitive priorities: price, quality, customer service, time, and flexibility, five critical facets areas that have been frequently discussed in logistics management literature (Sanders & Premus, 2002). These competitive priorities have been well established in literature beginning with the work of Van Dierdonck and Miller (1980) and Hayes and Wheelwright (1979, 1984). *Price* and *quality* are competitive priorities that focus on organizational resources to compete on the basis of either low price or quality leadership. *Time* is a competitive priority that refers to a focus on faster production and delivery times, while customer service focuses on providing highly individualized services, such as high-performance design and customization. Researchers in the past have also recognized the strategic importance of time-based performance (Nahm, Vonderembse, Rao, & Ragu-Nathan, 2006). Subsequently, they have considered various aspects of time-based performance relative to different stages of the overall value delivery cycle and have proposed several measures to evaluate them (Droge et al., 2004). The frequent appearance of the measures including delivery speed and delivery reliability/dependability (Handfield, 1995) suggests the important effect of delivery performance on agility. In addition, the advent of time based competition has elevated the strategic importance of customer responsiveness (Stalk & Hout, 1990).

Customer responsiveness describes a firm's ability to respond in a timely manner to customers' needs and wants. Thus, a firm's ability to respond promptly to customers' needs can be a source of enduring competitive advantage (Cusumano & Yoffie, 1998). The last competitive priority, *flexibility*, refers to a company's agility, and can take on a number of forms (Vickery, Calantone, & Droge, 1999). *Product flexibility* is the ability of a company to offer a large number of product features and options and to rapidly add or delete these features based on market competition. *Vol-*

Figure 2. Motivational process

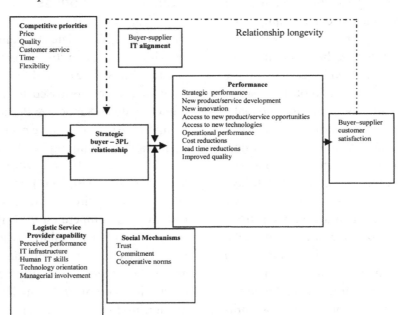

ume flexibility is the ability to speed up production to meet peak demands or cut production in slack periods, a feature especially important in industries characterized by extreme demand fluctuations. *Launch flexibility* is the ability to reduce time from idea conceptualization to product design, production, and final delivery. This capability is important in industries characterized by rapid rates of technological change and in business segments where style changes are frequent. Finally, *access flexibility* refers to the ability of a firm's distribution system to deliver products to multiple market segments, which can change at quick notice. Being quicker than other firms at getting products to new market segments can be an important competency to allow the firm to stay ahead of competitors.

Logistics Service Provider Capability

IT capability is cited as one of the seven capabilities that a logistics firm needs to achieve world-class business performance (Closs et al., 1997). It is

therefore believed that 3PL firms with superior IT capability can achieve a strong advantage over competitors by reducing costs and providing wider connectivity and strategic links with their business partners and customers, thus allowing the variety and quality of the services they provide to be expanded and improved. Consistent with previous studies (e.g., Kathuria, 2000), we investigate the competitive advantage of 3PL firms in three areas: cost advantage, service variety advantage and service quality advantage. Service variety advantage comes from a 3PL firm's ability to provide and customize a variety of 3PL services and products and meet the special service requirements of its customers. Service quality advantage includes the ability to provide a fast and reliable delivery of products and services, maintain superior order accuracy, and respond to customer inquiries, complaints, and claims in a prompt manner. However, according to the RBV (Barney, 1991), technological resource commitment is not sufficient by itself to develop IT capability in 3PL firms, and previous studies have found that accelerated expenditure on IT does not always result in increased firm productivity. Bharadwaj (2000) highlights the importance of developing human capital, such as technical and managerial IT skills, in enhancing IT capability. Boynton, Zmud, and Jacobs (1994) find that the effective implementation of IT depends on the business knowledge of IT managers, the IT knowledge of business managers, and the exchange of knowledge between the two. Top management involvement is also found to be critical to the successful assimilation of IT within a firm once resources have been invested (Chatterjee, Grewal, & Sambamurthy, 2002). All of these ideas indicate that investment and resource development in a 3PL firm are necessary for the development of IT capability.

Buyer-Supplier Integration and Social Mechanism

While IT is a critical element of SCM, IT is not a source of value by itself. Indeed, as explained by Mentzer et al. (2001) effective SCM also requires some antecedents. Antecedents to SCM are the factors that enhance or impede the implementation of an SCM philosophy. The above review of the literature documents shows that integration between supply chain partners improves firms' performance (Sanders, 2005; Vickery et al., 2003). Buyer-3PL integration is a concept defined as an effective, mutually shared process where two or more organizations work together, have mutual understanding, have a common vision, share resources and achieve collective goals (Sanders, 2005). Strategic buyer-3PL relationships involve informal processes based on trust, mutual respect and information sharing, the joint ownership of decision making and collective responsibility for outcomes. Researchers have empirically documented how relationship commitment and trust foster greater cooperation, reduce functional conflict and enhance integration (Paulraj & Chen, 2007). Within solid relationships, reciprocity is the norm, and it serves an important function. Duck

(1991) suggests that the relationship between two companies is most often defined by what the companies (in it) provide for one another, the resources that they distribute, and the exchanges that takes place. The increased focus on reciprocity within 3PL relationships is highlighted by the use of cost-sharing as one of the preferred approaches for their deal structure(s). Prior studies have suggested a positive relationship between cooperative norms and buyer-supplier performance, and has been considered a critical aspect of successful supply chain management (Cay & Yang, 2008). As argued by Heide and Jhon (1992), the purpose of cooperative norms is to structure an economically efficient relationship. In the same vein, Cannon et al. (2000) maintains that cooperative norms could "provide a general framework of reference, order, and standards against which to guide and assess appropriate behavior in uncertain and ambiguous situations", and thus are able to enhance the supplier's performance.

Research Method

A case study was conducted in order to explore the hypotheses of research. The case under consideration is Laziale Distribuzione s.p.a. (LD), a logistics service provider located in Pomezia, a city about 35 km from Rome, the capital City of Italy. The data mainly consists of primary data collected through qualitative explorative and semi-structured interviews. Secondly, data such as reports and other material on e-services development provided by the LD personnel and other material retrieved on the web were also used. Seven face to face qualitative interviews were conducted. The interviews lasted about 2 hrs; all interviews were recorded and fully transcribed. The sampling was purposeful (Patton, 1990). For our study, the literature review presumed that we interviewed managers working in the e-service development process at top management level and that they were involved in customer satisfaction process. Interview questions were open-ended.

Although the number of interviews may be considered relatively small, they were related to the key role that the respondents had in the planning and development of e-services, they provide a high level of reliability and validity to the research findings as in the case of Ozdemir et.al. Through our reference to Yin, the data was carefully analyzed by following the general strategy of theoretical orientation of the case study. In order to boost the validity the article, partial reports of the study were introduced and discussed with one of the LD's top managers.

Laziale-Interlaziale Logistic Group

Laziale Distribuzione was founded in 1978 in order to offer Italian companies modern shipment and logistic services. It counts approximately 182 employees. Over the years it has succeeded in consolidating its position in the market: by differentiating its offer; by expanding its field of action (just) to international markets; by investing in structures, training and IT infrastructures, becoming one of the most important logistics & distribution centres in Italy nowadays. The Logistic Group LD is composed of 6 companies: Laziale Distribuzione, Interlaziale, Laziale express, LOG.DI. srl, Logigraf s.r.l. and inLog. The Logistic Group LD has been nominated as the most innovative logistic Italian business by Assologistica (National Association of Italian Logistics Operators), Euromerci (a leading logistics magazine) and SITL Italia (International Freight & Logistics Verona Fair). In the Italian market, the Logistic Group is between the first 100 enterprises of logistics services providers (Confetra, 2005). In 1997 the Laziale-Interlaziale Logistic Group obtained the certification of own Quality System based on UNI EN ISO 9002:1994 rules. In 2003, they renewed the certification of Quality System in agreement to UNI EN ISO 9001:2000 (Vision, 2000). This important result permits them, to consolidate the business process of continuous improvement, guarantee proven level of service

reliability and professionalism that (it) answers to the needs of Customers. At present, LD Group is starting a new project: creation of Santa Palomba's logistic HUB, the first (logistic HUB) in the middle-south of Italy.

The LD may be classified as *integrator logistic operator*, because of two factors, the number of sectors managed and the number of services supplied. Indeed, LD Group's work in different sectors[1], each one has got a specific market with rules and different techniques. With reference to numbers and type of services supplied, the LD Group offers different services that can be classified in two different principal classes: *distribution* and *logistics services*. The LD Group offers distribution services fit for needs[2], thanks to their widespread network that allows fast and customized deliveries. The LD Group Network is composed of 15 platforms located throughout Italy. These platforms distribute any kind of goods (raw materials, semi-finished products, and finished products) both at a national and an international scale, according to appointed times, thanks to their selected carriers. Over the years, the LD Group has become skilled in logistics, by creating purpose-built plants. Thanks to skilled technical and computer structures, in Italy the company is a competitive reality in terms of costs and services. The added value of logistic solutions derives from the concentration in a single production unit of all operational activities: from directly receiving the products to handling and distribution. This system architecture allows logistic costs to be reduced by about 25%, thereby determining huge savings in management costs. Thanks to skilled technical and computer structures the logistic services allow the customers to: manage bigger volumes (getting thereby economies of scale); reduce handling of products; reduce internal management costs; improve performance efficiency; respect lead time; offer its own customers new services; greater flexibility; reducing communication errors.

Analysis and Results

LD have a formal, explicitly defined innovation strategy. An important aspect of LD Group's strategic orientation is the philosophy of leveraging IT to create superior products and services and to build solutions to respond to customer and partner requirements. In other words, the provider customises its services to the buyer's needs, and the relevant services are provided in a consistent and reliable manner: "The LD culture leads us to listen and understand the reasons and needs of our beloved customers, discussing with them the diagnosis and the relative solution to adopt. A progressive engagement made of attention and empathy. What they get undoubtedly is an overwhelming satisfaction to be recognized as catalysts of process of amazing steps ahead towards the street of success". One of the first visible effects associated with the increasing dissemination of ICT in the LD logistics service is the integration of traditional services (transportation and warehousing) with information-based services (e.g., tracking and tracing (T&T), booking, freight rate computation, routing and scheduling). Although LD may not be considered leaders in the field of technological innovation, over the last few years this organization has made significant progress in the adoption of new technologies, particularly those linked to the Internet. The type of IT known as an interorganizational system (IOS) lies at the heart of the ability of IT to support the logistic integration between buyers and LD. The logistics integration activity typically involves the sharing of very timely and very sensitive demand and sales data, inventory data, and shipment status data. Data sharing often involves a firm giving direct access to its computerized data bases to its supply chain partners. LD use AS/400, developed by IBM, (is) an information system that links the provider with more members of the supply chain, such as manufactures, transportation firms, retailers, or customers. This link automates some element of the logistics workload, such as order processing,

order status inquiries, inventory management, or shipment tracking. Vendor Managed Inventory (VMI), Quick Response (QR), Collaborative Planning, Forecasting and Replenishment (CPFR), Efficient Consumer Response (ECR), and postponement are among some of the initiatives that LD use for product and information flow. VMI allows the supplier to monitor the retailer's inventory levels and make periodic replenishment decisions involving order quantities, delivery mode, and the timing of replenishments. QR focuses on building a collaborative partnership between manufacturers and retailers by shortening the manufacturer's replenishment lead-time and giving the retailer a chance to place a small order at the beginning of the season, observe early demand, and choose an optimal replenishment quantity to maximize profits based on observed demand. The manufacturer gains from the collaboration by improving forecast accuracy and revising production schedules based on early demand. CPFR automates and improves sales forecasting and replenishment between trading partners, enabling participants to share improvements in inventory costs, revenue, and customer service. ECR decreases time and costs in the core, value-adding processes through four specific strategies: efficient store assortment, efficient replenishment, efficient promotion, and efficient product introduction. Postponement as a coordination mechanism attempts to reduce risk and uncertainty of operations by delaying operational commitment (form, place, and time) until final customer commitments have been obtained. All of these applications help to achieve service innovation and meet the level of customization required by logistic users, thus conferring a service variety advantage. Well-developed IT capability in LD improves the efficiency, effectiveness and productivity of business operations and thus provides a cost advantage. For example, the adoption of e-business by LD may help it to expand its customer base and save costs through automated online transactions. The results of this study show that the implementation of information networks through IOS improves data and information sharing and lowers coordination costs.

In the e-service innovation process we found that the most important customers were involved. We also found that customers had an important role in the initiation and implementation phases of the innovation process. We found very little evidence of customer as co-creator. Users were also involved in the implementation stage of the new e-services, thus contributing mainly to incremental changes with suggestions or improvements. This collaboration with customers has been very crucial for setting the vision and strategic goals for e-service innovation at LD. On an informal level, new ideas and suggestions are collected by LD managers and discussed in formal or informal meetings.

Superior IT capability in terms of diverse and sophisticated IT infrastructure and other intangible resources (as trust, commitment and cooperative norms) help LD to integrate with different logistics users and with relationship longevity.

DISCUSSION, CONCLUSIONS AND LIMITATIONS

This paper has presented the preliminary results of a case study about coordination and the role of 3PL. To summarize we found that the exchange of information through customers and 3PL is an essential condition for realizing the potential benefits of collaborative relationships (performance). The increased speed and flexibility of information and knowledge transfer allowed for more efficient coordination, and eventually higher revenues and profits, for all members of the supply chain, reducing communication errors, facilitating information knowledge sharing and increasing integration between the supplier and buyer firms. Especially, the relationship and collaboration with customers is very important for e-service innovation.

The findings of our study also have implications on managers both of 3PL providers and logistics users. The first is that 3PL managers must realize the importance of IT capability, as superior IT capability is a source of competitive advantage. The critical role of IT in saving costs, improving service quality and providing a variety of services cannot be ignored. Our study clearly shows the importance of technology orientation in the development of IT capability. 3PL firm managers are therefore recommended to dedicate themselves to the development of an organizational culture that favors IT. A stronger technology orientation in a 3PL firm helps top business managers to facilitate resource allocation to IT investment and encourages their involvement in IT strategies and planning. Well-developed IT capability in a 3PL firm can improve the efficiency, effectiveness and productivity of business operations and thus provide a cost advantage. For example, the adoption of e-business by 3PL firms may help them to expand their customer base and save costs through automated online transactions. Our findings are consistent with other empirical studies that find support for the positive effect of IT capability on cost advantage (e.g., Ravichandran & Lertwongsatien, 2005).

Finally, we recommend that the managers of logistics users consider the IT capability of 3PL firms in making outsourcing decisions. Especially, such a consideration is more important for logistics users which aim at long-term relationships with 3PL firms (Knemeyer & Murphy, 2004; Cox et al., 2005). The developing of strong, long-term partnerships requires designed managerial components and relationship activities that support the development of intangible connections (trust, commitment, cooperative norms, etc.) with partners.

We acknowledge some limitations of this study that might provide opportunities for future research. A limitation of this study was the sample population we only conducted in depth interviews in one Logistic service provider. The future research aim will include a broader population.

Finally, this study focused on the buyer-supplier dyad as the unit of analysis, and assumed the supplier firm's perspective. Also it would be interesting to investigate the customers' and competitors' perspective. We plan to do so in a second phase of the study. Despite these limitations, this study provides interesting insights in about the remarkable role of 3PL in Supply Chain coordination.

REFERENCES

Achrol, R. S., & Gundlach, G. T. (1999). Legal and social safeguards against opportunism in exchange. *Journal of Retailing, 75*(1), 107–124. doi:10.1016/S0022-4359(99)80006-2

Andersen, M. G., & Katz, R. B. (1998). Strategic sourcing. *International Journal of Logistics Management, 9*(1), 1–13. doi:10.1108/09574099810805708

Andraski, J. C. (1998). Leadership and the realization of supply chain collaboration. *Journal of Business Logistics, 19*(2), 9–11.

Autry, C. W., Griffis, S. E., Goldsby, T. J., & Bobbitt, L. M. (2005). Warehouse management systems: Resource commitment, capabilities, and organizational performance. *Journal of Business Logistics, 26*(2), 165–182. doi:10.1002/j.2158-1592.2005.tb00210.x

Barney, J. B. (1991). Firm resources and sustained competitive advantage. *Journal of Management, 17*(2), 99–120. doi:10.1177/014920639101700108

Bharadwaj, A. S. (2000). A resource-based perspective on information technology capability and firm performance: An empirical investigation. *Management Information Systems Quarterly, 24*(1), 169–196. doi:10.2307/3250983

Bowersox, D. J., & Closs, D. C. (1996). *Logistical management: The integrated supply chain process.* New York, NY: McGraw-Hill.

Boynton, A. C., Zmud, R. W., & Jacobs, G. C. (1994). The influence of IT management practice on IT use in large organizations. *Management Information Systems Quarterly, 18*(3), 299–318. doi:10.2307/249620

Cachon, G. P. (2004). The allocation of inventory risk in a supply chain: Push, pull, and advance-purchase discount contracts. *Management Science, 50*(2), 222–238. doi:10.1287/mnsc.1030.0190

Cannon, J. P., Achrol, R. S., & Gundlach, G. T. (2000). Contracts, norms, and plural form governance. *Journal of the Academy of Marketing Science, 28*(2), 180–195. doi:10.1177/0092070300282001

Carr, A. S., & Pearson, J. N. (1999). Strategically managed buyer-seller relationships and performance outcomes. *Journal of Operations Management, 17*(5), 497–519. doi:10.1016/S0272-6963(99)00007-8

Cay, S., & Yang, Z. (2008). Development of cooperative norms in the supply-supplier relationship: The Chinese experience. *Journal of Supply Chain Management, 44*(1), 60.

Chatterjee, D., Grewal, R., & Sambamurthy, V. (2002). Shaping up for e-commerce: Institutional enablers of the organizational assimilation of web technologies. *Management Information Systems Quarterly, 26*(2), 65–89. doi:10.2307/4132321

Chen, I. J., & Paulraj, A. (2004a). Understanding supply chain management: Critical research and a theoretical framework. *International Journal of Production Research, 42*(1), 131–163. doi:10.1080/00207540310001602865

Cheng, L., & Grimm, C. M. (2006). The application of empirical strategic management research to supply chain management. *Journal of Business Logistics, 27*(1), 1–57. doi:10.1002/j.2158-1592.2006.tb00240.x

Child, J., & David, F. (1998). *Strategies of cooperation.* Oxford, UK: Oxford University Press.

Christiaanse, E., & Kumar, K. (2000). ICT-enabled coordination of dynamic supply webs. *International Journal of Physical Distribution and Logistics Management, 30*(3-4), 268–286.

Closs, D. J., Goldsby, T. J., & Clinton, S. R. (1997). Information technology influences on world class logistics capability. *International Journal of Physical Distribution and Logistics Management, 27*(1), 4–17. doi:10.1108/09600039710162259

Cooper, M. C., Ellram, L. M., Gardner, J. T., & Hanks, A. M. (1997). Meshing multiple alliances. *Journal of Business Logistics, 18*(1), 67–89.

Cooper, M. C., Lambert, D. M., & Pagh, J. D. (1997). Supply chain management: More than a new name for logistics. *International Journal of Logistics Management, 8*(1), 1–14. doi:10.1108/09574099710805556

Cox, A. D., Chicksand, P. I., & Davies, T. (2005). Sourcing indirect spend: A survey of current internal and external strategies for non-revenue-generating goods and services. *Journal of Supply Chain Management, 41*(2), 39–51. doi:10.1111/j.1055-6001.2005.04102004.x

Crowley, A. G. (1998). Virtual logistics: Transport in the marketspace. *International Journal of Physical Distribution & Logistics Management, 28*(7), 547–574. doi:10.1108/09600039810247470

Cusumano, M. A., & Yoffie, D. B. (1998). *Competing on Internet time: Lessons from Netscape and its battle with Microsoft*. New York, NY: Free Press.

Dahlstrom, R., McNeilly, K., & Speh, T. (1996). Buyer-seller relationships in the procurement of logistical services. *Journal of the Academy of Marketing Science, 24*(2), 110–124. doi:10.1177/0092070396242002

Daugherty, P. J., Autry, C. W., & Ellinger, A. E. (2001). Reversel: The relationship between resource commitment and program performance. *Journal of Business Logistics, 22*(1), 107–112. doi:10.1002/j.2158-1592.2001.tb00162.x

Daugherty, P. J., Ellinger, A. E., & Dale, S. R. (1995). Information accessibility: Customer responsiveness and enhanced performance. *International Journal of Physical Distribution and Logistics Management, 25*(1).

Droge, C., Jayaraman, J., & Vickery, S. K. (2004). The effects of internal versus external integration practices on time-based performance and overall firm performance. *Journal of Operations Management, 22*(6), 557–573. doi:10.1016/j.jom.2004.08.001

Dubois, A., & Gadde, L. E. (2002). Systematic combining: An abductive approach to case research. *Journal of Business Research, 55*.

Earl, M. J. (1993). Experiences in strategic information systems planning: Editor's comments. *Management Information Systems Quarterly, 17*(3), 2–3.

Eisenhardt, K. M. (1989). Building theory from case study research. *Academy of Management Review, 14*(4).

Ellinger, A. E., Lynch, D. F., Andzulis, J. K., & Smith, R. J. (2003b). B-to-B e-commerce: A content analytical assessment of motor carrier websites. *Journal of Business Logistics, 24*(1), 119–220. doi:10.1002/j.2158-1592.2003.tb00037.x

Ellinger, A. E., Lynch, D. F., & Hansen, J. D. (2003a). Firm size, web site content and financial performance in the transportation industry. *Industrial Marketing Management, 32*, 177–185. doi:10.1016/S0019-8501(02)00261-4

Evangelista, P., & Sweeney, E. (2006). Technology usage in the supply chain: The case of small 3PLs. *International Journal of Logistics Management, 17*(1), 55–74. doi:10.1108/09574090610663437

Evans, P. B., & Wurster, T. (1997). Strategy and the new economies of information. *Harvard Business Review, 75*(5), 15–21.

Fisher, R. J., Maltz, E., & Jaworski, B. J. (1997). Enhancing communication between marketing and engineering: The moderating role of relative functional identification. *Journal of Marketing, 61*(3), 54–70. doi:10.2307/1251789

Forrester, J. W. (1958). Industrial dynamics: A major breakthrough for decision makers. *Harvard Business Review, 38,* 37–66.

Fugate, B., Sahin, F., & Mentzer, J. T. (2006). Supply chain management coordination mechanism. *Journal of Business Logistics, 27*(2), 129–134. doi:10.1002/j.2158-1592.2006.tb00220.x

Galliers, R. (1992). *Choosing information system research approaches, information systems research: Issues, methods and practical guidelines.* Oxford, UK: Blackwell Scientific.

Giunipero, L., Handfield, R. B., & Eltantawy, R. (2006). Supply management's evolution: Key skill sets for the supply manager of the future. *International Journal of Operations & Production Management, 26*(7), 822–844. doi:10.1108/01443570610672257

Global Logistics Research Team at Michigan State University. (1995). *World class logistics: The challenge of managing continuous change.* Oak Brook, IL: Council of Logistics Management.

Grover, V., & Malhotra, M. (1997). Business process re-engineering: A tutorial on the concept, evolution, method, technology and application. *Journal of Operations Management, 15,* 192–213. doi:10.1016/S0272-6963(96)00104-0

Gundlach, G. T., Ravi, S. A., & Mentzer, J. T. (1995). The structure of commitment in exchange. *Journal of Marketing, 59*(1), 78–92. doi:10.2307/1252016

Gustin, C. M., Daugherty, P. J., & Stank, T. P. (1995). The effects of information availability on logistics integration. *Journal of Business Logistics, 16*(1), 1–21.

Handfield, R. B. (1995). *Re-engineering for time-based competition.* Westport, CT: Quorum Books.

Hayes, R. H., & Wheelwright, S. C. (1979). Linking manufacturing process and product life cycles. *Harvard Business Review, 5*(1), 133–140.

Hayes, R. H., & Wheelwright, S. C. (1984). *Restoring our competitive edge: Competing through manufacturing.* New York, NY: John Wiley & Sons.

Hayes, R. H., Wheelwright, S. C., & Clark, K. (1988). *Dynamic manufacturing.* New York, NY: Free Press.

Heide, J. B., & John, G. (1992). Do norms matter in marketing relationships? *Journal of Marketing, 56*(2), 32–44. doi:10.2307/1252040

Houston, M. B., & Johnson, S. A. (2000). Buyer-supplier contracts versus joint ventures: Determinants and consequences of transaction structure. *JMR, Journal of Marketing Research, 37*(1), 1–15. doi:10.1509/jmkr.37.1.1.18719

Hunt, S. D., & Davis, D. F. (2008). Grounding supply chain management in resource-advantage theory. *Journal of Supply Chain Management, 44*(1), 10–21. doi:10.1111/j.1745-493X.2008.00042.x

Jayaram, J., Vickery, S. K., & Droge, C. (1999). An empirical study of time-based competition in the north American automotive supplier industry. *International Journal of Operations & Production Management, 19*(10), 1010–1033. doi:10.1108/01443579910287055

Karoway, C. (1997). Superior supply chains pack plenty of byte. *Purchasing Technology, 8*(11), 32–35.

Kathuria, R. (2000). Competitive priorities and managerial performance: A taxonomy of small manufacturers. *Journal of Operations Management, 18*(6), 627–641. doi:10.1016/S0272-6963(00)00042-5

Knemeyer, A. M., & Murphy, P. R. (2004). Evaluating the performance of third-party logistics arrangements: A relationship marketing perspective. *Journal of Supply Chain Management, 40*(1), 35–51. doi:10.1111/j.1745-493X.2004.tb00254.x

Kotabe, M., Martin, X., & Domoto, H. (2003). Gaining from vertical partnerships: Knowledge transfer, relationship duration, and supplier performance improvement in the U.S. and Japanese automotive industries. *Strategic Management Journal, 24*(4), 293–316. doi:10.1002/smj.297

KPMG. (2003). *Logistica integrata ed operatori di settore: Trend e scenari evolutivi del mercato Italiano.* Milan, Italy. KPMG Business Advisory Services.

Lai, F., Li, D., Wang, Q., & Zhao, X. (2008). The information technology capability of third party logistics providers: A resource based view and empirical evidence from China. *Journal of Supply Chain Management, 44*(3), 22. doi:10.1111/j.1745-493X.2008.00064.x

Lambert, D. M., Stock, J. R., & Ellram, L. M. (1998). *Fundamentals of logistics.* New York, NY: McGraw-Hill.

Lee, H. L., & Billington, C. (1992). Managing supply chain inventory: Pitfalls and opportunities. *Sloan Management Review,* 65–73.

Lee, H. L., Padmanabhan, V., & Whang, S. (1997). Information distortion in a supply chain: The bullwhip effect. *Management Science, 43*(4), 546–558. doi:10.1287/mnsc.43.4.546

Lewis, M. W. (1998). Iterative triangulation: A theory development process using existing case studies. *Journal of Operations Management, 16,* 455–469. doi:10.1016/S0272-6963(98)00024-2

Lusch, R. F., & Brown, J. R. (1996). Interdependency, contracting, and relational behavior in marketing channels. *Journal of Marketing, 60*(4), 19–39. doi:10.2307/1251899

Lynagh, P. M., Murphy, P. R., Poist, R. F., & Grazer, W. F. (2001). Web-based informational practices of logistics service providers: An empirical assessment. *Transportation Journal, 40*(4), 34–45.

Mahoney, J. T., & Pandain, J. R. (1992). The resource-based view within the conversation of strategic management. *Strategic Management Journal, 20*(10), 935–952.

Maloni, M., & Benton, W. C. (2000). Power influences in the supply chain. *Journal of Business Logistics, 21*(11), 49–73.

Maloni, M. J., & Carter, C. R. (2006). Opportunities for research in third-party logistics. *Transportation Journal, 45*(2), 23–38.

McDuffie, J. M., West, S., Welsh, J., & Baker, B. (2001). Logistics transformed: The military enters a new age. *Supply Chain Management Review, 5*(3), 92–100.

Menon, M. K., McGinnis, M. A., & Ackerman, K. B. (1998). Selection criteria for providers of third party logistics services: An exploratory study. *Journal of Business Logistics, 19*(1), 21–37.

Mentzer, J. T., DeWitt, W., Keebler, J. S., Min, S., Nix, N. W., & Smith, C. D. (2001). Defining supply chain management. *Journal of Business Logistics, 22*(2), 1–25. doi:10.1002/j.2158-1592.2001.tb00001.x

Merlino, M., & Testa, S. (1998, May 21-22). L'adozione delle tecnologie dell'informazione nelle aziende fornitrici di servizi logistici dell'area genovese-savonese: i risultati di un'indagine empirica. In *Proceedings of the 2nd Workshop I Processi Innovativi Nella Piccola Impresa,* Urbino, Italy.

Minguzzi, A., & Morvillo, A. (1999, June 20-23). Entrepreneurial culture and the spread of information technology in transport firms: First results on a Southern Italy sample. In *Proceedings of the 44th ICSB World Conference Innovation and Economic Development: The Role of Entrepreneurship and Small and Medium Enterprises,* Naples, Italy.

Nahm, A. Y., Vonderembse, M. A., Rao, S. S., & Ragu-Nathan, T. S. (2006). Time-based manufacturing improves business performance - results from a survey. *International Journal of Production Economics*, *101*(2), 213–229. doi:10.1016/j. ijpe.2005.01.004

Novack, R. A., Langley, C. J., & Rinehart, L. M. (1995). *Creating logistics value: Themes for the future*. Oak Brooks, IL: Council of Logistics Management.

Ozdemir, S., Trott, P., & Hoecht, A. (2007). New service development: Insight from an explorative study into the Turkish retail banking sector. *Innovation: Management. Policy & Practice*, *9*(3-4), 276–289. doi:10.5172/impp.2007.9.3-4.276

Patton, M. Q. (1990). *Qualitative evaluation and research methods* (2nd ed.). Newbury Park, CA: Sage.

Paulraj, A., & Chen, I. (2007). Strategic buyer-supplier relationships, information technology and external logistics integration. *Journal of supply Chain Management*, 4.

Porter, M. E. (1985). Technology and competitive advantage. *The Journal of Business Strategy*, *5*(3), 60–78. doi:10.1108/eb039075

Prahalad, C. K., & Krishnan, M. (1999). The meaning of quality in the information age. *Harvard Business Review*, *77*(5), 15–21.

Prahinksi, C., & Benton, W. C. (2004). Supplier evaluations: Communication strategies to improve supplier performance. *Journal of Operations Management*, *22*(1), 39–62. doi:10.1016/j. jom.2003.12.005

Ravi, K., Anandarajan, M., & Igbaria, M. (1999). Linking IT applications with manufacturing strategy: An intelligent decision support system approach. *Decision Sciences*, *30*(4), 959–992. doi:10.1111/j.1540-5915.1999.tb00915.x

Ravichandran, T., & Lertwongsatien, C. (2005). Effect of information resources and capabilities on firm performance: A resource-based perspective. *Journal of Management Information Systems*, *21*(4), 237–276.

Richey, R. G., Daugherty, P. J., & Roath, A. S. (2007). Firm technological readiness and complementarity: Capabilities impacting logistics service competency and performance. *Journal of Business Logistics*, *28*(1), 195–228. doi:10.1002/j.2158-1592.2007.tb00237.x

Sahin, F., & Robinson, P. (2002). Flow coordination and information sharing in supply chains: Review, implications, and directions for future research. *Decision Sciences*, *33*(4), 505–536. doi:10.1111/j.1540-5915.2002.tb01654.x

Sahin, F., & Robinson, P. (2005). Information sharing and coordination in make-to-order supply chains. *Journal of Operations Management*, *23*(6), 579–598. doi:10.1016/j.jom.2004.08.007

Sanders, N. R. (2005). IT alignment in supply chain relationships: A study of supplier benefits. *Journal of Supply Chain Management*, *41*(2), 4–13. doi:10.1111/j.1055-6001.2005.04102001.x

Sanders, N. R., & Premus, R. (2002). IT applications in supply chain organizations: A link between competitive priorities and organizational benefits. *Journal of Business Logistics*, *23*(1), 65–83. doi:10.1002/j.2158-1592.2002.tb00016.x

Sauvage, T. (2003). The relationship between technology and logistics third-party providers. *International Journal of Physical Distribution and Logistics Management*, *33*(3), 236–253. doi:10.1108/09600030310471989

Selviaridis, K., & Spring, M. (2007). Third party logistics: A literature review and research agenda. *International Journal of Logistics Management*, *18*(1), 125–150. doi:10.1108/09574090710748207

Sinkovics, R. R., & Roath, A. S. (2004). Strategic orientation, capabilities, and performance in manufacturer-3PL relationships. *Journal of Business Logistics*, *25*(2), 43–64. doi:10.1002/j.2158-1592.2004.tb00181.x

Stank, T. P., Keller, S. B., & Daugherty, P. J. (2001). Supply chain collaboration and logistics service performance. *Journal of Business Logistics*, *22*(2), 29–47. doi:10.1002/j.2158-1592.2001.tb00158.x

Van Hoek, R. (2002). Using information technology to leverage transport and logistics service operations in the supply chain: An empirical assessment of the interrelation between technology and operation management. *International Journal of Information Technology and Management*, *1*(1), 115–130. doi:10.1504/IJITM.2002.001191

Vickery, S. K., Calantone, R., & Droge, C. (1999). Supply chain flexibility: An empirical study. *Journal of Supply Chain Management*, *35*(3), 16–24. doi:10.1111/j.1745-493X.1999.tb00058.x

Yin, R. K. (1994). *Case study research design and methods*. Newbury Park, CA: Sage.

ENDNOTES

[1] Pharmaceutical, Promotional, Publishing, Automotive I.T. Electronics and Telecommunications, Banking Insurance, Fashion.

[2] Ordinary, Express, Intermode, Door to doord, Special delivery.

This work was previously published in the International Journal of E-Services and Mobile Applications, Volume 3, Issue 4, edited by Ada Scupola, pp. 21-36, copyright 2011 by IGI Publishing (an imprint of IGI Global).

Section 3
Models for Innovative E–Government Services

Chapter 9

E-Service Research Trends in the Domain of E-Government:
A Contemporary Study

M. Sirajul Islam
Örebro University, Sweden

Ada Scupola
Roskilde University, Denmark

ABSTRACT

Government 'e-service' as a subfield of the e-government domain has been gaining attention to practitioners and academicians alike due to the growing use of information and communication technologies at the individual, organizational, and societal levels. This paper conducts a thorough literature review to examine the e-service research trends during the period between 2005 and 2009 mostly in terms of research methods, theoretical models, and frameworks employed as well as type of research questions. The results show that there has been a good amount of papers focusing on 'e-Service' within the field of e-government with a good combination of research methods and theories. In particular, findings show that technology acceptance, evaluation and system architecture are the most common themes, which circa half of the studies surveyed focus on the organizational perspective and that the most employed research methods are case studies and surveys, often with a mix of both types of methodologies.

INTRODUCTION

E-services, intended as services provided through the use of information and communication technologies (ICTs) are a recent technological innovation, which is designed to provide real-time, anyplace, 24/7 accessibility and high quality value added services at individual, organizational and societal levels. Until now, however, in the literature there has not been a universally accepted definition of e-services (e.g., Scupola et al., 2009).

Although the term 'e-service' is generally used in relation to e-services provided in different sectors among which the private and public

DOI: 10.4018/978-1-4666-2654-6.ch009

sectors (e.g., Scupola et al., 2009), in this paper we refer only to government e-services, that is services provided by the government to the citizens through the use of information and Communication Technologies. For the purpose of this paper, we define e-services within the e-government domain as "the electronic delivery of government information, programs, and services often (but not exclusively) over the Internet" (Dawes, 2002).

In recent years, e-government has become both an important research domain especially in the context of public policy and has gained strategic importance in public sector modernization (Wimmer et al., 2008). However, despite the growing demand for accessing government services through modern information and communication technologies, Wimmer et al. (2008) reported that there has been a deficiency in e-government research concerning the future government and ICT with specific focus on e-services. This study has been the main motivation to investigate the status quo of recent e-service research within the e-government domain, thus leading to the main research question of this paper: What are the methodological and theoretical trends of 'e-service' research within the e-government research domain in the last few years? In order to investigate the research question a thorough literature review of circa 150 papers (Webster & Watson, 2002) published over the period 2005-2009 has been conducted. The papers have been mainly analyzed according to the types of research question investigated; the theories used as well as the unit of analysis (perspective) and research methods employed (Webster & Watson, 2002). The major contribution of this article lies therefore in providing a thorough and updated overview of e-services research within the e-government domain over the last few years.

The paper is structured as follows. In this introduction, the background, motivation and research question of the study have been provided. The following section describes the research method with special focus on the search process and criteria for information source selection, data

collection and analysis. The next section presents and discusses the results, while the last section provides some concluding remarks and suggestions for future research.

RESEARCH METHODOLOGY

Selection of Papers

This paper is based on a systematic literature survey of papers published within the period between 2005 and 2009. As Figure 1 shows, in order to make the research process rigorous, thus increasing the validity of the study, Webster and Watson (2002) guidelines for literature review and Grönlund and Andersson (2006) guidelines for paper selection and analysis have been adopted. According to Webster and Watson (2002, p. 4), "the major contributions are likely to be in the leading journals. It makes sense, therefore, to start with them. ..... You should also examine selected conference proceedings, especially those with a reputation for quality". In order to find the leading journals, the guidelines of AIS's 'Senior Scholars' Basket of Journals, as listed in Figure 1 had been explored. This search method also helped to frame the boundary of the literature review and to limit the content of the analysis.

Furthermore, other relevant journals within the e-services field (See Figure 1) were also selected. Being e-services a relatively young field of research, it was assumed that newer journals within the field contained also relevant and interesting research. Finally for identifying the conferences related to e-government, Grönlund and Andersson's (2006) suggestions have been mostly followed (See Figure 1), but also some other relevant conferences have been included.

Search Procedure

Webster and Watson (2002, p. 4) suggest that "a systematic search should ensure that you accumulate a relatively complete census of relevant

Figure 1. The Information source selection model

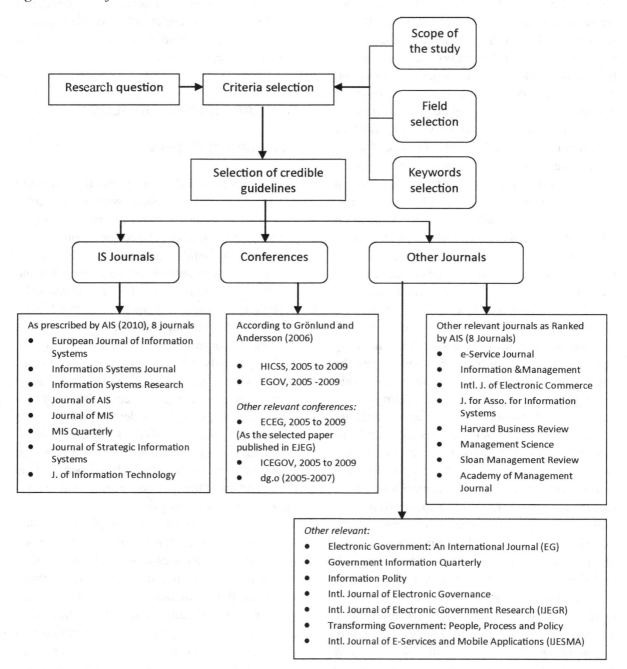

literature". Here the literature search has been conducted through an iterative process mainly based on the analysis of the contexts as advised by Walsham (1995, p. 76). In particular, the steps

suggested by Webster and Watson (2002) were followed, which can be shortly summarized as follows: Locating papers published in leading journals initially from major journal database (e.g., ABI/Inform); Examining relevant reputed conference proceedings; Going backward by reviewing the citations for the articles identified. Going forward to other relevant sources (e.g., Web of Science); Discussing the scope of the research with colleagues or friends interested in the subject matter either prior to or after the completion of the paper; Developing a concept Matrix which may be augmented with Units of Analysis. To conduct the search, four major keywords (e-service, electronic service, e-government, public administration) were used alternatively or sequentially based on the situation. This technique was used both to narrow down the number of papers and to find and select the most relevant papers for the study.

Data Collection

Refining the data was the most critical part of the data collection process as inappropriate data would mislead the research outcome. Though initially 16 journals (Figure 1) from the AIS's 'Senior Scholars' basket were used, only 10 journals had come up with relevant contents. For example, searching with the keyword 'e-service' in 'e-Service Journal' had produced 167 matches. However, only 4 of them were initially found relevant to the domain of e-government which had been later reduced to 2 during the sorting of the content according to the study's research objectives. Furthermore for example, Harvard Business Review did not provide any result on e-services and e-government in the given time period.

The data analysis process was based on a database especially developed for this paper by one of the authors as discussed by Grönlund and Andersson (2006). This database helped to analyze the data according to the study's research question and provided flexibility in generating partial reports in relation to any desired criteria. The fields of this database were filled with themes emerging from the research methods, research questions, theoretical frameworks and models and unit of

analysis of the papers selected. In the analysis of the data, a number of tables and matrices were built and presented (e.g., Webster & Watson, 2002). As search stopping criteria, a systematic saturation point of data collection was used.

This data collection method has initially produced about 150 articles, which have finally been reduced to 95. Please see the list of selected conferences and journals with corresponding number of papers in Appendix 1.

RESULTS

As it is showed in Table 1, most of the papers analyzed focus on assessing the needs on the demand side of e-government services such as e-service acceptance, trust in e-services, project management and the crucial factors for e-services quality.

In particular, the theme 'acceptance' is particularly dominating in e-service research, followed by evaluation, architecture and trust (Table 1). Adoption of e-services (AlAwadhi & Morris, 2007; Tung & Rieck, 2005; Lee & Lei, 2007; Arendsen et al., 2008, Carter & Schaupp, 2009), barriers to access (Vassilakis et al., 2005) and e-tax filing (Fu et al., 2006; Hung et al., 2006) are some of major research sub-themes that are mainly covered under the 'acceptance' research theme. Project management, channel management and benchmarking are some of the new themes that emerged in 2009.

In addition, Tang (2006) distinguished three perspectives research can focus on: individual, organizational or society.

As Table 2 shows, 'e-service' research mainly focuses on the organizational perspective (53%), followed by the individual (27%) and societal (20%) perspectives. This trend is more evident in the papers published in the conferences than journals. As shown in table 2 in fact, 27% conference papers focus on the organizational perspective. This percentage is more than three times

Table 1. E-service themes found in the period 2005-2009

Themes/keywords	2005	2006	2007	2008	2009	Total
Acceptance	4	2	4	2	3	15
Evaluation		3	4	1	7	15
Architecture	1		2	6	2	11
Trust	1		1	2	1	5
Challenges & Success factors	1	1		1	1	4
Service quality			3	1		4
Impact assessment		1		2	1	4
Need assessment		1	1	2		4
E-service (general)		1	1	1		3
Stakeholder			2		1	3
Strategies				1	2	3
Benchmarking					2	2
Effectiveness	2					2
Usability	1	1				2
Participation				2		2
Accessibility					1	1
Administrative burden			1			1
Availability				1		1
Business model			1			1
Channel management					1	1
Collaboration		1				1
Cross-agency integration			1			1
Democracy	1					1
Digital divide			1			1
Dynamic taxonomies			1			1
Infrastructure	1					1
Privacy			1			1
Project management					1	1
Return on Investment		1				1
Usefulness			1			1
Values				1		1
Total	**12**	**12**	**25**	**23**	**23**	**95 papers**

higher than the conference papers focusing on the individual perspective (8%). However this trend is quite balanced in case of journal publications with 26% of the papers focusing on the organizational perspective and 19% on the individual perspectives.

RESEARCH APPROACHES

In regard to the types of research approaches, the study found that, "descriptive research" (45%) and "theory generation" (40%) type of research are the most dominating ones, followed by the

Table 2. Perspectives of e-service research (n=95)

Venues	Individual	Organization	Society
Conference	8%	27%	12%
Journals	19%	26%	8%
Total	27%	53%	20%

"theory utilization" (15%) approach (See Table 3 and Table 4). While the 'descriptive' approach explains the context and phenomena of an entity, the theory generation approach is mainly focusing on either developing new or refining existing theories. The latter type aims both at explaining certain phenomena and at reducing the research gap not adequately addressed by previous research. Finally, the studies that fall within the category "theory utilization" are mainly studies that examine the applicability of certain theories in a different setting.

As shown in Tables 5 and Table 6, regarding the research methodologies used, the case study and survey are significantly dominating the e-service research field. In fact, they together represent more than 85% of the prevailing research methods employed in the sample of papers analyzed here. It also appears that papers published in journals rely more on the case-study method in comparison to papers published in conference proceedings. However in many instances it has been found that the use of various forms of 'survey' methods is mixed with other methods, where the use of 'case study' is mostly common.

TYPE OF RESEARCH QUESTIONS

To conduct an analysis of the concepts contained in the surveyed papers, the research questions of each paper have been grouped based on four thematic categories – explorative, develop, evaluative and causal. These categories have been empirically determined based on the frequency of occurrences of similar types of research questions. The

findings show (Table 7 and Table 8) that most of the research questions are explorative in nature followed by research questions trying to develop new research frameworks or ideas. Explorative research questions mainly investigate issues like status and functionalities (Zhenyu, 2007), adoption perception and process (Arendsen et al., 2009; Tung et al., 2005; Belanger et al., 2008; Papadomichelaki, 2006), identification of barriers and success factors (Vassilakis et al., 2005; Islam & Grönlund, 2007), research gaps (Phang et al., 2005), and administrative literacy requirements (e.g., Grönlund et al., 2007), and understanding and explaining of new insights about certain practices, relationship or perceptions (Axelsson & Melin, 2007; Fu et al., 2006; Hypponen et al., 2005; Chan & Pan, 2008).

The following research questions or objectives are more or less common in such category: "To investigates the real driving forces concerning the 'demand' side of egovernment and the take-up of public e-services" - (Kunstelj et al., 2007); "Which factors influence the adoption of high impact governmental e-services" (Arendsen et al., 2008); "Why did the system fail, and what, if anything can be done to improve it" (Islam and Grönlund, 2007); "Understanding of citizens' needs regarding public e- services" (Axelsson & Melin, 2007); "Analyses the Finnish electronic prescription system against the ramifications given for a national infrastructure "(Hypponen et al., 2005); and "Elaboration on e-government systems implementation with a focus on user engagement" (Chan & Pan, 2008).

Table 3. Research approaches (n=95)

Venues	Descriptive	Theory generation	Theory utilization
Conference	19%	17%	12%
Journals	26%	23%	3%
Total	45%	40%	15%

Table 4. Types of research approaches and corresponding studies

Research Approach	Authors
Descriptive	Grönlund et al. (2007); Salhofer et al. (2008); Cullen & Reilly (2007); Anthopoulos et al. (2006); Vassilakis et Al. (2005); Zhenyu (2007); Chen et al. (2009); Gibson et al. (2009); Pelly & Sia (2007); Horan et al. (2006); Melin & Axelsson (2009); Lourdes et al. (2005); Pardhasaradhi & Ahmed (2007); Kaliannan et al. (2009); Chan & Pan (2008); Wendy & Leela (2007); Buccella & Cechich (2009); Balci et al. (2008); Janssen & Feenstra (2008); Roy, J. (2009); Kaaya, J. (2009); Asgarkhani (2005); Arendsen & Hedde (2009); Mitra, A. (2005); Sarikas & Weerakkody (2007); Yang & Paul (2005); Janssen & Klievink (2009); Ask et al.(2008); Kariofillis-Christos & Economides (2009); Lee at el. (2008); Stoica & Ilas (2009); Islam & Grönlund (2007); Anthopoulos et al. (2007); Deursen (2007); Hypponen et al. (2005); Kunstelj et al. (2007); Charalabidis et al. (2006); Connolly (2007); Carratta et al (2006); Tan et al. (2005); Furuli & Kongsrud (2007); Scupola et al. (2009); Axelsson & Melin (2007)
Theory generating	Anastasios & Vasileios (2008); Sehl & Faouzi (2009); Luka (2009); Verdegem & Verleyea (2009); Gouscosa et al. (2007); Verdegem & Hauttekeete (2008); Gasmelseid (2007); Hu et al. (2008); Golubeva & Merkuryeva. (2006); Janssen & Kuk (2007); Sahu & Gupta (2007); Carter & Schaupp (2009); Lepouras et al. (2008); Leben et al. (2006); Park (2008); Papadomichelaki & Mentzas (2009); Corradini et al. (2008); Velsen et al. (2008); Yu (2008); Axelsson & Melin (2008); Andersen & Medaglia (2008); Papadomichelaki et al. (2006); Wang et al. (2005); Belanger & Carter (2006); Benjamin & Whitley (2009); Boyer-Wright & Kottemann (2008); Pinho & Macedo (2008); Gallant et al. (2007); Shachaf & Oltmann (2007); Tung & Rieck (2005); Belanger & Carter (2008); Pentafronimos et al. (2008); Lee & Lei (2007); Carter & Belanger (2005); Chen et al. (2006); Mike & Anthony (2007); Kanat & Özkan (2009); Welch & Pandey (2007)
Theory utilization	Sacco, G. M. (2007); Fu et al. (2006); Kraussl et al. (2009); Klischewski & Ukena (2008); McLeod & Pippin (2009); Schaupp et al. (2009); AlAwadhi & Morris (2008); Arendsen et al. (2008); Chee-Wee et al. (2008); Magoutas et al. (2007); Phang et al. (2005); Magoutas & Mentzas (2009) ; Hung et al. (2006);

On the other hand, the main objectives of what here is called 'develop' type of research are either to develop or refine a new or existing theory, model or framework that can be used subsequently to explain a phenomenon under investigation. Examples include: "To develop a constructive, value-based approach to aid the realization of e-customs initiatives in real-life setting" (Kraussl et al. 2009); "Proposes a cost-benefit model for evaluating front-end e- government services" (Andersen & Medaglia, 2008).

The evaluative research approach generally tries to assess the impacts and expected returns on implemented electronic services in various contexts. Examples include: "Reviews the existing literature on public return on investment

(ROI) and presents an assessment conducted on an Italian circuit of eGovernment services" (Carratta et al., 2006); "To evaluate the level of satisfaction derived by citizens while utilizing government-led ATIS services for trip planning" (Horan et al., 2006).

'Causal' research studies explore and experiment with the causal relationships and the post and pre-implementation effects of e-services on individual, organizational and societal perspectives. In this case, 'e-service' is considered an independent variable, whiler dependent variables are the associated outcomes or likely effects. The following are some examples of research questions: "Do e-services provide equitable online services to the public?" (Shachaf & Oltmann, 2007); "Do High Quality Websites matter for building Citizen Trust towards E-Government Services?" (Chee-Wee et al. 2008); "Explores the potential effects of the digital divide on e- government by surveying a diverse group of citizens to identify the demographic characteristics that impact use of e-government" (Belanger & Carter, 2006).

Table 5. Research methods (n=95)

Venues	Case study	Survey	Interpretive
Conference	18%	24%	5%
Journals	31%	14%	8%
Total	49%	38%	13%

Table 6. Types of research methods and corresponding studies

Research Methods	Authors
Case Study	Chen et al. (2006) ; Lee at el. (2008) ; Asgarkhani (2005); Connolly (2007); Furuli & Kongsrud (2007); Mike & Anthony (2007); Kraussl et al. (2009); Chen et al. (2009); Sarikas & Weerakkody (2007); Benjamin & Whitley (2009); Grönlund et al. (2007); Pardhasaradhi & Ahmed (2007); Balci et al. (2008); Buccella & Cechich (2009); Chan & Pan (2008); Fu et al. (2006); Roy, J. (2009); Sehl & Faouzi (2009); Luka (2009); Verdegem & Verleyea (2009); Gouscosa et al. (2007); Golubeva & Merkuryeva. (2006); Lepouras et al. (2008); Kanat & Özkan (2009); Kaliannan et al. (2009); Pinho & Macedo (2008); Kaaya, J. (2009); Janssen & Kuk (2007); Ask et al.(2008); Janssen & Klievink (2009); Melin & Axelsson (2009); Pelly & Sia (2007); Hu et al. (2008); Tan et al. (2005); Corradini et al. (2008); Velsen et al. (2008); Islam & Grönlund (2007); Kunstelj et al. (2007); Charalabidis et al. (2006); Carratta et al (2006); Wang et al. (2005); Phang et al. (2005); Hypponen et al. (2005); Klischewski & Ukena (2008); Janssen & Feenstra (2008)
Survey	Hung et al. (2006); Yu (2008); Carter & Belanger (2005); Gallant et al. (2007); Leben et al. (2006); Anthopoulos et al. (2007); Boyer-Wright & Kottemann (2008); McLeod & Pippin (2009); Deursen (2007); Papadomichelaki & Mentzas (2009); Arendsen et al. (2008); Kariofillis-Christos & Economides (2009); Arendsen & Hedde (2009); Anastasios & Vasileios (2008); Yang & Paul (2005); Chee-Wee et al. (2008); Magoutas & Mentzas (2009); Welch & Pandey (2007); Belanger & Carter (2006); Park (2008); AlAwadhi & Morris (2008); Gibson et al. (2009); Schaupp et al. (2009); Lee & Lei (2007); Horan et al. (2006); Axelsson & Melin (2008) ;Tung & Rieck (2005); Andersen & Medaglia (2008); Cullen & Reilly (2007); Stoica & Ilas (2009); Axelsson & Melin (2007); Sahu & Gupta (2007); Carter & Schaupp (2009); Shachaf & Oltmann (2007); Belanger & Carter (2008); Vassilakis, et al. (2005)
Interpretive	Mitra, A. (2005); Lourdes et al. (2005); Gasmelseid (2007); Papadomichelaki et al. (2006); Wendy & Leela (2007); Scupola et al. (2009); Salhofer et al. (2008); Zhenyu (2007); Magoutas et al. (2007) ; Verdegem & Hauttekeete (2008); Pentafronimos et al. (2008) ; Anthopoulos et al. (2006) ; Sacco, G. M. (2007)

As in most of the IS research, also the research in government e-services mainly uses theories and frameworks with origins in other disciplines such as marketing, behavioral or social sciences. In fact, as shown in Table 8, although the analytical foundation of around 85% of the papers is based on some models or frameworks, 40% of these rely on self-designed research frameworks, many of which are extensions of existing theories. In addition, around 17 per cent of the papers are fully descriptive and the arguments are validated mainly by empirical evidences or literature reviews. For example, the Technology Acceptance Model (TAM) (Davis et al., 1989), which explains the factors influencing the behavior of an individual

to accept and use a new technology, is the most influential model in the studies of technology acceptance (Gefen & Straub, 2000) also in the case of e-service. Other technology acceptance theories, such as the Theory of Planned Behavior (TPB) (Ajzen, 1985) and Unified Theory of Acceptance and Use of Technology (UTAUT) (Venkatesh et al., 2003) are also frequent in the study of e-service adoption. Finally, regarding e-service quality assessment, the service quality model – SERVQUAL, which originated from strategic business management domain and 'e-Government service quality model (e-GovQual)' in particular have been found to be quite used.

CONCLUDING DISCUSSION

This paper has conducted a literature review to examine the e-service research trends during the period between 2005 and 2009 mostly in terms of research methods, theoretical models and frameworks employed as well as the type of research

Table 7. Type of Research addressed in the papers (n= 95)

Venues	Explorative	Develop	Evaluative	Causal
Conference	19%	19%	6%	3%
Journals	25%	7%	13%	13%
Total	44%	26%	19%	16%

Table 8. Theories and models used in different research approaches

RQ	Theories/models	Authors
Explorative	Self-designed models	Anthopoulos et al. (2007) Furuli & Kongsrud (2007); Golubeva & Merkuryeva (2006); Hu et al. (2008); Papadomichelaki et al. (2006); Kunstelj et al. (2007); Deursen (2007); Arendsen et al. (2008); McLeod & Pippin (2009); Welch & Pandey (2007); Belanger & Carter (2008) ; Verdegem & Hauttekeete(2008); Janssen & Kuk (2007); Chen et al. (2009); Pelly & Sia (2007)
	e-Government stage model	Stoica & Ilas (2009); Zhenyu (2007); Sarikas & Weerakkody (2007)
	Technology Acceptance models (TAM, TPB, UTAUT)	Lee & Lei (2007); Phang et al. (2005); Tung & Rieck (2005); Kanat & Özkan (2009); Fu et al. (2006); AlAwadhi & Morris (2008) ; Gallant et al. (2007)
	Intermediation theory	Janssen & Klievink (2009)
	Users perspective problem solving process	Grönlund et al. (2007)
	Stakeholder theory	Islam & Grönlund (2007); Chan & Pan (2008)
	Dynamic taxonomies	Sacco, G. M. (2007)
	Web information system's implementation plan	Anthopoulos et al. (2006)
	Descriptive (no model followed)	Scupola et al. (2009); Axelsson & Melin (2007); Cullen & Reilly (2007); Kaliannan et al. (2009); Kaaya, J. (2009); Arendsen & Hedde (2009); Ask et al.(2008); Tan et al. (2005); Mitra, A. (2005); Yang & Paul (2005); Gibson et al. (2009)
Develop	Self-designed models	Charalabidis et al. (2006); Buccella & Cechich(2009); Carter & Schaupp (2009); Gasmelseid (2007); Andersen & Medaglia (2008); Janssen & Feenstra (2008); Wang et al. (2005); Yu (2008); Velsen et al. (2008); Sehl & Faouzi (2009); Corradini et al. (2008); Boyer-Wright & Kottemann (2008); Chen et al. (2006); Kraussl et al. (2009); Axelsson & Melin (2008); Mike & Anthony (2007); Mike & Anthony (2007);
	Systems development life cycle	Melin & Axelsson (2009)
	Activity theory	Klischewski & Ukena (2008)
	Technology Acceptance models (TAM, TPB, UTAUT)	Carter & Belanger (2005); Schaupp et al. (2009)
	Value Theory (Value-Focused Thinking Approach)	Park (2008)
	Government Enterprise Architecture – Public Administration (GEA-PA) service model	Salhofer et al. (2008)
	Descriptive (no model followed)	Balci et al. (2008)
Evaluative	Self-designed models	Leben et al. (2006); Gouscosa et al. (2007); Pentafronimos et al. (2008); Horan et al. (2006) ; Anastasios & Vasileios (2008)
	e-GovQual/SERVQUAL/QeGS	Papadomichelaki & Mentzas (2009) ; Kariofillis-Christos & Economides (2009); Connolly (2007); Magoutas et al. (2007); Magoutas & Mentzas (2009)
	The reference model	Lee at el. (2008)
	Three-phase evaluation model	Hypponen et al. (2005)
	Technology Acceptance model	Verdegem & Verleyea (2009)
	Martin Heidegger's etymological enquiry	Benjamin & Whitley (2009)
	Service maturity and delivery maturity	Lourdes et al. (2005)
	XML Model	Vassilakis, et al. (2005)
	Descriptive (no model followed)	Roy, J. (2009); Pardhasaradhi & Ahmed (2007); Carratta et al (2006); Asgarkhani (2005)

continued on following page

Table 8. Continued

RQ	Theories/models	Authors
Causal	Self-designed models	Pinho & Macedo (2008), Belanger & Carter (2006); Sahu & Gupta (2007); Lepouras et al. (2008)
	e-GovQual/SERVQUAL	Shachaf & Oltmann (2007) Chee-Wee et al. (2008)
	Stakeholder theory	Luka (2009)
	Technology Acceptance models	Hung et al. (2006)
	Enid Mumford's concepts	Wendy & Leela (2007)

questions and perspectives of the research. The results have shown that the number of publications focusing on 'e-service' within the e-Government research domain seems to have increased since 2005 with a peak in 2007. In particular the findings show that technology acceptance, evaluation and system architecture are the most common themes. Service or technology adoption and acceptance, quality assessment, stakeholder analysis and trust are the main subjects investigated. As for adoption and acceptance studies, TAM (Davis et al., 1989) is the most frequently used model. Furthermore, among the various quality assessment models, SERVQUAL is widely used in the studies dealing with the e-services quality assessment. Most of the studies focus on the 'organizational' perspective, while the research approaches used are mainly descriptive or intend to generate new theory. However, given the growing use of government e-services in practice, there is a lot of unexplored potential for e-services research in the future especially addressing issues such as eGov 2.0, data security and data privacy. As the e-service topic is relatively new in the domain of e-government and IS in particular, the study found a lack of established models or theories in the field. In fact, most of the papers analyzed use self-designed models which are derived from or are combination of well known theories taken from other disciplines, such as service marketing or stakeholder theories. Some papers are mainly theoretical and are based on literature reviews (as it is the case for this study as well) and explain

the e-service phenomenon in a descriptive way. In fact, regarding the type of research approach, the interpretive research is highly dominating. Regarding the research methods, this study found that case study and survey are more or less equally dominating research methods and in most cases both is used together with support from brief literature reviews.

Overall it can be concluded that within the e-government research domain, there has been over the period 2005-2009 a good amount of studies particularly addressing e-services, with a reasonable combination of research approaches, theories and methods.

LIMITATIONS AND SUGGESTIONS FOR FUTURE RESEARCH

It has to be acknowledged that the above findings are indicative in nature as they are only based on a limited amount of papers and on a relatively short period of time. This is indicative in the sense that there might be more journal and conference papers that have been missed here due to the search criteria adopted as discussed in the method section. Nevertheless, the findings of this study help to provide a picture about the contemporary research trend in 'e-service' research within the e-government domain over the last 5 years.

Finally, regarding future research it is suggested here that more attention should be paid to the individual and societal perspectives of e-service research. More focus should also be put to emerging issues, such as eGov 2.0, data security and data privacy. These considerations are made in light of the spectacular advancement of information and communication technologies and their equal diffusion in all three levels: the individual, organizational and societal. The limitation of this study as discussed above, in turn, calls for future research focusing on more papers with extended span of time.

REFERENCES

AIS - Association for Information Systems. (n.d.). *Senior Scholars' Basket of Journals*. Retrieved from http://home.aisnet.org/displaycommon. cfm?an=1&subarticlenbr=346

Ajzen, I. (1985). From intentions to actions: A theory of planned behavior. In Kuhl, J., & Beckmann, J. (Eds.), *Action control: From cognition to behavior*. Berlin: Springer Verlag.

AlAwadhi, S., & Morris, A. (2008). The Use of the UTAUT Model in the Adoption of E-Government Services in Kuwait. In *Proceedings of the 41st Annual Hawaii International Conference on System Sciences (HICSS'08)* (p. 219).

Anastasios, E., & Vasileios, T. (2008). Evaluating tax sites: an evaluation framework and its application. *Electronic Government*, *5*(3), 321–343. doi:10.1504/EG.2008.018878

Andersen, K. V., & Medaglia, R. (2008). e-Government Front-End Services: Administrative and Citizen Cost-Benefits. In *Proceedings of the 7th International Conference (EGOV 2008, TORINO, ITALY)* (Vol. 5184).

Anthopoulos, L., Siozos, P., Nanopoulos, A., & Tsoukalas, I. A. (2006). The Bottom-up Design of e-Government: A Development Methodology based on a Collaboration Environment. *e-Service Journal*, *4*(3).

Anthopoulos, L. G., Siozosa, P., & Tsoukalas, I. A. (2007). Applying participatory design and collaboration in digital public services for discovering and re-designing e-Government services. *Government Information Quarterly*, *24*(2), 353–376. doi:10.1016/j.giq.2006.07.018

Arendsen, R., Engers, T. M. V., & Schurink, W. (2008). Adoption of High Impact Governmental eServices: Seduce or Enforce? In *Proceedings of the 7th International Conference (EGOV 2008)*, Torino, Italy (Vol. 5184).

Arendsen, R., & Hedde, M. Jt. (2009). On the Origin of Intermediary E-Government Services. In *Proceedings of the 8th International Conference (EGOV 2009)*, Linz, Austria (Vol. 5693).

Asgarkhani, M. (2005). The Effectiveness of e-Service in Local Government: A Case Study. *Electronic. Journal of E-Government*, *3*(4), 157–166.

Ask, A., Hatakka, M., & Grönlund, Å. (2008). The Örebro City Citizen-Oriented E-Government Strategy. [IJEGR]. *International Journal of Electronic Government Research*, *4*(4), 69–88.

Axelsson, K., & Melin, U. (2007). Talking to, Not About, Citizens – Experiences of Focus Groups in Public E-Service Development. In *Proceedings of the 6th International Conference (EGOV 2007)*, Regensburg, Germany (Vol. 4656).

Axelsson, K., & Melin, U. (2008). Citizen Participation and Involvement in eGovernment Projects: An Emergent Framework. In *Proceedings of the 7th International Conference (EGOV 2008)*, Torino, Italy (Vol. 5184).

Balci, A., Kumaş, E., Taşdelen, H., Süngü, E., Medeni, T., & Medeni, T. D. (2008). Development and implementation of e-government services in Turkey: issues of standardization, inclusion, citizen and satisfaction. In *Proceedings of the ICEEGOV 2008* (Vol. 351, pp. 337-342).

Belanger, F., & Carter, L. (2006). The Effects of the Digital Divide on E-Government: An Empirical Evaluation. In *Proceedings of the 39th Annual Hawaii International Conference on System Sciences (HICSS'06)*.

Belanger, F., & Carter, L. (2008). Trust and risk in e-government adoption. *The Journal of Strategic Information Systems*, *17*(2). doi:10.1016/j. jsis.2007.12.002

Benjamin, M., & Whitley, E. A. (2009). Critically classifying: UK e-government website benchmarking and the recasting of the citizen as customer. *Information Systems Journal*, *19*(2).

Boyer-Wright, K. M., & Kottemann, J. E. (2008). High-Level Factors Affecting Global Availability of Online Government Services. In *Proceedings of the 41st Annual Hawaii International Conference on System Sciences (HICSS'08)*.

Buccella, A., & Cechich, A. (2009). A semantic-based architecture for supporting geographic e-services. In *Proceedings of the 3rd International Conference on Theory and Practice of Electronic Governance (ICEGOV2009)* (Vol. 322, pp. 27-35).

Carratta, T., Dadayan, L., & Ferro, E. (2006). ROI Analysis in e-Government Assessment Trials: The Case of Sistema Piemonte'. In *Proceedings of the 5th International Conference (EGOV 2006)*, Krakow, Poland (Vol. 4084).

Carter, L., & Belanger, F. (2005). The utilization of e-government services: citizen trust, innovation and acceptance factors. *Information Systems Journal*, *15*(1). doi:10.1111/j.1365-2575.2005.00183.x

Carter, L., & Schaupp, L. C. (2009). Relating Acceptance and Optimism to E-File Adoption. *International Journal of Electronic Government Research*, *5*(3), 62–74.

Chan, C. M. L., & Pan, S. L. (2008). User engagement in e-government systems implementation: A comparative case study of two Singaporean e-government initiatives. *The Journal of Strategic Information Systems*, *17*(2). doi:10.1016/j. jsis.2007.12.003

Charalabidis, Y., Askounis, D., Gionis, G., & Lampathaki, F. (2006). Organizing Municipal e-Government Systems: A Multi-facet Taxonomy of e-Services for Citizens and Businesses. In *Proceedings of the 5th International Conference (EGOV 2006)*, Krakow, Poland (Vol. 4084).

Chee-Wee, T., Benbasat, I., & Cenfetelli, R. T. (2008). Building Citizen Trust towards E-Government Services: Do High Quality Websites Matter? In *Proceedings of the 41st Annual Hawaii International Conference on System Sciences (HICSS'08)*.

Chen, A. J., Pan, S. L., Zhang, J., Huang, W. W., & Zhu, S. (2009). Managing e-government implementation in China: A process perspective. *Information & Management*, *46*(4). doi:10.1016/j. im.2009.02.002

Chen, C. C., Wu, C. S., & Wu, R. C. F. (2006). e-Service enhancement priority matrix: The case of an IC foundry company. *Information & Management*, *43*(5). doi:10.1016/j.im.2006.01.002

Connolly, R. (2007). Trust and the Taxman: A Study of the Irish Revenue's Website Service Quality. *Electronic. Journal of E-Government*, *5*(2), 127–134.

Corradini, F., Angelis, F. D., Polini, A., & Polzonett, A. (2008). Improving Trust in Composite eServices Via Run-Time Participants Testing. In *Proceedings of the 7th International Conference (EGOV 2008)*, Torino, Italy (Vol. 5184).

Cullen, R., & Reilly, P. (2007). Information Privacy and Trust in Government: a citizen-based perspective from New Zealand'. In *Proceedings of the 40th Annual Hawaii International Conference on System Sciences (HICSS'07)*.

Davis, F. D., Bagozzi, R. P., & Warshaw, P. R. (1989). User Acceptance of Computer Technology: A Comparison of Two Theoretical Models. *Management Science*, *35*, 982–1003. doi:10.1287/mnsc.35.8.982

Dawes, S. (2002). *The future of e-government*. Albany, NY: University at Albany/SUNY. Retrieved from www.ctg.albany.edu/publications/reports/future_of_egov/future_of_egov.pdf

Deursen, A. V. (2007). Where to Go in the Near Future: Diverging Perspectives on Online Public Service Delivery. In *Proceedings of the 6th International Conference (EGOV 2007)*, Regensburg, Germany (Vol. 4656).

Fu, J. R., Farn, C. K., & Chao, W. P. (2006). Acceptance of electronic tax filing: A study of taxpayer intentions. *Information & Management*, *43*(1). doi:10.1016/j.im.2005.04.001

Furuli, K., & Kongsrud, S. (2007). Mypage and Borger.dk - a Case Study of Two Government Service Web Portals. *Electronic. Journal of E-Government*, *5*(2), 165–176.

Gallant, L. M., Culnan, M. J., McLoughlin, P., Bentley, C., & Waltham, M.A. (2007). Why People e-File (or Don't e-File) Their Income Taxes. In *Proceedings of the 40th Annual Hawaii International Conference on System Sciences (HICSS'07)*.

Gasmelseid, T. M. (2007). A Multiagent Service-oriented Modeling of E-Government Initiatives. *International Journal of Electronic Government Research*, *3*(3), 87–106.

Gefen, D., & Straub, D. (2000). The Relative Importance of Perceived Ease of Use in IS Adoption: A Study of E-Commerce Adoption. *Journal of the Association for Information Systems*, *1*(8).

Gibson, A. N., Bertot, J. C., & McClure, C. R. (2009). Emerging Role of Public Librarians as E-Government Providers. In *Proceedings of the 41st Annual Hawaii International Conference on System Sciences (HICSS'09)*.

Golubeva, A., & Merkuryeva, I. (2006). Demand for online government services: Case studies from St. Petersburg. *Information Polity*, *11*(3-4), 241–254.

Gouscosa, D., Kalikakisa, M., Legalb, M., & Papadopouloub, S. (2007). A general model of performance and quality for one-stop e-Government service offerings. *Government Information Quarterly*, *24*(4), 860–885. doi:10.1016/j.giq.2006.07.016

Grönlund, Å., & Andersson, A. (2006). e-Gov Research Quality Improvements Since 2003: More Rigor, but Research (Perhaps) Redefined. In *Proceedings of 5th International Conference (EGOV 2005)*, Krakow, Poland (LNCS 4084, pp. 1-13). Berlin: Springer.

Grönlund, Å., Hatakka, M., & Ask, A. (2007). Inclusion in the E-Service Society – Investigating Administrative Literacy Requirements for Using E-Services. In *Proceedings of the 6th International Conference (EGOV 2007)*, Regensburg, Germany (Vol. 4656).

Horan, T. A., Abhichandani, T., & Rayalu, R. (2006). Assessing User Satisfaction of E-Government Services: Development and Testing of Quality-in-Use Satisfaction with Advanced Traveler Information Systems (ATIS). In *Proceedings of the 39th Annual Hawaii International Conference on System Sciences (HICSS'06)*.

Hu, G., Zhong, W., & Mei, S. (2008). Electronic Public Service (EPS) and its implementation in Chinese local governments. *Int. J. of Electronic Governance*, *1*(2), 118–138. doi:10.1504/IJEG.2008.017900

Hung, S. Y., Changa, C. M., & Yu, T. J. (2006). Determinants of user acceptance of the e-Government services: next term the case of online tax filing and payment system. *Government Information Quarterly*, *23*(1), 97–122. doi:10.1016/j.giq.2005.11.005

Hypponen, H., Salmivalli, L., & Suomi, R. (2005). Organizing for a National Infrastructure Project: The Case of the Finnish Electronic Prescription. In *Proceedings of the 38th Annual Hawaii International Conference on System Sciences (HICSS'05)*.

Islam, M. S., & Grönlund, Å. (2007). Agriculture Market Information E-Service in Bangladesh: A Stakeholder-Oriented Case Analysis. In *Proceedings of the 6th International Conference (EGOV 2007)*, Regensburg, Germany (Vol. 4656).

Janssen, M., & Feenstra, R. (2008). Socio-technical design of service compositions: a coordination view. In *Proceedings of the 2nd International Conference on Theory and Practice of Electronic Governance (ICEGOV 2008)* (Vol. 351, pp. 323-330).

Janssen, M., & Klievink, B. (2009). The Role of Intermediaries in Multi-Channel Service Delivery Strategies. [IJEGR]. *International Journal of Electronic Government Research*, *5*(3), 36–46.

Janssen, M., & Kuk, G. (2007). E-Government Business Models for Public Service Networks. [IJEGR]. *International Journal of Electronic Government Research*, *3*(3), 54–71.

Kaaya, J. (2009). Determining Types of Services and Targeted Users of Emerging E-Government Strategies: The Case of Tanzania. [IJEGR]. *International Journal of Electronic Government Research*, *5*(2), 16–36.

Kaliannan, M., Awang, H., & Raman, M. (2009). Electronic procurement: a case study of Malaysia's e-Perolehan (e-procurement) initiative. *Int. J. of Electronic Governance*, *2*(2/3), 103–117. doi:10.1504/IJEG.2009.029124

Kanat, I. E., & Özkan, S. (2009). Exploring citizens' perception of government to citizen services: A model based on theory of planned behaviour (TBP). *Transforming Government: People, Process and Policy*, *3*(4).

Kariofillis-Christos, C., & Economides, A. A. (2009). A holistic evaluation of Greek municipalities' websites. *Electronic Government: An International Journal*, *6*(2), 193–212. doi:10.1504/EG.2009.024442

Klischewski, R., & Ukena, S. (2008). An Activity-Based Approach towards Development and Use of E-Government Service Ontologies. In *Proceedings of the 41st Annual Hawaii International Conference on System Sciences (HICSS'08)*.

Kraussl, Z., Yao-Hua, T., & Gordijn, J. (2009). A Model-Based Approach to Aid the Development of E-Government Projects in Real-Life Setting Focusing on Stakeholder Value. In *Proceedings of the 41nd Annual Hawaii International Conference on System Sciences (HICSS'09)*.

Kunstelj, M., Jukić, T., & Vintar, M. (2007). Analysing the Demand Side of E-Government: What Can We Learn From Slovenian Users? In *Proceedings of the 6th International Conference (EGOV 2007)*, Regensburg, Germany (Vol. 4657).

Leben, A., Kunstelj, M., Bohanec, M., & Vintar, M. (2006). Evaluating public administration e-portals. *Information Polity*, *11*(3-4), 207–225.

Lee, C. B. P., & Lei, U. L. E. (2007). Adoption of E-Government Services in Macao. In *Proceedings of the 1st International Conference on Theory and Practice of Electronic Governance (ICEGOV 2007)*, Macao, China (pp. 217-220).

Lee, H., Irani, Z., Osman, I. H., Balci, A., Ozkan, S., & Medeni, T. D. (2008). Toward a reference process model for citizen-oriented evaluation of e-Government services. *Transforming Government: People, Process and Policy, 2*(4).

Lepouras, G., Vassilakis, C., Sotiropoulou, A., Theotokis, D., & Katifori, A. (2008). An active blackboard for service discovery, composition and execution. *Int. J. of Electronic Governance, 1*(3), 275–295. doi:10.1504/IJEG.2008.020450

Lourdes, T., Vicente, P., & Basilio, A. (2005). E-government developments on delivering public services among EU cities. *Government Information Quarterly, 22*(2), 217–238. doi:10.1016/j.giq.2005.02.004

Luka, S. C. Y. (2009). The impact of leadership and stakeholders on the success/failure of e-government service using the case study of e-stamping service in Hong Kong. *Government Information Quarterly, 26*(4), 594–604. doi:10.1016/j.giq.2009.02.009

Magoutas, B., Halaris, C., & Mentzas, G. (2007). An Ontology for the Multi-perspective Evaluation of Quality in E-Government Services. In *Proceedings of the 6th International Conference (EGOV 2007),* Regensburg, Germany (Vol. 4657).

Magoutas, B., & Mentzas, G. (2009). Refinement, Validation and Benchmarking of a Model for E-Government Service Quality. In *Proceedings of the 8th International Conference (EGOV 2009),* Linz, Austria (Vol. 5693).

McLeod, A. J., & Pippin, S. E. (2009). Security and Privacy Trust in E-Government: Understanding System and Relationship Trust Antecedents. In *Proceedings of the 41nd Annual Hawaii International Conference on System Sciences (HICSS'09).*

Melin, U., & Axelsson, K. (2009). Managing e-service development – comparing two e-government case studies. *Transforming Government: People, Process and Policy, 3*(3).

Mike, G., & Anthony, M. (2007). e-Government information systems: Evaluation-led design for public value and client trust. *European Journal of Information Systems, 16*(2).

Mitra, A. (2005). Direction of electronic governance initiatives within two worlds: case for a shift in emphasis. *Electronic Government, an Int. J., 2*(1), 26-40.

Papadomichelaki, X., Magoutas, B., Halaris, C., Apostolou, D., & Mentzas, G. (2006). A Review of Quality Dimensions in e-Government Services. In *Proceedings of the 5th International Conference (EGOV 2006),* Krakow, Poland (Vol. 4084).

Papadomichelaki, X., & Mentzas, G. (2009). A Multiple-Item Scale for Assessing E-Government Service Quality. In *Proceedings of the 8th International Conference (EGOV 2009),* Linz, Austria (Vol. 5693).

Pardhasaradhi, Y., & Ahmed, S. (2007). Efficiency of Electronic Public Service Delivery in India: Public-Private Partnership as a Critical Factor. In *Proceedings of the 1st International Conference on Theory and Practice of Electronic Governance (ICEGOV 2007),* Macao, China.

Park, R. (2008). Measuring Factors that Influence the Success of E-Government Initiatives'. In *Proceedings of the 41st Annual Hawaii International Conference on System Sciences (HICSS'08).*

Pelly, P. K., & Sia, S. K. (2007). Challenges in delivering cross-agency integrated e-services: The OBLS project'. *Journal of Information Technology, 22*(4).

Pentafronimos, G., Papastergiou, S., & Polemi, N. (2008). Interoperability testing for e-government web services. In *Proceedings of the 2nd International Conference on Theory and Practice of Electronic Governance (ICEGOV 2008),* Egypt (Vol. 351, pp. 316-321).

Phang, C. W., Li, Y., Sutanto, J., & Kankanhalli, A. (2005). Senior Citizens' Adoption of E-Government: In Quest of the Antecedents of Perceived Usefulness. In *Proceedings of the 38th Annual Hawaii International Conference on System Sciences (HICSS'05)*.

Pinho, J. C., & Macedo, I. M. (2008). Examining the antecedents and consequences of online satisfaction within the public sector: The case of taxation services. *Transforming Government: People, Process and Policy, 2*(3).

Roy, J. (2009). E-government and integrated service delivery in Canada: the Province of Nova Scotia as a case study. *Int. J. of Electronic Governance, 2*(2/3), 223–238. doi:10.1504/IJEG.2009.029131

Sacco, G. M. (2007). Interactive exploration and discovery of e-government services. In *Proceedings of the 8th annual international conference on Digital government research*, Philadelphia, PA (Vol. 228, pp. 190-197).

Sahu, G. P., & Gupta, M. P. (2007). Users' Acceptance of E-Government: A Study of Indian Central Excise. [IJEGR]. *International Journal of Electronic Government Research, 3*(3), 1–21.

Salhofer, P., Tretter, G., Stadlhofer, B., & Joanneum, F. H. (2008). Goal-oriented service selection. In *Proceedings of the 2nd International Conference on Theory and Practice of Electronic Governance (ICEGOV2008)*, Egypt (Vol. 351, pp. 60-66).

Sarikas, O. D., & Weerakkody, V. (2007). Realizing integrated e-government services: a UK local government perspective. *Transforming Government: People, Process and Policy, 1*(2).

Schaupp, L. C., Carter, L., & Hobbs, J. (2009). E-File Adoption: A Study of U.S. Taxpayers' Intentions. In *Proceedings of the 41st Annual Hawaii International Conference on System Sciences (HICSS'09)*.

Scupola, A., Henten, A., & Nicolajsen, H. W. (2009). E-Services: Characteristics, Scope and Conceptual Strengths. [IJESMA]. *International Journal of E-Services and Mobile Applications, 1*(3), 1–16.

Sehl, M., & Faouzi, B. (2009). Multi-agent based framework for e-government. *Electronic Government: An International Journal, 6*(2), 177–192. doi:10.1504/EG.2009.024441

Shachaf, P., & Oltmann, S. M. (2007). E-Quality and E-Service Equality. In *Proceedings of the 40th Annual Hawaii International Conference on System Sciences (HICSS'07)*.

Stoica, V., & Ilas, A. (2009). Romanian Urban e-Government. Digital Services and Digital Democracy in 165 Cities. *Electronic. Journal of E-Government, 7*(2), 171–182.

Tan, C. W., Pan, S. L., & Lim, E. T. K. (2005). Towards the Restoration of Public Trust in Electronic Governments: A Case Study of the E-Filing System in Singapore. In *Proceedings of the 38th Annual Hawaii International Conference on System Sciences (HICSS'05)*.

Tang, L. (2006). Group Effectiveness: An Integral and Developmental Perspective. In *Proceedings of the Annual meeting of the International Communication Association*, Dresden, Germany.

Tung, L. L., & Rieck, O. (2005). Adoption of electronic government services among business organizations in Singapore. *The Journal of Strategic Information Systems, 14*(4). doi:10.1016/j.jsis.2005.06.001

Vassilakis, C., Lepouras, G., Halatsis, C., & Lobo, T. P. (2005). An XML model for electronic services. *Electronic Government, an Int. J., 2*(1), 41-55.

Velsen, L. V., Geest, T. V. d., Hedde, M. t., & Derks, W. (2008). Engineering User Requirements for e-Government Services: A Dutch Case Study. In *Proceedings of the 7th International Conference (EGOV 2008)*, Torino, Italy (Vol. 5184).

Venkatesh, V., Morris, M. G., Davis, G. B., & Davis, F. D. (2003). User acceptance of information technology: Toward a unified view. *Management Information Systems Quarterly, 27*(3), 425–478.

Verdegem, P., & Hauttekeete, L. (2008). The user at the centre of the development of one-stop government. *Int. J. of Electronic Governance, 1*(3), 258–274. doi:10.1504/IJEG.2008.020449

Walsham, G. (1995). Interpretive case studies in IS research: nature and method. *European Journal of Information Systems, 4,* 74–81. doi:10.1057/ejis.1995.9

Wang, L., Bretschneider, S., & Gant, J. (2005). Evaluating Web-Based E-Government Services with a Citizen-Centric Approach. In *Proceedings of the 38ᵗʰ Annual Hawaii International Conference on System Sciences (HICSS'05).*

Webster, J., & Watson, R. (2002). Analyzing the past to prepare for the future. *Management Information Systems Quarterly, 26*(2).

Welch, E. W., & Pandey, S. (2007). Multiple Measures of Website Effectiveness and their Association with Service Quality in Health and Human Service Agencies. In *Proceedings of the 40th Annual Hawaii International Conference on System Sciences (HICSS'07).*

Wendy, O., & Leela, D. (2007). Citizen Participation and engagement in the Design of e-Government Services: The Missing Link in Effective ICT Design and Delivery. *Journal of the Association for Information Systems, 8*(9).

Wimmer, M., Codagnone, C., & Janssen, M. (2008). Future of e-Government Research: 13 research themes identified in the eGovRTD2020 project. In *Proceeding of the 41ˢᵗ Hawaii International Conference on System Sciences.*

Yang, J., & Paul, S. (2005). E-government application at local level: issues and challenges: an empirical study. *Electronic Government, an Int. J., 2*(1), 56-76.

Yu, C. C. (2008). Building a Value-Centric e-Government Service Framework Based on a Business Model Perspective. In *Proceedings of the 7ᵗʰ International Conference (EGOV 2008),* Torino, Italy (Vol. 5184).

Zhenyu, H. (2007). A comprehensive analysis of U.S. counties' e-Government portals: development status and functionalities. *European Journal of Information Systems, 16*(2).

APPENDIX

LIST OF JOURNALS AND CONFERENCES USED IN THE STUDY

Table 9.Conferences – 45 papers

International Conference on Theory and Practice of Electronic Governance (ICEGOV)	7
1st Intl. Conference on Theory and Practice of Electronic Governance (ICEGOV2007) – 2	
2nd Intl. Conference on Theory and Practice of Electronic Governance (ICEGOV2008) – 4	
3rd Intl. Conference on Theory and Practice of Electronic Governance (ICEGOV2009) - 1	
International Conference (EGOV)	18
5th International Conference (EGOV 2006) – 3	
6th International Conference (EGOV 2007) – 6	
7th International Conference (EGOV 2008) – 6	
8th International Conference (EGOV 2009) - 3	
Digital Government Society (dg.o 2007)	1
Annual Hawaii International Conference on System Sciences (HICSS)	19
Proceedings of the 38th Annual Hawaii Intl. Conference on System Sciences (HICSS'05) – 4	
Proceedings of the 39th Annual Hawaii Intl. Conference on System Sciences (HICSS'06) - 2	
Proceedings of the 40th Annual Hawaii Intl. Conference on System Sciences (HICSS'07) - 4	
Proceedings of the 41st Annual Hawaii Intl. Conference on System Sciences (HICSS'08) - 5	
Proceedings of the 42nd Annual Hawaii Intl. Conference on System Sciences (HICSS'09) - 4	

Table 10. Journals - 50 papers

International Journal of Electronic Government Research (IJEGR)	7
Electronic Government, an International Journal	6
Government Information Quarterly	6
Int. J. of Electronic Governance	5
Transforming Government: People, Process and Policy	5
Electronic Journal of e-Government	4
Information & Management	3
Information Systems Journal	2
Journal of strategic information systems	3
e-Service Journal	2
European Journal of Information Systems	2
Information Polity	2
International Journal of E-Services and Mobile Applications (IJESMA)	1
Journal of Information Technology	1
Journal of the Association for Information Systems	1

This work was previously published in the International Journal of E-Services and Mobile Applications, Volume 3, Issue 1, edited by Ada Scupola, pp. 39-56, copyright 2011 by IGI Publishing (an imprint of IGI Global).

Chapter 10
Proposing a Knowledge Amphora Model for Transition towards Mobile Government

Tunc D. Medeni
Turksat, Turkey

İ. Tolga Medeni
Turksat, Turkey

Asim Balci
Turksat, Turkey

ABSTRACT

As an important project for Turkey to achieve Information/Knowledge Society Strategic Goals, the e-Government Gateway currently focuses on the delivery of public services via a single portal on the Internet. In later stages, other channels such as mobile devices will be available for use, underlying a transition towards mobile and ubiquitous government services. In order to provide a supportive base for this transition, the authors develop a modeling of knowledge amphora (@), and link this conceptual model with the e-government gateway. Based on Knowledge Science concepts such as ubiquity, ba (physical, virtual, mental place for relationship-building and knowledge-creation), ma (time-space in-between-ness), reflection and refraction, the modeling of Knowledge Amphora incorporates the interactions @ the Internet and mobile devices that contribute to cross-cultural information transfer and knowledge creation. The paper presents recent electronic and mobile government developments of E-Government Gateway Project in Turkey as an application example of this philosophical and theoretical modeling. The contributed Ubiquitous Participation Platform for Policy Making (UbiPOL) project aims to develop a ubiquitous platform allowing citizens to be involved in policy making processes (PMPs). The resulting work is a practical case study as that develops new m-government operations.

DOI: 10.4018/978-1-4666-2654-6.ch010

INTRODUCTION

The e-Government Gateway Project is a major milestone for achieving the Information/Knowledge[1] Society Strategic Goals set by Turkey. Here, e-government basically means provision of public services through electronic means, which implies faster and cheaper access to these services. In order to facilitate access to electronic public services by citizens and enterprises, it will be ensured that these e-government services are reached from a single portal and via multiple channels such as mobile phones. Users will be able to access the system with smart cards or imprinted digital certificates for a secure transaction.

In this paper, we aim to provide a modeling of ubiquitous knowledge amphora (@) that can be linked with the e-Government Gateway Project in Turkey. This model can then provide the theoretical base and philosophical vision to pave the way for the necessary transition from e-government to m-government in Turkey. The outline of the paper is, mainly, first the development of the conceptual modeling for knowledge amphora, then the background information about the e-Government Gateway Project, finally a discussion about mobile government and citizen-oriented e-government initiatives as a part of e-Government Gateway Project. So the paper is a mix between theory/philosophy and a practical case study, aiming to bridge these two. Thus, this piece of work can be considered as not only both academic and practical, but also neither fully academic nor fully practical. One of the main contributions of this paper is then providing a novel cognitive lens covering the knowledge creation process and philosophy, applying and demonstrating to a real case example of e-government and mobile government development in Turkey.

The paper is structured as follows. After a brief introduction on knowledge in e-government services, the next section provides a philosophical discussion on knowledge creation, the following section briefly covers Turkish e-government activity in Turkey, and the final section provides an application of the theoretical framework into an actual case, interlinking a philosophy with practice.

KNOWLEDGE IN E-GOVERNMENT SERVICES

Within the scope of "e-services and mobile applications"; generally, importance of knowledge and knowledge management is recognized in the literature on e-government and public transformation. For example, Reid, Bardzki and McNamee (2004) underline the importance of communication and culture in addition to knowledge-sharing processes and appropriate infrastructure establishing a knowledge-enabled environment to effect (local) government reform. More generically, Cooper, Lichtenstein, Smith (2009) highlight the challenging nature of knowledge transfer among stakeholders to consider and resolve various needs and concerns for success of Internet-based (Information and Communication Technologies (ICT) support) services. Meanwhile, interoperability of different e-government initiatives actually refers (internally or externally) "to the process of ensuring that information systems, procedures and culture of an organization are managed with the aim to maximize opportunities for the exchange and re-use of information" (Brusa, Caliusco, & Chiotti, 2007, p. 35). Furthermore, benefiting from common understandings for the stages of e-government services (such as Layne & Lee, 2001), Fraser et al. (2003)[2] apply a perspective of knowledge into these common stages. (p.14)

1. **Publishing—One-Way Communication:** Knowledge is needed for how to present information clearly online and manage its publication, and how citizens, businesses or government agencies can use the information. Knowledge may be needed about the design, completion and processing of forms.

Thus, for instance, if a template is offered for use, then knowledge about how to guide and constrain its usage is needed.

2. **Interacting — Two-Way Communication:** Knowledge is needed for how to react "electronically" to requests from citizens. For instance knowledge may be needed for how information can be searched and received, how to user information can be accepted and maintained, and how information security can be established.

3. **Transaction — Exchange of Resources of Nominally Higher Value Than Information:** Knowledge is needed for how to ensure secure online exchange of items other than information such as taxes, registration fees and licenses, as well as how to efficiently and smoothly interface the online system with back-office processing systems. Awareness raise on issues of trust and details of engaged processes.

4. **Integration — All Aspects:** Knowledge is needed for how to streamline and coordinate the design and delivery of services with the attributes from the previous stages. This integration blurs the distinctions such as which unit provides a particular service or holds particular data, where one service ends and another begins, in the eyes of the user.

Accordingly, knowledge units can be directly or indirectly associated with service components. The structure of these associations can then define the basic concepts and relationships of a domain map, or ontology, use of which can be considered as an application of knowledge management into e-government services (ibid.). Furthermore, ontologies can be seen as a (semantic) model that forms the basis of a "Model Driven Architecture (Miller et al., 2001)" approach to e-Government (Salhofer, Stadlhofer, & Tretter, 2010).

In addition to a common understanding of content as information or knowledge, generally forms can also be considered instruments to transfer information, as well as to communicate between citizens and public agencies (Axelsson, Melin, & Persson, 2007). As suggested by Henten (2010), in fact (commercial or non-commercial) e-services encompass all informational services (data, information, and knowledge that also incorporate content) and software delivered on digital networks to users. Codifiability, digitization and interpretation (a common interpretative context) of knowledge can be seen as among the conditions for (and implications of) the development of e-services.

As e-government services and initiatives aim to benefit from the use of leading-edge, most innovative ICT in improving governments' fundamental functions, recent mobile and wireless technologies also create a "mobile government" direction for e-government. Mobile government (m-government) can then provide more convenient accessibility and availability (power of pull), better precision and personalization in targeting users and delivering content (power of push), and larger and wider user base (power of reach) (Kuscu, Kushchu, & Yu, 2007) M-Government can then not only provide ubiquitous access but also can take upon the movement in favour of mobile technology usages, and initiate new ways that ultimately better benefits, and even empowers the users and citizens in their various flexibility needs (Rossel, Finger, & Misuraca, 2006). For instance, location-based context aware services have the ability to utilize information about the user's context and adapt services to a user's current situation and needs, enhancing the utilization of mobile data services that so far lag behind mobile voice communication services (de Vos, Haaker, Teerling, & Kleijnen, 2008). As a consequence, integration of e-government and m-government initiatives can play a significant role in public administration reengineering and transformation (Avdagić, Šabić, Zaimović, & Nazečić, 2008). In addition to provision of new, direct interaction channels between government and citizen, more efficient means of work are also operationalised, thanks to use of mobile technologies in public sector such as hand held terminals for postal, traffic information system for police, vehicle monitoring by forestry administration (Yener, 2008). Meanwhile, for instance, Choi and Kim (2006) introduce an ontology-based context model in ubiquitous computing environment, modeling context metadata as well as context information (in a home domain). Finally, Ekong and Ekong (2010), among others, also recommend m-voting as a solution for enhanced e-participation and e-governance.

All these works can then provide a base for approaching e-government from a perspective that enables discussing and reinterpreting certain essential aspects of "knowledge" for public transformation and transition towards mobile service provision. With respect to this, the next section provides a philosophical discussion that underlies issues such as space, time, reflection and refraction for knowledge creation and management.

CONCEPTUALIZING SPACE-TIME (BA, MA), KNOWLEDGE AMPHORA (@), AND REFLECTION AND REFRACTION FOR UBIQUITY AND MOBILITY

Since the ancient times, knowledge has been an ultimate destination that is impossible to reach. For instance, according to legends, Adam and Eve were punished, as they ate the forbidden fruit of the knowledge tree. In the legend of Babylon Tower, then, the desire of the humans to reach their knowledge made Gods angry and they destroyed the tower and separated humans to different tribes with different languages.

Still, the human nature strives for knowledge and knowing, aware of the importance of the sophistication journey for the quest of reaching Sophia or ultimate knowledge. The humans, then, continuously furthered their knowledge. With the invention of the computer, and the revolution of the Internet, finally, the world has again seemed to find a common language with 0 s and 1s to share the culture and bring humans again together.

Finding a common language with the coding of 0 and 1, on the other hand, does not ensure the common understanding across cultures. Generally such series of codes in the forms of numbers or letters are recognized as data with little meaning, value or use for a common sense. As pieces of meaning, information can be derived from data. Then, knowledge can be constructed as an action-oriented system or network of valuable information. Knowledge "that is time-tested and proven to be truthful and useful can finally be considered as wisdom". Similarly, Aktas (1987) defines the data, information and knowledge under the topic of meaning. In this definition data has the lowest level of meaning; they are raw facts and opinions. Information is above the data with higher level of meaning, and it is useful for the present decision situation. In this trio, knowledge has the highest level of meaning, because it represents information that can be potentially useful in future decision situations.

Following the knowledge-creating spiral of Nonaka and Takeuchi (1995), Umemoto (2004) discusses that knowledge can be created spirally as a result of continuous interaction among these different types of episteme. For simplicity, currently knowledge is used to represent data, information, knowledge and wisdom (as all types of episteme).

This hierarchical interlink among data, information, knowledge and wisdom constitutes Sophia, as the ultimate outcome (knowledge). This can be compared with Sophistication, a development and maturation journey that emphasizes the process (knowing) that also underline knowing each other and together (Medeni & Umemoto, 2010) This notion of knowing each other and together highlights the importance of cross-cultural interaction and knowledge management. Accordingly, knowledge has to be understood and agreed upon (known) by all the different entities involved with the process of knowledge creation and management, if ever it deserves to be called knowledge (Medeni et al., 2008; Medeni et al., 2009)[3].

With respect to mobility and knowledge management, Derballa and Pousttchi (2006) discuss that ubiquity, "the possibility to send and receive data anytime and anywhere, and thus eliminates any spatiotemporal restriction" (p. 647), adds value to the processes of mobile knowledge management. As Derballa and Pousttchi (2006, p. 647) continues, ubiquity is originated "not only in the technical possibility but also in the typical usage of mobile devices." This discussion about the ubiquity and typology of mobile devices can then be extended to include the philosophy and theology of (ubiquitous or not) all computing, communication, and information technologies, which can all be considered as virtual and networked organizations. A conceptualization of these technologies, and virtual and networked organizations that is based on knowledge science concepts such as Far-East concepts of *ba* and *ma* as well as Mediterranean conceptualization of Amphora (@), or theories about reflection and refraction can then contribute to such discussion about ubiquity, mobility and spatio-temporal issues.

Ba and Ma

Utilizing space and time to explain social phenomena has been one of the major concerns in history. However, we rarely pay attention to how a spatio-temporal function is utilized. This common ignorance takes place in the context of having an objective view of self and objects without noticing space and time (Medeni, Iwatsuki, & Cook, 2008).

Ships, computer devices, and so on, have led us to utilize space and time even more in condensed and efficient manner. This is the process of what Cooper (1998) describes as "a case of reterritorializing space and retemporalizing time" (p. 113), the double function of space and time. As a way of human orientation, according to Elias (1992), the spatio-temporal functions do not affect us

independently but act as: "Every change in 'space' is a change in 'time'; every change in 'time' is a change in 'space'" (pp. 99-100). At the same time, this dual function carries ontological characteristics. As Whitehead (1967) claims: "things are separated by space, and are separated by time: but they are also together in space and together in time, even if they be not contemporaneous" (p. 64). As Massey (1994) suggests, space is not static nor is time spaceless. Although spatiality and temporality are different from each other, neither can be conceptualized in the absence of the other.

According to knowledge science in Japan (for example, Medeni, Iwatsuki, & Cook, 2008) "*Ba*" is recognized as the 'place', the shared context for relationship building and knowledge creation for which trust among stakeholders is very important. It does have a physical, a relational, and a spiritual dimension. Moreover, it can be physical, mental, or technological (in the sense of information-communication technology) (Nonaka, Toyama, & Scharmer, 2001).

According to Medeni, Iwatsuki, and Cook (2008), "*Ma*" is the in-between-ness, or 'interval' that conveys both time and space as a conceptual and perceptual unity. It is a tension between things allowing for different patterns of interpretation, a constant flow of possibilities, awaiting or undergoing transformation by the availability of physical components and potential uses. Moreover, it is expected to be recognized in relationships, as degree of formality is articulated by measuring *ma* in place, time, social position, and age (Kerkhove, 2003; Hayashi, 2004).

As a significant social phenomenon, *ma* conceptualizes and perceives the interval and in-between-ness that comes with the unity of time and space, thus capturing also the spatial emphasis of *ba*. Depending on the context, *ma* works as a mediator for space and/or time, and refines a dichotomous situation, such as a sender and a receiver, in a communication setting. However, *ma* not only catalyses the dichotomies, but creates a meaning as it is utilized. Berque (1982) also points out that *ma* works as a free zone in a communication, where a sender puts consecutive signs and a receiver finds meanings out of the signs, whose discrepancies are considered as a connection between the both.

Ba and *ma* are useful concepts to address issues about space and time, thus mobility and ubiquity. With respect to such conceptualization of *ba* and *ma* that contributes to the understanding of ubiquity, mobility and spatio-temporal issues, next we present an idea of "knowing-ship" and "knowledge amphora (@)".

Knowledge Amphora (@)

The following discussion explains the major concepts and perspectives regarding this understanding of knowledge amphora, benefiting from authors previous work elsewhere[4]. Knowledge amphora brings together various concepts and perspectives regarding the concepts of knowledge and amphora. Firstly, here knowledge and knowing should be understood as "knowing-ship." Here, "knowing-ship" can be considered a carrier that brings knowledge within different forms of entrepreneurship, creatorship, partnership, and leadership, among others. This metaphor of "knowing-ship" can be considered as a means that makes cross-cultural interaction more mobile, in comparison to more stable means, such as bridges or boundaries that provide space-times for such interaction. Such a notion of "knowing-ship" also conceptualizes "knowing together and each other," as "everyone navigates on the same ship through the journey," which can highlight a general model of cross-cultural interaction, as well as mobile, ubiquitous information and knowledge carriers. Such mobile means could then be considered

- An anaphora as "a carrying up or as back" in Ancient Greek, and
- An amphora "a whirling eddy" (which fits well with the current understanding of

knowledge creation as a spiral) in current Turkish (anafor),
- Or an amphora as "a carrying craft" in a more common sense with its all mental, real and virtual aspects of time-space, the last of which are already being represented by the well-known sign for the Internet navigation, @.

In fact, @ symbol is originally used for amphora. For instance, it used to represent amphorae, as a measurement of mapping sign (calculating the sale's value of the mercantile items or showing the location of shipwrecks carrying amphorae on the maps). @ can then be a good concept to integrate natural, social, organizational, technological aspects for transfer of material, data and information and creation of knowledge. This conceptualization of amphorae for mobile and ubiquitous interaction of knowing and knowledge is illustrated in Figure 1.

Reflection and Refraction Theories for Ba and Ma

While reflection can be understood as seeing reality as it is, refraction complements this understanding as reconceiving and changing reality (Wankel-DeFillippi, 2006). While reflections can then be understood as deeply dwelling upon existing knowledge as an intra-cultural interaction that relies on understanding, harmony, open idea and consensual decision; refractions can be understood as generating new knowledge as an inter-cultural interaction that can also rely on different understandings, misunderstandings, conflicts and confusions (Medeni & Umemoto, 2010).

Thus refractions matches well with critical reflections; furthermore, from a different but related perspective, they can also contribute to creativity and cross-cultural communications. In that sense, the physical phenomenon of refraction provides interesting metaphorical connections to social phenomenon of refraction (Uno, 1999; Amar,

Figure 1. Knowledge Amphora (@)

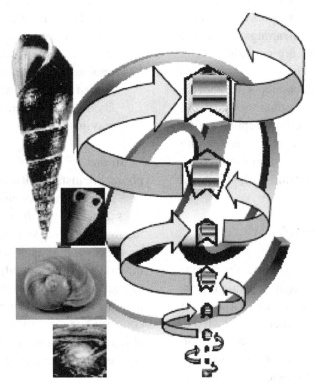

2002; Drucker & Maciariello, 2004; Medeni & Umemoto, 2008). For instance,

- Due to optical refraction the eyes perceive a spoon in a glass of water broken, which warns us that our perceptions and understandings can mislead us, thus we have to critically reflect for the reality;
- Again, the refractions in prism that turns white incident ray into different colors can be associated with more critical, cross-cultural creative ways of thinking, reminding us that even for one incident there could be various ways of interpretations. [5]

The importance of reflection and refraction for boundary-crossing interactions can then be argued (Medeni-Cook-Elwell, 2007), as well. With respect to this, *ba* and *ma* can be considered as technologies for reflective and refractive practice.[6]

Tihon (2006) asserts that knowledge emerges from the information system whose *attractor* is the *ba*, the *shared* place, context, or *ba*sin. This assertion is supported by Tihon's findings that the implicit and explicit domains of the organization's information system generally echo each other. Using Tihon's (ibid.) approach and analysis, it can also be suggested that beside the attractor of sharing, to a certain extent, the attractors of tension and rupture are also needed. As the implicit and explicit domains' echoing each other is not a one-to-one but refracted reflection due to tensions as well as contextual difficulties, Tihon's approach highlights domains that not only reflect each other and share a common basin, but also refract each other and differentiate an in-between-ness bound.

In addition to *ba*, such an approach would incorporate *ma*, which supports the context for dealing with tensions, refractions and ruptures that occur in the progress of time, besides the relations and reflections cultivated by the shared space. Working together and managed thoughtfully, the attractors of sharing, as well as tension and rupture, can turn the negative elements into positive aspects so that useful knowledge can emerge from the information system. The temporal and spatial in-between-ness, provided by *ma*, would provide suitable conditions for refraction, especially for those that can assume a role of bystanders, as discussed by Drucker and Maciariello (2004, p. 48).

To sum up the above discussion, while the real, virtual and mental space of *ba* matches very well with the concept of reflection, the in-between-ness and interval of time and space that *ma* provides can be used for the facilitation of refraction, as another important concept that complements reflection. In return, we can consider these concepts of reflection, refraction, *ba* and *ma* from the perspective of cross-cultural interaction and mobile knowledge management.

Knowledge Amphora as Reflections and Refractions @ Expanding Ba and Ma

We also think the reflections and refractions @ expanding *ba* and *ma*, which address the spatial and temporal aspects of systemic knowledge creation, can also be reinterpreted as a *knowledge amphora*. This knowledge amphora incorporates not only information transfer but also knowledge creation, which, according to Nonaka and Takeuchi's (1995) model, occurs basically as a knowledge-creating spiral. Accordingly, these reflective and refractive interactions are perceived as important dynamics of time and space expansion that can be visualized as refractive and reflective *ba-ma*s, which, as a whole, construct a knowledge amphora. With respect to this knowledge amphora that expands by reflections and refractions, *ma*-boundaries are the intersection intervals of two *ba*-buildings.

The explanation of these discussions about knowledge amphora as reflections and refrac-

tions @ expanding *ba* and *ma* can be found in the illustration of Figure 2. Further explanation on understanding these refractive *ba* and refractive *ma* as Space-Time Curvature embedded in Hyperspace can be found in the work of Medeni et al. (2008).

The resulting modeling of Knowledge Amphora provides a conceptual base for developing not only the operations in the e-Government Gateway project in Turkey but also mobile, ubiquitous services and (electronic) interactions in today's and tomorrow's knowledge society in general. However, our attention will be on e-government services and e-Government Gateway project in Turkey, as a case study, which will be presented next.

E-Government Initiatives in Turkey, E-Government Gateway

According to the recent report of Brown University, U.S.A. (West, 2007), the e-government services provided in Turkey are considered to be in the top ten among the 198 countries of the world. However, according to other reports disseminated by the UN, OECD or EU, the performance of Turkey in e-government can be considered to be about average. For instance, according to the latest report of the UN (2008), Turkey has the ranking of 76 among 182 countries in the world in terms of e-government readiness. Evaluating for the EU, Capgemini (2007, p. 15) reaches similar conclusions for Turkey with a 69% performance ranking in general, regarding online sophistication maturity.

In Turkey, in order to facilitate access to electronic public services by citizens (and enterprises), these e-government services should be reached from a single portal, called e-Government Gateway, and via multiple channels. Users will be able to access the system with smart cards or imprinted digital certificates for secure transactions. Finally, "integration standards" will be adapted to the es-

Figure 2. Knowledge Amphora of Expanding Reflective Ba and Refractive Ma

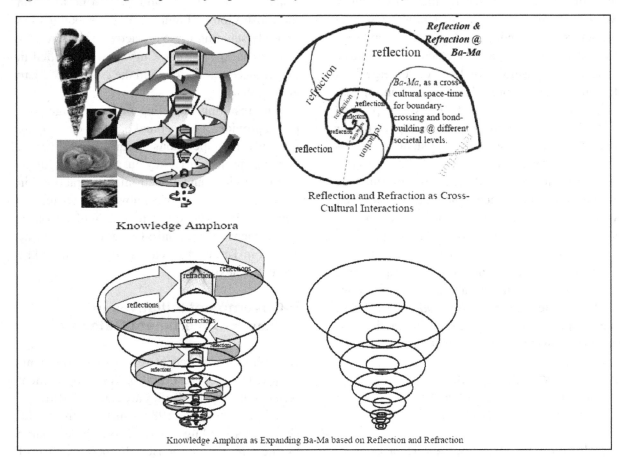

The modules of presentation layer do involve minimum number of business rules so that injecting a different kind of presentation domain into this layer such as wap or mobile java applications reside as an easy task in terms of development. The modules of this layer are mainly the portlet applications for web and jsp/servlet applications for mobile implementations. The aim of ESB of e-Government Gateway is to orchestrate the connections and transactions between e-Government Gateway, and web service publishers or web service clients. These clients are government organizations. The main purpose of this orchestration is to connect necessary web services of the government organizations to another one. (Yildirim et al., 2010, p. 55)

tablishment of an interoperability framework. For interoperable cross-border interactions, ubiquity, "the possibility to send and receive data anytime and anywhere, and thus eliminate any spatiotemporal restriction", is also a key concept, which paves the way toward mobile government services.

As explained in Yildirim et al. (2010), e-Government Gateway is formed of three main layers that are both separated and collaborating: 1. the gateway layer, 2. the backbone-repository layer, 3. and the presentation layer (a combination of view and controller layers consumes the related web services on the gateway layer, called as e-Government Gateway Enterprise Service Bus, ESB).

With this approach, it is aimed that the citizens can interact with a user-friendly Portal interface without witnessing the complexity of the Gateway system behind. Meanwhile, e-Government Gateway functions in the middle to establish communication lines to establish interoperability between different government organizations, using web services.

For transactions and processes concerning web service usage, ESB offers services like those it offers to citizens. At this point, the ESB is also used to connect the repository layer which consists of the database and the e-mail servers of e-Government Gateway to the presentation layer so that the full orchestration of the infra-structure is completed. (Yildirim et al., 2010, p. 55)

Integration and Interoperability Framework for Citizen-Oriented Transformation of E-Government Services in Turkey

In recent years e-government services have brought about a more collaborative mindset, owing to the tremendous opportunities for sharing information and aligning (if not integrating) service offerings across different providers. Then, a core challenge for e-government's enterprise architecture is that a more seamless governance be nurtured through collaborative opportunities between units (i.e., departments and agencies), or more aggressively pursued through a single, central service provider. One centralizing force is the pursuit of greater interoperability across enterprise-wide architectures (important elements of a platform for service delivery) for the public sector as a whole. Increased integration in service delivery based on commonality of infrastructures, data, and business processes, and service innovation achieved by multi-channel service delivery and *smaller and smarter* use of back-end processes and systems to support *bigger and better* front-end

operations encourage more collaborative models of service delivery (UN, 2008).

These models of "connected or networked government" request government agencies to rethink their operations, to move towards a chain-oriented paradigm with respect to structure, culture, knowledge and management, and to look towards technology as a strategic tool and an enabler for public service innovation and productivity growth (UN, 2008). According to the same UN report (2008), following a systemic approach to collection, reuse, and sharing of data and information, networked government is based upon interoperability as the ability of government organizations to share and integrate information by using common standards. Potential common standards, policies, and frameworks should be flexible enough to respond to changing conditions and varying requirements. Networked governance encourages creative and collective societal action to advance the public good, influencing and incorporating the strategic actions of multi-stakeholders regionally and internationally.

Going back to Turkish case; in general in Turkey, for integration and interoperability, database management and online accessibility for the following issues are necessary (Altınok, 2008[7]; Acar & Kumaş, 2008).

1. Real Person Entities
2. Address Information
3. Legal Institutional Entities
4. Movable Tangible Assets
5. Land register and real estate
6. Geographic Information System (GIS)

In Turkey, objectives for the Real Person Entities are mostly accomplished by MERNIS (Central Population Affairs System), Address Information system is also considered to be completed. The following items (#3 - #6), however, can still be considered as work-in-progress in order to achieve integration and interoperability of e-government services in Turkey. As one step forward, recent workshop, conference and seminars highlight integration and interoperability issues at interorganizational, intraorganizational and technical levels.

One of the next steps for integration and interoperability is the development of the legal entity system. Currently there is a searchable online company registration database, and a single application form, from which information is distributed by post/courier. The process provides a one-stop shop for registration with trade registry, tax, labour and insurance authorities.[8] However, there are still unsettled political, administrative and legal issues regarding the improvement and implementation of the project. For instance, for registration of a new company, the responsibility currently belongs to the Union of Chambers and Commodity Exchanges, and Trade Registry Offices of the Chamber of Commerce. Delegation of responsibility over business registration, however, was transferred from the Ministry of Trade and Industry to the Trade Registry Office.[9] Such unsettled issues with respect to authority and responsibility have an inevitable influence on targets and projects such as the development of a Knowledge System for Legal Entities on the way to become an Information Society.

In general, in 2011, 70% of all the e-government services will be ready according to the Information Society Strategy in Turkey (DPT, 2006). In the 2008 progress report by DPT (2008), it is noted that among the 111 actions defined in the strategy document, only 3 are concluded, 51 are work-in-progress, 34 are at the beginning stage, and 23 are yet to start. In the report DPT (2008) highlights that the priorities and objectives of the Strategy still need to be appreciated and owned by all stakeholders, responsible and interested entities in the society. Problems experienced in the implementation of the Strategy are also underlined by the report under the headings of Legislation (and Legal) Issues, Financial Issues, Personnel (and Human Resources) Issues, Issues of Intra- Institutional Coordination, Issues of Inter-Institutional Coordination, and Other

Issues (DPT, 2008). According to our findings based on, overall, problems of Legislation Issues and Issues of Inter-Institutional Coordination are evaluated to have the highest (negative) impact (21%), followed by Personnel Issues (19%), Financial Issues (16%), Issues of Inter-Institutional Coordination (13%) and Other Issues (which also highlight coordination issues) (%10) in the implementation of the Strategy. Similarly, problems of Legislation Issues and Issues of Inter-Institutional Coordination are highlighted the most (19%) by individual actions of the strategy, followed by Personnel and Financial Issues (18%), Issues of Inter-Institutional Coordination (17%) and Other Issues (9%). These analyses, we think, underline the necessity of cross-cultural interaction among various societal entities for integration and interoperability of e-government services in Turkey.

Officially launched with approximately 20 services in December 2008 in Ankara as a major milestone of the specific action plan of the strategy, e-Government Gateway currently (as of June 2010) provides around 180 e-services and new services are under development. Besides these services, e-Government Gateway provides links to hundreds of electronic services provided by individual agencies from their own websites. Currently e-Government Gateway services do not "take over" agency web site services, however it provides a trustworthy, user-friendly choice with a single sign-up mechanism to securely access all services in one place. Identity management using personal password and e-signature are integrated into the system, and limited financial transactions and higher levels of personalization are planned for the near future.

The e-Government Gateway project will also be under continuous development in response to the arising needs of integration and interoperability from citizens, business enterprises and public institutions, which could also be resembled with a spiral model for project management. Accordingly, the project can be divided into phases (Analysis, Design, Implementation, Testing) that build upon the previous one and with a running release of software produced at the end (Ariadne Training, 2001) (Figure 3). In collaboration with front office, the back office can then work on the entire lifecycle, within which testing and learning from experience is embedded. First, the main modules are developed together with a limited number of services available, upon which new modules and services are integrated in time. Meanwhile, Continuous, incremental improvements on available services with respect to user feedback are utilized.

With respect to these continuous improvements resulting from citizens' feedback so far, certain operational examples can be given. For instance, regarding single sign-in (identification and authentication) module, # button has been added to the virtual keyboard, and automatic log-out has been extended, following user feedbacks. Soon, it will also become possible to fill-out forms for contact one-time only, as the necessary information will be automatically provided from the records of the registered users. Moreover, for example it has become possible for users to choose the default location for the weather forecast information, while it used to be Ankara, as the political capital city, by default, at the beginning. Again, in the near future, thanks to citizen feedback and demands, it will also be possible to expand the operations to the citizens living abroad. Fi-

nally, adding necessary parameters to print-outs as confirmations of the information transactions conducted via e-Government Gateway has been done for certain services (Citizens request these hard-copies to be recognized as official document, even if the whole concept of the e-Government Gateway aims to eliminate the need for citizens to provide these hard-copies to public institutions, enabling flow of information virtually among the public institutions themselves).

By completion of the project, for instance, a conceptual integrity called "integration standards" will also be ensured where security standards and data-sharing of public institutions at inter-organizational level is provided under one roof. These integration standards then contribute to the establishment of an interoperability framework. Interoperable systems working in a seamless and coherent way across the public sector will then be a turnkey for providing better, tailored, cheaper services for citizens.

To underline the importance and necessity of such integration and interoperability, the e-Government Gateway project is publicized as a publicly-owned project. Created out of a national competition for logo (and motto) design, the logo (Figure 4) and identity of the project, rather than of the company, is mostly used for publicity

Figure 3. Spiral Model (Adapted from Ariadne Training, 2001, p. 12)

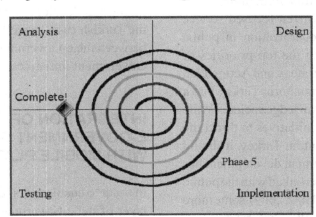

Figure 4. The Logo of e-Government Gateway, Turkey

purposes, such as in ICT fairs, or e-government conferences, among other related events.

Actually, the e-Government Gateway project is under the responsibility of Ministry of Transportation. The convergence of transportation service and technologies with those of information and communication is an interesting point to highlight the analogy between transfer of tangible materials (transportation) and transfer of data, information and knowledge (information and communication). Here, even data can be considered as raw material for creation of knowledge. This convergence of information, communication and transportation technologies also matches well with the understanding of mobility.

In conclusion; "social transformation", "ICT adoption by business", "citizen focused service transformation", and "modernization in public administration" are among the top priorities of the Information Society Strategy and Action Plan (DPT, 2006) that aims to transform Turkey into a developed information/knowledge society. While the Strategy incorporates initiatives to direct the e-government development, in Turkey, it should be reminded that e-government development has not been able to be fully interlinked with the public sector reform (Balci & Kirilmaz, 2009), some more basic problems need to be solved beforehand. Going back to the basics, for instance, according to

Rehan and Koyuncu (2009, p. 385), in short, the main problem of Turkey towards e-government developments is the weakness of ICT usage and ICT infrastructure. Therefore, by all means certain necessary actions to improve ICT usage and further develop the current infrastructure should be realized within a short period of time.

As stated by the e-Government Factsheet of Turkey[10], "as of June 2008, the total number of fixed subscribers is around 18 million with a penetration rate of roughly 25%. The total number of mobile subscribers reached to 63.6 million with a penetration rate of 90%. The number of Internet subscribers reached to approximately 5.4 million, 5.3 million of which are broadband (ADSL) subscribers." Thus, for outreaching citizens throughout the country, use of mobile devices holds a profound place, which is both an opportunity and necessity for e-government operations, underlining the incentive for the development of mobile government initiatives.

Having initially presented a theoretical perspective that can provide a conceptual base for the practical e-government operations as well as guide the transition from e-government to m-government, we have then provided a brief background and current situation of the e-Government Gateway in Turkey, underlying some specific issues about integration and interoperability, citizen-orientation and public transformation. Next we will provide another discussion on the mobile-government extension of the e-Government Gateway, applying the developed theory and philosophy to the Turkish case, as well as establishing the link between the e-Government Gateway and mobile Government initiatives.

INTEGRATION OF E-GOVERNMENT GATEWAY WITH MOBILE PLATFORMS

In order to enable users reaching the presentation layer of the e-Government Gateway via mobile devices in addition to their computers, a mobile application and a wap portal of e-Government Gateway was launched in October 2009.

Mobile Government Application is a Java 2 Platform, Micro Edition (J2ME) application which is available for the mobile devices with MIDP 2.0 and CLDC 1.1 configuration. Once the citizen downloads the application, 10 services which have been chosen among the frequently used ones in web portal, are available for use as well as various information related with and content provided by the public agencies in cooperation.... the citizens can use the provided services by validating their identities with their e-Government Gateway passwords or mobile signatures. Security of the application is provided by the encryption and server-side computations. (Yildirim et al., 2010, p. 56)

Thanks to this Mobile Government Application that uses the web services on the e-Government Gateway ESB in order to provide a service to the user, users do not need to re-download the application when a new patch or version of the application is provided. While the mobile application runs on the mobile phone of the user, it dynamically connects to the related web services to download the necessary, updated information (Yildirim et al., 2010).

Beside this Mobile Government Application, the necessity of a Wap portal arises from the need to reach different user groups, as in Turkey still widely used are primitive types of mobile phones with low-levels of configurations, browsers of which can only support wap pages. Implementation of the wap portal is handled by changing only the presentation level while keeping all the mechanisms and business logic behind unchanged, confirming once again the strength of service-oriented architecture. As a result, although the mobile government extension is very much new, it is promising to have broad accessibility. Still, the mobile extension is its early stages, and there is much space for improvement. For instance, the

most frequently used feature of the wap portal is the search mechanism (32%), which could be due to the difficulty of navigation in both wap and web sites on mobile platforms. (Yildirim et al., 2010) (Figure 5)

One of International Ubiquity and Mobility Projects at Türksat can be given as another example. In more practical terms, development of a Ubiquitous Participation Platform for Policy Making (UbiPOL) project, into which Turkish partners including Türksat are also involved, has just started. UbiPOL project aims to develop a ubiquitous platform that allows citizens be involved in policy making processes (PMPs) regardless of their current locations and time. As a result of this project, a proof of concept and a pilot application will be exhibited that practically citizens will be able to use their mobile devices to take part in Municipality policy making and implementation processes in their neighborhoods (Figure 6).

It is suggested that the more citizens find connections between their as-usual life activities and relevant policies, the more they become pro-active or motivated to be involved in the PMPs. For this reason, UbiPOL aims to provide context aware knowledge provision with regard to policy making. That is citizens using UbiPOL will be able to identify any relevant policies and other citizen's opinion whenever they want wherever they are according to their as-usual life pattern. (UbiPOL, 2009)

The major innovative aspects of the UbiPOL project are the following:

- **Ubiquitous Governance Model:** UbiPOL will enable a new governance model in which policy making processes are shared between policy makers and citizens. Furthermore, citizens will be able to interact with each other to share their opinions based on their contexts (locations). This is

Figure 5. Sample Screenshots from Mobile Government Application (main menu and sign-in with mobile signature) and Wap Portal of e-Government Gateway (services and institutions) as well as Service Usage Statistics in e-Government Gateway Wap Portal (Adapted from Yildirim et al., 2010, pp. 57, 59, 60)

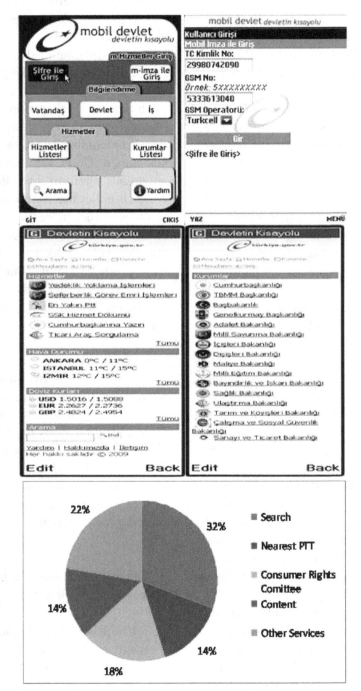

Figure 6. Ubipol infra and supra-structure (Adapted from Irani et al., 2010, pp. 80, 88)

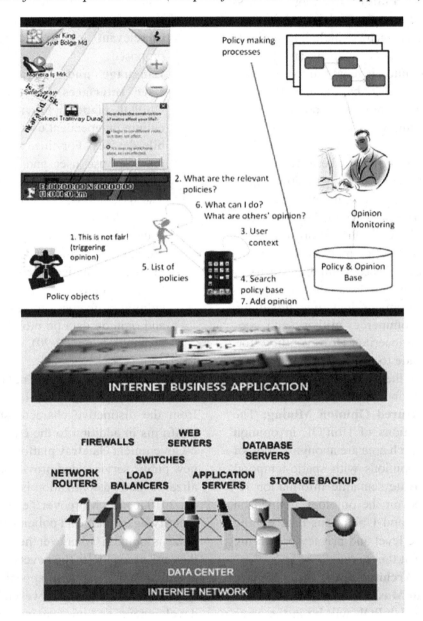

an ideal governance model that incorporate G2C (Government to Citizen), C2G (Citizen to Government) and C2C (Citizen to Citizen) interactions for policy making.

- **Mobile eParticipation Tools:** UbiPOL will increase citizen's motivation to participate to policy making processes by linking policy issues with geographical objects which are confronted by citizens in their daily life. Citizens now can link site objects on the street with relevant political issues and the probability to participate to the issues becomes higher.

- **Executable Policy Making Process Model:** The process model needs to consider various factors including citizens and policy makers' interests and be able to penetrate into citizens' life to overcome possible resistance by citizens to being involved in policy making processes. The policy making workflow model (PMWF model) which is one of the deliverables of UbiPOL project will be executable by computer system according to citizen opinion and feedbacks from policy makers.

- **Location Based Policy-Map Services:** UbiPOL will be the first effort to provide location based policy map services to citizens. Location based computing technology has been widely applied to various application domains including travelling, education, commerce, military and geographical engineering. However, no effort has been made to provide policy map services on handheld devices based on their current location and relevant site objects.

- **Privacy Ensured Opinion Mining:** The main innovations of UbiPOL in opinion mining research area are anonymization of text based opinions with spatio-temporal context; private semantic information retrieval based on the ontology defined on the policies; and Ubiquitous data mining at the device level and privacy preserving data mining at the server.

- **Scalable Architecture for Service Provision to Mass Citizens:** The main innovations of UbiPOL will be in the areas of load balancing, novel communications with very low overhead and latency, evaluating state of the art algorithms and data structures for various application domains, caching techniques and parallel programming models.

- **Security Over Citizen Identity and Opinion:** UbiPOL will propose a security mechanism in delivering policy and opinion data to mobile devices from server side and vice versa. For this, selective encryp-

tion mechanism which considers the citizens role and the level of confidentiality of relevant opinion and policies will be proposed.

- **Language and Culture Independent User Interfaces For Mobile Devices:** UbiPOL platform provides language and culture independent user interfaces on mobile devices. For this, an intelligent user manager use user and device profile to render the user interface mobile devices. The user and device profiles will be given for developing application in different countries during the deployment time and the intelligent user manager will interpret them to dynamically compose a user interface according to given language and cultural preference on the display of policy and opinion data on mobile devices (Irani et al., 2010, pp. 89-90).

In summary, e-Government Gateway project aims to develop mobile services "which benefit from the distinctive characteristics of mobile platforms in addition to the enhancement of all e-Government Gateway platforms by integrating new public services and providing more personalization and customization issues" (Yildirim et al., 2010, p. 60). Moreover, "enabling citizens in identifying any relevant policies along with other citizens' opinion 'whenever they want' 'wherever they are' according to their everyday life pattern", is anticipated to be more appealing for citizens to get involved with the newest relevant policies developments and implementations (Irani et al., 2010, p. 78). As a result, e-government operations will outreach more citizens for their demand and use.

CONCLUDING DISCUSSION

E-Government Gateway in Turkey aims to provide 75 million citizens and business entities with a single point of access, one-stop-shop, to e-government services. The implementation process of e-Government Gateway is based on service transformation. The main principle of this service transformation is the consideration of user satisfaction issues when re-designing the business processes of public institutions and services delivered to citizens and enterprises. The priority in services transformation is not merely to transfer available business processes to electronic channels without making any improvements; on the contrary, the aim is to deliver these electronic government services, as processes which are re-designed according to user needs for integration or simplification when necessary, in an effective, uninterrupted, fast, transparent, reliable way. Currently mobile government initiatives complement and extend such initiated e-government operations in the Turkey case.

The theoretical perspective presented in this paper that underline concepts such as *ba, ma, reflection, refraction and knowledge amphora* can provide a conceptual base for the practical e-government operations as well as guide a transition from e-government to m-government. Reflections at the physical, virtual or mental place of *ba* underline intra-cultural interactions within a societal entity that contributes to the development of relations, understanding of each other, and comprehension of existing knowledge. Meanwhile, refractions at the time-space in-between-ness of *ma* points out the inter-cultural interactions among different societal entities that contributes to the creation of new knowledge, accompanied by mobility and transfer, which can incorporate misunderstandings, conflicts and confusions, as well.

These conceptual and philosophical perceptions apply and explain well certain phenomena in e-government operations. First of all, we should realize the use of Internet or any other ICT as for not only transferring data and information but also creating new knowledge. Moreover, the accompanying concepts of mobility and ubiquity request a readjustment of our practical relations with time and space, especially for exchange of existing and creation of new knowledge in-

dividually and collectively. Then, relations and interactions across societal entities at different levels need to address issues such as interoperability and integration carefully. Especially for the success of e-government initiatives, fulfilling trust-based interactions among different institutional stakeholders, or improving relations with citizens and other users is crucial. The continuous improvement of e-government services with respect to various stakeholder and user feedback can eventually lead to public transformation. User participation is considered to be at high levels, when ubiquitous devises are used for their involvement in local policy-making initiatives that will have direct impact on their daily life cycles. Meanwhile, continuous development and sustainable operations of mobile systems that can function at low-tech mobile devices is crucial for outreaching various citizen groups, thus diminish digital divide. As a result, the suggested conceptual and philosophical framework can lead to a transition towards ubiquitous government, starting first as a mobile government extension to the existing e-government operations and then continue to expand mobile and ubiquitous initiatives until reaching to a certain sophistication level to become operational on its own - with respect to a spiral model of project management and development of e-government operations under the umbrella of converging information, communication (and transportation) technologies.

Discussing the operations in practice and developing our perspective in theory, we also underline the inter-link between the theory and practice with respect to the present situation and future prospects in e-government. At the end, the Figure 7 illustrates this inter-link between the ubiquitous knowledge amphora and e-Government Gateway Project in Turkey. Rather than being specific to Turkish case, however, we hope this work leads to further works, giving inspirations and insights to other researchers that can be useful for electronic and mobile government services in general, or for certain specific cases in other parts of the world. Meanwhile, this inter-link between

Figure 7. Concept of Knowledge Amphora (@), and E-Government Gateway Project @ turkiye.gov.tr

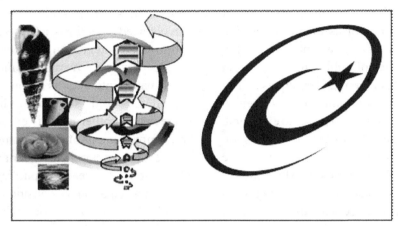

theory and practice of e-government and mobile government needs to be supported and verified by further practical and academic work.

SUGGESTIONS FOR FURTHER RESEARCH

The resulting work is an initial attempt to propose a not only theoretical but also philosophical framework that is applicable to the development of electronic and mobile government operations – specific (if not limited) to Turkish experience. As a future outlook, contributing to the conceptualizations of knowledge-creating spiral, Knowledge Amphora can be a useful framework for explaining the emerging ubiquitous interactions of cross-cultural information transfer and knowledge creation that informally happen or formally occur via fixed and mobile devices @ the Internet, or any other networked and virtual organizations in today's global knowledge society. Moreover, understanding amphora as a mobile, ubiquitous device for the carriage of not only material but also information and knowledge can also contribute to our understanding and development of mobile services and products for the increasingly-sophisticating knowledge society and economy. Even if it is still an initial (and

limited) attempt and open to further development, knowledge amphora can then provide not only a conceptual base for the e-government operations but also guide the transition from e-government to m-government. Similarly, knowledge amphora can help us improve the e-Government Gateway project in Turkey. Knowledge Amphora as reflections and refractions @ expanding *ba* and *ma* (space-time) can be useful for modeling mobility and ubiquity, underlying cross-institutional, citizen-satisfaction-and public-transformation-oriented-interactions as important parts of the e-Government Gateway Project of Turkey.

ACKNOWLEDGMENT

This work is partially supported by the Ubiquitous Participation Platform for POLicy Makings (UbiPOL), EU FP7-ICT-2009-4 STREP project. We also would like to thank Türksat, Turkey for supporting this work. As not only our part of fulfilling the responsibility to contribute to future work, but as part of our affiliations with these above-mentioned projects and institutions, as well, we also try to disseminate the findings of this academic work by presenting certain parts of this paper in various platforms.

REFERENCES

Acar, M., & Kumaş, E. (2008). Türkiye'nin Dönüşüm Sürecinde Anahtar bir Mekanizma Olarak E-Devlet, E-Dönüşüm ve Entegrasyon Standartları. In *İAS2008*. Turkey: TÜİK.

Aktaş, Z. (1987). *Structured Analysis and Design of Information Systems*. Upper Saddle River, NJ: Prentice-Hall.

Amar, A. D. (2002). *Managing Knowledge Workers. Unleashing Innovation and Productivity*. London: Quorum Books.

Ariadne Training Limited. (2001). *Engineering Software - Applied Object Oriented Analysis And Design Using The Uml*. Ariadne Training Limited.

Avdagić, M., Šabić, Z., Zaimović, T., & Nazečić, N. (2008, September). eGovernment and mGovernment Integration: Role in Public Administration Reform. In *Proceedings of the Third International Conference & Exhibitions on Mobile Government (mLife 2008), Mobile Government Consortium International (mGCI) UK*, Turkey.

Axelsson, K., Melin, U., & Persson, A. (2007). Communication Analysis of Public Forms - Discovering Multi-Functional Purposes in Citizen and Government Communication. *International Journal of Public Information Systems*, *3*, 161–181.

Balcı, A., & Kirilmaz, H. (2009). Kamu Yönetiminde Yeniden Yapilanma Kapsaminda E-Devlet Uygulamalari. [Reorganization of Public Administration and e-Government Applications]. *Türk İdare Dergisi*, *81*, 463–464.

Berque, A. (1982). *Vivre L'espace au Japon* (Miyahara, M., Trans.). Tokyo, Japan: Presses Universitaires de France.

Brusa, G., Caliusco, M. L., & Chiotti, O. (2007). Enabling Knowledge Sharing within e-Government Back-Office Through Ontological Engineering. *Journal of Theoretical and Applied Electronic Commerce Research*, *2*(1), 33–48.

Capgemini. (2007). *The User Challenge Benchmarking: The Supply of Online Public Services, 7th Measurement*. Retrieved July 28, 2008, from http://ec.europa.eu/information_society/eeurope/i2010/docs/benchmarking/egov_benchmark_2007.pdf

Cooper, R. (1998). Assemblage Notes. In Chia, R. C. H. (Ed.), *Organization Worlds: Exploring in Technology and Organization with Robert Cooper* (pp. 108–129). London: Routledge.

Cooper, V., Lichtenstein, S., & Smith, R. (2009). Successful Web-Based IT Support Services: Service Provider Perceptions of Stakeholder-Oriented Challenges. *International Journal of E-Services and Mobile Applications*, *1*(1), 1–20. doi:10.4018/jesma.2009092201

de Vos, H., Haaker, T., Teerling, M., & Kleijnen, M. (2008). Consumer Value of Context Aware and Location Based Mobile Services. *International Journal of E-Services and Mobile Applications*, 36–50.

Derballa, V., & Pousttchi, K. (2006). Mobile Knowledge Management. In Schwartz, D. G. (Ed.), *Encyclopedia of Knowledge Management*. Hershey, PA: IGI Global.

DPT. (2008). *Bilgi Toplumu Stratejisi ve Eylem Planı 1. Değerlendirme Raporu*. Devlet: Planlanma Teşkilatı.

Drucker, P. F., & Maciariello, J. A. (2004). *The Daily Drucker*.

Ekong, U. O., & Ekong, V. E. (2010). M-Voting: A Panacea for Enhanced E-Participation. *Asian Journal of Information Technology*, *9*(2), 111–116. doi:10.3923/ajit.2010.111.116

Elias, N. (1992). *Time: An Essay*. Oxford, UK: Blackwell Publishers.

Fraser, J., Adams, N., Macintosh, A., McKay-Hubbard, A., Lobo, T. P., & Pardo, P. F. (2003). Knowledge Management Applied to e-Government Services: the Use of an Ontology. In *Knowledge Management in Electronic Government*. Berlin: Springer. doi:10.1007/3-540-44836-5_13

Hayashi, T. (2004). Captured Nature and Japanese Way of Tolerance. *MAJA Estonian Architectural Review*. Retrieved from http://www.solness.ee/majaeng/index.php?gid=44&id=453

Henten. (2010). Services, E-Services, and Non-services. *Electronic Services: Concepts, Methodologies, Tools and Applications* (Vol. 3, pp. 1-9). Hershey, PA: IGI Global.

Irani, Z., Lee, H., Weerakkody, V., Kamal, M., Topham, S., & Simpson, G. (2010). Ubiquitous Participation Platform for POLicy Makings (Ubi-POL): A Research Note. *International Journal of Electronic Government Research*, *6*(1), 78–106. doi:10.4018/jegr.2010102006

Kerkhove, D. D. (2003). *NextD Journal: Re-Rethinking Design Issue Two, Conversation 2.3*. Retrieved from http://www.nextd.org/02/pdf_download/NextD_2_3.pdf

Kim, E., & Choi, J. (2006). An Ontology-Based Context Model in a Smart Home. In M. Gavrilova et al. (Eds.), *ICCSA 2006* (LNCS 3983, pp. 11-20). Berlin: Springer Verlag.

Kuscu, M. H., Kushchu, İ., & Yu, B. (2007). Introducing Mobile Government. In Kuschu, I. (Ed.), *Mobile Government: An Emerging Direction in e-Government* (pp. 1–11). Hershey, PA: IGI Global. doi:10.4018/978-1-59140-884-0.ch001

Layne, K., & Lee, J. (2001). Developing fully functional E-government: A four stage model. *Government Information Quarterly*, *18*(2), 122. doi:10.1016/S0740-624X(01)00066-1

Massey, D. (1994). *Space, Place, and Gender*. Open University.

Medeni, T., Elwell, M., & Cook, S. (2007). "Digitally Deaf" into Games for Learning: Towards a Theory of Reflective and Refractive Space-Time for Knowledge Management. In *BEYKON 2007*. Turkey: Immersing.

Medeni, T., Iwatsuki, S., & Cook, S. (2008). Reflective *Ba* and Refractive *Ma* in Cross-Cultural Learning. In Putnik, G. D., & Cunha, M. M. (Eds.), *Encyclopedia of Networked and Virtual Organizations*. Hershey, PA: IGI Global. doi:10.4018/978-1-59904-885-7.ch178

Medeni, T., Medeni, İ. T., Balci, A., & Dalbay, Ö. (2009). *Suggesting a Framework for Transition towards more Interoperable e-Government in Turkey: A Nautilus Model of Cross-Cultural Knowledge Creation and Organizational Learning*. Ankara, Turkey: ICEGOV.

Medeni, T., Tutkun, C., Medeni, İ. T., Kumas, E., & Balci, A. (2008). Proposing a Modeling of Ubiquitous Knowledge Amphora for the Transition from e-Government to m-Government in Turkey. In *Proceedings of the E-Government Gateway Project mLife Events 2008,* Turkey.

Medeni, T., & Umemoto, K. (2008). An Action Research into International Masters Program in Practicing Management (IMPM): Suggesting Refraction to Complement Reflection for Management Learning in the Global Knowledge Economy. *Eurasian Journal of Business and Economics*, *1*(1), 99–136.

Medeni, T., & Umemoto, K. (2010). *Educating Managers for the Global Knowledge Economy*. VDM Publications.

Miller, C., Mukerji, J., Burt, C., Dsouza, D., Duddy, K., & El Kaim, W. (2001). *Model Driven Architecture (MDA) (Document No. ormsc/2001-07-01). Architecture Board ORMSC* (Miller, C., & Mukerji, J., Eds.). OMG.

Nonaka, I., & Takeuchi, H. (1995). *The Knowledge-creating Company: How Japanese Companies Create the Dynamics of Innovation*. Oxford, UK: Oxford University Press. doi:10.1016/0024-6301(96)81509-3

Nonaka, I., Toyama, R., & Scharmer, O. (2001). *Building Ba to Enhance Knowledge Creation and Innovation at Large Firms*. Retrieved from http://www.dialogonleadership.org/Nonaka_et_al.html

OECD. (2006). *OECD e-Government Studies*. Turkey: OECD.

Rehan, M., & Koyuncu, M. (2009). *Towards e-government: A survey of Turkey's progress*. Ankara, Turkey: ICEGOV.

Reid, V., Bardzki, B., & McNamee, S. (2004). Communication and Culture: Designing a Knowledge-enabled Environment to Effect Local Government Reform. *Electronic. Journal of E-Government, 2*(3), 197–206.

Rossel, P., Finger, M., & Misuraca, G. (2006). Mobile e-Government Options: Between Technology-driven and User-centric. *Electronic. Journal of E-Government, 4*(2), 79–86.

Salhofer, P., Stadlhofer, B., & Tretter, G. (2009). Ontology Driven e-Government. *Electronic. Journal of E-Government, 7*(4), 415–424.

Tihon, A. (2006). The Informational Attractors: A Different Approach of Information and Knowledge Management.

UbiPOL. (2009). *Project Document 2009*. Retrieved from http://www.ideal-ist.net/Countries/UK/PS-UK-3051

Umemoto, K. (2004). Practicing and Researching Knowledge Management at JAIST. In *Proceedings of the Technology Creation Based on Knowledge Science: Theory and Practice JAIST Forum*, Japan.

United Nations. (2008). *E-Government Survey 2008 - From E-Government to Connected Governance*. Geneva, Switzerland: UN.

Uno, Y. (1999). Why the Concept of Trans-Cultural Refraction Necessary. *NEWSLETTER: Intercultural Communication,* (35).

Wankel, C., & DeFillippi, R. (2006). *New Visions of Graduate Management Education*. CT: Information Age Publishing.

West, D. M. (2007). *Global E-Government*. Providence, RI: Center for Public Policy, Brown University.

Whitehead, A. N. (1967). *Science and the Modern World*. New York: The Press.

Yener, E. (2008). Adoption of Mobile Technologies by the Turkish Public Sector. Plenary Talk. In *Proceedings of the Third International Conference & Exhibitions on Mobile Government (mLife 2008)*, Antalya, Turkey.

Yildirim, G., Medeni, T., Aktaş, M., Kutluoğlu, U., & Kahramaner, Y. (2010). *M-Government as an Extension of E-Government Gateway: A Case Study*. Antalya, Turkey: ICEGEG.

ENDNOTES

[1]   In Turkish, the same word, "Bilgi", is used for information and knowledge interchangeably.

[2]   http://www.accessegov.org/acegov/uploadedFiles/webfiles/cffile_3_24_06_3_08_35_PM.pdf last access 20.06.2010

[3]   Rather than particularly distinguishing information and knowledge, Turkish language and culture signify knowing, knowing-yourself, knowing-each-other, and knowing-together ("bilişmek"). Sophism, tasavvuf, tefekkür (reflection) are among the issues that could be investigated elsewhere for further information.

[4]   This subsection is written based on authors previous works elsewhere such as Medeni et al. 2008, Medeni et al 2009, Medeni & Umemoto 2010.....

[5]   The incident ray refracts moving from one environment to a different environment,

for which an analogy can be drawn with the knowledge that flows from one societal entity to another societal entity that ranges from an individuals' thinking-out-of-the-box to two institutions' inter-cultural exchange.

6   As a not only personal but organizational and social process, reflection is a significant concept for creating knowledge. It plays an important role not only for the epistemological product-ion of sophia, but also the process of sophistication. Accordingly, we conceptualize that reflection and refraction are interlinked as social and natural phenomena. This meta-conceptualization also highlights the spiritual importance of such phenomenon in faith systems such as Christianity, Sophism, Taoism or Zen Buddhism.

7   Prime Ministry, Meeting in June 2008.

8   (eupractice website factsheet, http://www.epractice.eu/, accessed on 10 June 2010)

9   (eupractice website, http://www.epractice.eu/, accessed on 10 June 2010)

10   (http://www.epractice.eu/en/document/288413, accessed on 10 June 2010)

This work was previously published in the International Journal of E-Services and Mobile Applications, Volume 3, Issue 1, edited by Ada Scupola, pp. 17-38, copyright 2011 by IGI Publishing (an imprint of IGI Global).

Chapter 11
Technical Audit of an Electronic Polling Station:
A Case Study

Hector Alaiz-Moreton
Universidad de Leon, Spain

Luis Panizo-Alonso
Universidad de Leon, Spain

Ramón A. Fernandez-Diaz
Universidad de Leon, Spain

Javier Alfonso-Cendon
Universidad de Leon, Spain

ABSTRACT

This paper shows the lack of standard procedures to audit e-voting systems and also describes a practical process of auditing an e-voting experience based on a Direct-recording Electronic system (D.R.E). This system has been tested in a real situation, in the city council of Coahuila, Mexico, in November 2008. During the auditing, several things were kept in mind, in particular those critical in complex contexts, as democratic election processes are. The auditing process is divided into three main complementary stages: analysis of voting protocol, analysis of polling station hardware elements, and analysis of the software involved. Each stage contains several items which have to be analyzed at low level with the aim to detect and resolve possible security problems.

Elections are the most important processes in a democratic country; as citizens must accept not only the results but also whole process. Hence, only if both, security and the transparency are guaranteed, e-voting systems will be acceptable tools in the elections processes.

There are two types of electronic voting systems (Puiggali, 2007) remote e-voting.and face e-voting. Main difference between these kinds of methods is the physical situation of the elector. In the first one the user uses TCP/IP support to vote. In the second one, the user goes to his electoral

DOI: 10.4018/978-1-4666-2654-6.ch011

district to vote and make his election, aided for a digital voting system (Ruth & Mercer, 2007). This system is called polling station based on D.R.E (Direct-recording Electronic). D.R.E. station saves the vote of the users. When the Election Day finishes, results can be obtained soon with little effort.

The three common pillars of any election process are: confidentiality, integrity and availability (Morales, 20099). E-voting systems must, therefore, fulfill these fundamental pillars as any other traditional voting system. These three pillars are implicit the main requisites of a D.R.E. system (Fujioka, Okamoto, & Ohta, 1992; Indrajit, Indrakshi, & Natarajan, 2001):

1. There is a secure authentication method to access the system.
2. Only the people authorized to vote can do it.
3. The anonymity of the voter and the privacy of its vote are guaranteed and preventing from any kind of coerciveness.
4. There exists some protocol that allows the authorities to test, verify and certify the system.
5. The system must be useful, easy to use and accessible to people with disabilities
6. Intruders must be detected and potential attacks prevented.
7. The system must be robust, fault tolerant and available during the whole Election Day.
8. The whole process must be auditable.

It might appear that auditing a simple machine for counting votes would be an easy task, but nothing could be further from the truth. A technical audit of an electronic polling system, does not only need to check every single component but also must inspire confidence in potential users (Helbach & Schwenk, 2007). It should be an open process, clear and based on fixed standards. In practice this is dependent upon the laws and

electoral authorities in the various countries and states involved. This leads to extensive and complicated debates. Some scholars even talk of the impossibility of performing a successful audit of an electoral procedure based on technology or propose a number of methods for indirect verification (Schoenmakers, 2000).

This paper describes a real audit of a real electronic voting machine, Figure 1. It is the system of the State of Coahuila in Mexico, which has been used on a whole range of occasions since 2004 and recently in binding electoral processes.

The first difficulty was that of establishing the list of items to carry out the requirements described. In fact, there is no clear and comprehensive documentation on which to rely. In the section "Relation between secure items and requirements of the audit process" it can see like this list of items has impact on the eight requirements defined.

Every country, state and even electoral district has its own opinions on the topic. There are few published standards to act as a basis, and even when they exist, it is not always clear how to use them (Barrat, 2008). In the case of Europe there are certain recommendations made by the Council of Minister of the Council of Europe, designated Rec (Electronic Frontier Finland, 2008), but certain E.U. countries quite definitely do not observe them (Cohen, 2005). In the U.S.A. there are some developed proposal for standardization, thanks to the larger number of experiments, even successful, that have taken place there (Fischer & Coleman, 2006).

It was a big mistake on the part of the system and program developers and researchers to assume that the level of security for electronic voting would be similar to what is needed by financial institutions (Cox & Rubin, 2004). For the latter, the confidential operation may be made known to authorized third parties, while, in contrast, in electronic voting anonymity is an essential part of the process. Hence, nobody can be permitted to obtain information about how anybody votes

Figure 1. Polling station (© Observatorio Voto Electrónico)

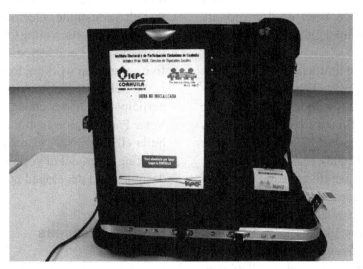

other than in the final count, and then only for the purpose of totalling the votes cast. These kinds of drawback demand the use of a range of techniques for indirect checking of votes, such as V.V.A.T., or voter-verified audit trial (Mercury, 2007), which complicates and slows down the use of machines involved in electronic voting. Despite everything, there are obvious limitations on the security of this equipment unless appropriate measures are taken from the very start of the process of designing it (Armen & Morelli, 2005; Kohno, Stubblefield, Wallacj, & Rubin, 2004).

Another negative aspect is the apparent ease with which it would be possible to perpetrate fraud with D.R.E. polling system. In the present case, there are rigorous studies (Di Franco, Petro, Shear, & Vladimirov, 2004), that demonstrate that minor tinkering with the master copy of the voting software would be enough to allow electoral fraud on a grand scale. Moreover, a considerable part of the electronic voting systems market is in the hands of private companies that are reticent about opening up their systems to audit by third parties (McGaley & McCarthy, 2004). All of this means that extreme precautions must be taken during the

audit process, which must leave no aspect of the security of the device and its software unchecked. There are other sorts of audit to cover features of the accessibility and usability of electronic voting equipment (Falcão, Cunha, Leitão, Faria, Pimenta, & Carravilla, 2006).

Other examples of e-voting experiences, which have been cancelled due to the fact that the audit process had not been well defined, are Netherlands, and Ireland, where the Independent Commission on Electronic Voting (I.C.T.E.) decided not to use the D.R.E due to lack of one organization and one standard dedicated to ensured the correct work of the electronics polling systems (Commission on Electronic Voting, 2010). On the other hand, Brazil, where the vote of the citizens is compulsory, the D.R.E. system based on biometric security primitives, are being very accepted thanks to voting specifications defined by Brazil's Superior Electoral Tribunal (TSE).

In countries such as Spain (Gutiérrez-Rubí, 2009) or United Kingdom (Open Rights Group, 2007), several experimental experiences have been done with not binding results; due to laws of country do not permit e-voting systems in official government elections. However some binding e-voting processes have been used in private companies such as football clubs or banks (Real Madrid, 2009).

There are many manufactures of D.R.E. systems such as INDRA (Electoral Advisory Systems for Citizens, 2004; Indra, 2009), Bharat Electronics Limited, (Bharat Electronics, 2009), Hart InterCivic (Ladendorf, 2008), Microvote General Corporation (2008), to mention only the most important. These manufactures develop polling station with custom software highly secure but not easy to audit in a standard process.

Other manufactures like Premier Elections Solutions have suffered several audit processes to analyze and study their security features (Feldman, Halderman, & Felten, 2007; Wertheimer, 2004; Lamone, 2003; Kohno, Stubblefield, Wallacj, & Rubin, 2004). Thanks to these audit processes each manufacture includes new features in its systems to insure the security and make easier the own audit process, however a critical question appears. Which is the standard audit process procedure? And more important, which is the standard where the security elements to be evaluated are collected?

In this context, the decision of using former experience was taken in order to draw up a document listing the basic requirements that had to be checked in the I.E.P.C.C.'s voting machine prior to undertaking the actual audit. This paper does not present results, since they are copyrighted by the I.E.P.C.C. and, therefore, subjected to confidentiality clauses, at least until they are officially published. For this reason, only generic security aspects of the audit are detailed, whilst the conclusions concerning the security of the audited system are avoided.

- **Case Study:** Electoral and Voter Participation Institute of the State of Coahuila

The Organizer of the Elections is the "Instituto Electoral y de Participación Ciudadana de Coa-

huila" (IEPC, 2008), an independent organization that also constitutes electoral authority.

The Electronic Voting Observation Unit of the University of Leon in Spain (OVE, 2005) has for five years been working from an academic angle in the field of technology-based voting and voter participation. It was chosen by the Electoral and Voter Participation Institute of the State of Coahuila (IEPC, 2008) to resolve the problems of auditing its model of voting machine, which was designed and developed by the Institute itself.

Election Process

For each electronic polling station, an electoral table supervises the voting process, including both, the casting of the ballots and the counting of the votes. Citizens who live in the constituency where the polling station is installed compose the electoral table. That means that a President, a Technical Secretary, a person in charge of counting, and a Substitute to replace any of the other three in case of absence.

The people of the Electoral Table have to do several tasks such as: open and close the Electoral Table, install, open and close the polling station, use the control codes, receive the votes, close the voting, close the polling station, count the votes, send the results and finally publishing the results at the Electoral College. All steeps have perfectly defined by the Electoral and Voter Participation Institute of the State of Coahuila.

The polling station of our case study is based on a printed code to operate the e-Polling Station. Two kinds of codes are available to operate the e-polling station: control codes and access election codes. In our station, these codes consist of printed barcode cards that slide through a reader. These codes allow the handling of the station, and must be used only by the president of the electoral table:

- **Verification Code:** To check the right function of the station prior to opening it.

- **Open Code:** Used just once each Election Day to start the voting.
- **Close Code:** To be used only once in order to finish the voting.
- **Restore Code:** In case of power failure, the station must be restarted and data recovered by means of this code.
- **Reprint Code:** To get a second copy of the printed ballot in case of malfunction during its printing.

Electoral Day: Opening

The Electoral Day, under the supervision of both, representatives of the political parties and electoral observers, the people of the electoral table will handle the polling station and guide all the process of casting the votes.

First, they will set the e-polling station up, checking also that the ballot box is empty. After that, they will count and register the number of available access codes so that they can certify, at the end of the day that the results from the e-polling station are valid. The president will install the station and the screen in order to guarantee the secrecy of the votes.

After installing the station, the president will open the station and check its function by sliding its verification code card. Once this task is finished, the Open Code will be used to open the station. The result of this operation is a printed report that must show that the database is empty and, therefore, no votes at all. The number of access votes is also printed, and must match the number of cards received. Finally, date and hour of opening are also printed. This printed report must be attached to the minutes taken during the whole Election Day.

In case of failure of the electronic polling station, the ballots will be cast with a traditional ballot box, being this fact written in the minutes (and also in the incident registration form). After this previous work, the president will open the Electoral College.

Changing the Station's Address

Each electronic Polling Station has to be placed in the address the Electoral Authority decides. It can only be changed due to the following reasons: The selected place does not exist; it is closed; it is a forbidden place (a factory, a church, a bar, etc.); it does not allow the privacy, secrecy of the vote; there are no easy ways to access it; in case of force majeure. The new location must be placed in the same constituency and as close as possible to the original one, being its address published in the outside of that original location.

Reception of the Votes

Each voter will show his/her credential to the president, the technical secretary will check that his/her name appears in the Electoral Register and, eventually, the president will give the voter its access code. The voter will go to the station, touch the display and follow the audiovisual instructions that will appear. After a vote is cast, a ballot is printed in order the voter to check it and cast it in a traditional box.

If a ballot is not properly printed, the president will use de reprint code to allow the voter to get a new copy of the ballot. The reprint code will also be used after changing the printer paper roll, in case it would be necessary.

The Opinion of the Users

During the design of the audit process two simulated voting experiences were carried out to know the opinion of the users about electronic voting systems.

The first experience was based on a fictitious election process. A set of users used the polling station to select their favourite's icons to represent an unreal company. After that, users filled up a questionnaire, extracting the following conclusions (score from 1 to 5):

- **Information About Operation System:** 3,884
- **Confidence in the Voting System:** 3,186
- **Security of Election:** 3,186
- **Simplicity in the Voting Procedure:** 3,163
- **Speed of the Voting Process:** 4,8

Analyzing results it can be say that DRE system is a good way to develop an election process, thank to its speed and the generous package of information showed along election process. However features such as usability and the perception of safety should be improved.

Other important fact is that the 64% of the respondents preferred the e-voting than traditional procedure, according with this quantitative data, is necessary to ask itself, how to convince the other 36 percent. The answer is: "improving perception of safety".

The final score of the election process developed with an electronic polling was 8.3 over 10 a good calcification but have to be improved the perfection of safety feature if a country want to use e-voting system in an official election process.

The second experience was developed in TEC-NOMEDIA (2009), a technologic fair. Thanks to special setup of the polling station, users filled up a questionnaire using the polling station to select each option of the questionnaires with the following result:

- 43% of the users would use this system if previously a security audit process is made.
- 27% of the users would use this system in any case.
- 14% of the users would use this system to no official elections; this is in not binding democratic processes.
- 10% of the users never use this system.
- 6% of users do not have an explicit opinion.

One more time the perception of safety is a big handicap due to a high percent of users prefers a DRE system audited due to they see a polling station like a black box.

Scenario of the Testing Experiences

First experience was developed in the University of Leon. The number of participants was 172, divided in different groups of students: fourth course of biology science, second course of mining engineering and a course of a special program of the University of Leon oriented to people with more 65 years old who want to study and to acquire generic knowledge in new technologies.

Second experience was developed in a regional fair called TECNOMEDIA 09, which main goal is the divulgation of new technologies. The number of participants in the experience was 210. The users profile was very heterogeneous, from young students of primary schools, to computer science professionals and people without knowledge in informatics.

DESIGN OF THE AUDIT PROCEDURE

The audit procedure presented in this document was divided into a series of stages. This was done in order to simplify the tasks that had to be performed and thus set up, an optimally efficient auditing method that would guarantee security and viability of the electronic voting system under real-life conditions.

It is worthwhile emphasizing that the group of assessable components that will be outlined are those which for one reason or another proved to have some weak point during the experiences of electronic voting that are covered by this paper. The analysis of these components can be extrapolated to experimental uses of other different electronic voting systems. It will serve as a complement to future auditing processes in such as way as to define a generic procedure applicable to any system.

The division of the auditing procedure into three groups of features for analysis was intended to facilitate the task of listing and assessing all the components involved in an electronic voting system. These three groups of features may be summarized as comprising study of the following:

- The protocol for voting by means of electronic polling systems (electronic voting machines or ballot boxes).
- The hardware components of the electronic voting machine.
- The software components involved in the voting process.

The reason that the first set of features to be studied are those included in the voting protocol is the fact that certain weak points in this might in some instances be remedied. This would be by means of changes and improvements to the hardware and software components, which would logically follow proposals for solutions relating to the voting protocol. Figure 2 shows the auditor process with its three stages associated and how solutions proposed after the first stage (analysis of the voting process) may involve modifying hardware and software elements).

As occurs on the ISO standard safety, items discussed in the audit process designed will be described at low level and although these items might seem trivial, is necessary to revise all of them to guarantee the polling station security.

Analysis of the Voting Protocol

This section reviews a number of features involved in the voting procedure, some external, some internal. An attempt has been made to pre-empt the problems of security and efficiency which are critical in a context as complex as that of electronic voting.

Hence, an assessment was made of a range of aspects involving logistics, environment and protocols that will define the quality and security of the process of voting. The various elements intrinsic to the voting process are considered in depth below. A description of what is evaluated

Figure 2. Scheme of the audit procedure (© Observatorio Voto Electrónico)

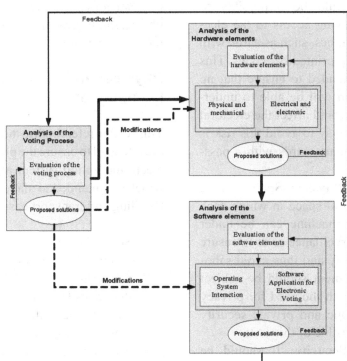

and the requirements that should be met is given in each case.

- **User Operations:** The officials assigned to a given polling station should not carry out any action individually. This would ensure that all actions involving validation, resetting, closing or opening will not be undertaken by only a single user. To avoid this risk, users with personal secret PINs (Personal Identification Number) must be added.

- **Procedure for Initializing and Powering Down Machines.** In no circumstances should it be possible during the polling process to turn an electronic voting machine on or off, other than in such a way that all the registers or meters recording the results of voting are re-initialized to zero. Software of the polling station must be supported this types of operations whether controlled or not.

- **Administrator Access to Voting Machines:** It should be impossible to duplicate items giving access to the electronic voting machines, such as cards with magnetic stripes or printed bar codes. This requirement is intended to ensure that no undesirable person will be able to undertake actions that would affect the electoral process. The best way to avoid this risk is to use personal smarts cards with a password associated.

- **Security of the Physical Space:** There should be a physical space in which each electronic voting machine can be under constant visual surveillance so as to ensure its physical integrity and thus avoid sabotage. This requirement should be subject to compatibility with the need for privacy when voting. Public and private security officers are responsible of this task.

- **Environmental Security:** It should be made certain that the positioning of voting machines is such that they will not be exposed to external risks such as those arising from weather phenomena. To minimize this risk, the vote process should be developed in institutional buildings.

- **Maintenance and Support:** The voting machines should have a capacity to remain operational during the entire period of voting, from the start of polling until the downloading of data to allow compilation of the results. A technical expert has be stay in the Electoral College to solve any problem.

- **Availability of Languages:** The voting machines should be able to handle a range of languages or dialects. This is a software developed task.

Analysis of the Hardware Components of Machines

In order to cover all the components in the electronic voting system, these were grouped into two categories. On the one hand, there were the physical and mechanical components, on the other the electrical and electronic components.

Physical and Mechanical Components

This section refers both to those components which are intended to ensure the physical integrity of the electronic voting system and to those elements which are intended to make its use, accessibility, handling and transport easier.

- **Seals:** There is a need for a system of seals that would reveal any potential undesirable tampering. This should be able to cover all the stages from transporting the voting machine to the polling station up to the point when results are reviewed and validated. This item must be studied in the physical design stage of the polling station by the manufacture.

- **Locks:** A set of high-quality locks should be fitted to those parts of a voting machine which need to be accessible for carrying out repair or maintenance operations on the polling day. Manufactured is the responsible of the quality of these items.

- **Ventilation Systems and Operating Temperature:** The temperature inside the machine where its central processing unit is sited must remain in the acceptable range for the correct operation of the system. This problem can be solved by a low power fan.

- **Physical Accessibility:** The system must have appropriate arrangements to allow adjustments to the position of the voting machine. This is in order that electors with any sort of handicap can still get access to it so as to cast their votes.

- **Mobility:** The system should be easily transportable to polling stations. For this purpose, various factors should be taken into account, such as the system's total weight, its ruggedness of construction and the ease with which it can be handled when being moved.

- **Confidentiality:** There must be appropriate physical components to ensure the confidentiality of voting. To avoid this risk is very important the correct situation of the polling station in the Election Day.

- **Enclosure:** The physical structures within which the computer system is enclosed must be such as to protect it against potential sabotage. Thus, all hardware devices incorporated in the electronic voting machine that are not directly involved in the action of casting a vote must be totally out of reach to any user other than repair and maintenance staff.

Electrical and Electronic Components

This section assesses the features relating to components such as computer equipment, systems providing electric current (power supplies, protection, cabling) and peripheral devices

- **Power Supply Surge Protection:** There should be systems to protect against surges caused by variations in the mains electricity supply that might damage the equipment. A UPS (Uninterruptible Power Supply) is highly recommended.

- **Cabling:** The set of cables supplying power to the various devices should be totally hidden. If, for good cause, this is not possible, they must be shielded from any possible tampering by some protective fixture. Manufactures must take into account this problem in the design process of DRE system.

- **Identity Document Reader:** The system for reading the identity documents that allow to voters access should be robust enough to prevent impersonation by duplicating voter credentials. The best solution is to utilize reusable smart cards and a reader for these to discard the printed credentials, illegally copied.

- **Battery Life:** The capacity of the batteries feeding the electronic polling system must be sufficient to keep the electronic voting machine operational throughout the polling period, even in the case of a power cut. With this in mind, consideration should be given to installing an uninterruptable power supply, to the use of high-capacity batteries and to attempting to incorporate low-drain devices. If the polling station is a based on a tablet-pc or laptop, the battery should have a minimum of six cells. If the polling station is based in a simple PC system is recommended the use of an UPS (Uninterruptible Power Supply) system.

- **Heat Sinks:** The system must have fitments that will dissipate heat in those places where an overheating problem could occur. Furthermore, all unnecessary mechanical parts should be eliminated in order to avoid increased power consumption by the system, which would lead to greater heating, as well as a reduced ability to run independently of a mains power supply.

- **Printer Electricity Supply:** The printing devices must have a power supply independent from that of the computer system. This is so that it can print the voting docket necessary for the audited voting arrangements. One more time, using an UPS (Uninterruptible Power Supply) is much recommended.

- **Wireless Interfaces**: Access to wireless interfaces of types such as Bluetooth, IrDa, the WiFi 802.11 family and Wimax should be physically restricted. The aim of this is to ward off malicious attacks such as buffer overflows or the feeding of malicious code into the system. Wireless interfaces locked are an action that should be done by software and hardware mechanism.

- **Input and Output Devices:** Access to laser disk drives and U.S.B. interfaces should be physically restricted, with the same intention as mentioned in the previous paragraph.

- **Hard Disk Storage:** Measures should be in place to avoid the loss of data filed on internal storage units. This assessment looks at measures such as the implementation of some level of redundant arrays of independent disks (RAID), or storage based on solid-state memory, which is more stable and consumes less power than conventional hard disks.

Analysis of Software Components

The sections below cover the logical elements going to make up the electronic polling system. They refer to those aspects relating to the software application for voting and the operating system under which this application runs.

Interaction with the Operating System

The operating system is a key item when it comes to determining the overall security of the polling system. Good administration of the operating system is crucial in ensuring the proper functioning of the electronic polling system as a whole, since it controls the hardware peripherals and the software applications for voting.

- **Restriction of Access To The Operating System:** Execution of the applications program that enables voting must be in full-screen mode, not allowing any interaction with the operating system under which the application runs, particularly at critical points such as an unexpected rebooting of the system.

- **User Management:** There should be a hierarchy of users with the following structure:
  - **Elector:** A person using a voter identity document to cast a vote.
  - **Polling Officer:** Any of the users who are responsible for the actions involved in managing voting machines during polling.
  - **Administrator:** A user who sets up and configures the voting machine. Nomination of this person should be the responsibility of the manufacturer.

In this way a second line of defense is established if the requirements expressed in the previous section are not properly fulfilled.

- **Log Files:** There must be a set of mechanisms to record the actions carried out on the system, with the intention of certifying that the electronic polling has functioned correctly. These mechanisms must ensure checking and logging of:
  - Signing on and off by each of the various groups of users.
  - Exceptions and faults.
  - The actions performed by each generic elector user, always with the caveat that voting confidentiality should be maintained.
  - Monitoring of the processes of start-up and close-down of the voting machine.
  - All this information must be stored in an encrypted form. It should allow a security audit procedure to be performed if there is any suspicion that undesirable activities have been taking place.
- **System Files:** The operating system should ensure some form of security back-up copy is made so as to protect the data involved in the voting process.

Software Applications for Electronic Voting

These items are responsibility of the software manufacture. These enterprises have to apply specific quality controls oriented to general purpose software. Thanks to these controls the proper functioning is insured as well as the user interfaces adequacy.

The paragraphs below present those aspects that were analysed that refer to the quality and security of the software devoted to the voting process covered by this document.

- **Comments and Documentation:** The source code should include the comments and formal documentation necessary for incorporating improvements and thus giving a capacity for scaling up the system.
- **Graphic Interface Display:** Care should be taken that the resolution of the screen on which the software runs can be adjusted to any format required.
- **Programming Language:** The language used to implement the application must allow for adaptation to new platforms in an efficient and secure way.
- **Software Structures:** The structure of the software should be adapted to new object-oriented methodologies and hence facilitate scaling up by means of improvements affecting different modules separately.
- **User Interface:** A user-friendly graphic interface should be provided in which the sequence of steps to be taken in order to cast a vote should be clear and concise. Each step in the procedure should be accompanied by an audible explanation.

The set of items covered above refer to aspects concerning quality that must be kept in mind when the software is being written. Below, a definition is given of several major aspects that should be taken into account so as to ensure the security of the software.

- **Access to Management Codes:** The set of codes needed to perform voting machine management actions should be protected, encrypted and outside the execution environment of the software application.
- **Access to Elector Codes:** The voter codes associated with each identity document should be protected in the same way as the management codes, as in the previous paragraph.
- **Constituency Configurations:** It should be ensured that each electronic voting machine does not contain all possible configu-

rations for all the constituencies in which it might be used. Each machine should be configured only for a single constituency, depending exclusively upon the initialization code. This aspect is critical because all the voting machines are identical, each one being "personalized" in accordance with the initialization code. The code in question is visible on the machine, and hence may easily be duplicated. In this way, it would be possible to produce multiple machines for one and the same place that would generate valid results but with different final outcomes.

- **Protection of Electoral Information:** There must be some means for protecting the electoral information stored within the machine. This may be achieved either by means of the application and management of read and write permissions, or through some encryption protocol.
- **Diagnostics:** Logs should be kept by the software application so as to ensure that the program is being executed and is functioning without any unforeseen events.
- **Information on Progress:** The is a need to present information such as which constituency a machine is assigned to, what maximum number of electors it permits, how many votes have been cast on it, among other relevant data, so as to permit a process of dynamic auditing to be carried out while voting is still going on. This information, together with the diagnostic logs mentioned in the previous paragraph and in paragraph, will ensure the consistency of the information referring to the process of voting.
- **Security of the Electoral Information Within the Machine:** This point refers to the fact that the electoral software should guarantee that the information stored in the machine cannot be extracted by using particular software tools.

- **Security of Communications between the Machine and Other Devices:** It should be clarified how the software will guarantee that other pieces of equipment do not communicate with it during polling and are not able to change the electoral information previously stored.
- **Security and Authenticity of the Machine's Applications:** The software and the process of production and configuration of the machine should guarantee that the software is authentic during all stages in the polling process.
- **Security of the Secrecy of the Ballot:** It should be clear how the software guarantees that there is no way of identifying what choice each individual voter has made.

Relation Between Secure Items and Requirements of the Audit Process

The Table 1 shows how the items explained before affect in the eight requirements explained in the section "Technical audit of an electronic polling station: a case study".

Table 1 shows that the most important requirement is the number 7, "The system must be robust, fault tolerant and available during the whole election day" because the number of items influencing this, is quite big in comparison with others requirements. This requirement is strongly relational with the pillars integrity and availability.

It can observe that the rest of requirements are quite distributed along the list of items, this mean that all items are important to guarantee three pillars of e-voting and therefore the eight requirements defined.

CONCLUSION

Thanks to two testing experiences made before audit, it can be concluded that 43 percent of the users would use the electric polling station analysed

Table 1. Items-requirements relation

Items	Requirements							
	1	2	3	4	5	6	7	8
User operations	x	x	x	x	x	x		
Procedure for initializing and powering down machines							x	
Administrator access to voting machines	X	x		x				x
Security of the physical space	X	x	x				x	
Environmental security						x	X	
Maintenance and support								x
Availability of languages					x		X	
Seals					X	x	x	
Locks					X	x	x	
Ventilation systems and operating temperature							x	
Physical accessibility		x			x			
Mobility					x			
Confidentiality	X							
Enclosure					X		x	
Power supply surge protection							x	
Cabling							x	
Identity document reader	X	x		x				x
Battery life							x	
Heat sinks							x	
Printer electricity supply				x			x	x
Wireless interfaces		x				x	x	
Input and output devices		x				x	x	
Hard disk storage							x	x
Restriction of access to the operating system	X	x	x	x	x			
User management	X	x	x					X
Log files				x				x
System files							X	X
Comments and documentation								x
Graphic interface display					x			
Programming language								x
Software structures	X	x						

continued on following page

Table 1. Continued

Items	Requirements							
	1	2	3	4	5	6	7	8
User interface					x			
Access to management codes	X	x	x					X
Access to elector codes	X	x	x					
Constituency configurations							x	
Protection of electoral information			x					
Diagnostics							x	
Information on progress				x			X	X
Security of the electoral information within the machine			X	x			x	
Security of communications between the machine and other devices		x					x	
Security and authenticity of the machine's applications							x	x
Security of the secrecy of the ballot	X						X	

in this work if and only if this system is audited by an external organization of experts in security, also can be concluded that the confidence level of the users is 8.3 over 10. Therefore is a main priority to improve the perception of safety, a big handicap if a country wants to utilize e-voting system to develop an official election process.

The outlined method is no more than a first step towards a definition of a document to serve as a basis for other technological audits of electronic voting machines due to the lack of formal auditing methods for this type of systems, where the application of ISO 27001, ISO 27002 and ISO 17799 is not possible, as these standards do not take into account specific aspects of systems designed for electronic voting purposes. Therefore the correct implementation of the items assessed defines the features of an accurate, secure polling system.

After applying the proposed methodology, and once its results are known, the system should be improved by interpreting the data and, hence, being applied again in order to refine the procedure.

Future work is, therefore, ensured with the aim of optimizing the methodology so that a standard can be proposed to set the security control items required for Direct-recording electronic voting systems.

ACKNOWLEDGMENT

We are very grateful to I.E.P.C.C. for its support.

REFERENCES

Armen, C., & Morelli, R. (2005). E-voting and computer science: Teaching about the risks of electronic voting technology. In *Proceedings of the Tenth Annual Conference on Innovation and Technology in Computer Science Education*, Bologna, Italy (pp. 227-231).

Barrat, J. (2008). Electronic voting certification procedures. Who should carry out the technical analysis? In Barrat, J. (Ed.), *E-voting: The last electoral revolution* (1st ed.). Ann Arbor, MI: ICPSR.

Bharat Electronics. (2009). *Electronic voting machines*. Retrieved from http://www.bel-india.com/index.aspx?q=&sectionid=237

Cohen, S. (2005). *Auditing technology for electronic voting machines*. Cambridge, MA: MIT Press.

Commission on Electronic Voting. (2010). *Independent commission on electronic voting and counting at elections*. Retrieved from http://www.cev.ie/

Cox, C., & Rubin, A. (2004). *Is the U.S. ready for electronic voting?* Retrieved from http://teacher.scholastic.com/scholasticnews/indepth/upfront

Di Franco, A., Petro, A., Shear, E., & Vladimirov, V. (2004). Small vote manipulations can swing elections. *Communications of the ACM, 47*(10), 43–45. doi:10.1145/1022594.1022621

Electoral Advisory Systems for Citizens. (2004). *Indra*. Retrieved from http://94.126.241.45/web-electa/electa_indra_EN.htm

Electronic Frontier Finland. (2008). *Incompatibility of the Finnish e-voting system with the council of Europe e-voting recommendations*. Retrieved from http://www.effi.org/

Falcão, J., Cunha, M., Leitão, J., Faria, J., Pimenta, M., & Carravilla, A. (2006). A methodology for auditing e-voting processes and systems used at the elections for the Portuguese parliament. In R. Krimmer (Ed.), *Proceedings of the International Workshop on Electronic Voting in Europe: Technology, Law, Politics and Society* (LNI 86, pp. 145-154).

Feldman, A., Halderman, J., & Felten, E. (2007). Security analysis of the Diebold AccuVote-TS voting machine. In *Proceedings of the USENIX Workshop on Accurate Electronic Voting Technology* (p. 2).

Fischer, E., & Coleman, K. (2006). *The direct recording electronic voting machine (DRE)* (Tech. Rep. No. RL33190). Washington, DC: The Library of Congress.

Fujioka, A., Okamoto, T., & Ohta, K. (1992). A practical secret voting scheme for large scale elections. In *Proceedings of the Workshop on the Theory and Application of Cryptographic Techniques: Advances in Cryptology* (pp. 244-251).

Gutiérrez-Rubí, A. (2009). *El voto electrónico llega a España con las elecciones europea*. Retrieved from http://www.gutierrez-rubi.es/2009/06/04/el-voto-electronico-llega-a-espana-con-las-elecciones-europeas/

Helbach, J., & Schwenk, J. (2007). Secure Internet voting with code sheets. In A. Alkassar & M. Volkamer (Eds.), *Proceedings of the 1st International Conference on E-Voting and Identity* (LNCS 4896, pp. 166-177).

IEPC. (2008). *Memoria electoral: Electoral and voter participation institute of the State of Coahuila*. Retrieved from http://www.iepcc.org.mx/

Indra. (2009). *Indra realizará el escrutinio de las Elecciones al Parlamento Europeo del 7-J.* Retrieved from http://www.indra.es/servlet/ContentServer?pagename=IndraES/SalaPrensa_FA/DetalleEstructuraSalaPrensa&cid=1243482316640&pid=1087577300456&Language=es_ES

Indrajit, R., Indrakshi, R., & Natarajan, N. (2001). An anonymous electronic voting protocol for voting over the Internet. In *Proceedings of the Third International Workshop on Advanced Issues of E-Commerce and Web-Based Information Systems* (p. 188).

Kohno, T., Stubblefield, A., Wallacj, D., & Rubin, A. (2004). Analysis of an electronic voting system. In *Proceedings of the IEEE Symposium on Security and Privacy* (pp. 27-40).

Ladendorf, K. (2008). *Casting its lot with e-voting.* Retrieved from http://www.hartic.com/news/77

Lamone, L. H. (2003). *State of Maryland Diebold AccuVote-TS voting system and processes.* Annapolis, MD: Maryland State Board of Elections.

McGaley, M., & McCarthy, J. (2004). Transparency and e-voting democratic vs. commercial interests. In R. Krimmer & Grimm, R. (Eds.), *Proceedings of the 1st International Workshop on Electronic Voting in Europe: Technology, Law, Politics and Society* (LNI 47, pp. 143-152).

Mercury, R. (2007). *Mercury's statement on electronic voting.* Retrieved from http://www.notablesoftware.com/RMstatement.html

Microvote General Corporation. (2008). *Election solutions.* Retrieved from http://www.microvote.com/products.htm

Morales, V. M. (2009). *Seguridad en los procesos de voto electrónico remoto: Registro, votación, consolidación de resultados y auditoria.* Unpublished doctoral dissertation, Universitat Politecnica de Catalunya, Barcelona, Spain.

Observatorio Voto Electrónico (OVE). (2005). *Electronic voting observation unit of the University of Leon.* Retrieved from http://www.votobit.org/ove/index.html

Puiggali, J. (2007). *Voto electrónico.* Paper presented at the 2nd Jornadas de Comercio Electrónico y Administración Electrónica, Saragossa, Spain.

Real Madrid. (2009). *El voto electrónico en la Asamblea.* Retrieved from http://www.realmadrid.com/cs/Satellite/es/1193040472656/1202766421991/noticia/Noticia/El_voto_electronico_en_la_Asamblea.htm

Ruth, S., & Mercer, D. (2007). Voting from the home or office? Don't hold your breath. *IEEE Internet Computing, 11*(4), 68–71. doi:10.1109/MIC.2007.94

Schoenmakers, B. (2000). Fully auditable electronic secret-ballot elections. *Internet Technology Magazine,* 5-11.

Wertheimer, M. (2004). *Trusted agent report: Diebold AccuVote-TS voting.* Columbia, MD: RABA Technologies.

This work was previously published in the International Journal of E-Services and Mobile Applications, Volume 3, Issue 3, edited by Ada Scupola, pp. 16-30, copyright 2011 by IGI Publishing (an imprint of IGI Global).

Section 4
Interoperability in
E–Government and E–Business

Chapter 12

Measuring Interoperability Readiness in South Eastern Europe and the Mediterranean:
The Interoperability Observatory

Ourania Markaki
National Technical University of Athens, Greece

Yannis Charalabidis
University of the Aegean, Greece

Dimitris Askounis
National Technical University of Athens, Greece

ABSTRACT

This paper introduces the Interoperability Observatory, a structured research effort for measuring interoperability readiness in the regions of South Eastern Europe and the Mediterranean, supported by the Greek Interoperability Centre. The motivation for this effort derives from the fact that, although interoperability is a key element for public administration and enterprises effective operation, and an important enabler for cross-country cooperation, a standard framework for benchmarking interoperability developments at country level is currently not in place. Interoperability-related information is highly fragmented in different ICT, e-Government and e-Business reports. In this context, in the core of the Interoperability Observatory lies the definition of a structured collection of metrics and indicators, associated with the dimension of interoperability-governance, and a mechanism for gathering with regard to the latter suitable information for a number of countries from various sources. The ultimate goal is the use of this information towards the directions of raising awareness on the countries' interoperability status, promoting best practice cases and benchmarking.

DOI: 10.4018/978-1-4666-2654-6.ch012

INTRODUCTION

During the last few years, interoperability has been recognized as a critical factor for achieving true one-stop service provision for citizens and businesses, fostering collaboration among organizations and achieving efficiency and productivity gains in both the public and private sector. Additionally, it has been established as an important enabler for cross-country G2G (Government-to-Government) cooperation and cross-border service delivery (Commission of the European Communities, 2003; Charalabidis, Askounis, & Gionis, 2007). Thus, it has emerged as one of the most vivid research areas in the fields of electronic governance and electronic business (Charalabidis, Panetto, Loukis, & Mertins, 2008) and has become a key issue in the agenda of the respective research and practice communities (Codagnone & Wimmer, 2007; Information Society Technologies, 2008), leading to the uptake of various projects and initiatives. More importantly, researchers and practitioners have started to realize that when it comes to attempts related with aligning organization and processes, tackling semantic and technical shortcomings, building architectures, achieving legal interconnection and co-operation of systems or even developing standardization frameworks, there exist common practices and knowledge to be shared (Charalabidis et al., 2008; Laudi, 2010). Once identified, such practices can lead to an enhanced exploitation and reuse of real life cases and paradigms by the stakeholders interested, facilitating thus endeavors for promoting interoperability.

On the other hand, following the uptake of relevant activities, interoperability research has been extended so as to deal - besides the actual interconnection of diverse services, systems or organizations – with the development of methods and frameworks for reviewing or assessing the status of interoperability practice and for guiding the implementation of future endeavors. In this context, several researchers have proposed frameworks for assessing interoperability readiness, based on interoperability maturity levels (De Soria, Alonso, Orue-Echevarria, & Vergara, 2009; Gottschalk, 2008; Kasunic & Anderson, 2004; Pardo & Burke, 2008; Sarantis, Charalabidis, & Psarras, 2008). Additionally, interoperability research and practice have been enriched with the creation of mechanisms enabling the dissemination of research outcomes and best practices, and the creation of awareness among the communities involved (IDABC, 2009; SEMIC, n.d.). In fact, such developments have also rendered clear the value and usability of interoperability-related information (e.g. statistics, implicit performance evaluation information, best practices, policy material etc.), for the purposes of awareness raising in the field, benchmarking, carrying out comparative analyses, providing recommendations to tackle possible weaknesses and challenges and enabling more informed decisions about the allocation of scarce resources to solve interoperability problems.

Yet, the relevant evaluation approaches have been addressing only specific interoperability dimensions (e.g. technical, semantic, organizational, etc.) and have been focusing mainly at system (Kasunic & Anderson, 2004) or enterprise/organization level (De Soria et al., 2009; Pardo & Burke, 2008; Sarantis et al., 2008). As a result, they have failed to provide a structured framework for reviewing interoperability advancements at country level and to examine interoperability from a wider institutional perspective. Such a perspective, coined by the European Public Administration Network (2004) as interoperability governance, should be concerned with the political, legal and infrastructural conditions that are relevant for developing and using interoperable systems that span both intra- and inter-organizational boundaries (MODINIS, 2007), and should be viewed as an issue that cuts across all other interoperability dimensions.

Systematic approaches that have indeed addressed interoperability at a country level have been provided within the frame of the MODINIS Study (2007) and the NIFO (National Interoperability Frameworks Observatory) project (ID-ABC, 2009), an accompanying activity to the IDABC projects of the European Interoperability Framework 2.0 and Architecture Guidelines. The subject of the MODINIS initiative has been the study on interoperability at local and regional level in terms of key success factors' and barriers' detection and recommendations' provision to the stakeholders involved. On the other hand, the NIFO project has targeted the creation of awareness on National Interoperability Frameworks (NIFs) and the speeding up of their development in Europe. Still, the project has focused mainly on the existence or status of current NIFs, when the latter, though important, are not the only aspect for benchmarking a country's performance with regard to interoperability readiness.

The absence of a framework for benchmarking national interoperability developments has been further complemented by the unavailability of explicit statistic or empirical information with regard to the issue in question at country level. The unavailability of such information has been however justified by the technical difficulties involved in collecting data at such a level. Yet, as interoperability, although well established as a research domain, has been mainly perceived, and thus assessed, under the prism of its application domains, i.e. mainly e-Government and e-Business, relevant information has been implicitly and fragmentarily provided in several different national or international ICT, e-Government or e-Business reports, being accordingly hard to discover. Similarly, interoperability research outcomes, activities and projects with a good practice potential have been repeatedly labeled according to the particular application domain, to which they fall into, being therefore difficult to point out.

As a consequence, and despite its usability and importance for the purposes stated above, interoperability-related information has been highly dispersed. This fact has pointed out the need for an organized approach to standardize a set of metrics and indicators on interoperability-related factors and combine information from multiple sources to address the dimension of interoperability governance and to shed more light on the status and progress of relevant developments at national level. Under these circumstances, this paper introduces the *Interoperability Observatory*, a structured research initiative for benchmarking interoperability readiness at country level, supported by the *Greek Interoperability Centre* (2008). Being the outcome of an extensive literature review, the Interoperability Observatory is based on the establishment of a series of indicators that are associated with the dimension of interoperability governance and touch upon the policy, research and practice domains, and the collection of relevant information. Its scope of application covers South-Eastern Europe and the Mediterranean, i.e. a region that has been attributed within the last years priority for the expansion of the European Research Area and the development of its research potential, as demonstrated by the European initiatives and the number of research projects carried out within the countries that it enumerates.

In this context, this paper is organized as follows: First we comment on the significance of interoperability-related information identifying possible cases of use, reviews research endeavors for benchmarking interoperability status and developments and exposes the scope of the paper. Next, we introduce the Interoperability Observatory and presents the types of information that are of interest within its context. We then define the collection of indicators applied within the framework of the Interoperability Observatory and link the latter with current available data sources. The next section exposes the methodology for gathering information, providing thus a more thorough view on the particular sources used, and describes the underlying processing and reporting system. Then we present sample data sheets for a set of countries selectively and discusses on the results. Finally, we summarize the ideas presented pointing out the added value of the Interoperability Observatory as well as the insufficiencies and limitations involved, and describes the research steps to follow.

INTRODUCING THE INTEROPERABILITY OBSERVATORY

As already suggested in the introduction of this paper, the Interoperability Observatory is a research initiative that aims at monitoring interoperability developments in the area of South-Eastern Europe and the Mediterranean. Its target region encompasses 15 countries (namely Albania, Bosnia and Herzegovina, Bulgaria, Croatia, Cyprus, F.Y.R.O.M., Greece, Hungary, Malta, Montenegro, Moldova, Romania, Serbia, Slovenia and Turkey), and constitutes an area that is, at least partially, yet immature with regard to interoperability developments (Western Balkans Network for Inclusive e-Government, 2008).

The goal of the Interoperability Observatory is to improve the stakeholders' understanding on interoperability at regional level by capitalizing on important interoperability-related knowledge that touches mainly upon the policy, research and practice domains. In this direction, it involves the standardization of a set of indicators, associated with the dimension of interoperability governance, and the detection and aggregation of relevant information. The types of information that are of interest in the context of the Interoperability Observatory are outlined hereafter as we expose the aspects to be taken into account within its frame and present its structure.

The acknowledgement of the significance of interoperability and its associated benefits from a policy perspective can be considered as an important precursor for a country's engaging in interoperability projects and activities. As a result, a critical element to be taken into account within

the frame of the Interoperability Observatory is a country's interoperability awareness. The latter is likely to be reflected in the potential incorporation of the concept of interoperability in national policy documents and strategic frameworks or more specifically in the existence of National Interoperability Frameworks, providing recommendations to guide the development of interoperable systems, services and organizational structures. Both the incorporation of interoperability in such documents and the gravity attributed to the latter within their context, as well as the existence of a National Interoperability Framework along with its level of detail, provide indications on the degree at which there are clearly defined goals and priorities and the appropriate stakeholder commitment or top management support.

Moving from the policy to the practice and research domain, the concrete steps taken by a country and also its actual progress towards the direction of achieving interoperability are depicted in the volume, intensity and size of national interoperability-related projects and activities as well as in the degree of its involvement in EU level relevant research. Consequently, these constitute further aspects for consideration. On an upper level, a measure of a country's success and also actual proof, that the activities undertaken have resulted in tangible benefits, is revealed through the identification of best practices and reusable solutions and the examination of their eventual impact.

Finally, specific aspects on the operation of the Public Administration or the enterprise sector, such as the degree of interoperability applied to the provision of public services or enterprise business processes respectively, are presumed to enhance the stakeholders' understanding on e-Government and e-Business interoperability at practical level and are therefore also taken into account.

Based on the types of knowledge described above, the emerging structure of the Interoperability Observatory consists of six thematic axes, shown in Figure 1, namely:

Figure 1. The axes of the Interoperability Observatory

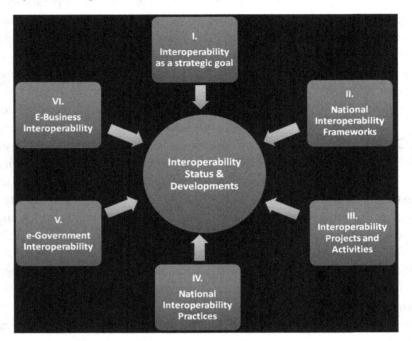

1.   Interoperability as a strategic goal
2.   National Interoperability Frameworks
3.   Interoperability Projects and Activities
4.   National Interoperability Practices
5.   E-Government Interoperability
6.   E-Business Interoperability

Axes I and II focus on the policy level and investigate the acknowledgement of interoperability as a strategic priority and the existence of a national interoperability framework respectively. Axes III to IV address a country's advancement towards interoperability in terms of relevant projects and practices, while axes V-VI examine specific aspects that reveal the level of e-Government and e-Business interoperability.

These axes are each further analyzed in the following section in a number of indicators and provide a generic view of the status of interoperability at regional level.

DESIGNING THE INTEROPERABILITY OBSERVATORY

This section presents the indicators that have been defined within the context of the Interoperability Observatory, as a result of an extensive literature review conducted within the frame of the Greek Interoperability Centre (2008). These indicators, being both quantitative and qualitative, are classified in the six thematic areas, outlined in section 2 and can be uniformly applicable in all countries under consideration, provided that the required information is publicly available. In the following, we provide for each metric a brief description, the list/range of its values and cite potential sources for information gathering. These sources constitute additionally part of the literature reviewed within the frame of the definition of the framework's indicators.

Interoperability As A Strategic Goal

Strategic Priority on Interoperability

- **Values:** Yes/No/Unknown (and justification).

Addressing interoperability issues at the policy level is an important precursor for engaging in interoperability activities and constitutes the first substantial formal step in the quest for managing this utility-like capability, even if there are several ad-hoc initiatives to be reported. As a result, this qualitative indicator examines whether Interoperability is recognized as a strategic priority, a fact which holds truth if the latter is explicitly mentioned in either one of the national e-Government, Interoperability, local e-Government, Digital Planning, IT or Information Society (etc.) strategies or other strategic framework of the country. In this case it is awarded the value "Yes"; otherwise the value "No". In case there is no relevant information, the indicator is granted the value "Unknown". A suitable justification is also provided depending on the case.

Potential sources of information include the aforementioned types of strategies, the e-Government Factsheets, available through the ePractice. eu portal (http://www.epractice.eu/en/factsheets/) for 34 European countries, the MODINIS Study (2007), as well as the latest version of the Capgemini Benchmark Measurement (2009).

National Interoperability Strategy Status

- **Values:** Not planned yet/Planned/Under development/Published/Unknown.

This indicator draws its inspiration from the European Interoperability Strategy – EIS, which is an action plan that addresses cross-border and cross-sectoral interoperability with the aim of facilitating the implementation of EU policies (Deloitte Consulting, 2009), and investigates therefore the existence of a similar action plan at national level. Such an action plan may be absent, and thus "Not planned yet", "Planned", "Under development" or already "Published". In case no information is available, the National Interoperability Status is attributed the value "Unknown". The information sources previously outlined, apply here as well.

National Interoperability Frameworks

National Interoperability Framework Status

Interoperability Frameworks appear as the instruments for a common approach to interoperability for organizations wishing to interact with each other. Within their scope and audience they specify a common vocabulary, concepts, principles, policies, guidelines, recommendations and practices, and form the basis for the assessment and selection of standards and technical specifications (Gartner, 2009). National Interoperability Frameworks prove that a country is interoperability-aware and pose as the cornerstone for the resolution of interoperability issues in the public sector and the provision of one-stop, fully electronic services to businesses and citizens (Charalabidis, Lampathaki, Kavalaki, & Askounis, 2010). Such frameworks outline the essential prerequisites for joined-up and web-enabled e-government and provide the necessary methodological support to an increasing number of projects related to the interoperability of information systems, in order to better manage their complexity and risk and to ensure that they will deliver the promised added value (Koussouris, Lampathaki, Tsitsanis, Psarras, & Pateli, 2007; Ralyte, Jeusfeld, Backlund, Kuhn, & Arni-Bloch, 2008). In this context, this indicator examines the status of a National Interoperability Framework by aggregating information on a series of relevant aspects, namely:

- **Status**
- **Values:** Not planned yet/Planned/Under development/Published/Unknown.
- **Title of NIF:** (If published, under development or planned)
- **Version:** (If published)
- **Release Date:** (If published)
- **Focus/Scope:** (If published, under development or planned)
- **Values:** Governance (G) / Conception (C) / Implementation (I) / Operation (O), and/or their possible combinations or Unknown
- **Audience:** (If published, under development or planned)
- **Values:** Government sector / Business sector, and/or their combination or Unknown.
- **Responsible Agency:** (If known)

The primary source of information for most aspects of this indicator is the Final report of the "National Interoperability Frameworks Observatory – NIFO" project (Gartner, 2009). Additional sources constitute the various' countries individual frameworks, in case they are available in English.

European Interoperability Framework Incorporation Status within National Interoperability Frameworks

- **Values:**
- *Low: The NIF demonstrates low compliance to the EIF or has been developed independently*
- *Moderate: The NIF follows the basic guidelines of the EIF*
- *High: The NIF is highly conformant with the EIF and incorporates all necessary recommendations*
- *Unknown*

This qualitative metric provides an indication of the level of importance and incorporation of the European Interoperability Framework (EIF), that has been issued by the IDABC ("Interoperable Delivery of European e-Government Services to Public Administrations, Businesses and Citizens") initiative, within National Interoperability Frameworks.

The idea is that, in the diversified forest of interoperability frameworks that have been produced as a result of the diverse needs and studies of different nations and organizations (Peristeras & Tarabanis, 2006), cross-border interoperability and collaboration is not only a challenge in terms of information system interoperability but also a challenge in the way interoperability frameworks are compatible with each other, and can thereby serve as a vehicle for communication between the parties involved in cross-border interoperability collaborations (Gøtze, Christiansen, Mortensen, & Paszkowski, 2009). As a result, this indicator uses the EIF as a reference point and measures the compatibility of National Interoperability Frameworks with the latter.

Interoperability Projects and Activities

National Interoperability-Related Activity

- **Values:**
- *Non-Existent: 0 projects*
- *Low/Limited: 1-5 projects*
- *Moderate: 6-20 projects*
- *High: Over 20 projects*

This indicator brings the evaluation of interoperability efforts down to the practical level and rates the National Interoperability-related Activity of a country by means of the number of relevant projects, funded from national resources. These are mainly e-Government projects, where interconnection, integration or interoperability have a central role, and target indicatively the development of portals/gateways to serve as single entry points to information and services, the implementation of integrated information systems, as well as the establishment of country-wide infrastructures, e.g.

networks, offering communication, interconnection of state-administrative resources and G2G/G2B/G2C services. Both ongoing and completed projects are taken into consideration. The indicator is measured in a qualitative four level scale, determined according to the authors' perception, and including the levels "Non-existent", "Low/limited", "Moderate" and "High". A quite descriptive but non-exhaustive list of national interoperability-related projects is also provided, in order to offer a more clear view on the relevant activities.

Indicative sources of information include the e-Government Factsheets and cases' descriptions available through the ePractice.eu portal (http://www.epractice.eu/), the MODINIS Study (2007), as well as the Capgemini 8th Benchmark Measurement (2009).

EU Interoperability Research Involvement

- **Values:**
- *Non-Existent: 0 projects*
- *Low/Limited: 1-5 projects*
- *Moderate: 6-20 projects*
- *High: Over 20 projects*

This metric captures the degree of a country's engagement with EU research and development (R&D) activities, and provides therefore an indication of the number of EU-funded interoperability-related projects in which the country participates. The latter are taken as a proxy for a country's willingness to link up with other countries and to support the promotion of interoperability at cross-border level and reveal moreover the maturity level of research conducted within its borders. Both ongoing and completed research projects are taken into consideration. The same four level qualitative scale is used in this case as well, while an indicative list of EU-funded projects is also included.

The CORDIS portal (http://cordis.europa.eu/home_en.html) of the European Commission serves as the main source of information for this indicator. Additional evidence is possible to be found in the sources outlined above.

National Interoperability Practices

Number of Interoperability Cases with a Good Practice Label

- **Values:**
- *No Cases At All: 0 cases*
- *Low: 1-5 cases*
- *Medium: 6-20 cases*
- *High: Over 20 cases*

The metric in question provides an indication of the number of interoperability cases with a good practice label that have been implemented by a country. These are projects and other activities that have resulted in the development of innovative, flexible and reconfigurable interoperability solutions with an appreciable impact in terms of users' uptake, a series of articulated benefits (e.g. managerial, financial, cultural etc.), and a number of patterns and components that may be either reused in other activities within the country or in other countries, or that can be exploited for educational and/or benchmarking purposes. This indicator is particularly important as it reveals the country's progress towards the achievement of interoperability as a result of the initiatives and activities undertaken.

Information on this indicator derives mainly from the descriptions of real life cases, submitted on the ePractice.eu portal and awarded by the ePractice.eu community the "Good Practice Label" or qualified as "e-Government European Awards' Nominees/Winners" etc.

Best Interoperability Practice

In the frame of this indicator, one interoperability case with a good practice label is selected as the most important or indicative one and is described with regard to the following aspects:

- **Title and Short Description**
- **Status**
  - **Values:** Pilot application
    - Operational since…
    - Launched in…
    - etc.
- **IOP aspects covered**
  - **Values:** Technical
    - Semantic
    - Organizational
    - Legal
    - Standardization
    - Assessment
    - Training
    - Subcategories of the above and/or their combinations according to the taxonomy of interoperability horizontal issues, developed by the Greek Interoperability Centre (2008).
- **Impact**
  - Brief description of Benefits, Reusable Components, Patterns and Lessons Learnt from the particular IOP case.

E-Government Interoperability

Interoperability Level of Core E-Government Services to Citizens

$$IOP\ Level_{citizens} = \frac{N_o\ of\ core\ services\ provided\ in\ stage\ 4\ or\ 5}{12}100\%$$

This indicator reveals the degree of interoperability that reaches the final recipients of public services and measures thus the interoperability level of core e-Government services to citizens, based on the 5-stage sophistication stage proposed by Capgemini (2007). "Core" covers the 12 public services most frequently used by households/citizens. The indicator is expressed as the percentage of the 12 basic services which are provided in the 4th or 5th stage of sophistication maturity, where stage 4 (transaction) corresponds to full electronic case handling, requiring no other formal procedure from the applicant via "paperwork", and stage 5 (targetisation) provides an indication of the extent, by which front- and back-offices are integrated, data is reused and services are delivered proactively.

Interoperability Level of Core E-Government Services to Businesses

$$IOP\ Level_{business} = \frac{N_o\ of\ core\ services\ provided\ in\ stage\ 4\ or\ 5}{8}100\%$$

Similarly, this metric measures the interoperability level of core e-Government services to businesses. In this case, "core" covers the 8 public services most frequently used by businesses. The percentages for both indicators are calculated using information on the sophistication stage of e-Government services, included in the eGovernment factsheets, available through the ePractice.eu portal (http://www.epractice.eu/en/factsheets/).

Connected Government Status

$$CGS = \frac{N_o\ of\ services\ provided\ in\ Stage\ V\ "Connected"}{Total\ number\ of\ services}100\%$$

The term "Connected Government" is used in the e-Government Survey of the United Nations (2008) within the frame of the Web Measure

Box 1.

$$Int.\ Level_{intra-organizational} =$$
$$\frac{N_o\ of\ enterprises\ where\ sales'\ \&\ purchases'\ information\ is\ shared\ electronically}{Total\ N_o\ of\ enterprises\ with\ at\ least\ 10\ persons\ employed}100\%$$

Box 2.

$$Int.\ Level_{cross-organization} =$$
$$\frac{N_o\ of\ enterprises\ using\ automated\ data\ exchange\ with\ external\ ICT\ systems}{Total\ N_o\ of\ enterprises\ with\ at\ least\ 10\ persons\ employed}100\%$$

Index in order to describe the situation in which governments transform themselves into a connected entity that responds to the needs of its citizens by developing an integrated back office infrastructure. In this context, the indicator "Connected Government Status" expresses the percentage of services, which are provided in Stage V "Connected", based on the information on Service Delivery by Stages 2008, included in the e-Government Survey as well.

E-Business Interoperability

Intra-Organizational Integration Level

Focusing on a typical aspect of the enterprise sector's operation i.e. information sharing, this metric provides an indication of the intra-organizational integration level that characterizes the latter in terms of the percentage of enterprises in which information on sales and purchases is shared electronically among the different internal functions (e.g. management of inventory levels, accounting, production or services management, distribution management etc.) (Box 1). Sharing information electronically and automatically is considered more specifically under the aspects of:

- Using one single software application to support the different functions of the enterprise;
- Data linking between the software applications that support the different functions of the enterprise;
- Using a common database or data warehouse accessed by the software applications that support the different functions of the enterprise;
- Automated data exchange between different software systems.

Cross-Organization Integration Level

In a similar context, the cross-organization integration level of the enterprise sector is expressed as the percentage of enterprises that use automated data exchange between their own and other ICT systems outside the enterprise group, in order to send orders to their suppliers or receive orders from their customers (Box 2).

Cross-Organization Application-to-Application Integration Level

This indicator goes beyond the aspect of information exchange and investigates the cross-organi-

Box 3.

$$A2A\ Int.\ Level_{cross-organization} =$$
$$\frac{N_o\ of\ enterprises\ with\ linked\ business\ processes\ at\ cross-organizational\ level}{Total\ N_o\ of\ enterprises\ with\ at\ least\ 10\ persons\ employed}100\%$$

Box 4.

$$e-Invoicing\ Status = \frac{N_o\ of\ enterprises\ sending\ or\ receiving\ electronic\ invoices}{Total\ N_o\ of\ enterprises\ with\ at\ least\ 10\ persons\ employed}100\%$$

zation application-to-application integration level in the enterprise sector, based on the percentage of enterprises, whose business processes are automatically linked to those of their suppliers and/or customers (Box 3). This entails electronic and automated sharing of information on the supply chain management and encompasses the aspects of

- Exchanging all types of information with suppliers and/or customers, in order to coordinate the availability and delivery of products or services to the final consumer;
- Sharing information on demand forecasts, inventories, production, distribution or product development;
- Using computer networks; not only the Internet but also other connections between computers of different enterprises.

E-Invoicing Status

Considering e-Invoicing as another aspect of e-Business Interoperability, this metric measures the percentage of enterprises that send and/or receive electronic invoices (Box 4).

All four metrics are based on the i2010 benchmarking indicators of Eurostat (2009), the Statistical Office of the European Communities.

All enterprises with at least 10 persons employed are included. Complementary material derives from the reports of the European e-Business Market Watch.

Figure 2 summarizes the indicators defined above and links them with potential information sources.

INFORMATION GATHERING, DATA POPULATION AND PRESENTATION

In order to fulfill its purpose, the theoretical framework of the Interoperability Observatory is complemented by actual data, and thus an information collection methodology, and a web-based processing and reporting system. These are described in the following paragraphs.

Data Collection

The method for detecting and gathering information relies on desktop research and employs both Internet-based simple and advanced keyword searching. The latter is a well-established method for collecting data and constitutes within the context of the Interoperability Observatory an ongoing task. Particular emphasis is given to the material being accurate and up-to-date, while objectivity

Figure 2. Linking Interoperability Indicators with potential information sources

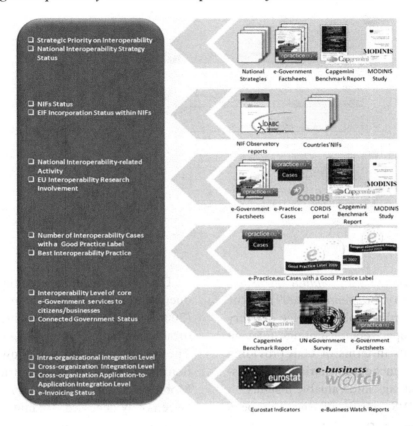

of data is ensured through exploitation of external benchmarkings and cross-checking of information from multiple sources. The resources being accessed are publicly available and include reports, surveys, papers, policy documents and a wide range of portals and web sites of miscellaneous institutions, projects and authorities. Among the sources, already outlined in the previous section, the following are to be distinguished:

- The *Final Report of the NIFO project* (Gartner, 2009), summarizing the results of "setting up the National Interoperability Frameworks Observatory". That is an on-line observatory with the objective of improving interoperability of public services' delivery by means of raising awareness about the rules of collaboration and the interoperability layers addressed in different NIFs, and providing recommendations to tackle possible incompatibilities of the latter.

- The *ePractice.eu portal* (ePratice, n. d.), acting as the central hub for all e-Government activity in the EU and associated states, and operating as a service that merges the e-Government Observatory with the Good Practice Framework, and within its context:
  - The *e-Government Factsheets*, exposing among others a country's main strategic objectives and principles, and
  - A number of *real life cases' descriptions* posted by the portal's users.

Figure 3. Interoperability observatory – indicators presentation (screenshot)

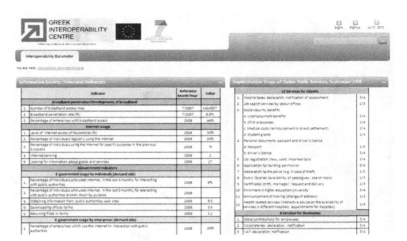

- The *most recent (8th annual measurement) report of Capgemini* (2009) on the progress of online public service delivery across Europe, demonstrating relevant improvements in terms of both providing hard-wired benchmark metrics and transforming rankings into valuable insights.
- The *CORDIS portal* of the European Commission, serving as the gateway to European research and development projects.
- The *MODINIS Study* (2007), providing status reports on interoperability at local and regional level for 25 EU member states.

Data Population and Presentation

Data population and presentation takes place using a processing and reporting system. This supports simple data entry and management through standardized data forms, interconnection with the Observatory's database, record keeping and automatic generation of reports. Data forms are designed in accordance to the axes and the indicators adopted, while reports are code-generated, their generation being triggered automatically every time a data form is updated, as shown in Figure 3.

Dissemination of the resulting reports is conducted through the Greek Interoperability Centre (2008). Advanced features include the generation of tables, charts, comparative figures and statistics. The system is scalable and allows the incorporation of more indicators or countries upon demand. Moreover, it supports user management and can be remotely accessed, enabling access and data entry or validation from multiple authorized users, and offering as a result the potential for other entities or organizations to contribute to the Interoperability Observatory by adding or editing information on the country they represent.

RESULTS AND DISCUSSION

In this section we provide a series of indicative data sheets, generated within the framework of the Interoperability Observatory, and analyze the respective results. The analysis is deliberately scaled down to a sample of five countries (Bulgaria, Croatia, Cyprus, Greece and Hungary), so that representative aspects from all axes are exposed for all five countries. The results are indicative and aim primarily at presenting the rationale and the philosophy of the Interoperability Observatory. For more information and a thorough view

Table 1. Interoperability as a strategic goal (Axis I)

Country	Strategic Priority on Interoperability	National Interoperability Strategy Status
Bulgaria	Yes. General interoperability guidelines are defined in the Bulgarian National Interoperability Framework for Governmental information Systems (6/2006) and the Ordinance on the General Requirements for Interoperability and Information Security (11/2008). There is also the strategic goal of developing centralized e-government systems (provision of a centralized integrated information environment for public services, delivery of centralized services by proposing standardized solutions; activities related to the security of centralized information and systems). Such tasks imply drawing up uniform standards to be used in the communications and exchange of data.	Not planned yet.
Croatia	Yes. Interoperability stands as a key area in the eCroatia Programme, which defines the main directions of the Croatian e-Government Strategy.	Not planned yet.
Cyprus	Yes. Interoperability is part of the national e-Government strategy. Moreover, within the scope of the project for the revision of the Cyprus e-Government Strategy there is the plan of preparing a National Interoperability Framework based on the guidance provided by the European Interoperability Framework (EIF). Strategic priority on interoperability is further reflected in the Information Systems Strategy. Additionally, the e-Government vision of the country is to create efficient and effective public services that will be accessible to external users/organizations on line through electronic means (without the need of visiting any government department) and as such implies a high degree of interoperability.	Not planned yet. An overall individual strategy on Interoperability has not been officially established, yet it is embedded in the Information Systems Strategy.
Greece	Yes. The Greek e-Government Interoperability Framework is part of the overall design of the Greek Public Administration aiming to provide e-Government services to enterprises and citizens. It is the cornerstone of the Digital Strategy for the period 2006-2013, and it is also directly related to the objectives and guidelines of EU Policy 2010, European Information Society 2010. The Framework aims to support e-Governance at central, regional and local level and to achieve interoperability at the information systems level, processes and data by defining the standards, specifications and rules for the development and deployment of web-based front and back-office systems.	Not planned yet.
Hungary	Yes. Hungary has a National Interoperability Framework since 2008. Furthermore, one of the four strategic fields of the "E-Public Administration 2010" Strategy is the introduction of integrated services for the governmental institutions, back offices in order to promote an interoperable, transparent and effective Public Administration.	Not planned yet.
	[3]Established within the e-Government Framework System project that determines the standards, requirements and regulations covering unified technical, semantic and IT-security aspects, methodological application development and project management, as well as the monitoring of the platform for the development and operation of e-Government in order to guarantee that the development of the independent sectoral and municipal sub-systems will result in the establishment of an interoperable, safe and modern e-Government. [4] Available only in Hungarian	

on the values of the metrics in question for each country, the reader is prompted to the interoperability country factsheets, available through the Greek Interoperability Centre (2008) website.

In this context, Tables 1 and 2 summarize the findings for the axes "Interoperability as a strategic goal" and "National Interoperability Frameworks". Figure 4 provides an overview on the volume of interoperability-related activities, while Figures 5 and 6 illustrate respectively the

countries' rankings with regard to the axes of e-Government and e-Business Interoperability.

As seen in Table 1, it is quite remarkable that all five countries examined are interoperability-aware and have recognized interoperability as a strategic priority: even if the latter may not appear in the national strategies (e-Government strategies, digital strategies etc.) as a key field, ambitious goals have been set in all five countries for developing centralized, integrated systems and

Table 2. National interoperability frameworks (Axis II)

National Interoperability Framework Status & EIF Incorporation Status within NIF							
Coun-try	Status	Title of NIF	Ver-sion	Re-lease date	Focus/Scope	Audi-ence	Responsible Agency
Bul-garia	Pub-lished	Bulgarian National Interoperability Framework for Governmental Information Systems	-	June 2006	Concep-tion	Gov-ern-ment	Ministry of State Administration and Administrative Reform http://www.mdaar.government.bg/emanagement.php
	EIF Incorporation Status within NIF	Moderate. The Bulgarian NIF has been developed in compliance with the "European Interoperability Framework for pan-European e-Government Services" - version 1.0, published in November 2004 and adopts its basic guiding principles.					
Croatia	Under devel-opment	Croatian framework for electronic government interoperability (http://www.e-hrvatska.hr/sdu/en/EGovernment/Interoperability.html)	N/A	N/A	Unknown	Gov-ern-ment	Central State Administrative Office for e-Croatia http://www.e-hrvatska.hr/sdu/en/e-hrv.html
	EIF Incorporation Status within NIF	Unknown. The Framework is still under development.					
Cyprus	Under devel-opment	Unknown	N/A	N/A	Un-known[1]	Gov-ern-ment	Department of Information Technology Services, Ministry of Finance, http://www.mof.gov.cy/mof/dits/dits.nsf/
	EIF Incorporation Status within NIF	Envisaged to be high, since the National e-Government Interoperability Framework will be prepared based on the guidance provided by the European Interoperability Framework.					
	[1]The Framework is going to include standards for security (authentication/authorization), encryption, digital signatures, etc.						
Greece	Pub-lished	Greek e-Government Interoperability Framework http://www.e-gif.`gov.gr/	v3.0	Janu-ary 2009	Concep-tion - Imple-mentation –Opera-tion[2]	Gov-ern-ment	Greek Ministry of Interior, Public Administration and Decentralization - General Secre-tariat of Public Administration and Electronic Government http://www.gspa.gr/%2864042249117262l5%29/eCHome.asp?lang=1
	EIF Incorporation Status within NIF	High. The Greek Interoperability Framework has been based on the outcomes of relevant European and international initiatives and is thus highly conformant with the European Interoperability Framework (EIF).					
	[2]The Framework defines standards, specifications and rules for the development and deployment of web-based front and back office systems for the Greek Public Administration at National and Local level. Although it refers to vision and strategy it is not a systematic approach. The conception, the implementation and the operation are analyzed adequately.						
Hun-gary	Pub-lished	Hungarian National Interoperability Framework[3]	-	De-cem-ber 2008	Imple-menta-tion-Op-eration	Un-known[4]	EKK- Senior State Secretariat for Infocommunication http://ekk.gov.hu/hu
	EIF Incorporation Status within NIF	Unknown					

standardized solutions. Still, neither of the countries has established a "pure" national interoperability strategy, indicating that the development of a National Interoperability Framework, which in general defines standards, specifications and rules for the development of interoperable front- and back-office systems, currently emerges as the most critical step towards the achievement of interoperability. In fact, three out the five countries presented (Bulgaria, Greece and Hungary) have

Figure 4. Interoperability-related activity (Axes III & IV)

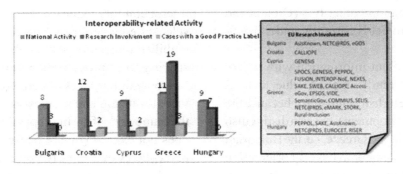

Figure 5. e-Government Interoperability (Axis V)

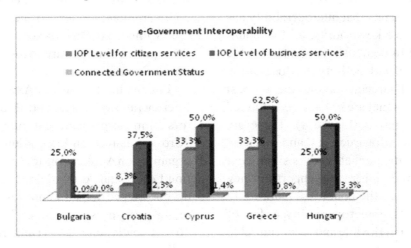

Figure 6. e-Business Interoperability (Axis VI)

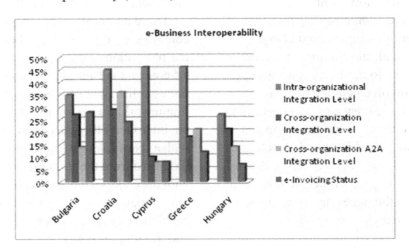

already established a National Interoperability Framework, while in the other two (Croatia and Cyprus), such a project is currently under development (Table 2).

Although important information with regard to the NIFs is currently missing (due to non-availability in the English language or because the project is under development), one can distinguish the cases of Bulgaria and Greece, i.e. the Bulgarian and the Greek NIFs that both have an orientation towards the government sector and have been developed in conformance with the European Interoperability Framework (EIF).

As a result of the fact that interoperability has been recognized as a key priority, all five countries are in place to demonstrate a considerable interoperability-related activity ("Moderate"), with Croatia and Greece having the highest numbers of projects that are or have been funded from national resources (Figure 4). These are mainly large-scale infrastructures' and services' projects that do indicate that there is significant governmental and administrative support, but they do not reflect as well the countries' maturity to harness the benefits of interoperability, as few of them have resulted in flexible and reconfigurable interoperability solutions, i.e. good practice cases, that are transferable and reusable: Greece, Croatia and Cyprus have each a low number of cases with a good practice label, while similar information cannot be reported for Bulgaria and Hungary.

On the other hand, the findings, reported in Figure 4 with regard to the European Interoperability Research Involvement are not that favorable, since with the exceptions of Greece and Hungary that participate or have participated in a remarkable number of EU-funded activities and projects ("Moderate" involvement), the rest of countries present quite limited involvement and need to create the necessary conditions for interoperability research to evolve further.

On another level, an indication of the countries' interoperability maturity is provided through the selected e-Government and e-Business interoperability indicators. Although all countries are striving to improve online public services, the interoperability level of core e-Government services to citizens and businesses, as defined within the framework of the Interoperability Observatory, does not exceed (Figure 5) for all countries the percentages of 33,3% (Cyprus and Greece) and 62,5% respectively (Greece), proving that there is still room for more advancements. Still, shifting the focus from the limited number of core e-Government services to the broader scope of the "Connected Government Status" indicator, the reported findings are extremely unfavorable for all five countries.

Finally, the findings of Axis VI ("e-Business Interoperability"), depicted through the indicators "Intra-organizational Integration Level", "Cross-organization Integration Level", "Cross-organization Application-to-Application Integration Level" and "e-Invoicing Status" in Figure 6 do not exceed for all countries the percentages of 46% (Cyprus and Greece), 29% (Croatia), 36% (Croatia) and 28% (Bulgaria) respectively, indicating that significant effort has to be put on bringing the benefits of interoperability down to the practical level in the business sector as well.

The data sheets exposed prove that all five countries need to intensify their efforts towards the promotion and adoption of interoperability. Of course, the first step towards this direction is the establishment of a National Interoperability Framework; however once a consistent strategy on interoperability has been formulated along with a set of relevant guidelines, the focus has to be shifted from the strategic to the practical level with the undertaking and implementation of suitable initiatives and projects as well as with the evaluation of their outcomes for benchmarking and training purposes.

CONCLUSIONS AND FURTHER STEPS

This paper introduced the Interoperability Observatory, a research initiative for benchmarking interoperability at regional level, supported by the Greek Interoperability Centre (2008). The paper discussed on the motivation behind this initiative, lying in both the absence of a standardized framework for monitoring interoperability developments and the unavailability of relevant information at country level, and indicated how the Interoperability Observatory builds on multiple different sources and several aspects in order to provide through a set of indicators a composite view of the dimension of interoperability governance at regional level. Results generated within this frame were also presented and discussed for a sample of countries, indicating that efforts towards the promotion and adoption of interoperability need to be intensified.

Besides the innovativeness of its conceptualization, the added value of the Interoperability Observatory lies in the collection-combination, filtering and refinement of interoperability-related knowledge, which is inherently scattered in different sources. Such knowledge can be thereafter exploited among others for the conduction of comparative analyses, awareness raising and the provision of suitable recommendations. Additionally, the significance of this research initiative lies in its regional scope, i.e. in the fact that it focuses on a region that is yet immature with regard to interoperability developments.

Still, the Interoperability Observatory and its supporting framework are also subject to a number of insufficiencies, the most crucial being the absence of a suitable validation mechanism to guarantee the accuracy and reliability of the information provided. Further limitations are imposed by the rarity of reporting information outside the EU as well as well as by the language barrier hindering the exploitation of appreciable amounts of information.

In this context, future steps target among others the healing of the former insufficiencies and include thus, the development of a flexible information collection and validation framework that will be based on the establishment of a network of contact points in the countries monitored. The latter will both serve as information sources and will validate the resulting reports. To this end, means such as interviews or suitable questionnaires will be employed, in order to gather first-hand experience from each country. Such a data collection and validation mechanism is additionally envisaged to be supported by the web-based reporting system, already in place.

In order to ensure the continuity and accuracy of the research framework, subsequent steps will include moreover the proliferation and refinement (if applicable) of the indicators adopted, the upgrading of the underlying reporting system accordingly, and the investigation and use of additional sources. The enrichment of the research framework is intended to be directed towards both the consideration of more specific interoperability dimensions, e.g. technical, semantic, organizational etc., and the incorporation of extensive benchmarking research on the current status of e-Government and e-Business maturity in each country under investigation.

Finally, future plans encompass the development of an interoperability maturity model that is going to consist of a series of levels, describing the evolvement of a country from a stage of lack of awareness and/or planning on interoperability, to a stage of interoperability activities and projects being value-adding and resulting in tangible and measurable benefits. Such a model will analyze the different characteristics related to each level and will draw connections between each stage of interoperability maturity and potential actions and measures to be taken, in order to advance the status of interoperability per country. The model will serve the classification, based on the indicators adopted, of the countries monitored in a series of interoperability maturity levels and the suggestion, depending on the case, of the activities that each country should undertake, in order to move upwards in the hierarchy outlined, complementing thus the Observatory's functionality. Extension of the covered area, to include US and Asia countries can then be based on the above common model.

REFERENCES

Capgemini, R. E. IDC, Sogeti, & DTi. (2009). *Smarter, faster, better egovernment* (8th benchmark measurement). Retrieved from http://ec.europa. eu/information_society/eeurope/i2010/docs/ benchmarking/egov_benchmark_2009.pdf

Capgemini. (2007). *The user challenge: Benchmarking the supply of online public services* (7th measurement). Retrieved from http://ec.europa. eu/information_society/eeurope/i2010/docs/ benchmarking/egov_benchmark_2007.pdf

Charalabidis, Y., Askounis, D., & Gionis, G. (2007). A model for assessing the impact of enterprise application interoperability in the typical European enterprise. In Doumeingts, G., Müller, J., Morel, G., & Vallespir, B. (Eds.), *Enterprise interoperability: New challenges and approaches* (pp. 287–296). London, UK: Springer. doi:10.1007/978-1-84628-714-5_27

Charalabidis, Y., Lampathaki, F., Kavalaki, A., & Askounis, D. (2010). A review of electronic government interoperability frameworks: Patterns and challenges. *International Journal of Electronic Government, 3*(2), 189–221. doi:10.1504/ IJEG.2010.034095

Charalabidis, Y., Panetto, H., Loukis, E., & Mertins, K. (2008). *Interoperability approaches for enterprises and administrations worldwide.* Electronic Journal for e-Commerce Tools and Applications.

Codagnone, C., & Wimmer, M. A. (Eds.). (2007). *Roadmapping egovernment research. Visions and measures towards innovative governments in 2020: Results from the EC-funded project eGovRTD2020.* Retrieved from http://www. egovrtd2020.org/EGOVRTD2020/FinalBook.pdf

Commission of the European Communities (CEC). (2003). *Linking up Europe: The importance of interoperability for egovernment services.* Retrieved from http://www.csi.map.es/csi/pdf/ interoperabilidad_1675.pdf

Consulting, D. (2009). *Supporting the European interoperability strategy elaboration.* Retrieved from http://ec.europa.eu/idabc/servlets/Doc9cff. pdf?id=32455

CORDIS. (n. d.). *European Commission homepage.* Retrieved from http://cordis.europa.eu/ home_en.html

De Soria, I. M., Alonso, J., Orue-Echevarria, L., & Vergara, M. (2009). Developing an enterprise collaboration maturity model: research challenges and future directions. In *Proceedings of the 15th International Conference on Concurrent Enterprising*, Leiden, The Netherlands. ePractice (n. d.). *Meet, share, learn.* Retrieved from http:// www.epractice.eu/en/factsheets/

European Public Administration Network (EPAN). (2004). *eGovernment working group: Key principles of an interoperability architecture.* Retrieved from http://www.epractice.eu/ document/2963

Eurostat. (2009). *i2010 Benchmarking indicators.* Retrieved from http://epp.eurostat.ec.europa.eu/ portal/page/portal/statistics/search_database?_pi ref458_1209540_458_211810_211810.node_ code=tin00115

Gartner. (2009). *NIFO project – final report: A report for European commission directorate general for informatics* (Version 130). Retrieved from http://ec.europa.eu/idabc/servlets/Doc?id=32120

Gottschalk, P. (2008). Maturity levels for interoperability in digital government. *Government Information Quarterly*, *26*, 75–81. doi:10.1016/j.giq.2008.03.003

Gøtze, J., Christiansen, P. E., Mortensen, R. K., & Paszkowski, S. (2009). Cross-national interoperability and enterprise architecture. *Informatica*, *20*(3), 369–396.

Greek Interoperability Centre. (2008). *Interoperability guide* (Version 1). Retrieved from http://www.iocenter.eu/

IDABC. (2009). *European interoperability framework for pan-European e-government services*. Retrieved from http://ec.europa.eu/idabc/en/document/2319/5644

IDABC. (2009). *National interoperability frameworks observatory (NIFO)*. Retrieved from http://ec.europa.eu/idabc/en/document/7796

Information Society Technologies. (2008). *Enterprise interoperability research roadmap* (Version 5.0). Retrieved from ftp://ftp.cordis.europa.eu/pub/fp7/ict/docs/enet/ei-research-roadmap-v5-final_en.pdf

Kasunic, M., & Anderson, W. (2004). *Measuring systems interoperability: Challenges and opportunities: Software engineering measurement and analysis initiative* (Tech. Rep. No. CMU/SEI-2004-TN-003). Pittsburgh, PA: Carnegie Mellon University.

Koussouris, S., Lampathaki, F., Tsitsanis, A., Psarras, J., & Pateli, A. (2007). A methodology for developing local administration services portals. In P. Cunningham & M. Cunningham (Eds.), *Proceedings of the eChallenges conference: Expanding the knowledge economy: Issues, applications, case studies*. Amsterdam, The Netherlands: IOS Press.

Laudi, A. (2010). The semantic interoperability centre Europe – reuse and the negotiation of meaning. In Charalabidis, Y. (Ed.), *Interoperability in digital public services and administration: Bridging e-government and e-business* (pp. 144–161). Hershey, PA: IGI Global. doi:10.4018/978-1-61520-887-6.ch008

MODINIS. (2007). *Study on interoperability at local and regional level* (Final version). Retrieved from http://www.epractice.eu/files/media/media1309.pdf

Pardo, T. A., & Burke, G. B. (2008). *Improving government interoperability: A capability framework for government managers*. Albany, NY: University at Albany, SUNY.

Peristeras, V., & Tarabanis, K. (2006). The connection, communication, consolidation, collaboration interoperability framework (C4IF) for information systems interoperability. *International Journal of Interoperability in Business Information Systems*, *1*(1), 61–72.

Ralyte, J., Jeusfeld, M., Backlund, P., Kuhn, H., & Arni-Bloch, N. (2008). A knowledge-based approach to manage information systems interoperability. *Information Systems*, *33*, 754–784. doi:10.1016/j.is.2008.01.008

Sarantis, D., Charalabidis, Y., & Psarras, J. (2008). *Towards standardising interoperability levels for information systems of public administrations*. Electronic Journal for e-Commerce Tools and Applications.

SEMIC. (n. d.). The semantic interoperability centre. *Europe*. Retrieved from http://www.semic.eu/semic/view/snav/About_SEMIC.xhtml.

United Nations. (2008). *eGovernment survey 2008: From eGovernment to connected governance*. Retrieved from http://unpan1.un.org/intradoc/groups/public/documents/UN/UNPAN028607.pdf

Western Balkans Network for Inclusive eGovernment. (2008). *Roadmap for inclusive eGovernment in the Western Balkans: Building e-services accessible to all.* Retrieved from http://e-society.org.mk/portal/download/Roadmap-for-inclusive-eGovernment-in-the-Western-Balkans.pdf

This work was previously published in the International Journal of E-Services and Mobile Applications, Volume 3, Issue 2, edited by Ada Scupola, pp. 73-91, copyright 2011 by IGI Publishing (an imprint of IGI Global).

Chapter 13
Policy Cycle–Based E–Government Architecture for Policy–Making Organisations of Public Administrations

Konrad Walser
Bern University of Applied Sciences, Switzerland

Reinhard Riedl
Bern University of Applied Sciences, Switzerland

ABSTRACT

This article outlines a business and application architecture for policy-making organisations of public administrations. The focus was placed on the derivation of processes and their IT support on the basis of the policy-cycle concept. The derivation of various (modular) process areas allows for the discussion of generic application support in order to achieve the modular structure of e-government architectures for policy-making organisations of public administrations, as opposed to architectures for operational administration processes by administrations. In addition, further issues and spheres of interest to be addressed in the field of architecture management for policy-making organisations of public administrations will be specified. Different architecture variants are evaluated in the context of a potential application of the architecture design for policy-making organisations of public administrations. This raises questions such as how the issue of interoperability between information systems of independent national, state, and municipal administrations is to be tackled. Further research is needed to establish, for example, the level of enterprise architecture and the depth to which integration in this area must or may extend.

DOI: 10.4018/978-1-4666-2654-6.ch013

INTRODUCTION

Motivation for the Article and Statement of the Problem

Enterprise architecture management in e-government has been frequently discussed and operated at a highly technical level so far. An extremely compelling method for deriving e-government application landscapes appears to come from business. In order to understand business in the administrative context, it is necessary to record and to differentiate business process areas and organisational correlations. In a business process model for the e-government area, it is possible to differentiate the following process areas according to Walser (2008) and Walser and Riedl (2009): policy-making processes, operational business processes, strategic business processes for the two aforementioned process categories, and support processes and processes in the area of interoperability (extending business processes across administrative units in hierarchical, vertical or network form). The discussion of political or policy-making processes is therefore difficult and problematic, because political activities are less transparent, less straightforward, and more complex, than operational administration processes, for example. Moreover, until now there has been no clear and reliable model for explaining policy-making processes and procedures, which may vary depending on national state systems. In order to discuss the architecture, therefore, it was necessary to find a concept or model which is as simple as possible, and the components of which can be converted into enterprise architecture. This is the case with the policy cycle concept.

It is very likely that enterprise architectures for organisational units which are specialised in making policies on a federal level are more dominant and more differentiated than those on the member state or municipality level. However, the mechanisms among administrations, executive, legislature, and stakeholders as well as voters on all three levels, federal, (member) state, and municipality, can be considered to be similar, even if – from an institutional point of view – they are not as extensively developed. From this perspective, it may seem obvious to consider an independent generic architecture model for policy-making organisations of public administrations which involves all possible stakeholders, based – for instance – on a stakeholder model of a policy domain. An interoperability concept should be implemented between operational administration information systems and policy administration information systems. Policies may be (but do not have to be) based on information input from the operational administration level.

Thus the notion of forming the architectural concept in conjunction with the policy cycle is a new subject, as is addressing the generic enterprise architecture topic in administrations. Little literature is available. Only few convincing solutions for the issues to be addressed have been visualised or realised in practice. Apart from the policy cycle concept (Lasswell, 1956, 1971; Héritier, 1993; Everett, 2003; Howard, 2005) – which is considered to be controversial due to its practicability – no empirically verified and unique concept of policy-making organisations exists that could serve as a basis for the specification of architectures. In addition, the stakeholder concept (participants of the political process) needs to be considered in order to distinguish political processes of an administration in terms of cooperation. Currently, only a few aspects of the policy cycle are discussed via certain keywords in e-government: e.g. e-participation, e-voting, e-citizenship. All these concepts need to be properly distinguished from the operational administration work through an appropriate architecture discourse and must be put in a binding framework. However, interfaces do exist between operational administration and policy-making, e.g., in the data area. (Electronic) elections and votes require citizen data which is managed and maintained by operational administrations. Thus, the architectures of both policy-making and operational administration are explicitly linked with one another, and so the understanding of the political process, its participants and involved institutions, as well as the dependencies among the system elements, forms the focus for analyzing the requirements for an enterprise architecture for policy-making organisations of public administrations. The advantage of the policy-cycle framework is that, thanks to its simplicity, it is relatively easy to implement in applications. This is a good reason to adopt it as a template for discussing enterprise architecture management in the policy-making administration domain. Some essential and influential parameters must be considered with regard to the creation of architectures for policy-making organisations of public administrations. Namely, the political process as a whole has no clearly defined owner; the different phases of the policy cycle are owned rather by various and changing participants. It is therefore difficult to determine an "owner of the enterprise architecture". The political process is characterised by a kind of "free floating" or power play of different interest groups and by various policy aggregation and specification stages. This impedes the clear assignment of responsibility for the process or parts thereof. The same applies to responsibilities for possible applications, with the exception of core administration applications supporting the political process. Thus, the entire architecture-addressing infrastructure does not necessarily have to be provided by federal government. Instead, different infrastructures are possible and indeed more feasible; however, it should be possible to link them in an intelligent way with the focus on interoperability in stakeholder networks. An initial conclusion is that the architectural concept should be open, similar to the organisational setting, and with facilities for ad-hoc interoperability. Enterprise architecture for a policy-making organisation of a public administration may also be considered from the perspective of the four views or levels of the TOGAF framework of the Open Group (2004):

business architecture, application architecture, system architecture and data architecture.

The question of modularising the diverse areas for processes, data, and applications as well as for systems forms the focus of the four TOGAF architecture domains, for the purpose of overcoming intricacy issues. Another question is whether modularisation should be primarily based on domains. This can be realised in compliance with the different business, activity, process or application areas of the policy-making organisation.

Objectives and Content

On the basis of the above descriptions, the present article will pursue a number of objectives. Firstly, the policy cycle concept and dissociation of its (modular) domains and process characteristics are presented briefly. Then the distinction is outlined between the different stakeholder groups, and their involvement and participation in the fields of communication, collaboration and documentation in the various process areas of the policy cycle. This is followed by the systematic derivation and specification of modular e-government architectures for policy-making organisations of public administrations, taking the aforementioned TOGAF architecture levels into account. Next there is a description of modularisation or domain-building principles for elements of e-government architectures. The structuring principles of the domains are then evaluated on the business and application architecture level. Reference is then made to supporting empirical research based on a proposition of an "architecture prototype" which can be empirically validated. The following questions may therefore be asked: to what extent does the deployment of architectures for political administrations in the context of federation, federal member states and municipalities differ and to what extent is it equally developed? To what extent may different argumentations be required? Which clarifications of the structure model of the political administration (architecture) and its specification need to be looked at separately? The architecture prototype is followed by the derivation and discussion of alternative designs for business and application architectures based on the modular framework. Finally, the article investigates potential future research projects in this field on the basis of a "prototype architecture" to be proposed, which may be empirically validated and/or further developed theoretically. The following article is an extended version of Walser and Riedl (2010). An important additional section about several design variants for business and application architectures was added.

Systematic Approach

A number of aspects are significant for the systematic approach of this article. The procedure is clearly conceptual. This approach enables modular differentiation and design principles and alternatives to be developed for business and application architectures for policy-making organisations. Business and application architectures are derived systematically on the basis of real (conceptual and theoretical) derivations and considerations as well as concepts taken from literature. Various policy-making process areas are derived from business analysis, which is based on the policy-cycle concept, and from the stakeholder approach and its application in the policy-cycle area. These form the basis for the modular business and application architecture. The methodical procedure is based on the aforementioned enterprise architecture management framework TOGAF. Unlike in TOGAF, which distinguishes between four different views, in this case only the business and application layers are examined in greater detail. The conceptual approach also encompasses a modularisation concept for a basic enterprise architecture for policy-making. A number of variant business or application architectures are derived for policy-making and its processes, which may extend beyond the executive state authority. Principles for architectural work and guidelines for further empirical research into the specifica-

tion of architectures for the policy-making part of public administrations are developed, and a "building proposal" is formed, with variants, for an architecture for policy-making parts of public administrations. Alternative options for the design of business and application architectures in policy-making organisations of public administrations are developed and discussed. Due to space constraints, the question of how various political systems influence the architecture design must remain open.

Stakeholder Models in E-Government

For a better understanding of the approach, the stakeholder concept must be examined. This may consist of the components shown in Figure 1.

Stakeholders serve as participants of the opinion-forming process, which is ultimately a political process. However, the term 'institutions' means state-run institutions, which occupy certain recipient roles in the political opinion-forming process and discourse (Linder, 2005). Ministers, as law-enforcement officers, lead stakeholders from within the administration; the administration itself is a traditional, state-run institution. Others involved are executive officers as employees of an administration, civil servants (of policy-making organisations and at the operational process level), as well as experts from within the administration, experts from outside the administration, and the press. Others are parties, associations and unions, which – as well as stakeholders and/or lobbyists – may have various directly-linked roles in the policy-making process, parliamentary chambers and services as traditional state-run institutions, chancelleries or departments as brokers or hubs for parliamentary information (distribution and collection), ministries, chancellery offices as collection points for referendum petitions as well as votes and elections, citizens, enterprises, entrepreneurs or clients of an administration in general, and parliamentarians as representatives of the people. The stake-

Figure 1. Possible links between state-run institutions and stakeholders

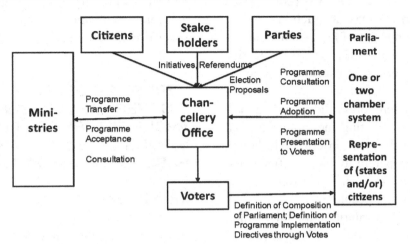

holders and their interaction permit the derivation of activities which guide the creation of the e-government architecture. The relationship between the different stakeholders can be displayed using a few examples as shown in Figure 1. For the purposes of this article, it is essential to remember that stakeholder and expert groups differ depending on the field of policy, and probably also on the policy programme which is discussed in a policy field. Different communication, collaboration, and internal processes may thereby evolve for the various participants. This in turn affects the enterprise architecture of the corresponding area of the policy cycle. It is the objective of the present paper to provide a generic approach across all fields of policy and to discuss policy-cycle-phase-specific differences.

Positioning of the Process Framework for Policy-Making Organisations of Public Administration in the Context of an Administration Process Framework

Policy-making processes must first be placed in a larger context. We differentiate between five generic types of administrative processes: political

processes, operational processes (including front-office and back-office processes; daily contact with customers), management processes, support processes, and intergovernmental (inter-agency; inter-organisational) processes. Furthermore, the above explanations raise some questions. Namely: which modularisation criteria already exist from the perspective of the management of e-government business (process) architectures? To what extent can these modularisation criteria be further developed with regard to the application architectures and the resulting data and system architectures on the level of hardware, software, and network architecture (Niemann, 2005)? How can architectural structuring criteria be mapped in compliance with the aforementioned modularisation criteria according to Aier (2007), Frick-Marre (1995), and Walser (2008) specified an e-government process framework which can be used as a model for the creation of appropriate business-process or architecture modules.

A possible solution for dealing with architectural intricacy is the aforementioned modularisation of architectures or parts thereof, e.g. of the aforementioned business, application, data and system architectures. These modules, or the ones to be created, possess certain characteristics which can be referred to as follows by Aier (2007) based on Frick and Marre (1995): abstraction from the implementation (in terms of IT systems; the following criteria were defined from the perspective of software engineering), encapsulation in terms of hiding internal modes of operation, exchangeability, reusability, temporal validity, orthogonality – in terms of not affecting one another, mutual exclusivity, exhaustiveness – in terms of isolation, universality, interoperability, well-defined and minimal interfaces, and generic as well as hierarchic structures if applicable.

On the basis of the criteria of modularisation defined here, and on the basis of Walser (2008), an attempt is made to bundle processes and to unite these bundles via domains, components or modules. E-government process areas are modularised in a first step in Table 1, in the left-hand column. In a second step, further process categories are distinguished in the right-hand column. The overall number of process modules, as well as the actual sub-process modules in Table 1, must be further specified. This is done in Table 2 with regard to the following criteria: participants, objectives, input-output relations, clients, degree of structuring and standardisation, IT support options, development opportunities (with regard to various administration departments), and intricacy of processes. The process categories addressed never

emerge exclusively or separately, but always in a federal government-specific or organisation-specific combination.

DERIVATION OF AN E-GOVERNMENT-BUSINESS ARCHITECTURE FOR POLICY-MAKING ORGANISATIONS OF PUBLIC ADMINISTRATIONS

Introduction

In Figure 2, in the framework of the reference process model for public administrations on the level of political processes suggested by Walser (2008), the prevailing policy cycle model is recommended as the basis for the structuring of architecture and business processes. Since there are currently no other reference models available for this field, it seems appropriate to use this model for the mapping of policy-making processes. It can also be used for initial analysis of business processes and the extent to which these are covered in terms of applications. On this basis, a proposal may be made for the implementation of a business and application architecture. The proposed enterprise architecture should be as generic as possible to ensure that various administration

Table 1. E-government process areas and appropriate modularisation

E-government processes according to Walser (2008) (supersets)	Possible process module developments (according to Walser (2008), Walser & Riedl (2009))				
Strategic policy-making management processes	Policy cycle planning	Management of organisational aspects in the policy making organisation	Management of personnel aspects in the policy making organisation	Management of leadership of policy cycle processes and organisations as far as possible	Controlling of policy making organisation	---
Policy cycle processes (of policy-making organisations of public administrations)	Policy initiation	Policy estimation	Policy selection	Policy implementation	Policy evaluation	Policy termination

Table 2. Characterisation of process supersets

	Strategic political administration management processes	Policy-cycle processes (of policy-making organisations of public administrations)
Participants	Top-ranking officials of policy-making organisations of public administrations	Policy-cycle stake-holders and policy field audience
Objectives	Objectives of policy pro-grammes (impact and outcome)	Increase of impact and outcome
Input-output rela-tions	Reason for policy programme; suc-cessfully realised policy programme	Input for ini-tialisation of pol-icy programmes, achievement of objectives in terms of the outcome across all policy programmes over a period of time
Clients	Audience of the field of policy and the policy programme	Audience of the fields of policy
Degree of struc-turing	Medium	Mixed
IT support options	Low; usu-ally information evaluation systems (data-warehouse based)	Mixed; support through collabora-tive information systems (web 2.0; social software)
Degree of stan-dardisation	Low	Mixed
Process character-isation (regarding various fields of policy)	Indicator-oriented; information compression and review; usually deterministic	Very different, in-secure, stochastic
Process intricacy	High	High

units and employees on a superordinate level are involved. However, it must be emphasised that the proportion of political processes differs de-pending on the level of government (federal level, state-level, municipality level). Taking the entire administration volume into account, the share of policy-making on each level decreases from the federation to the municipality level, whilst the proportion and importance of operational level processes in turn increases. This affects the defi-nition, specification and generalisability of the statements made here.

Policy-Cycle Model as Basis for Architecture Discussions

The policy cycle as a basic element for the purposes of this article and for the creation of an appropriate IT architecture will be examined in greater detail below. The focus is on the following aspects. The policy cycle consists of the following process phases: policy initiation or problem perception, policy estimation or policy programme formula-tion, policy selection or decision, policy imple-mentation or realisation of the policy programme, policy evaluation, and policy termination. Each phase of the policy-making process includes dif-ferent inputs and outputs, various sub-tasks and stakeholders. Different objectives may be defined for each phase, and the extent to which the ad-ministration is the process owner in corresponding networks, is open to consideration. On the basis of the process analysis, it is necessary to decide upon information system support in terms of a comprehensive architecture management in order to support the whole range of political processes.

In Table 3, processes of the policy cycle are distinguished on the basis of the graphic represen-tation in Figure 2, for each the following phases: input, throughput, and output, as well as a legisla-tion example from Switzerland for each phase.

Firstly, considerations relating to the pro-cesses are explained on the basis of Figure 2 and Table 3. Next, the derivation of applications for supporting policy-making processes of public administrations is discussed (compare Table 4). The interaction of the various participants in the policy-cycle processes can therefore also be re-garded as an intricate process which is shaped by different collaboration and communication net-works and interested parties, and which is not

Figure 2. Political processes based on the policy cycle (Lasswell, 1956, 1971; Héritiér, 1993; Howard, 2005)

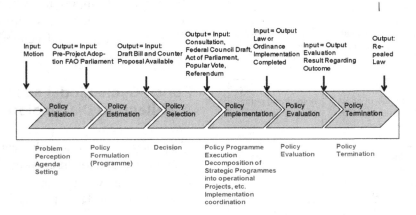

necessarily controlled by the administration. These participants include policy-making bodies, operational administration (e.g. as the implementing body or data supplier), (chambers of) parliament, coordination bodies for administration bodies, parliament, government agencies, cabinet, stakeholders (e.g. federal or state chancelleries), etc. as displayed in Figure 1.

Figure 3 also indicates the diversity of the aforementioned relationships between operational administration and policy making. This diversity may include data supply by the operational administration units and applications for

Table 3. Distinction of the processes of the policy cycle

Process area of policy cycle	Input	Throughput	Output	Swiss legislation example
Initiation	Drafting of motion	Discussion of motion in parliament	Adoption as pre-project of the department	Parliamentary motion, popular or canton initiative
Estimation	Initiation of pre-project of department	Setting up of departmental pre-project or creation of expert draft	Bill or draft made available, as well as counterproposal	Pre-project of department, expert draft, counterproposal
Selection	Bill or draft as well as counterproposal	Creation and consultation of, for example, draft legislation of the federal council, creation of draft parliamentary legislation, etc., definition of draft legislation for votes	Creation of request for consultation, creation of federal council draft legislation, creation of parliament draft legislation, popular vote initiated, referendum	Consultation, federal council draft legislation, parliamentary draft legislation, popular vote, referendum
Implementation	Adoption of legislation or regulation	Enactment processes, enforcement processes	Implementation of law or regulation completed	Enactment, enforcement
Evaluation	Evaluation task and research questions as well as evaluation design	Evaluation processes	Evaluation outcome	Verification of the effect of the legislation by internal and external bodies
Termination	Decision regarding the termination of the policy programme	Termination processes	Terminated policy programme with potential motivation to re-start the policy programme	

Table 4. Distinction of policy cycle processes and possible application support

Phase designation	Core activities	Possible application support
Policy initiation	The focus regarding communication and collaboration between stakeholders and administrations; integrated business administration and document management, for example for processing queries to parliament, etc.	Through communication and collaboration systems, electronic document management systems, citizen relationship management systems and policy-making-body-specific applications.
Policy estimation	Expert work and comprehensive communication of the expert work, programme draft, law and ordinance texts, etc., possibly also campaign preparatory work regarding referendum and initiative management.	Through document management systems and citizen relationship management systems with campaign (planning and execution) functionality, Office environment, expert systems and database access, etc.
Policy selection	Comprehensive communication during consultations and consideration of consultation results in parliamentary work, campaigns regarding referendum and initiative management, initiative and referendum votes; preparation of implementations, etc.	Through communication and collaboration platforms; document management systems, citizen relationship management systems (including campaign planning and execution functionality); deployment of simulation or "model building" tools, depending on the policy-making body.
Policy implementation	Policy implementation on the operation-level architecture and adjustment of IT support, if necessary, according to amendments pursuant to process adjustments at operation level. If possible, the deployment of IT must allow for structured evaluations on the spot, provided the policy-making body approves.	Through specific administration applications, preparation of data warehouses (data collection and aggregation), operational administration processes of high-performance administrations.
Policy evaluation	Sharing of information from various IT systems of the execution of operational processes, but also of systems in the justice area which similarly support operational processes in jurisdiction. Communication of results for the alignment of policy, the policy programme, if necessary, etc.	Through business intelligence based on data warehouses, integration of specific administration applications of the execution of operational processes towards citizens and enterprises.
Between **policy implementation** and **policy evaluation** the following applies	o The policy programme then progresses to the implementation (operation level) phase. o Thus, the policy programme is implemented on an operational process or administration level (focus: architecture for operational administration units): implementation of laws, ordinances, adjustments or new definition of operational administration processes based on the programme. o If possible, special focus should be placed on the IT support because data for the evaluation of the programme can be easily generated through appropriate IT implementations; however, this is seldom the case and depends on the policy-making body (see therefore Figure 3).	

evaluations or votes and elections, measures for the implementation of policies in operational administration, etc.

For the operational administration process to be able to supply data for the evaluation of policies in the policy-making process, the policy programme must be planned, formulated and decided, etc. in as much detail as possible with regard to the information technology support and its implementation. On this basis, information systems can be specifically built. These, in turn, may be used for measuring the success of policies and, thus, for deriving information-based conclusions regarding the effect of policies on the audience. This task is facilitated if, for example, a

reduction of carbon dioxide emissions is measured using scientific methods and as a result of certain policy programmes. Social behaviour that is expected to change as a result of a policy programme is harder to measure.

Requirements of Business and Application Architectures for Policy-Making Organisations of Public Administrations

The following sections are an attempt to describe the various requirements of business and application architectures for policy-making organisations of public administrations. Appropriate and detailed

Figure 3. A model displaying the relationships between Operational administration processes and policy-cycle processes: Data aggregation, e.g., for Policy evaluation purposes

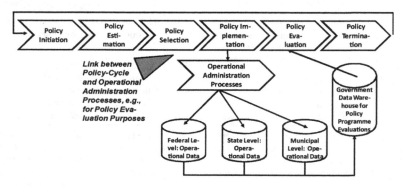

network analyses would actually be required, but this is an attempt to create a generic requirement profile. In addition, applications may be assigned to support these process requirements. The need to deploy an integrated information system and to practise generic management of policy-making is crucial. The interaction of stakeholder groups and institutions of the political system, that are affected by the processes in various phases and different network configurations, requires a number of functionalities, for example communication functionality via various communication media and various contact points, collaboration functionality allowing experts, parliamentarians, employees of an administration, etc. to work together, documentation system functionalities, and contact and campaign management (planning and execution) functionalities. Also required are citizen relationship management logic (Walser, 2006) for private sector customer relationship management and King (2006) for public sector CiRM), e.g. for consultations, votes and elections etc., and knowledge-retrieval functionality for experts who create draft legislation for various policy-making bodies and who explore or try to discover interdependencies, etc.

These functionality requirements also permit the derivation of the most important architecture domains, which are at the centre of architectures

for policy-making units of public administrations. The functionalities must be available for use in an open (external) or closed way (internal, e.g. administration, stakeholders). Policy-making organisational units work independently from the rest of the administration for the most part (Page & Jenkins, 2005); however, interfaces to other process areas exist which must be considered when building the architecture. These include access to citizen data for the addresses of voters, as well as access to data of operational administration units and applications for evaluating policy programmes, etc. Based on these requirements, information systems may be derived that cover the needs appropriately: communication applications, collaboration applications, documentation management applications for policy-making processes, citizen relationship management systems with contact, and campaign management functionalities, and knowledge management databases and applications. CRM logic for supporting operational administration in the front office and processes is different because of the numbers of partners and the way in which collaboration takes place in the policy-cycle domain. In policy-making organisations of public administrations, the communication type used is reminiscent rather of public relations in private industry instead of one-to-one communication, as in operational administration. Here, therefore, integrated communication via electronic means has a different meaning than on the operational level of the administration. However, due to the increasingly electronically supported environment, integrated electronic communication scenarios must also be considered by the political area. Participants may be the aforementioned stakeholders of the political process. In the policy-cycle programme area, the focus is placed on communication within the administration, as well as between the various participants and stakeholders of the political process who define and determine a policy programme. For this purpose, integrated communication and collaboration structures can be used to facilitate communication among the various participants.

Specification of a Component-Based Architecture for Policy-making Organisations of Public Administrations

On the basis of the requirements outlined in this article with regard to the architecture of policy-making organisations of public administrations, the following process components or domains

(including appropriate information system support) can be named: communication component including appropriate platforms (phone, e-mail, web, face-to-face), collaboration component including appropriate platforms (e.g. project-oriented collaboration allowing various experts or stakeholders to work together to support the creation of policy programmes) as well as a lean citizen relationship management logic (platform) for intelligent relationship management (campaign planning and execution) in administrations and for other stakeholders, document management component with appropriate information system support (for document creation, implementation, and evaluation of policy programmes), operation-level-specific components and applications of the various administration departments and information systems of all kinds as the basis for the creation, specification and implementation of policy programmes in policy-making organisations of public administrations.

Figure 4 shows the various components or domains which can be bundled for each policy cycle phase instead of working with different layers as shown in Figure 5. The bundling option takes account of the consistent handling of security

Figure 4. Policy cycle architecture: Process and application components or domains

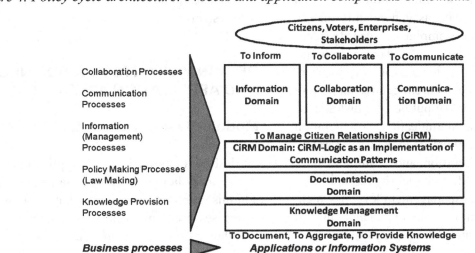

CiRM: Citizen Relationship Management (Tools)

Figure 5. Possible domain or component architecture for sub-areas of the policy cycle

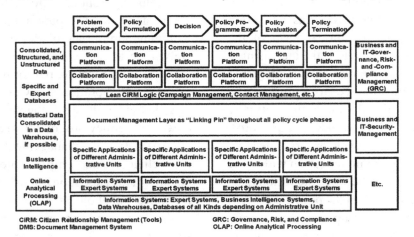

CIRM: Citizen Relationship Management (Tools)
DMS: Document Management System

GRC: Governance, Risk, and Compliance
OLAP: Online Analytical Processing

and compliance aspects. A temporary solution between Figure 4 and Figure 5 is to implement continuous layers for documentation management, for example. This needs to be considered for the management of security limits between the upper and lower architecture area in Figure 5. Figure 5 is an attempt to show a comprehensive architecture for policy-making organisations of public administrations based on the policy cycle. Various layers can be distinguished as follows: communication, collaboration, citizen relationship management (logic), document management, specific applications for policy-making organisations of public administrations, information, monitoring, and expert systems as well as general information systems. In addition, compliance, security (identity and access management) and data warehouse systems are required as domains or components in business and application architectures for policy-making organisations of public administrations.

The fact that communication, collaboration and specific applications are looked at separately shows that collaboration and communication systems can have different security requirements (identity and access management levels), which, in turn, results in more or less open and, thus, in more or less integrated information systems. This distinction also reflects and adopts the various configuration requirements of the different phases of the policy cycle as well as the fact that the first phases or processes of the cycle are not necessarily led by the administration but by the stakeholders. Thus, different requirements regarding the provision and management of exchange or communication platforms must be met. Also, when dealing with personal data in the decision-making process, for example in the event of popular votes or elections, special attention must be paid to security management, which, in turn, results in special requirements for citizen relationship management, collaboration and communication systems.

DESIGN VARIANTS FOR BUSINESS AND APPLICATION ARCHITECTURE

Module-Based Architecture Variants

On the basis of the business and application architecture for policy-making organisations of public administrations described in Chapter 4, which comprises a communication, a documentation, a collaboration and an information domain or corresponding components, we will now look at the resulting possible architecture design variants for policy making.

The features of the political process, the prototype of which is set out in the policy cycle, give rise to the discussion of different variants of the design, so that expanded design options for the architecture, i.e. business and application architecture, may be discussed on this basis. As mentioned, the political process as a whole has no clearly defined owner. The different phases of the policy cycle are owned rather by sometimes various, changing parties or stakeholders. It is therefore necessary to define a single "architectural owner" for the entire business and application architecture. Architectural modules or domains must be formed, which may be linked together flexibly and depending on the situation. The aim should be to create areas of focus in loose or closely connected networks (Van Waarden, 1992), which may be configured quite differently depending on the political area. A clear allocation of responsibility or overall responsibility for policy-cycle processes, along with responsibilities for application components, is thereby made more difficult or does not exist at all.

It appears inconceivable that the state should make infrastructures extensively available also for further stakeholders in the policy process, with the exception of citizens and voters. Far more likely to be considered are infrastructures that are likewise modular, compatible with one another, and which may be interlinked. This means it must be possible for interoperability to be established in stakeholder networks, and also for collaborative networks to be loosely connected. It is therefore necessary to define a blueprint for interoperability between the information system components to be integrated in the form of a network. Essentially, the architectural concept should be open in a similar manner to the organisational setting and should

also permit ad-hoc interoperability on the basis of defined standards.

Overview of Possible Design Variants

Now that the problem of designing business and application architectures for policy-making organisations of public administrations has been specifically defined, different variants may be considered which, in turn, are dependent on context factors that are to be differentiated in further research. The aforementioned variants are discussed and examined in greater depth below. The following solution alternatives may be defined in specific terms:

- **Variant 1:** A core module with different components as shown in Figure 4 is created for each institutional participant (including stakeholders).
- **Variant 2:** A uniform architecture with the defined components is designed where feasible for governmental participants (executive, chancellery, legislative; state institutions). This means that document management, for example, is universally integrated.
- **Variant 3 (Supplementary to Solution Variant 2):** Individual application components are made available by the state for external stakeholders. At the very least, only clearly defined interfaces can be made available, which the state provides to facilitate connection to external systems (participants) of a similar type as illustrated in Figure 4.
- **Variant 4:** Phase-specific modules are provided with reference to the policy cycle (analogous to Figure 4). External offices that are not part of the executive or legislative structure could obtain access to the

corresponding modules, e.g. by registering. This would enable the IT security policies for identity and access management to be designed with different levels of protection for each phase of the policy cycle (differentiation of more open and more closed architectural zones).

- **Variant 5:** Combinations of the solution alternatives described above are possible. In this way features of the policy cycle may be taken into account in specific policy areas or in different political systems. Likewise, specific features of certain network configurations which require corresponding adaptations may be taken into account more effectively.

The further differentiations of solution variants do not refer completely and precisely to the solution variants explained here. They are rather varieties of them and they are not evaluated in detail. Any evaluation should be carried out by means of further qualitative, empirical research.

Allocation of Module Clusters to Policy-Cycle Phases

Analogously to the variants described above, we shall look first at the variants for the allocation of domain clusters to policy-cycle phases (Figure 6).

The modularisation options documented here correspond on a first level to the principle of similarity. On a second level, certain independent applications of the module packages or combinations, as illustrated in Figure 4, may be discussed from specific perspectives as follows:

- From the perspective of an administrative unit (a government office, directorate or ministry)
- From the perspective of the state (incl., e.g., parliament, ministries, chancellery)
- From the perspective of external stakeholders which are not part of the closer circle of the areas specified above (e.g. parties, lobby groups, experts, certain population or corporate groups)
- From the perspective of parliament (various chambers, parliamentary coordination and services, etc.)
- From the perspective of the chancellery as coordinator between national administrations.

These different perspectives determine the architectures to the effect that they result in different requirements such as security, interfaces, openness, collaborativeness, and data access. It is necessary to specify, in greater detail, which requirements here lead to encapsulations (modules mentioned here or other ones) of partial architectures, for example from the above-mentioned perspectives.

Figure 6. Policy-cycle-phase-specific superdomains as a proposed architecture (Variant 4)

Allocation of Module Clusters to National Administration

A further means of structuring or arranging the modules into an integrated policy-making organisation may be illustrated as shown in Figure 7. Module or domain clusters are assigned to the different national administrations and a uniform architecture is differentiated for these. Interface issues are of particular interest here. Interfaces must be implemented between the elements of modules for policy-making management. These may be used to define only partial architectures for policy-making management. A federal, state or municipal chancellery acts in this case as intermediary between parliament and administrative units. Particular attention must be paid to data protection (identity and access management), since certain data of the executive or of parliament are confidential and must, if necessary, be locked away in separate architectural zones.

Discussion of Interoperability Issues on the Business and Application Architecture Level

Issues of interoperability from the business and application perspective are discussed below. Thus it is necessary to differentiate between interfaces on the policy-making level and those on the operational administration level. Informations and data are required from operational administrations to enable policies and therefore programme outputs to be improbe and adapted (Figure 3). In the context of policy evaluation, data is again required from the operational administration or from law enforcement, on the basis of which it is possible to check whether the policy programme is having the desired effect (outcome). A great variety of information exchanges are therefore possible, for example on the basis of the data repositories or data warehouses illustrated in Figure 3, in which data from various applications is integrated in order to facilitate information exchanges appropriate to the matters being addressed. From the architectural perspective, this data or information is to be transferred from one architectural nucleus to another, as shown in Figure 6. Furthermore, there is the issue of business transactions which are relevant for the various domains on the policy-making level. Internal and external business transactions need to be taken into account. The corresponding phase also includes dealing with or looking after voters with regard to elections, electoral lists, etc. In future, consultations with citizens may even be included in earlier phases of the policy cycle, for example to help citizens become more involved in the political process (e.g. as is already part of the

Figure 7. Integration of process and application modules between different state institutions (variant 2)

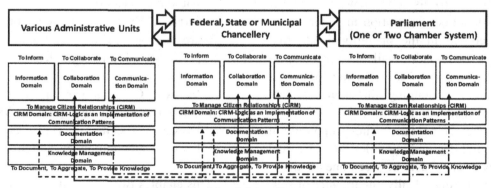

legislative process for the purposes of extensive e-participation). Thus the networks of stakeholders, as illustrated by Van Waarden (1992), for example, would be extended into earlier phases of the policy cycle. In other words, the networks may also be expanded, for example, to include opportunities for citizens to have an influence in the phases in which, until now, policy has been handled by the administration only. This gives rise to the question as to which interfaces are to be formed between operational administration and policy-making organisations, and what this means in specific terms for the architecture and architecture management for the policy-making organisation. The following different interfaces are possible:

- Interfaces to be used repeatedly for transferring citizens' addresses from operational administration for elections and votes. There must be no integration at all in the reverse direction, for example with regard to data on voting behaviour, etc. At any rate, however, the integration of data on non-addressable citizens, etc., should be announced in order to enable the operational administration to make corrections to address data.
- Interfaces to be used repeatedly for transferring statistical data for evaluation where possible from systems of the operational administration involved; at any rate, comments by citizens about new drafts or the operational administration in general may be recorded in the operational administration, and made available for the policy-making organisation in the form of a repository containing the actual citizen comments.
- Interfaces for transferring input/output from architectural modules of individual policy-cycle phases between one another
- Non-recurring: ideally, the transfer of statutory implementation repositories/rule sets

for implementation of the law in the operational administration (e.g. on the basis of information systems).
- Consolidation of messages/information etc. from front offices, i.e. Web or Contact Center environments, in order to bring "policy making" closer to the pulse of stakeholders. Contacts may be created directly between the policy-making organisation and the population, or matters may be presented to the population in the form of lists. This is illustrated on the basis of a study by Schellong of the 311 Contact Center Initiatives in major cities of the USA (Schellong, 2008, 2009).

Application-Oriented Discussion of Interoperability Using Electronic Business Administration as an Example

The discussion of different types of interoperability and integration may be illustrated using the example of electronic business administration (document management systems) as follows (Figure 8 which includes the module for documentation/DMS and business administration).

Figure 8 shows three different interoperability variants (others are possible):

- **Variant A:** Separate subcomponents or modules are used for each institutional unit without connection.
- **Variant B:** Subcomponents or modules are used across the entire administration, possibly with different clients, but without connection problems.
- **Variant C:** This variant must be considered as being analogous to Variant A. However, specific interfaces are defined between subcomponents and modules.

This enables two different settings to be discussed for the implementation of the functional

Figure 8. DMS or business administration integration options

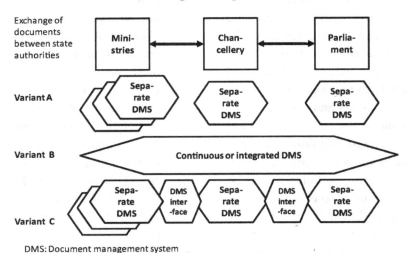

DMS: Document management system

modules, which may be combined but may also be complementary:

1. Implementation of the architecture for each phase of the policy cycle (depending on the institutional boundaries; at all events with client capability). This means that the module clusters or applications are applied independently of government authorities, institutions and stakeholders. This raises issues of ownership and financing.
2. Implementation of the architecture for each institution within or beyond the administration. This implementation is independent of the policy-cycle phases. This solution raises issues of coordination problems and data reconciliations across different institutions.

The two fundamental perspectives shown here each have different features for the management of architectures for the policy-making organisation as a whole and for the management of the integration technologies or interoperability infrastructures. It indeed appears novel or innovative to connect clusters (of functions or applications) from information systems in series repetitively in similar form as components, in order to take

into account the different network and collaboration conditions – e.g. in different stakeholder environments – and the security of the relevant data/system accesses.

CONCLUSION AND OUTLOOK

First of all, architecture management was placed in the context of the policy-making organisation of a public administration, policy-making processes, and their application support. Architecture management in the context of policy-making processes was looked at from the perspective of the comprehensive management of administration business architectures. The focus was placed on the derivation of activities in the policy cycle network, which is used for assigning supporting system types to the various activities and permits the creation of component-based e-government architectures for policy-making organisations of public administrations. An attempt was also made to illustrate the architecture for policy-making organisations of public administrations on the basis of the component concept and the highly generic framework of the policy cycle. The main focus was on deriving architecture models that may be regarded overall as clusters of components, starting with the derivation of activities in the policy-cycle network and of corresponding applications. The modules in the modular structure may be used either individually, or in a universally integrated way, for a great variety of target groups.

In addition, critical areas and specific areas of architecture management in policy-making organisations of public administration were specified on the basis of the policy cycle. A question that remains is whether the modules should be aligned with the policy cycle and its phases, or whether the alignment should be based on the similarities of the applications that are intended to support policy-making organisations of public administrations in their core areas.

Starting with the differentiation of business and application architectures and a corresponding component or module framework, possible variants of architectures for policy-making organisations were discussed as follows:

- **According to Variant 1:** A module combination analogous to Figure 4 is provided for each institutional participant.
- **According to Variant 2:** Those involved in the political decision-making process (executive, chancellery, and parliament) are each provided with a separate module combination as shown in the diagram. Between the module combinations, a loose or even close connection is to be achieved between individual modules where feasible. A particularly central example of this is document management (cf. integration variants shown in Figure 8).
- **Variant 3 (Supplementary to Variant 2):** Individual application components are made available by the state for external participants (e.g., collaboration platforms to recommend issues and to exchange ideas). A minimum variant of this is the specific definition of interfaces. The state

could make these available for stakeholders, citizens, or voters.

- **Variant 4:** Phase-specific modules are provided on the basis of the policy cycle as shown in Figure 6. External offices that are not part of the executive or legislative structure could obtain access to the corresponding components (combinations), e.g. in return for registration and more or less rigid identification and authentication rules. This means that the (IT) security policies may be controlled and implemented differently for each phase of the policy cycle.
- **Variant 5:** Combinations of the above solution alternatives are possible. In this way features of the policy cycle may be taken into account in specific policy areas.

Different integration variants were also specifically defined for Variant 2 above, using examples. The outlook for the research is as follows. The solutions proposed are to be further differentiated and empirically validated. An initial validation has been carried out by means of a series of expert interviews. However, since the framework enters new territory that has never been studied in this respect, further validation of this approach is needed. As with operational administration, it makes sense to differentiate between the front and back office, as is briefly addressed in this article, for example, on the basis of the citizen relationship management approach, i.e. front office application system, including communication, collaboration, and citizen relationship management components. Adaptations to existing applications used in e-government for policy-making are to be discussed in practical, day-to-day administration, and together with potential solution providers. A further issue that also needs to be discussed relates to the interoperability requirements between modules of the business and application architecture in relation to the different variants of the architecture design, as well as to other business process areas, as mentioned at the beginning of the article. Equally interesting is the extent to which the modular framework may be extended to include applications for supporting strategic administration management in the policy environment. Various issues addressed in the text reveal the need for further research, both empirical and conceptual. The research carried out here raises indirect questions, such as how to deal with interoperability between information systems of national administrations, which – for constitutional reasons – are independent of one another. Further research will therefore focus on clarifying, for example at the enterprise architecture level, the extent to which such integration is permitted or possible.

Finally the different variants of enterprise architectures for policy making institutions may also serve administrations for the purposes of enterprise architecture management or softwares developers to develop integrated software suites for policy making institutions and administrations.

REFERENCES

Aier, S. (2007). *Integrationstechnologien als basis einer nachhaltigen enterprise architecture – abhängigkeiten zwischen organisation und informationstechnologie.* Berlin, Germany: Gito Verlag.

Aier, S., & Schönherr, M. (2004). Flexibilisierung von organisations- und IT-architekturen durch EAI. In Aier, S., & Schönherr, M. (Eds.), *Enterprise application integration* [Flexibilisierung komplexer enterprise architekturen]. (pp. 1–60). Berlin, Germany: Gito Verlag.

Everett, S. (2003). The policy cycle – democratic process or rational paradigm revisited? *Australian Journal of Public Administration, 62*(2), 65–70. doi:10.1111/1467-8497.00325

Frick, A., & Marre, R. (1995). *Der software-entwicklungsprozess.* Munich, Germany: Hanser.

Hach, H. (2005). *Evaluation und optimierung kommunaler e-government-prozesse.* Unpublished doctoral dissertation, Universität Flensburg, Flensburg, Germany.

Héritier, A. (1993). *Policy-analyse: Kritik und neuorientierung.* Opladen, Germany: Westdeutscher Verlag.

Howard, C. (2005). The policy cycle: A model of post-machiavellian policy making? *Australian Journal of Public Administration, 64*(3), 3–13. doi:10.1111/j.1467-8500.2005.00447.x

King, S. F. (2007). Citizens as customers: Exploring the future of CRM in UK local government. *Government Information Quarterly, 24*(1), 47–63. doi:10.1016/j.giq.2006.02.012

Krcmar, H. (2005). *Informationsmanagement.* Berlin, Germany: Springer-Verlag.

Lasswell, H. D. (1956). *The decision process: Seven categories of functional analysis.* College Park, MD: University of Maryland.

Lasswell, H. D. (1971). *A pre-view of policy sciences.* New York, NY: Elsevier.

Linder, W. (2005). *Schweizerische demokratie – institutionen, prozesse, perspektiven.* Bern, Germany: Haupt.

Niemann, K. D. (2005). *Von der enterprise architecture zur IT-governance.* Braunschweig, Germany: Vieweg.

Page, E. C., & Jenkins, B. (2005). *Policy bureaucracy: Government with a cast of thousands.* Oxford, UK: Oxford University Press.

Picture. (2009). *Final results from the picture project.* Retrieved from http://www.picture-eu.org/

PriceWaterhouseCoopers. (2002). *Gesamtschweizerische strategie zur dauerhaften archivierung von unterlagen aus elektronischen systemen (Strategiestudie) – appendix III.* Retrieved from http://www.vsa-aas.org/uploads/media/d_strategie_anh_3.pdf

Schellong, A. (2008). *Citizen relationship management – a study of CRM in government*. Frankfurt, Germany: Peter Lang Verlag.

Schellong, A. (2009). Calling 311: Citizen relationship management in Miami-Dade county improving access to government information and services. In Rizvi, G., & De Jong, J. (Eds.), *The state of access: Success and failure of democracies to create equal opportunities* (pp. 191–206). Washington, DC: Brookings Institution Press.

Schönherr, M. (2004). Enterprise architecture frameworks. In Aier, S., & Schönherr, M. (Eds.), *Enterprise application integration – serviceorientierung und nachhaltige architekturen* (pp. 3–48). Berlin, Germany: Gito Verlag.

The Open Group. (2004). *TOGAF 8.1: Certification for practitioners version 1.xx by architecting-the-enterprise*. Retrieved from http://www.opengroup.org/togaf/cert/protected/certuploads/6853.pdf

Van Waarden, F. (1992). Dimensions and types of policy networks. *European Journal of Political Research*, *21*(1-2), 29–52. doi:10.1111/j.1475-6765.1992.tb00287.x

Walser, K. (2006). *Auswirkungen des CRM auf die IT-integration*. Lohmar, Germany: Eul-Verlag.

Walser, K. (2008). Umrisse eines e-government-prozess-referenzmodells. *eGov-Präsenz*, *1*, 61-63.

Walser, K., & Riedl, R. (2009). Skizzierung transorganisationaler modularer e-government-geschäftsarchitekturen. In *Proceedings der 9 Internationalen Tagung für Wirtschaftsinformatik Business Services: Konzepte, Technologien, Anwendungen* (pp. 565-574).

Walser, K., & Riedl, R. (2010, July 1-2). Outline of a generic e-government architecture for political administrations – based on the policy cycle concept. In *Proceedings of the 4th International Conference on Methodologies, Technologies, and Tools Enabling e-Government, Olten, Switzerland* (pp. 1-10).

This work was previously published in the International Journal of E-Services and Mobile Applications, Volume 3, Issue 3, edited by Ada Scupola, pp. 49-68, copyright 2011 by IGI Publishing (an imprint of IGI Global).

Chapter 14
Architectural Guidelines and Practical Experiences in the Realization of E–Gov Employment Services

Elena Sánchez-Nielsen
Universidad de La Laguna, Spain

Daniel González-Morales
Universidad de La Laguna, Spain

Carlos Peña-Dorta
ARTE Consultores Tecnológicos, Spain

ABSTRACT

Today's Public Administration faces a growing need to share information and collaborate with other agencies and organizations in order to meet their objectives. As agencies and organizations are gradually transforming into "networked organizations," the interoperability problem becomes the main challenge to make possible the vision of seamless interactions across organizational boundaries. Today, diverse architectural engineering guidelines are used to support interoperability at different levels of abstraction. This paper reviews the main guidelines' categories which support aspects of architecture practice in order to develop interoperable software services among networked organizations. The architectural guidelines and practical experiences in the domain of e-Gov employment services for the European Union member state Spain are described. The benefits of the proposed solution and the lessons learned are illustrated.

INTRODUCTION

Nowadays, the Public Administration (PA) must be agile and responsive in order to be operative and efficient. One of the major changes has taken place in the way PA cooperates with agencies and organizations by forming networked organizations in order to optimally collaborate towards the provision of software services in a service oriented environment. In this context, interoperability is explicitly identified as one of the key bottlenecks. Legacy systems/applications often

DOI: 10.4018/978-1-4666-2654-6.ch014

hinder collaboration endeavors, since, in many cases the applications are not even designed to interoperate with other applications. Diverse architectural engineering guidelines have been used to support interoperability from IT architect perspective: enterprise architecture, software architecture design, service oriented architecture paradigm, frameworks and maturity models and infrastructures architecture. Over the last years a number of organizations and individual researchers have developed and documented techniques, processes, guidelines and best practices for the different viewpoints of architectures proposed. In the European Union (EU), the European Interoperability Framework (EIF, 2009) defines a set of recommendations and guidelines for the development of PA e-Gov services. However, there is not a consensus about what architecture guide use in order to design and implement interoperable systems for PA.

The goal of this paper is twofold. First, we provide a review of practice architectures, frameworks and models of maturity with architectural guidelines for the problem of interoperability between networked organizations. Second, we describe as case study about how the interoperability problem is addressed at national, regional and local level for Spaninish e-Gov employment services. Our practical experiences and lessons learned during the software project development for the Public Administration and collaborative organizations in the context of e-Gov employment services are illustrated using different architectural guidelines at national, regional and local level. The European Interoperability Framework (EIF, 2009) is adopted as interoperability framework guidelines at national, regional and local level. At regional and local level, the following guidelines are addressed: Zachman Framework (Zachman, 1987) as enterprise architecture, UML and BPML enterprise modeling language as process language to describe business processes and their executions, Service oriented Architecture (OASIS SOA, 2006) as paradigm to implement a service oriented

environment, and open source middleware as infrastructure interoperability.

The remainder of the paper is structured in the following way. First we review the main architectural guidelines from a top-down approach that require special attention in order to develop interoperable services. Next, we focus on the information strategy for Spanish e-Government and the domain scenario for developing e-Gov employment services. We describe how the interoperability problem is addressed at national, regional and local level using the European Interoperability Framework as main interoperability guideline for the Spanish employment scenario. Our practical experiences using the Canary Islands Community as use case to validate the provision of centric-citizen services in the employment domain are illustrated. The functionalities of the services developed, the approach used, the results and lessons learned with the implementation of the architectural guidelines are described. Finally, we provide concluding remarks and future work.

RELATED WORK

This section presents relevant literature associated with the software services development in an interoperability scenario from a top-down viewpoint. Different areas need to be covered in order to provide a solution to problems of interoperability between institutions and organizations. These areas are:

- **Enterprise Modeling:** The goal is to make explicit knowledge that adds value to the enterprise or can be shared by business applications and users for improving the performance of the enterprise.
- **Interoperability Frameworks:** Which provide a set of standards and guidelines that describes the way in which organizations have agreed, or should agree, to interact with each other.

- **Software Architecture Design and Service Technologies:** Provide a way to describe the functionality of software modules and to facilitate their interoperability.
- **The Open Source Model:** Provides means for implementing the standard and non-proprietary solutions required with new opportunities to software entrepreneurs.
- **Middleware Solutions:** Offer application development platform functionalities.

Each of the areas is described in the following sections.

Enterprise Modelling

Enterprise models have the goal to reduce the complexity and give a representation of the structure, activities, processes, information, resources, people, behaviour and goals of an enterprise and the dependencies between them (Knothe et al., 2007). However, the models often do not fit users' requirements, e.g. the model is not detailed enough or the level of formalization is not appropriate. The enterprise modelling (EM) approach is different for each enterprise, depending on its current practices, systems, knowledge and culture. A path to adopt an enterprise modelling approach towards interoperability is required in order to improve competitiveness in a more and more complex enterprise environment.

Two European projects ATHENA (n. d.) and INTEROP (Panetto, Scannapieco, & Zelm, 2004) have delivered diverse publications with a detailed discussion of state-of-the-art about techniques and technologies concerning enterprise modelling. INTEROP (Interoperability Research for Networked Enterprise Applications and Software) is the nucleus mainly of the university research community and defines the conceptual as well as the technical integration of business by means of reference models. ATHENA (Advanced Technologies for Interoperability of Heterogeneous Enterprise Networks and their Applications) is an IT industry platform.

Enterprise modeling approaches can be structured in three parts: (1) enterprise frameworks and architectures, (2) industrial initiatives and standardization bodies working on enterprise interoperability and, (3) enterprise modeling languages. Each of the approaches is outlined at the three following sections.

Enterprise Frameworks and Architectures

An enterprise framework can be defined as a fundamental structure which allows defining the main sets of concepts to model and to build an enterprise. Enterprise architecture (EA) research has resulted in a number of elaborated architecture proposals with a general scope. The main elaborated architecture proposals are: (1) The Zachman Framework (Zachman, 1987), (2) The GRAI Framework from GRAI Laboratory on GRAI and GIM (Doumeingts et al., 1998), (3) the CIMOSA Framework (Computer Integrated Manufacturing Open System Architecture) (CIMOSA, n. d.), (4) PERA (Purdue Enterprise Reference Architecture), (5) The GERAM Framework (Generalized Enterprise Reference Architecture and Methodology), an overall definition of a generalized architecture based on existing enterprise architectures such as CIMOSA, GRAI and PERA, (6) ARIS, (7) the DoDAF Architecture Methodology, called formerly C4ISR (n. d.) is being developed by the FEAC Institute of Washington DC in close cooperation with the Air Force, The Navy, The Army and Pentagon (8) TOGAF Architecture Methodology that is property of the Open Group, an international interest organization, (9) the TEAF Methodology from the US Department of Commerce, (10) the AKM technology, and (11) ISO 15745 as framework elaborated for application integration.

Industry Initiatives and Standardizations Bodies

The state-of-the-art concerning with the European and international standards relevant to enterprise interoperability are: EN/ISO 19439 (framework for enterprise modelling), EN/ISO 19400 (constructs for enterprise modelling), CEN TS 14818 (decisional reference model), ISO 15704 (requirements for enterprise architectures and methodologies), ISO 14258 (concepts and rules for enterprise models) and ISO/IEC 15414 (Open Distributed Processing – reference model).

Enterprise Modelling Languages

Enterprise Modelling Language (EML) allows building the model of an enterprise according to various point of view such as function, process decision and economic in an integrated way. The main EML are: (1) IEM – Integrated Enterprise Modelling based on the object-oriented modeling approach for modeling business processes and related organizational structures (Spur, Mertins, & Jochem, 1996), (2) ITM, a commercial template and modeling language available from Computas, (3) BPM, template that implements most of the BPMN language standard for Busines Process Modelling, (4) UML, (5) MEML (Monesa Enterprise Modelling Language), (6) Petri Nets, formal, graphical and executable technique for the specification and analysis of concurrent, discrete-event dynamic systems, (7) CIMOSA, (8) GRAI, (9) IDEF (Integrated DEFinition methodology), (10) PSL (process specification language), a neutral representation for manufacturing processes, (11) WPDL (workflow process definition language), (12) BPML (Business Process Modelling Language), (13) EDOC – UML Profile for Enterprise Distributed Object Computing Specification, and (14) ebXML (Electronic Business using eXtensible Markup Language).

Frameworks with Architectural Guidelines for Support Interoperability

The IEEE Standard Glossary of Software Engineering Terminology defines interoperability as "the ability of two or more systems or components to exchange information and to use the information that has been exchanged" (IEEE Computer Society, 1990). Interoperability problems occur at three different levels: (1) at the organizational or business level, it is concerned with defining business goals, modeling business processes and bringing about the collaboration of administrations that wish to exchange information and may have different internal structures and processes; (2) at the semantic or knowledge level, it is related to ensure that the precise meaning of exchanged information is understandable by any other application that was not initially developed for this purpose. Generally, different formats, schemas, and ontologies are implied; and (3) at the technical level, it is concerned with the underlying information and communication technologies and systems. It includes aspects such as open interfaces, standards, interconnection services, data integration and exchange, accessibility and security services.

In the following section, it is described the interoperability framework used as interoperability guideline in our case study of employment services.

European Interoperability Framework

The European Interoperability Framework (EIF, 2009) defines a set of recommendations and guidelines for eGovernment services so that Public Administrations, enterprises and citizens can interact across borders, in a pan-European context. The objectives of this framework are: (1) to support the European Union's strategy of providing user-centred eServices by facilitating the interoperability of services and systems between PA's, PA's as the public (citizens and enterprises), at a pan-European level; (2) to supplement national interoperability frameworks in areas that cannot be adequately addressed by a purely national approach and (3) to help achieve interoperability both within and across different policy areas, notably in the context of the IDABC programme (IDABC, n. d.). As a result, the European Interoperability Framework shows how services and systems of administrations throughout Europe should interrelate in order to serve, supplement and enrich each other with a view to providing pan-European eGovernment services. The considerations and recommendations of the EIF are based on the following principles: (1) accessibility, related to ensure access for disabled persons and offer support in a language understood by the user; (2) multilingualism, the underlying information architectures should be linguistically neutral, so that multilingualism does not become an obstacle to the delivery of eGovernment; (3) security, from the user perspective, identification, authentication, non-repudiation, confidentiality should have a maximum level of transparency, involve minimum effort and provide the agreed level of security; (4) privacy, appropriate information regarding the data processing activities should be made available to the concerned individuals. Full compliance with the existing European and national data protection legislation should be ensured; (5) subsidiarity, in line with the principle of subsidiarity, the guidance does not interfere with the internal workings of administrations and EU Institutions; (6) Use of open standards; (7) assess the benefits of open source software, where the use of open source software (OSS) should be assessed and considered favorably alongside proprietary alternatives and (8) use of multilateral solutions, where the interoperating partners adopts the same set of agreements for interoperability solutions.

Software Architecture Design

Enterprise architecture defines frameworks that describe the main sets of concepts to model and build an enterprise focusing on the attainment of the business objectives and concerning with items such as business agility and organizational efficiency while software architecture design is related to the structure or structures of the software system, which comprise software elements, the externally visible properties of those elements, and the relationships among them (Bass, Clements, & Kazman, 2003).

Over the last years a number of organizations and individual researchers have developed techniques, processes, guidelines and best practices for software architecture design (Hofmeister, Kruchten, Nord, Obbink, Ran, & America, 2006; Bass et al., 2003; Bosch, 2000; Clements et al., 2002; Dikel, Kane, & Wilson, 2001; Gomaa, 2000). Since many of the design methods were developed independently, their descriptions use different vocabulary and appear quite different from each other. Architecture design methods that were developed in different domain exhibit domain features and emphasize different goals.

Service Oriented Architecture

The notion of Service Oriented Architecture (SOA) has received significant attention with the software design and development community. The result of this attention is the proliferation of many conflicting definitions of SOA. In this context, OASIS (a not-for-profit consortium that drives the development, convergence and adoption of open standards for the global information society) provides the OASIS Reference Model for Service Oriented Architecture 1.0 (OASIS SOA, 2006). This reference model describes an abstract framework for understanding the significant entities and their relationships within a service-oriented environment. The reference model focuses on the field of software architecture and defines SOA as "a paradigm for organizing and utilizing distributed capabilities that may be under the control of different ownership domains." Visibility, interaction and effect are key concepts for describing the SOA paradigm. The value of SOA is that it provides a scalable paradigm for organizing large networks of systems that require interoperability to realize the value inherent in the individual component.

Free, Libre Open Source Software

Free, libre and open source software referred to as acronym FLOSS is a serious force that is shifting the traditional economic paradigm of software development by the emergence of non-proprietary software. The term open source is a combination of two properties: (1) access to source code and (2) freedom to modify and give it (or part of it) away (Panetto et al., 2004; EIF, 2009). The importance of this new paradigm is the opportunity to use existing software systems as a base for a new design. In this context, FLOSS denotes the freedom to modify and redistribute the software product, and not its monetary cost. In fact, most of the widely used open source license models do not prohibit charging money for the product being distributed.

Nowadays, FLOSS is a tremendous opportunity to European software entrepreneurs. However, the attitude towards FLOSS represents their significant weakness. Europe has a solid basis for Open Source Software development, but the commercial aspect of Open Source Software happens rather in North American than in Europe (Panetto et al., 2004).

Infrastructure Architecture - Middleware Solutions – Open Source SOA

Equally, it is fully recognized that open source development is a key issue for interoperability (Vitvar, Kerrigan, Overeem, Peristeras, & Tarabanis, 2006; Ley, 2007). It has been suggested that open source reference implementation of standards has a positive impact on the validation and adoption of the standards, as well as the robustness of the standards-based software. Moreover, development based on FLOSS creates new business opportunities for small and medium sized enterprises (SMEs) in the Enterprise Interoperability markets (Knothe et al., 2007).

However, no significant results have been done about practical experiences that combine the use of SOA and FLOSS in SMEs in order to provide a solution to the interoperability problem for the development of software services.

SPANISH E-GOVERNMENT INFORMATION STRATEGY

In the EU, the e-government information strategy can be seen at two levels (Vitvar et al., 2006) as (1) a global strategy driven by the European Commission to enable e-government services across the EU member states and (2) national strategies to form a national e-government available within a particular EU member state. The initiative which aims to develop a global strategy at the EU level is called IDABC program (IDABC, n. d.). Based on the fundamental principles of the EU, one of the goals of this program is to develop a European Interoperability Framework (EIF, 2009). Different national initiatives related to IDABC have been started such as GovTalk (Cabinet Office, 2010), ADAE (2011), and Digitalisér.dk (Digitalisér. dk, n. d.).

In the e-Employment services domain, the SEEMP project (IST World, 2008) is focused on designing and implementing in a prototypal way an EIF*-compliant architecture to further enhance the interoperability of existing national/ local job e-Employment services in the scenario of job-seekers.

In the Spanish context, the Act 11/2007, of 22 June, regulating the electronic access to Public Services from citizens (Ley, 2007), states in its explanatory statement that *"Public Administrations will use information technologies in their relationships with the diverse administrations and citizens, using informatics, technological, organizational and security resources which guarantee an appropriate level of technical, semantic and organizational interoperability and avoid the discrimination of citizens by their technological election"*.

Spanish Employment Services Scenario

Spain is made up by 17 autonomous communities (regions) in which the Spanish State is organized, with powers of self-government for managing its own interests. To achieve real efficiency in the design of electronic services oriented towards Spanish citizens' requirements, a high level of communication between the PA's and collaborative organizations in central, regional and local level is needed. However, one of the main problems is that Spanish country is characterized by a highly decentralized PA, as a result of which the closest local authority to the citizen will generally be his or her interlocutor for diverse matters. But there are a great many situations where competencies are shared by several levels of PA's.

The employment's competences in Spain started to emerge with an approach focused on a national (central) PA information system that covered the basic functionalities. However, at the present time, employment's competences have been transmitted to every autonomous community. This new conceptual model of Spanish Public Services has represented the need and challenge of restructuring the organizational and technical environment of every autonomous community in order to take into account the coexistence of diverse PA information systems which need to be integrated at national, regional and local level with the purpose of handling active employment policies used by the different Autonomous Public Employment Services.

ADDRESING INTEROPERABILITY IN SPANISH E-GOV EMPLOYMENT SCENARIO

In this section, we describe how the interoperability problem for Spanish e-Gov Employment services is addressed. The Canary Islands community is selected as use case to describe the interoperability problem with the stakeholders implied: (1) the national employment PA information system (SISPE), (2) the regional PA information system for the Canary Islands Community (SISPECAN) and (3) information systems of the local collaborative organizations. Interactions between information systems are supported at three different levels: (1) interactions between SISPE (national PA information system) and SISPECAN (regional PA information system), (2) interactions between SISPECAN and local information systems of collaborative organizations and (3) interactions between SISPE and local information systems of collaborative organizations.

Interoperability at National Level

The SISPE PA information system aims to solve the problems of integration and interoperability at national level among the diverse information systems that made up the different autonomous communities. At the same time, SISPE aims at promoting the cooperative and collaborative relationships between the different Employment Public Services, with the purpose of enhancing the working mobility. Two operating levels are enabled: (1) autonomous communities that make direct use of a single PA national information system and (2) autonomous communities that have developed their own PA information systems that make use of the national information system.

The three dimensions of the interoperability problem are addressed as follows:

- **Organizational Dimension:** The business model of SISPE has to be accessible for every autonomous community in order to share active employment policies, employment services and data and information exchange. These processes are oriented to satisfy the citizens' needs.
- **Semantic Dimension:** A set of syntactic assets are defined based on common data structures for all the autonomous com-

munities in order to ensure that data elements were interpreted in the same way by communicating parties. This common terminology includes information about CV summary, job search, process management of job-seekers candidates and information about job contract.

- **Technical Dimension:** The need for flexible collaboration between the different autonomous communities and the national PA information system is addressed. A set of communication services are designed to implement exchange of common data and execution of common processes. The middleware Tuxedo (Oracle, n. d.) was selected as component to handle the technical level of interoperability. Communications services are designed with fine granularity, which has permitted enabling services' calls from high level of the applications and so, providing functionalities of major importance. As a result, the four levels of interactions, defined by EIF, have been achieved: (1) stages 1, 2: services of interaction where there is not any automatic processing of information; (2) stages 3, 4: services with transactions between administrations, enterprises and citizens have been fully automated. This flexibility has introduced the administrative modernization for the Spanish Employment Public Services by means of implementing front-office services (stages 1 and 2) and back-office services (stages 3 and 4). Data synchronization at transactional level represented an important issue, since at data layer, different databases were used: Oracle database in the information systems of some autonomous communities and Adabas database in the national PA information system. Figure 1 shows an overview of the SISPE approach. In the top of the figure is represented the different autonomous communities and at the down of

the figure is represented the national (state) employment PA. Intermediate components and adaptors are used to handle data access to Adabas database from legacy information systems of the different autonomous communities.

Interoperability at Regional and Local level

In the Spanish context, at regional level, every autonomous community is responsible of handling its own PA information system for Public Employment services.

In this section, we describe the strategic guidelines of the Canary Islands community in order to fulfill SISPE PA information system (at national level) and how to expand this strategy to incorporate collaborative organizations (at local level).

The strategic guidelines of the employment PA information system of the regional Canary Islands community (SISPECAN) are based on the following design principles:

- **Integrated System:** The different subsystems that made up SISPECAN are collaborative among them, sharing data and processes.
- **e-Administration:** All processes will be supported in an electronic way.
- **Proactive Administration:** The Administration provides a model which predict the citizens' needs and not wait for the citizens' requests.
- **Collaborative Administration:** Many employment services are supported by collaborative organizations in an indirect way. A relationship based on partners is defined in order to improve the delivery of services to citizens.
- **Closest and Multi-Channel Administration:** The personal attendance of citizens to the buildings of the

Figure 1. Overview of SISPE approach

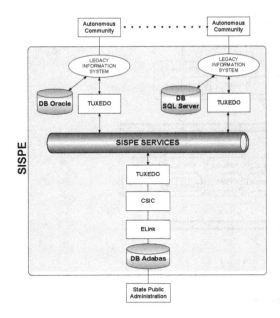

PA is changed by a multi-channel supply based on mobile services (m-government), e-government employment services (MyEmployment Portal), interactive TV, etc.

Figure 2 shows the architecture and functionalities of SISPECAN. SISPECAN is made up by the following subsystems:

- **Intermediation:** It is the main component of SISPECAN. It computes job-search process for job seekers and employers. This component implements the SISPE protocol of interoperability.
- **Training:** This module is responsible of all the aspects related to the training systems for job-seekers such as training courses, seminars, training in business, etc.
- **Guidance:** This module carries out the guidance to unemployed workers, describing the different and possible itineraries by means of individual orientation measures.
- **Funds:** This subsystem is responsible of the program of funds related to improve

the employability feature and skills of job seekers as well as job prospects.

- **Agreements:** This module is a specific component of the grants subsystem related to funds with local institutions.
- **Certification:** This module achieves the certification of the funds coming from the European Social Fund (ESF) and the State Employment Public Service (INEM).
- **Statistics:** this element is based on a data warehouse for data mining.
- **MyEmployment Portal:** This portal provides to citizens electronic and one access to employment services by means of a multi-channel approach.

Addressing Networked and Collaborative Organizations

The collaborative organizations play an important role in the delivery of better services to unemployed and employed workers.

As main guideline, the Canary Islands PA Employment (CIEPA) designs a cooperative model whose aim is to structure group of functions which are achieved by different collaborative entities. This model is a dynamic instrument that allows incorporating new organizations and therefore new electronic services to citizens in a progressive way. Thus, organizations will be gradually transformed into networked organizations, where interoperability will become the main challenge to make possible the vision of seamless business interactions across organizational boundaries. To this end, a new subsystem of SISPECAN called SISPECAN-WS was developed (Figure 2). This subsystem allows interoperating the different information systems of the diverse collaborative entities with the different subsystems of SISPE-CAN. At the same time, the SISPE protocol of interoperability was improved for sharing data between CIEPA, national PA, and PA's corresponding to other autonomous communities.

Figure 2. Overview of SISPECAN architecture

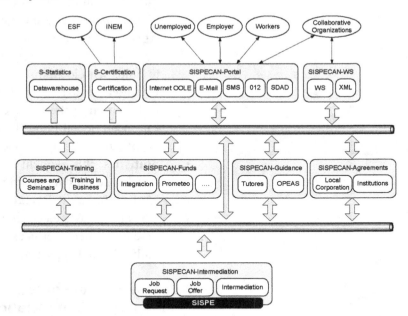

Addressing Architectural Guidelines for Collaborative Organizations

Different architectural guidelines are used to address the interoperability framework between SISPECAN (the PA regional information system) and the information systems of collaborative organizations at local level. The guidelines used are:

- The Zachman Framework (Zachman, 1987) as enterprise architecture for classifying and organizing the descriptive representation of CIEPA.
- UML and BPML enterprise modeling language as process language to describe business processes and their executions.
- The European Interoperability Framework (EIF, 2009) as main guideline to enable seamless integration of e-gov employment services.
- Service oriented architecture (SOA) (OASIS SOA, 2006) as paradigm to implement a service oriented environment.
- Open source middleware as technical infrastructure interoperability.

PRACTICAL EXPERIENCES

The use case implemented to validate the provision of centric-citizens services in the Canary Islands Community in the employment domain is focused on enhancing the training of unemployed workers with the accomplishment of new training courses and seminars according to his/her profile. These training courses are carried out by different and diverse collaborative organizations that have developed their own legacy systems/applications and have not electronic interaction with PA employment information systems at national and regional level.

Objective

The following functionalities are introduced by SISPECAN in order to fulfill the accomplishment of training courses by the different organizations: (1) to corroborate the conditions of unemployed workers at the beginning of the training course, in order to validate unemployed candidates can access to the training course, (2) to register and program the course of the collaborative organization responsible of the training course, (3) to register the unemployed workers who accomplish the training courses, (4) to annotate the results of the training course to curriculum summary of the unemployed workers, (5) to cancel courses if conditions of the collaborative organization are not validated and (6) to authorize and authenticate users who can access to the system.

Three different interaction types are necessary to cover these functionalities: (1) exchange of data between the national PA employment information system (SISPE) and the regional PA information system (SISPECAN), (2) exchange of data between SISPECAN and collaborative organizations information systems and (3) exchange of data between collaborative organizations information systems and SISPE. This last interaction is computed by means of SISPECAN.

Services

The approach proposed for the provision of centric-citizens services relies on the concept of services. With this aim, open source service oriented architecture is designed and implemented. The different functionalities described at previous section have been wrapped as services and they have been implemented by Web services technology on SOA. Open source software is used to perform the interoperability framework.

The proposed SOA approach is three-layer architecture: (1) the technical integration layer consists of different connectors, API's and wrappers establishing a technical integration between the SOA and the legacy and external applications of every collaborative organization; (2) an enterprise service bus (ESB) that enables interoperability between the Web services exposed by SISPE-CAN and the collaborative organizations; and (3) orchestration of the different services by means of a workflow engine based on BPEL (Business Process Execution Language).

ServiceMix (n. d.) is used as enterprise service bus. This ESB is based on the standard JBI

(Java Business Integration - JSR 208) (Vinoski, 2005). This solution includes a framework for the deployment and monitoring of JBI compliant components, as well as for the routing between these components (JSR 208) (Java Community Process, 2005). This ESB connects and routes the messages from collaborative organizations to the corresponding systems/applications of SISPECAN.

The security model is developed on two levels: (1) transport level, SSL is used for protection, integrity and confidentiality of exchanged data; and (2) authentication level, Web services security standard (OASIS, 2006) is used with the purpose of authenticating and authorizing the user of the system.

Figure 3 shows interoperability framework for e-Gov employment in the Spanish context, focusing on the use case of Canary Islands autonomous community.

A functional model based on two levels of interactions was achieved between three different entities in a networked way: collaborative organizations, autonomous community (in this case, Canary Islands PA Employment, represented by CIEPA) and INEM (that represents the national PA). As a result of using this framework of interoperability, the information system of every collaborative organization interoperates with the service oriented architecture of SISPECAN supported on a ServiceMix ESB and at the same time SISPECAN interoperates with SISPE (State Public Services of Employment) by means of the middleware Tuxedo (Oracle, n. d.).

This framework of interoperability closes the gap of interaction between different organizations and ensures that new organizations can be incorporated in a dynamic way, where the only prerequisite is to develop a client application on its legacy system.

From the citizen perspective, new services are accessed by the portal of Canary Islands Employment (MyEmployment Portal). To be precise, the physical interaction with civil servants in the

Figure 3. Interoperability Framework for Spanish e-Gov Employment. Canary Islands PA Employment (CIEPA) is selected as use case.

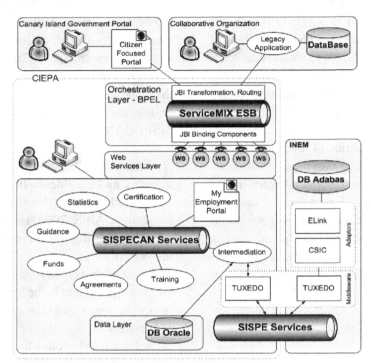

administrative buildings is being replaced with virtual assistances through the portal. In the future, all services provided by different administrations will be unified by the citizen-focused Portal of Canary Islands Government Portal.

Results and Lessons Learned

Our practical experience has demonstrated that working at national level with proprietary middleware infrastructure is more complex and limited. The main reason is that Tuxedo is not based on open standards and therefore it limits the use of different ICT technologies. On the other hand, the use of open source software is the opportunity to use existing software systems as a base for the SOA design. As a result, source code availability allows us to improve, fix and support the reused components. This situation improves our software development because new components were de-

veloped for the ServiceMix ESB. To be precise, new components were developed for integration services which used complex business logic.

With the proposed approach, the access from the collaborative organizations to the distributed services supported by the Canary Islands Employment Public Services is rerouted through the ESB and its orchestration component. This design model decouples the interface of the service calls from their implementation. That is, client access to services remains unchanged, independently of eventual changes in the underlying sub-systems at the Canary Islands PA Employment information system.

The underlying principles in the use case have been accomplished in the following way: (1) The Web Accessibility guidelines have been adopted in order to ensure accessibility feature; (2) multilingualism at the present time is not used in front office and web pages, only Spanish language is employed in the delivery of e-Employment services. At back-office level, XML-schemes are used; (3) security aspects and personal data protection have been carried out by means of LDAP, SSL protocol and Web security standard; (4) use of open standards and open software have been ensured; (5) the use of a SOA based on open standards and open software opens new changes to the integration towards the interoperability framework at a pan-European level.

We found that open source development tools for building, deploying and testing production quality work well together. However, we have found three drawbacks working with ServiceMix ESB: (1) insufficient documentation for software developers, (2) the technical complexity resulting from the combination of different technologies such as JBI, BPEL, SOAP, JMS, and (3) advanced knowledge is required in order to define services reactive to input messages and acting with diverse ways with different subsystems by means of different messages.

CONCLUSION AND FUTURE WORK

E-government employment services are defined in a complex architectural and technological scenario. The diverse Public Administrations, agencies and collaborative organizations that are involved are usually autonomously and independently managed. Therefore, sharing electronic services between them to reach a common overall goal is an important way to improve the delivery of services to citizens. From the citizen-centric viewpoint an important reduction of the administrative burden is achieved. The interoperability issue becomes one of the main challenges in order to provide efficient services to citizens from autonomous and networked institutions. Nowadays, diverse architectural engineering guidelines have been proposed to support the interoperability issue at different levels of abstraction. In this paper, we have reviewed the state of the art of the

interoperability problem and we have presented the motivating scenario of e-Gov employment services in the European Union member state, Spain. The Canary Islands community is selected as use case of interoperability with collaborative organizations at local level. The European Interoperability Framework is adopted as main guideline to supplement national interoperability framework and to support in future services at pan-European level.

ACKNOWLEDGMENT

This work has been supported by the Spanish Government under the project TIN2008-06570-C04-03 and the Consellería de Innovación e Industria - Xunta de Galicia under the project SEGREL: Semántica para un eGov Reutilizable en entornos locales (08SIN06322PR).

REFERENCES

ADAE. (2011). *Association des Dirigeants et Administrateurs d'Entreprise*. Retrieved from http://www.adae.asso.fr

ATHENA. (n. d.). *Advanced technologies for interoperability of heterogeneous enterprise networks and their application*. Retrieved from http://www.athena-ip.org/

Bass, L., Clements, P., & Kazman, R. (2003). *Software architecture in practice* (2nd ed.). Reading, MA: Addison-Wesley.

Bosch, J. (2000). *Design and use of software architecture: Adopting and evolving a product-line approach*. Reading, MA: Addison-Wesley.

BPEL. (2007). *Business process execution language for web services*. Retrieved from http://www.ibm.com/developerworks/library/specification/ws-bpel/

Cabinet Office. (2010). *GovTalk*. Retrieved from http://www.govtalk.gov.uk

CIMOSA. (n. d.). *CIM open system architecture*. Retrieved from http://www.pera.net/Methodologies/Cimosa/CIMOSA.html

Clements, P., Bachmann, F., Bass, L., Gralan, D., Ivers, J., & Litle, R. (2002). *Documenting software architectures: Views and beyond*. Reading, MA: Addison-Wesley.

Digitalisér.dk. (n. d.). *About Digitalisér.dk*. Retrieved from http://digitaliser.dk/resource/432461

Dikel, D. M., Kane, D., & Wilson, J. R. (2001). *Software architecture: Organizational principles and patterns*. Upper Saddle River, NJ: Prentice Hall.

Doumeingts, G., Vallespir, B., & Chen, D. (1998). Decisional modelling using the GRAI grid. In Bernus, P., Meritns, K., & Schmidt, G. (Eds.), *Handbook on architectures of information systems* (pp. 313–338). Berlin, Germany: Springer-Verlag.

EIF. (2009). *European interoperability framework for pan-european egovernment services*. Retrieved from http://ec.europa.eu/idabc/en/document/2319/5644.html

Gomaa, H. (2000). *Designing concurrent, distributed and real-time applications with UML*. Reading, MA: Addison-Wesley.

Hofmeister, C., Kruchten, P., Nord, R., Obbink, H., Ran, A., & America, P. (2006). A general model of software architecture design derived from five industrial approaches. *Journal of Systems and Software, 80*, 106–126. doi:10.1016/j.jss.2006.05.024

IDABC. (n. d.). *Interoperable delivery of European e-government services to public administrations, businesses and citizens*. Retrieved from http://europa.eu.int/idabc

IEEE. Computer Society. (1990). *IEEE standard glossary of software engineering terminology*. Washington, DC: IEEE Computer Society.

Java Community Process. (2005). *JSR 208: JavaTM Business Integration (JBI)*. Retrieved from http://www.jcp.org/en/jsr/detail?id=208

Ley. (2007). *Ley 11/2007 de acceso electrónico de los ciudadanos a los Servicios Públicos*. Retrieved from http://www.boe.es/aeboe/consultas/bases_datos/doc.php?coleccion=iberlex&id=2007/12352

OASIS. (2006). *Oasis web services security (WSS)*. Retrieved from http://www.oasis-open.org/committees/tc_home.php?wg_abbrev=wss

OASIS SOA. (2006). *OASIS reference model for service oriented architecture 1.0*. Retrieved from http://www.oasis-open.org/committees/download.php/19679/soa-rm-cs.pdf

Oracle. (n. d.). *Oracle tuxedo application runtime for CICS and Batch*. Retrieved from http://www.oracle.com/products/middleware/tuxedo/tuxedo.html

Panetto, H., Scannapieco, M., & Zelm, M. (2004). INTEROP NoE: Interoperability research for networked enterprises applications and software. In *Proceedings of the Workshops of On the Move to Meaningful Internet Systems* (pp. 866-882).

ServiceMix. (n. d.). *The software apache foundation*. Retrieved from http://servicemix.apache.org/home.html

Spur, G., Mertins, K., & Jochem, R. (1996). *Integrated enterprise modelling*. Berlin, Germany: Springer-Verlag.

Vinoski, S. (2005). Towards integration – java business integration. *IEEE Internet Computing, 9*(4), 89–91. doi:10.1109/MIC.2005.86

Vitvar, T., Kerrigan, M., Overeem, A., Peristeras, V., & Tarabanis, K. (2006, March). Infrastructure for the semantic Pan-European e-government. In *Proceedings of the Semantic Web and E-Government Conference*.

World, I. S. T. (2008). *SEEMP: Single European employment market-place.* Retrieved from http://www.ist-world.com/ProjectDetails.aspx?ProjectId=dba1d15b009e48ecad153f131c29120c&SourceDatabaseId=8a6f60ff9d0d4439b41532481 2479c31

Zachman, J. A. (1987). A framework for information systems architecture. *IBM Systems Journal, 26*(3). doi:10.1147/sj.263.0276

This work was previously published in the International Journal of E-Services and Mobile Applications, Volume 3, Issue 3, edited by Ada Scupola, pp. 1-15, copyright 2011 by IGI Publishing (an imprint of IGI Global).

Section 5
Applications for Innovation in E–Business and E–Government

Chapter 15

The WAVE Platform:
Utilising Argument Visualisation, Social Networking and Web 2.0 Technologies for eParticipation

Deirdre Lee
National University of Ireland, Ireland

Yojana Priya Menda
National University of Ireland, Ireland

Vassilios Peristeras
National University of Ireland, Ireland

David Price
ThoughtGraph Ltd., UK

ABSTRACT

The growth of Information and Communication Technologies (ICTs) offers governments advanced methods for providing services and governing their constituency. eGovernment research aims to provide the models, technologies, and tools for more effective and efficient public administration systems as well as more participatory decision processes. In particular, eParticipation opens up greater opportunities for consultation and dialogue between government and citizens. Many governments have embraced eParticipation by setting up websites that allow citizens to contribute and have their say on particular issues. Although these sites make use of some of the latest ICT and Web 2.0 technologies, the uptake and sustained usage by citizens is still relatively low. Additionally, when users do participate, there is the issue of how the numerous contributions can be effectively processed and analysed, to avoid the inevitable information overload created by thousands of unstructured comments. The WAVE platform addresses what the authors see as the main barriers to the uptake of eParticipation websites by adopting a holistic and sustained approach of engaging users to participate in public debates. The WAVE platform incorporates argument visualisation, social networking, and Web 2.0 techniques to facilitate users participating in structured visual debates in a community environment.

DOI: 10.4018/978-1-4666-2654-6.ch015

INTRODUCTION

Nowadays, Information and Communication Technologies (ICTs) are widely deployed across many sectors, for example the health sector, the education sector, and the government sector. The utilisation of ICT in the government sector offers opportunities to efficiently provide government information and public services to citizens in profound ways (Dutton, 2007; Anttiroiko, 2003). Democratic institutions now face the challenge of adapting to these modern methods of governance which catalyzed the emergence of eDemocracy (Hague & Loader, 1999). eDemocracy is defined as the use of ICTs to support the democratic decision-making processes, and may be further divided into two distinct areas: eParticipation and eVoting. eParticipation soon emerged as a research area in its own right, as it opens up greater opportunities for consultation and dialogue between government and citizens (Macintosh, 2004).

Participation generally implies 'joining in', either in the sense of taking part in some communal discussion or activity, as in deliberation, or in the sense of taking some role in decision making. The 'e' in eParticipation refers to the use of ICTs, in particular the Internet and Web 2.0 technologies, in the participation process, with the implication that technology has the ability to change and transform citizen involvement in deliberation and decision-making processes (Sæbø, Rose, & Skiftenes Flak, 2008). The focal point of eParticipation is the citizen, i.e. the purpose of eParticipation is to increase citizen's abilities to participate in the government's decision making process (Sæbø, Rose, & Skiftenes Flak, 2008; Tambouris, Liotas, & Tarabanis, 2007). However, the number of citizens joining and contributing to online participation sites is minimal. Possible explanations include a lack of awareness of the existence of these sites or a digital divide, which creates a gap between those that are familiar with technology and those that are not (Fang, 2002).

Therefore, of utmost importance is motivating citizens to engage in the government policy-making process. Merely providing typical ICT solutions is not the answer to improving citizen participation in government policy-making proceedings (Macintosh, 2007). In saying this, governments have already embarked upon a new wave of initiatives to actively engage citizens in the democratic processes, which is illustrated by the increasing number of eParticipation projects, tools and technologies (Tambouris, Liotas, & Tarabanis, 2007). Many of these existing sites use Web 2.0 technologies for easy publishing of information, information sharing and user collaboration. Some examples of popular Web 2.0 applications are blogs, wikis, podcasts, RSS feeds, tagging, social networks, search engines and multiplayer online games (Osimo, 2008).

However, the majority of existing eParticipation websites, are not successful in engaging large numbers of general public (Heeks, 2003). We believe that a major obstacle to sustainable user engagement is a lack of impact. A lack of impact refers to the absence of visible action resulting from eParticipation. eParticipation requires the committed contribution of all stakeholders in the decision making process. When citizens, businesses, and other organisations are willing to dedicate time and effort into voicing their opinions, it is the responsibility of the policy makers to acknowledge these opinions and respond accordingly. eParticipation websites that are not established by a government body, or that do not involve policy makers from the beginning are at an immediate disadvantage, as they need to first build up trust and acceptance from citizens for the content that is generated. But even in governmental eParticipation sites, the opinions expressed by citizens are rarely used in any visible way. This discourages citizens from further contributing to this site.

Another barrier to eParticipation is how to process and analyse the numerous user contributions effectively. Many opinions expressed on current eParticipation websites are structured in an informal, conversational format. Valid arguments, opinions, and disputes are lost in the sea of forums, blogs, wikis, and tweets. Technologies were designed to encourage discourse between people and to enable people to express opinions and reactions freely. However, debating on a particular issue requires a dedicated effort to formulate an argument and to evaluate how this argument compares with and addresses other arguments. Therefore to build a rich and expressive debate with clearly thought out arguments, a more structured format is required than those conversational based sites mentioned above.

In this paper we present the WAVE platform (http://www.wavedebate.eu/) which aims to improve the inclusiveness and transparency of EU decision making at the national and European level by incorporating argument visualisation, social networking, and Web 2.0 techniques. The use of highly integrated, state-of-the-art argument visualisation techniques to make the impact of complex European Union environmental legislation on climate change more accessible and easy to understand for citizens, special interest groups and decision makers alike.

The WAVE platform has been developed as part of the WAVE project (WAVE, 2009), an eParticipation Preparatory Action programme co-funded by the European Commission. The rest of this paper is structured as follows. First we present the current state of the art of eParticipation websites. Next we present the WAVE platform and its architecture. A preliminary evaluation is documented and the conclusions and future work are presented.

STATE OF THE ART

Many national and international eParticipation websites exist online today. In an extensive study carried out as part of the European eParticipation project, 258 eParticipation cases operating at European, national, regional, and local levels were identified and analysed (Panopoulou, Tambouris, & Tarabanis, 2009). Fifty eParticipation cases (namely 19% of the total eParticipation cases) were identified as having European scope. With regard to the eParticipation area that each case addresses, most European cases involve information provision services, followed by discussions and consultation facilities.

As part of the requirements-gathering phase for the WAVE platform, a study of existing European eParticipation websites was carried out (Lee et al., 2009). Although this study is not exclusive, in that it only provides information on a sub-set of all eParticipation sites available, these sites were chosen as they are representative of the kinds of eParticipation websites that are currently online. Also, the main aim of these sites are comparable to what the WAVE project is trying to achieve, as many incorporate deliberation and debating tools, and utilise Web 2.0 technologies. An overview of the use of Web 2.0 technologies by the analysed eParticipation sites are presented in Table 1.

Through this analysis we found that news updates, RSS news feeds, and pod/video-casting are prevalent features on eParticipation websites. As such, users of eParticipation websites will expect to have such features available to them. Blogging, discussion forums, calendar of events, inclusion of statistical information, and rating systems are also popular, with 50% of our websites including these features. While analysing why these technologies are not more prevalent, it should be noted that these technologies serve specific purposes and their inclusion may not have been in line with the goals of the particular website. The same may hold for the lesser-used technologies, such as bookmarking, social networking, tag clouds, micro-blogging, petition, and polling. Alternatively, these technologies may not be included in many of the eParticipation websites, as their benefit is not clear. For example, in our study, only ePractice.eu and Change.org have embraced social networking. Social networking is quite a recent phenomenon on the Internet and is seen as

Table 1. Overview of Web 2.0 technologies used in eParticipation websites

↓ Web 2.0 Technology / Website →	voice http://www.give-your-voice.eu	debate-europe http://europa.eu/debateeurope/index_en).htm	national-dialogue http://www.thenationaldialogue.org/ideas/)	ePractice http://www.epractice.eu/index.php?page=home	change.og http://www.change.org	Debatewise http://www.epractice.eu/index.php?page=home	africa-gathering http://africagathering.org	Ourclimate http://www.ourclimate.eu/ourclimate/default.aspx
Blog	✓			✓	✓		✓	
Discussion Forum	✓	✓	✓					✓
News	✓			✓	✓	✓	✓	✓
RSS Feed	✓			✓	✓		✓	✓
Podcast/ Videocast		✓	✓	✓	✓			✓
Bookmarking	✓			✓		✓		
Social Networking				✓	✓			
Tag Cloud			✓	✓			✓	
Micro-blogging	✓				✓		✓	
Petition					✓			
Poll	✓	✓	✓					
Calendar/Events	✓			✓			✓	✓
Search Engine	✓	✓	✓	✓	✓	✓		✓
Statistical Info			✓	✓	✓	✓		
Rating System			✓	✓	✓	✓		
Geo Map								

a social and informal meeting place for friends and strangers online. How informal social networking could benefit traditionally, formal government websites is yet to be researched. However, at the time of writing, ePractice.eu has 16,567 members, proving that the inclusion of such lesser used Web 2.0 technologies may be invaluable to the usage of eParticipation sites.

Engaging more people in the discussion and formulation of public policies and then being able to create meaningful summaries of their contributions is clearly a major challenge for eParticipation. First, technologies should lower the entry barriers to these types of discussions. This results in thousands of previously excluded people to now openly express and exchange their opinions and ideas, creating a vivid community of participation. Secondly, to process, present and query large vol-

umes of input generate mass collaborative public networks that emerge dynamically over the Web, there is a need to combine traditional intelligent technologies such as natural language processing with applications for argument representation, visualization and opinion processing (Peristeras, Mentzas, Tarabanis, & Abecker 2009). However, online collaborative policy mapping of this kind is a new field. Moreover, thinking deeply and constructively in a structured way as a community involves care, patience, discipline and tolerance for ambiguity: so the process of building, and to a lesser extent reading, the arguments place a greater intrinsic demand on participants than other familiar tools that work most naturally at the surface level of the conversation.

THE WAVE PLATFORM

In WAVE, we adopt a holistic approach to engaging users in eParticipation. Three pilot sites participate in WAVE: the UK pilot, the French pilot, and the Lithuanian pilot. Each pilot site has a national version of the WAVE platform; however the platform is only one aspect of the approach. The pilots are responsible for coordinating efforts to promote the tool, ensure take-up of the tool, and, if necessary, for guiding and assisting users in using the platform. The platform itself should engage users to participate on the site and even contribute to the live debates. The pilots then coordinate feeding the user contributions to policy-makers, which the policy-makers may consider when deciding on issues. This process of contribution and feedback is iterative, with the theory being that having direct impact on policy-makers encourages users to continue contributing to the platform.

The WAVE platform facilitates this approach. The goal of the WAVE platform is to engage users in sustainable eParticipation, with low technical entry-barriers. The WAVE platform achieves this goal by:

- Enabling all stakeholders, citizens and policy makers alike, to participate openly in a common community on their areas of interest.
- Involving policy makers from the beginning of the project, so that they can oversee the entire participation process, become actively involved, and use the results of debates in their governmental roles.
- Deploying an innovative argument visualisation platform for the sound and structured development of rich debates, in a multi-lingual, cross border context.

The WAVE platform is designed as an open, community-based, interactive website, providing a dynamic social environment that focuses users around particular issues and ultimately encourages them to participate in debates. In this way it overcomes the shortcomings of existing eParticipation web-sites. The WAVE platform aims to address the divide between government and social websites by providing a transparent and comfortable environment for a social community to grow. The WAVE Platform is shown in Figure 1.

While gathering requirements for the WAVE platform, many attributes continually appeared as being integral to the success of any eParticipation platform. We define these attributes as the fundamental requirements of the WAVE platform, meaning that they are the underlying principles

Figure 1. WAVE Platform (© NUIG, 2010, used with permission)

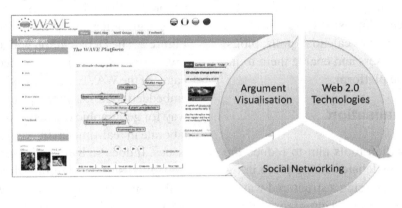

that guided the design, development, and implementation of the WAVE platform. These requirements emerged from an extensive literature review of eParticipation sites, online debating sites, and online communities, as well as collaboration with the pilot site teams, who have direct contact with the stakeholders. The requirements are:

- Usability
- Structured data
- Social networking
- Community awareness
- Multicultural support
- Security and transparency
- Interoperability
- Feedback and sustainability

In order to facilitate and absorb users from a wide and diverse demographic, with varying political and technical expertise, the WAVE platform incorporates the following functionality:

- **Argument Visualisation:** An innovative argument visualisation tool, Debategraph (http://www.debategraph.org/), allows users to engage in debates using a graphical mapping-interface, including a mechanism to rate and review the arguments and the structured dialogue
- **Social Networking:** Provision of personal space to create an identity in the community and to network with communities of interest by forming groups with other users
- **Web 2.0 Technologies:** Usage of Web 2.0 technologies such as calendar, polls, blogs, etc. to engage users and enable them to interact with the platform.

Argument Visualisation

Argument visualisation tools aim to reduce the complexity of online debates taking place in forums and blogs by enforcing a structured way of contributing and by employing intuitive graphical

user interfaces. An argumentative discussion starts with an initial proposition stated by a single creator and is then followed by supporting propositions or counter-propositions from other contributors (Lange, Bojārs, Groza, Breslin, & Handschuh, 2008). The initial proposition becomes the root node or parent node in the argument space, which in this case is the discussion under consideration, and the supporting or counter propositions become the children of the root/parent node. The interlinking between these nodes which represent the different propositions of an arguments, gives rise to well-structured graphical argumentation maps (Shum, 2003; Van Bruggen, Boshuizen, & Kirschner, 2003). Argumentation maps are in fact semantic networks of argument nodes connected by meaningful relations which provide a useful starting point to analyze dependencies between map parts and answer queries (De Moor, Park, & Croitoru 2009). Argumentation maps typically exhibit a high degree of formality and use incremental formalism, so that semantics is emergent and not predefined. This means that the argumentation tool permits a stepwise evolution of the argumentation space, through which formalization is not imposed by the system but is at the user's control (Tzagarakis, Karousos, Karacapilidis, & Nousia, 2009).

Debategraph

Debategraph is a standalone, argument-visualisation tool that allows users to engage in debates using a graphical mapping-interface, as shown in Figure 2. The Debategraph widget (which utilises JavaScript and Flash) is embedded into the relevant pages of the WAVE platform using iFrame HTML embedded code. Debategraph provides a powerful way for geographically distributed groups to collaborate in real-time in thinking through complex issues. It does so by enabling groups of any size to externalise, visualize, question, and evaluate all of the considerations that anyone thinks might be relevant to the issues at hand – and by facilitating

Figure 2. The Embedded Debategraph Widget (© ThoughtGraph, 2010, used with permission)

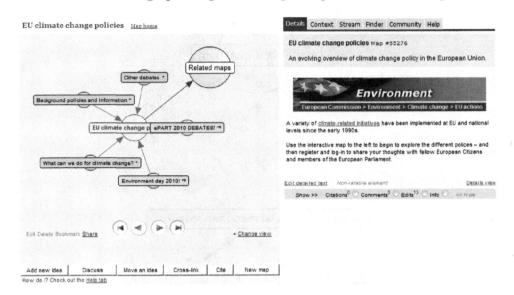

an intelligent, constructive dialogue around those issues. In essence, building the maps involves three steps:

1. Breaking down the subject into meaningful ideas;
2. Figuring out the relationships between those ideas; and,
3. Expressing the ideas and relationships visually.

The components of a Debategraph argument map are shown in Figure 3. Issues (or questions) are raised; Positions (or answers) suggested in response to these Issues, and Supportive and Opposing Arguments advanced for and against the Positions (and each other). Each building block has its own colour to make it easier to see the types of ideas and relationships at a glance. There's a wider set of building blocks beyond the three core ones, but the core set of Issues, Positions, and Supportive and Opposing Arguments can be combined and recombined many times to build rich maps on any scale.

The reasons for map-building are as follows:

- Complex debates can be mapped comprehensively, so that all of the relevant arguments (from all perspectives) can be woven together into a single, coherent and iteratively improvable visual structure;
- The process can be opened up transparently via the web so that many people can contribute to, visualize, learn from, and evaluate our collective understanding of the debates; and,
- These collaborative, visual maps can help us make sense of and act more efficiently and effectively in relation to these debates.

Social Networking

Social networking engages users by binding individuals or organisations together around a common theme/topic. Social networks are characterised by the principles of (Rose, Sæbø, Nyvang, & Sanford, 2007):

- Free access to information,
  ○ Self-organisation,
  ○ Mass collaboration,

Figure 3. Debategraph Map Components (© ThoughtGraph, 2010, used with permission)

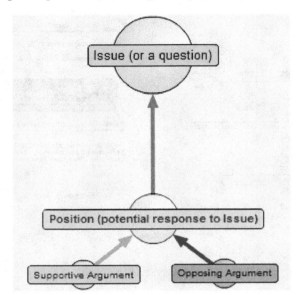

○   Non-exclusive services, and
○   User participation.

A recent survey conducted by the Semantic Web Company shows that social software, such as wikis, forums, blogs, and RSS feeds have already been widely introduced and are widely used in organisations today (Pellegrine, 2009). According to the results of this survey, the main benefits of social software are:

•   Access to information/knowledge quickly and
•   Networking with others easily and staying in touch.

The main barriers to social software are:

•   Amount of time that is required,
•   Redundancy with regard to other systems, and
•   Losing control over one's data/knowledge.

In eParticipation there is a need to take advantage of the widespread use and acceptance of social networking technology to improve citizen engagement. However additional challenges may occur in the context of social networking it the political domain, as there needs to be a high level of trust before users contribute. While government websites typically provide more formal and exposed eParticipation applications, social websites are perceived as offering a more trusted and intimate environment to express political views (Coleman, Macintosh, & Schneeberger, 2007).

The WAVE platform aims to address the divide between government and social websites by providing a transparent and comfortable environment for a social community to grow. Through user profiling and groups, a WAVE community exists on the WAVE platform, as shown in Figure 4. Each user completes a form on registration detailing their profile, e.g. gender, occupation, location, etc. While providing this information creates an atmosphere of openness and fuels statistical analysis of users, it may also be seen as a barrier to participation. Therefore the completion of all user information, apart from username, password, and email is optional. A user's profile also contains the groups that the user is a member of and activity

Figure 4. Social Networking in the WAVE Platform (© NUIG, 2010, used with permission)

information. The user's activity refers to any actions that the user has performed on the platform, e.g. created a group, event, or poll. Activity that has occurred in any of the groups that the user is a member of is also displayed.

A WAVE Group is dedicated to a particular issue and connects users with a common interest. The goal of a group is to facilitate users to become active on particular maps, rather than getting lost in issues that they have no real interest in. Each group page consists of a short description, a list of members, an activity stream, and an embedded map. The activity stream is similar to that on a user profile and shows activities that have occurred relating to the group. The root node of the embedded map corresponds to the group issue and is chosen on group creation.

Web 2.0 Technologies

Web 2.0 technologies are a new generation of online technologies that enable a user, not only to read information from the Internet, but also to contribute and participate online. These interactive tools facilitate user collaboration and co-operation.

- **RSS Map Updates:** Displays the stream of all activity on the Debategraph maps

- **Blog:** Enables the site moderator to post messages to the WAVE community, which any WAVE member can comment on.
- **Poll:** Displays the latest polls created by WAVE members.
- **Calendar Events:** Displays any events that have been created by WAVE members.
- **Tagging:** The tag cloud offers a great way to find and explore WAVE content that WAVE members have tagged across the platform. Click on a tag to open a page listing all of the current items tagged with a given tag.
- **Activity Notifications:** Shows you the stream of all activity on the WAVE platform, including news of any new WAVE members, new groups, new events, new polls, or new blog-posts.

WAVE PLATFORM ARCHITECTURE

The WAVE platform consists of three pilot sites (French, Lithuanian, and UK), which have a common layout and provide common functionality. Each of the pilot sites consist of a home page, user pages, and group pages, and includes the Debategraph tool. However, each of the sites will be presented in the national language and the content

Figure 5. Web 2.0 Tools in the WAVE Platform (© NUIG, 2010, used with permission)

of the site, including the Debategraph maps, will be country-specific. Figure 6 presents a deployment diagram of the WAVE platform, depicting the hardware for the system, the software that is installed on that hardware, and the middleware used to connect the disparate machines to one another.

As can be seen from the deployment diagram, all three websites are hosted on an Apache2 web server. Drupal is used to create the websites, while MySQL is used as the database management system. The main components of a Drupal website

are: modules, content, presentation, and user. Debategraph is embedded in the platform using iFrames and information is passed between the WAVE platform and Debategraph via REST Web Services and RSS Feeds.

In Figure 7, use-cases that capture the main system behaviour of the WAVE platform are displayed. As can be seen from the diagram, a user may interact with the platform as an anonymous user or a registered user. This is important as it means that an anonymous user may still browse the platform and explore the maps. However, in

Figure 6. WAVE Platform Deployment Diagram (© NUIG, 2010, used with permission)

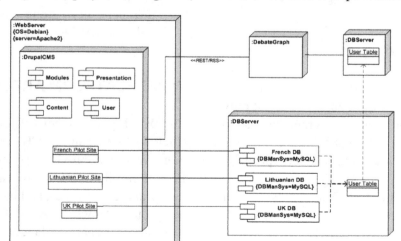

Figure 7. WAVE Platform Use-Cases (© NUIG, 2010, used with permission)

order to contribute to the platform, a user must register/login. A registered user may be further defined as a WAVE group member and/or a moderator. Any registered user may become a member of any WAVE group. A moderator is responsible for a pilot site and has control over all content on the platform.

The main components of the WAVE platform are depicted in Figure 8. The registration and authentication components enable the WAVE platform to keep track of its users. Moreover, registering users promotes a sense of trust and openness on the platform, as a user can be identified as the same person that made multiple contributions to maps, or as a member of multiple groups. If this is intimidating for a user wishing to express political opinions, they may still reg-

ister using an avatar and pseudonym. A user's profile contains user information as well as the groups that the user is a member of and activity information. The activity component handles this functionality. Another task carried out by the activity component is the displaying of recent map activity on the WAVE platform. This is achieved through the RSS notification stream that Debategraph publishes, handled by the RSS Feed component.

A WAVE Group is dedicated to a particular issue and connects users with a common interest. The root node of the embedded map corresponds to the group issue and is chosen on group creation. The Debategraph interface has been updated by Thoughtgraph, in order to integrate seamlessly with the WAVE platform and to improve the us-

Figure 8. WAVE Platform Component Diagram (© NUIG, 2010, used with permission)

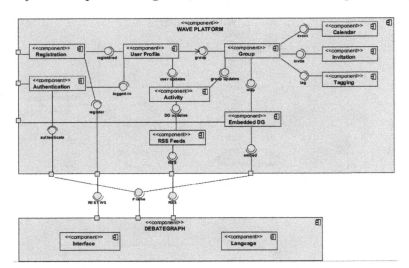

ability of the argument visualisation tool. Debategraph has also added a language component, so that the interface may be easily translated into many languages, hence widening the potential user catchment area.

EVALUATION

Preliminary Results

In Figure 9, a summary of the usage statistics of the WAVE platform for the closed user phase is displayed. This summary was generated from the Web server log files using WebLog Expert Lite (http://www.weblogexpert.com/) A hit represents a server request for any file (page, image, etc.), a page view represents a server request for a page, a visitor is determined from the IP address, and bandwidth represents the amount of traffic transmitted from the site. As can be seen, there has been a lot of activity on the pilot sites during this phase.

The main feedback gleaned from users during the closed user phase regarding the WAVE platform was that while the functionality of the platform was sufficient, there were general usability issues.

When the users were confronted with the homepage, they were unsure how to proceed, what was expected from them, and how they can interact with the platform. This was a major barrier to participation and needed to be addressed immediately. In response to this issue the WAVE workflow was developed.

WAVE Workflow

The WAVE workflow outlines the sequence of connected steps that a user may perform to navigate through the WAVE platform in a logical progression. The WAVE workflow consists of six steps and is displayed on each pilot site's front page, as shown in Figure 10. The first step, Explore, encourages users to browse around the maps so that they become familiar with the interface and see how arguments are structured. This learning process should be natural and subconscious as they find issues that are of interest to them. Users may also explore the Web 2.0 features on the WAVE platform, such as groups, other users' profiles, polls, etc. In order to contribute to the WAVE platform a user must Join. Rating is a simple way to contribute to a map, which a user

Figure 9. WAVE Platform Usage Statistics (© NUIG, 2010, used with permission)

Time range: 22/05/2009 20:27:38 - 25/03/2010 14:39:46 Generated on Thu Mar 25, 2010 - 15:57:49	
Hits	
Total Hits	1,124,426
Visitor Hits	685,980
Spider Hits	438,446
Average Hits per Day	3,650
Average Hits per Visitor	43.54
Cached Requests	251,425
Failed Requests	43,213
Page Views	
Total Page Views	179,792
Average Page Views per Day	583
Average Page Views per Visitor	11.41
Visitors	
Total Visitors	15,756
Average Visitors per Day	51
Total Unique IPs	5,859
Bandwidth	
Total Bandwidth	20.46 GB
Visitor Bandwidth	7.71 GB
Spider Bandwidth	12.75 GB
Average Bandwidth per Day	68.01 MB
Average Bandwidth per Hit	19.08 KB
Average Bandwidth per Visitor	512.80 KB

Figure 10. WAVE Workflow Front-Page (© NUIG, 2010, used with permission)

may perform while still becoming accustomed to the map structure. Sharing Ideas and Joining Groups are the next steps. Once a user is comfortable with the WAVE platform, they may decide to contribute more, e.g. adding their own ideas to the maps, joining groups, creating groups etc. At this stage a user is utilising most of the functionality of the WAVE platform, but because the user's participation on the site was done incrementally, following a clear workflow, the technical barriers to usage have been dramatically reduced. The final step of the workflow involves giving Feedback to the pilot site moderators regarding the WAVE approach. Feedback also refers to the onus on the pilot site moderators to distribute feedback stemming from users' contributions on the maps to policy-makers.

CONCLUSION

During the last few years a large number of eParticipation sites have been proposed. These platforms however have not achieved widespread adoption. With the WAVE platform, we address what we see as the main barriers to the uptake of eParticipation websites, by adopting the WAVE approach: a holistic and sustained approach of engaging users to participate in public debates. The WAVE platform incorporates argument visualisation, social networking, and Web 2.0 techniques to facilitate users participating in structured visual debates in a community environment. The WAVE platform provides a common space where all stakeholders can openly collaborate and deliberate over an issue (currently the focus in on climate change issues). As well as using argument visualisation tools, the WAVE platform incorporates many Web 2.0 technologies to maximise its accessibility, ease-of-use, and effectiveness. The Debategraph argument visualisation tool is at the core of the WAVE platform, enabling users to incrementally and collaboratively build debates around certain topics. The WAVE platform facilitates

users to form communities around these debates and focuses on current activity taking place in the debate map. All users are treated equally in the WAVE platform, so that citizens, business, NGOs, and policy makers can contribute and deliberate side by side. Each user has access to the same information and is aware of what other users are contributing, resulting in an open and transparent environment. The WAVE platform has been deployed and tested by a small number of users. Future work includes launching a large scale deployment of the platform anticipating its use by large number of citizens across Europe.

ACKNOWLEDGMENT

The work presented in this paper has been funded in part by Science Foundation Ireland under Grant No. SFI/08/CE/I1380 (Lion-2) and the European Union under Grant No. EP-08-01-002 (WAVE). The authors would like to thank all the WAVE project partners for the creative discussions and ideas.

REFERENCES

Anttiroiko, A. V. (2003). Building strong e-democracy: the role of technology in developing democracy for the information age. *Communications of the ACM, 46*(9), 121–128. doi:10.1145/903893.903926

Coleman, S., Macintosh, A., & Schneeberger, A. (2007). *D12.3: eParticipation research direction based on barriers, challenges and needs* (Tech. Rep. No. FP6-2004-ist-4-027219). Leeds, UK: University of Leeds.

De Moor, A., Park, J., & Croitoru, M. (2009). Argumentation map generation with conceptual graphs: The case for ESSENCE. In *Proceedings of the Fourth Conceptual Structures Tool Interoperability Workshop collocated with the 17th International Conference on Conceptual Structures.*

Dutton, W. H. (2007). *Through the network (of networks) – the fifth estate.* Oxford, UK: Oxford Internet Institute. doi:10.2139/ssrn.1134502

Fang, Z. (2002). E-Government in digital era: Concept, practice, and development. *International Journal of the Computer, the Internet and Management, 10*(2), 1-22.

Hague, B. N., & Loader, B. (1999). *Digital democracy: Discourse and decision making in the information age.* London, UK: Routledge.

Heeks, R. (2003). *Most eGovernment-for-development projects fail: How can risks be reduced?* Manchester, UK: Institute for Development Policy and Management.

Lange, C., Bojārs, U., Groza, T., Breslin, J. G., & Handschuh, S. (2008). Expressing argumentative discussions in social media sites. In *Proceedings of the Conference on Social Data on the Web at the International Semantic Web.*

Lee, D., Peristeras, V., Price, D., Kumar, S., Liotas, N., Ruston, S., et al. (2009). *D2.2: functional specification.* Retrieved from http://www.n4c. eu/Download/n4c-wp2-023-dtn-infrastructure-fs-12.pdf

Macintosh, A. (2004) Characterizing e-participation in policy-making. In *Proceedings of the 37th Annual Hawaii International Conference on System Sciences* (p. 10).

Macintosh, A. (2007). *Challenges and barriers of eParticipation in Europe?* Paper presented at the Forum for Future Democracy.

Osimo, D. (2008). *Web 2.0 in government: Why and how? (Tech. Rep. No. EUR 23358 EN).* Brussels, Belgium: European Commission.

Panopoulou, E., Tambouris, E., & Tarabanis, K. (2009). *D4.2b: eParticipation Good practice cases.* Retrieved from http://islab.uom.gr/eP/index. php?searchword=practice&option=com_search

Pellegrine, T. (2009). *A comparative study on approaches to social software and the semantic web.* Retrieved from http://www.semanticweb.at/ file_upload/1_tmpphpvuVU1T.pdf

Peristeras, V., Mentzas, G., Tarabanis, K. A., & Abecker, A. R. (2009). Transforming E-government and E-participation through IT. *IEEE Intelligent Systems, 24*(5), 14–19. doi:10.1109/MIS.2009.103

Rose, J., Sæbø, O., Nyvang, T., & Sanford, C. (2007). *D14.3a: The role of social networking software in eparticipation.* Retrieved from http://www.demo-net.org/what-is-it-about/research-papers-reports-1/demo-net-deliverables/RoseEtAl2007b

Sæbø, O., Rose, J., & Skiftenes Flak, L. (2008). The shape of eParticipation: Characterizing an emerging research area. *Government Information Quarterly, 25*(3), 400–428. doi:10.1016/j.giq.2007.04.007

Shum, S. B. (2003). The roots of computer supported argument visualization. In Kirschner, P. A., Buckingham Shum, S. J., & Carr, C. S. (Eds.), *Visualizing argumentation: Software tools for collaborative and educational sense-making.* Berlin, Germany: Springer-Verlag.

Tambouris, E., Liotas, N., Kaliviotis, D., & Tarabanis, K. (2007). A framework for scoping eParticipation. In *Proceedings of the 8th Annual International Conference on Digital Government Research: Bridging Disciplines & Domains* (pp. 288-289).

Tambouris, E., Liotas, N., & Tarabanis, K. (2007). A framework for assessing eparticipation projects and tools. In *Proceedings of the 40th Annual Hawaii International Conference on System Sciences* (p. 90).

Tzagarakis, M., Karousos, N., Karacapilidis, N., & Nousia, D. (2009). Unleashing argumentation support systems on the Web: The case of CoPe_it! *International Journal of Web-Based Learning and Teaching Technologies*, *4*(3), 22–38.

Van Bruggen, J. M., Boshuizen, H. P. A., & Kirschner, P. A. (2003). A cognitive framework for cooperative problem solving with argument visualization. In Kirschner, P. A., Buckingham Shum, S. J., & Carr, C. S. (Eds.), *Visualizing argumentation: Software tools for collaborative and educational sense-making* (pp. 25–47). Berlin, Germany: Springer-Verlag.

WAVE. (2009). *WAVE project portal*. Retrieved from http://wave-project.eu/

This work was previously published in the International Journal of E-Services and Mobile Applications, Volume 3, Issue 3, edited by Ada Scupola, pp. 69-85, copyright 2011 by IGI Publishing (an imprint of IGI Global).

Chapter 16
Protecting a Distributed Voting Schema for Anonymous and Secure Voting Against Attacks of Malicious Partners

Sebastian Obermeier
ABB Corporate Research, Switzerland

Stefan Böttcher
Universität Paderborn, Germany

ABSTRACT

A distributed protocol is presented for anonymous and secure voting that is failure-tolerant with respect to malicious behavior of individual participants and that does not rely on a trusted third party. The proposed voting protocol was designed to be executed on a fixed group of N known participants, each of them casting one vote that may be a vote for abstention. Several attack vectors on the protocol are presented, and the detection of malicious behavior like spying, suppressing, inventing, and modifying protocol messages or votes by the protocol is shown. If some participants stop the protocol, a fair information exchange is achieved in the sense that either all votes are guaranteed to be valid and accessible to all participants, or malicious behavior has been detected and the protocol is stopped, but the votes are not disclosed.

INTRODUCTION

Electronic voting is not only of interest for governmental elections; there are additional applications that motivate the development of a secure electronic voting schema that works without a trusted authority. An example is the energy market respective smart grids. In this scenario, end users use interconnected energy meters, which allows energy utilities to change prices due to demand. Within such scenario, an anonymous voting of smart meter users, for instance on their private

DOI: 10.4018/978-1-4666-2654-6.ch016

energy consumption, is an application for energy demand prediction, but such scenario clearly has to prevent a large variety of malicious behavior, which ranges from manipulating or suppressing the distribution of the voting result, to spying or manipulating single votes, i.e., suppressing, inventing or modifying votes, to spying or manipulating protocol messages, e.g. suppressing, inventing or modifying messages, to stopping cooperation or even stopping the protocol when an unwanted outcome of the electronic voting process is obvious.

Another different example where anonymous voting mechanism can be used is the computation of the union of different statements. In this case, participants do not choose from a fixed set of possible results, but are free to phrase the content of their vote. In this case, properties of an anonymous voting schema such as vote integrity and the limitation to a fixed set of participants that is allowed to vote motivate the use of such a voting schema.

Therefore, whenever participants may behave malicious, it is crucial that a voting protocol can detect such kinds of malicious behavior. In contrast to traditional paper based voting mechanisms, we consider detecting these kinds of malicious behavior after some votes have been disclosed as being not sufficient. When some votes have already been disclosed, other participants may vote differently in a repetition depending on the previously disclosed votes.

This is why we additionally require a secure anonymous voting protocol to meet the following requirements:

1.   The protocol prohibits disclosure of votes as long as malicious behavior can prevent a fair voting process, and even more, the protocol prohibits the disclosure of votes as long as the protocol execution can be stopped by the malicious behavior of a single participant.
2.   Whenever votes have been disclosed, even malicious behavior cannot prevent the voting

protocol to terminate "correctly". Informally speaking, Correct Termination means that every cooperative participant will get provable information about the correct voting result.

If only limited anonymity is required, i.e. if there is a trusted party which fulfills the following two requirements: every participant trust this third party (i.e. assumes that this trusted party does not act maliciously) and every participant allows this third party to know his vote (i.e. it does not require anonymity of his vote with respect to this party), there is the following straight forward solution. This trusted third party collects all votes, prevents duplicate votes, and distributes the voting result.

However, we argue that in many voting situations, this limited anonymity is not sufficient and a party trusted by every participant is not always given, i.e. when votes shall not be disclosed to any voting authority. For the smart grid scenario, competing energy utilities and an open market impede the establishment of a third party organization that collects sensitive data.

Therefore, we focus on the significantly more difficult situation where no trusted party is given, and full anonymity is required, i.e. the voter's identity shall not be disclosed to anybody, not even to a voting authority. Here, a special challenge is to achieve two goals which seem to be contradictory, i.e. guaranteeing that the final voting result contains exactly one uncorrupted vote of each participant and guaranteeing voter anonymity at the same time.

Problem Description and Requirements

Fair and secure anonymous voting has to solve a variety of requirements at the same time that seem to be contradictory.

According to Cranor and Cytron (1996), the following requirements have to be met by an anonymous voting schema:

- **Accuracy:** Requires that malicious behavior like inventing voters or votes, submission of multiple votes by the same voter, suppressing, stealing or modifying votes, stealing keys etc. cannot do any harm to the voting process.
- **Democracy:** Requires that only eligible voters can cast one single vote.
- **Privacy (or Anonymity):** Requires that the identity of the voter is not disclosed.
- **Verifiability:** Requires that every vote that is counted during the vote counting is submitted by a legal voter, and that legal votes are not duplicated.

However, we cannot prevent that participating voters or even voting authorities cheat. Thus, we have identified the following additional requirements:

- **Distribution:** Requires that the voting protocol should only involve one role, i.e. the voter's role. It should not rely on any trusted third party.
- **Provability:** Requires that each participant does not only receive the voting result itself, but can also construct a proof of the validity of the received voting result.
- **Fairness:** Means that malicious behavior, uncooperative behavior or even stopping the protocol does not lead to a situation where some, but not to all votes are disclosed. This includes preventing a fraudulent voter from stopping the vote counting when he can observe a voting result tendency by receiving some of the votes. Fairness has the consequence that either all votes are disclosed or no vote at all.
- **Correct Termination:** Requires that after all participants agreed to go to vote, the voting process cannot be stopped any more by malicious participants. More precisely, we do not attempt to prohibit that a non-cooperative participant stops the voting protocol, before the first vote has been disclosed and counted. However when disclosure of counted votes has started, the protocol has to guarantee that a malicious participant cannot stop the voting protocol anymore, i.e. the voting protocol must enable all remaining voters to reconstruct the final voting result even if malicious participants stop cooperation. This requirement includes that the voting result shall reach all participants that cooperate to get the vote.

Contributions

We present a voting protocol providing a unique combination of the following properties:

1. It is a fully distributed voting protocol that only needs one role – the voter's role, i.e. it does not need a trusted voting authority.
2. It does not disclose any voter's identity to any other party. Nevertheless, it guarantees that each counted vote is signed by one voter.
3. It does not disclose any vote before a fair termination of the protocol can be guaranteed.
4. It is a fair protocol, in the sense that non-cooperative participants cannot stop the voting protocol after the first vote has been uncovered.
5. When no cheating or manipulation of the voting preparations is detected, the voting process is started, and all the desired voting properties listed in Section "Problem description and requirements", i.e. Accuracy, Democracy, Anonymity, Verifiability, Protocol Distribution, Provability, Fairness, and Correct Termination are guaranteed.
6. Malicious participant behavior cannot prevent the non-malicious participants to achieve the desired protocol properties described above.

We consider the following kinds of malicious behavior. Malicious participants may try to: Invent or delete ballot papers; copy or steal valid ballot papers; duplicate counting of votes on legal bullet papers; manipulate votes or of the outcome of the voting; disclose other voters' identities; alter or invent voting protocol messages; suppress messages or change the protocol; stop the protocol at any time.

Beyond our previous publication Obermeier and Böttcher (2010), this article additionally

- Highlights and discusses potential attack vectors to the protocol that violate anonymity and fairness of the protocol
- Points out the protocol's resilience against the attack vectors and shows limitations of the protocol, e.g., in case a participant's machine is compromised
- Extends related work in order to set the contribution into a broader context

RELATED WORK

Blind signatures have been proposed by Chaum (1982) and can be used for anonymous voting as well, e.g. as described by Fujioka, Okamoto, and Ohta (1993). In this protocol, each voter signs a blinded ballot paper and sends it to a validator instance that again signs the vote and returns it to the voter. The voter then votes and returns the ballot paper to a vote compiler, which, in the end, is able to decrypt and verify the votes. However, as an entity that is able to verify a voter's signature is able to link a voter to a vote, the schema is not anonymous. Furthermore, the vote compiler, as a central instance, can publish a trend even while the vote progresses, influencing the incomplete vote.

Chaum (1988) has proposed a voting protocol that also provides anonymity of the voter's identity. The protocol uses an anonymous channel that is somehow related to our anonymous broadcast approach. However, within the approach of Chaum (1988), each voter must register at a certain instance to be able to vote. In contrast, our protocol allows all participants to agree in a distributed fashion on the voting community.

Although the protocol of Ray, Ray, and Narasimhamurthi (2001) does rely on third parties for special roles, these parties need not to be trusted. However, Ray, Ray, and Narasimhamurthi (2001) guarantee anonymity by setting up an FTP server to which all voters upload their votes. As a voter can be identified by its IP address, we do not consider this protocol as entirely anonymous.

The protocol proposed by Boyd (1989) also bases on an anonymous channel, which is used by the voters to cast their vote. However, it also requires a central instance, the vote Administrator. Although the protocol ensures that votes cannot be forged, this approach allows the vote Administrator to change votes and declare them as faked votes. In contrast, our protocol does not need a special vote Administrator that can declare votes as faked.

Data anonymity for voting and election mechanisms has also been studied by Cohen and Fischer (1985), Kiayias and Yung (2002), Groth (2004), and Benaloh and Tuinstra (1994).

These mechanisms are proposed for counting the number of votes for a previously fixed set of possible candidates, while our approach allows voting for a certain data item with any value.

Groth (2004) proposes a voting protocol in combination with a broadcast protocol. However, this protocol relies on the fact that the last participant must play fair. If he dislikes the result, he could stop the protocol execution and prevent all other participants from learning the result. To solve this problem, Kiayias and Yung (2002) propose a special trusted voting authority that always plays fair, which is in fact a trusted-third party. In comparison, our protocol works correctly without a trusted authority.

Approaches that guarantee anonymity by using zero-knowledge proofs by Cohen (1986), Benaloh and Yung (1986), and Cramer, Gennaro, and Schoenmakers (1997), guarantee privacy by certain computational problems. Although our voting schema uses cryptography for signing votes, the voter's anonymity is based on a special routing schema where the receipt of a vote does not allow concluding that the sender has been the voter.

Other approaches focus on receipt-freeness, which means that a voter must not be able to prove that a particular vote was casted by him in order to prevent "vote buying" or vote extortion (Benaloh & Tuinstra, 1994; Hirt & Sako, 2000; Okamoto, 1997). However, within our scenario, participants may act malicious and change other participants' data if they get in contact with them. Thus, each participant must be able to stop the complete protocol in case his encrypted data was changed or deleted before it is possible to decrypt the first vote.

To guarantee anonymity, mix-networks have been proposed by Abe (1997) and Desmedt and Kurosawa (2000). However, the voting authority and the mix networks could manipulate by aborting the complete voting process in case of an unpopular result. In comparison to all of these approaches, to the authors' knowledge, there is no secure and anonymous voting protocol that only requires one role and is thus fully distributed.

Another approach is the ThreeBallot voting system that has originally been proposed by Rivest (2006) and successfully been prototyped by Santin (2008). The goal of this voting system is to allow each participant to verify by means of a receipt whether their vote has been accepted, but to not allow conclusions what the actual vote has been. An application for this voting schema is the use in web sites that list the receipts. Although the ThreeBallot vote system guarantees a correct vote and anonymity if the voting authority plays fair, it does not solve the problem of having malicious participants directly in the vote authority.

Assumptions

Our anonymous voting schema relies on the following assumptions:

- The set of N participants joining the voting is known before the protocol starts and does not change during protocol execution. We consider this not being a limitation as each participant although being known still can vote with an abstention.
- Each participant provides one vote which may be an abstention.
- The number of malicious partners does not exceed a maximum number M.
- Every participant behaves opportunistically. Opportunistic behavior is defined as follows.

Definition of Opportunistic Behavior

1. The participant does not declare that he received a valid ballot paper when he did not receive a valid bullet paper. This includes the case that the participant detects that his bullet paper is invalid.
2. The participant does not deliberately disclose his votes, i.e., he does not disclose his votes in an unsafe situation. This includes situations in which not every participant has declared that he received a valid ballot paper.

Considered Attack Vectors

The following attack vectors could be identified as dangerous to voting protocols and the mitigation of these vectors is therefore discussed after the presentation of the protocol

1. **Invention of Ballot Papers:** Someone tries to invent a ballot paper. This attack vector includes malicious attackers who try to create a new ballot paper. For instance, if the

attacker knows how a valid ballot paper looks like, he might slightly change it in order to produce a new ballot paper.

2. **Copying Valid Ballot Papers:** Someone tries to copy a valid ballot paper

This attack vectors assumes a malicious user simply copies or duplicates a valid ballot paper in order to supply more votes than he is allowed to supply.

3. **Stealing Unused Valid Ballot Papers:** Someone tries to steal or to copy an unused valid ballot paper of a different participant. This attack vector assumes a malicious person copies or steals the unused valid ballot paper of another participant in order to cast a vote in the name of the person from which the ballot paper was stolen.

4. **Exchanging Used Valid Ballot Papers with Another Vote:** Someone tries to steal a used valid ballot paper of a different participant and to replace it with an own vote. This attack vector assumes a malicious person steals the used valid ballot paper of another participant by replacing the other participants vote by his own vote.

5. **Suppressing Valid Ballot Papers:** Someone tries to suppress the reception of a ballot paper for a certain participant. The attacker's goal here is to increase the weight of his own ballot paper, or to suppress potentially unwanted votes of other voters.

6. **Suppressing Casted Votes:** Someone tries to suppress the casted vote of another participant

Again, the attacker's goal is to increase the weight of his own ballot paper, or to suppress potentially unwanted votes of other voters.

7. **De-Anonymization of Votes:** Someone tries to read the content of a casted vote of a different participant. It would violate the anonymity of the voter if someone else is able to associate a certain vote with a known participant.

8. **De-Anonymization of Ballot Papers:** Someone tries to identify which ballot paper belongs to which participant. It would violate the anonymity of the voter if someone else is able to associate a certain ballot paper with a known participant.

9. **Spying Votes:** Someone spies the vote done on the ballot paper and later announces the vote to the public. The protocol should not allow constructing a proof of a vote.

10. **Cooperative Attacks Against the Safety Of The Protocol:** Two participants cooperate for undermining the safety of the protocol. Two participants may jointly try to betray other participants.

11. **Cooperative Attacks Against the Anonymity of the Protocol:** Two participants cooperate for undermining the anonymity of the protocol.

Of course, prevention of cooperative attacks is limited, e.g., if all participants but one cooperate in an anonymity attack, the vote of the last remaining participant can be disclosed.

OVERVIEW OF THE VOTING ALGORITHM

The Key Ideas Behind the Structure of the Overall Voting Algorithm

Our algorithm combines the following key ideas.

- To authorize ballot papers, a ballot paper becomes valid only when signed by each of the N participants.
- Security of signing ballot papers is guaranteed by a technique called verifiable decryption.

- Valid ballot papers are distributed in such a way that anonymity is preserved.

- Uniqueness and ownership of ballot papers is checked for preventing the copying, deleting, and stealing of valid ballot papers.

- The voting process, i.e. submission or disclosure of votes, is not started as long as a participant has not received a valid and unique ballot paper, i.e. each participant has to declare that he has received a valid ballot paper.

- When it is common knowledge that every participant has got a valid and unique ballot paper, the voting is done in a way that it cannot be stopped by malicious behavior.

- Provability of vote results and correct counting is guaranteed by uniqueness and provable validity of ballot papers.

- Anonymity of voters is guaranteed by using anonymous broadcast messages.

Overall Voting Protocol

The overall voting protocol for each participant Pi is summarized in the following Algorithm.

Algorithm 1: Voting algorithm

1. VotingAlgorithm (Participant Pi)
2. {
3. D1,...,DN = exchangePublicKeys(Di) ;
4. myBallotPaper = getMyBallotPaper (Pi) ;
5. localOK = locallyVerify(myBallotPaper) ;
6. Commit = globallyVerify(localOK) ;
7. if (Commit)
8. { Votes = anonymousBroadcast(Vote) ;
9. count(Votes) ; }
10. }

Let Dj be a public key of participant Pj that is used for sending secret messages to Pj. In the

procedure exchangePublicKeys (line (3) of Algorithm 1), the N participants agree on who are the other participants and exchange their public keys D1,...,DN in a safe way, e.g. by using certificates.

In the procedure getMyBallotPaper (in line (4) of the Algorithm), a distributed protocol (Phase 1) is executed such that each participant gets its ballot paper. In the procedure locallyVerify (line (5)), each participant Pi verifies whether his ballot paper is valid and unique, as described in Phase 2. The local verification results are exchanged among all participants in the procedure globallyVerify (line (6)). If the verification result of each participant is that his ballot paper is valid, this is the global decision to vote, i.e. *Commit* is *true* (line (6)), otherwise *Commit* is *false*. If the global decision is to vote (line (7)), votes are anonymously broadcasted (line (8)) and counted by every participant (line (9)). Otherwise, i.e. if *Commit* is *false*, the voting does not take place and no vote is disclosed. The procedures called in the Algorithm are described in more detail in the following sections.

From the Algorithm and the definition of opportunistic behavior, Lemma 1 can be concluded.

- **Lemma 1:** Assume every participant behaves opportunistically and that the procedure globallyVerify broadcasts the local verification result (*valid* or *invalid*) and returns *true* into the local variable *Commit* if and only if every participant has broadcasted the local verification result *valid*. Then, votes are disclosed only if every participant has locally verified that his ballot paper is *valid*.
  - **Proof Sketch:** Every participant first receives its ballot paper (line (4)), and then locally verifies that it is *valid* (line (5)). Opportunistic behavior requires that each participant declares to have a valid ballot paper only if the participant has a valid ballot paper. Within the global verification step (line (6)), localOK's value,

e.g. *valid,* is broadcasted by a signed message, such that each sender can be identified. The global verification step returns *true* into the local variable *Commit*, only if every participant has broadcasted that its local verification returned the result *valid*. And, the second property of opportunistic behavior requires that a participant only discloses his votes if he is in a safe situation (line (8)), i.e. only if he has got the information that *Commit* is *true* (line (7)).

MESSAGE TYPES

Individual and Broadcast Messages

There are three kinds of messages being used in our protocol: individual messages, broadcast messages, and anonymous broadcast messages (like in line (7) of the Algorithm).

We assume that each *individual message*, i.e. each message only intended for a single participant Pk, is encrypted with Pk's public key Dk and is signed by the sender for authentication purposes when needed. We do not mention this explicitly anymore.

A *broadcast message* is signed by the sender if necessary and is sent to every participant, i.e., the recipient knows who is the sender.

Anonymous Broadcast Messages

An *anonymous broadcast message* is sent by a sender S to each single participant Pk without uncovering S's identity to Pk. To ensure that each participant Pk receives the message, S sends it on multiple different anonymous paths to Pk. Within an anonymous path from S to Pk, each recipient only knows the prior and the next recipient of the message on the path. For example,

an *anonymous path* $S \rightarrow Pa \rightarrow Pb \rightarrow Pc \rightarrow \ldots \rightarrow Pk$ means that S sends an encrypted message to Pa, i.e. $Da(to(Pb),Db(to(Pc),\ldots(to(Pk),Dk(M))\ldots))$ such that Pa, after decrypting the message with his private key Ea which removes the encryption done by Da, knows that the message $Db(to(Pc),\ldots(to(Pk), Dk(M))\ldots)$ shall be forwarded to participant Pb. When Pb receives this message, he again decrypts the message to find out that the next recipient is Pc. As each non-malicious participant (e.g. Pb) suppresses who sent the message to him (i.e. Pa), the next participant (i.e. Pc) does not know about the previous participants (S, Pa) forwarding the message. Whenever the length of the path exceeds the maximum number M of malicious participants, at least one participant suppressing the sender exists.

S sends the anonymous broadcast message on multiple anonymous paths to each participant Pk in order to avoid that a single malicious participant can suppress the message by not properly decrypting and forwarding the message. Again, the message cannot be suppressed if it is sent on enough paths, i.e. on at least 1 path not containing a malicious participant. As anonymous broadcast messages are not signed, we need and will use other mechanism to verify that the message has not been changed or invented.

PHASE 1: DISTRIBUTION OF BALLOT PAPERS

The goal of the Phase 1 is that each participant Pi gets an anonymous ballot paper. Anonymity means that Pi when using this ballot paper cannot be identified by the other participants. Nevertheless, authentication of ballot papers must be supported in the sense that Pi's ballot paper is provably valid. Furthermore, malicious behavior like copying, inventing or stealing ballot papers shall be detected. This requires solving the challenging task of guaranteeing authentication and anonymity at the same time.

Accepting Only Signed Votes

The goal of our first contribution is guaranteeing authentication of votes and anonymity of voters at the same time. Our first main idea is to use an anonymous unique private key, called a *private key box*, for each voter to sign the vote, i.e., the vote on a ballot paper is only considered being valid, when it is signed with this anonymous private key box. Owning a valid ballot paper becomes equivalent to owning a valid private key box, as we can guarantee that a private key box can be used only once to sign a vote.

Using Private Key Boxes for Anonymous Signatures of Votes

The signature that guarantees authentication of votes must be safe, but nevertheless must be anonymous in the sense, that the identity of the voter must not be disclosed when the signature is checked for validity. To achieve both goals at the same time, anonymity of the signature of the vote and guaranteeing the authentication of a vote, the signature is done by encrypting the hash value $h(V)$ of the vote V with a *private key box* rather than with a single private key only. A *private key box* EBk that is used for anonymous encryption contains one private key EiBk contributed by each participant Pi.

As described in the corresponding subsections, private key boxes are generated and distributed in such a way that it is impossible to associate a private key box with a particular participant. In this sense, private key boxes support anonymity, and they can be used to sign votes.

For a vote V, the *signature given by a private key box EBk* containing the private keys E1Bk,…,ENBk is the value computed for E1Bk(…(EnBk(h(V)))…), where we assume that the pri-

vate keys E1Bk,…,ENBk are commutative and h is a publicly known collision free hash function.

The public keys D1Bk, …, DNBk corresponding to the private key box keys E1Bk,…,ENBk are collected in a public key box that corresponds to the private key box. The public key box is being used for checking the signature as follows. Given a value V, e.g. the vote on a ballot paper, and the signature S=E1Bk(…(ENBk(h(V)))…) computed by the owner of the private key box EBk, i.e. by the owner of the ballot paper EBk, the signature can be checked by applying all the public keys D1Bk, …, DNBk of the corresponding public key box to S. The signature S is considered being valid if and only if D1Bk(…(DNBk(S))…)=h(V).

Figure 1 illustrates the key box concept. Each key box pair consists of a public key box which contains N public keys and a private key box, which contains N corresponding private keys.

One key box pair is sufficient to sign and to check one vote. As there are N participants involved in the voting schema, our algorithm uses exactly one unique pair of a private key box and a corresponding public key box for each participant. In other words, for N participants, there are N different key box pairs with each private key box having N different private keys and each public key box having N different public keys, such that every participant Pi contributed one private key EiBk to each private key box EBk and one corresponding public key DiBk to each public key box DBk. Note that the N private keys contributed by a participant Pi to the N different private key boxes should be pair-wise different to each other and to the public key Di being used

Figure 1. Key box concept

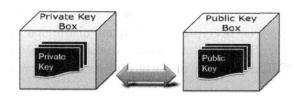

for secret messages to Pi or for authentication of Pi.

Getting the Voter's Ballot Paper

The first key part of the voting procedure is the generation of anonymous ballot papers, in other words, the key box pair generation and the anonymous private key box distribution. This is summarized in the following Boolean function getMyBallotPaper which is listed below. The main steps of the function getMyBallotPaper are the following.

Algorithm 2: The function getMyBallotPaper
1. getMyBallotPaper (i) // i is Pi's ID
2. {generateKeyPairs(N);//N participants
3. constructPublicKeyBoxes(N) ; // c.f. Section "Constructing public key boxes"
4. constructPrivateKeyBoxes(N) ; // c.f. Section "Generation of private key boxes"
5. myPrivateKeyBox = distributePrivateKeyBoxes(i,N) ;
6. } //c.f. sections "Verifiable Decryption" and "Safe distribution of key boxes"

In line (2), each participant Pi generates N pairs of keys, one pair of private key $EiBk$ and corresponding public key $DiBk$ for each pair (Ebk,DBk) of private key box and corresponding public key box. In line (3), the public key box keys are broadcasted and the public key boxes are constructed as described in Section "Constructing public key boxes". In line (4), private key box keys are N times encrypted and are used for constructing private key boxes as described in Section "Generation of private key boxes". Finally, encrypted private key boxes are step-wise decrypted, shuffled and distributed as described in sections "Verifiable decryption" and "Safe distribution of key boxes". As a result, each

participant (when reached line (6)) has received a decrypted private key box

Constructing Public Key Boxes

As mentioned, each participant Pi generates N pairs of keys, one pair $(EiBk,DiBk)$ of private key and public key for each key box pair (EBk,DBk). Then, Pi broadcasts each public key by sending a signed broadcast message (c.f. Section "Message Types") containing the pair $(DiBk,DBk)$, such that all participants know that they have to add the public key $DiBk$ to the public key box DBk. Broadcasting guarantees that all N*N public keys of all N public key boxes $DB1,...,DBN$ are publicly available.

Generation of Private Key Boxes

Our second goal is to avoid that the private keys which the participants deposit in a private key box are spied, copied or used by other participants than that participant that will become the owner of the private key box. The solution to meet this requirement combines the following ideas:

Example 1: Intermediate Decryption Results
1. Stealing a private key box key $EiBk$ can be detected and is prevented as after stealing a key from a private key box, this private key box does not correspond to a public key box anymore, and therefore, the private key box becomes useless.
2. Each participant Pi, before providing a private key $EiBk$ for a private key box EBk, encrypts $EiBk$ N times with the N public keys $D1,...,DN$ of all participants, i.e. Pi computes $XiBk = D1(...(DN(EiBk))...)$, and it contributes $XiBk$ to the private key box EB by anonymously broadcasting the pair $(XiBk,EBk)$. Thereby and by the way how the key $XiBk$ is decrypted to

EiBk, spying private keys and attacking anonymity by relating public keys to private keys are prevented.

1. $DO_i(...(DO_N(EiB)...)$, // added by participant O_i
2. $DO_{(i-1)}(...(DO_N(EiB)...)$, // added by participant $O_{(i-1)}$
3. ...
4. $DO_2(...(DO_N(EiB)...)$, // added by participant O_2
5. $DO_1(...(DO_N(EiB)...)$. // added by participant O_1

Verifiable Decryption

Verifiable decryption is used for the decryption of private key boxes. The goal of verifiable decryption is that a participant P_i decrypting a message can be sure about which message he decrypts. For this purpose, the *decryption history* is provided for each participant P_i, such that he can verify that what he is going to decrypt is what he wants to decrypt.

For example, assume that the algorithm requires each participant P in $\{P_1,...,P_N\}$ to decrypt one layer of an encrypted message $M=D_1(...(D_N(EiB)...)$. As the order of participants $P_1,...,P_N$ is different from the order in which they decrypt, we introduce the terms $O_1,...,O_N$ for participants, when regarding their decryption order. Let $\{O_1,...,O_N\} = \{P_1,...,P_N\}$ where $O_1,...,O_N$ is the order in which the participants remove their layer of decryption. When it is O_i's turn to remove a layer of decryption, i.e. to remove his public keys from the public box key DO_i, the decryption history is provided, i.e. participant O_i gets and increments a list of intermediate decryption results (shown in Example 1) of the decryption steps previously done by O_1, O_2, ..., $O_{(i-1)}$, where each previous participant removed one layer of encryption.

By step-wise applying the public keys $D_1,...,D_{(i-1)}$ to the elements of the list and by checking whether each computation result is equal to the next list element and that the last list ele-

ment $DO_1(...(DO_N(EiB)...)$ is one of the private key box keys, each participants O_i can check that he really is removing one decryption layer from the encrypted value $DO_i(...(DO_N(EiB)...)$, and he is not decrypting something else.

Since participant O_i knows all the public keys $D_1,...,D_{(i-1)}$ and can verify all the previous decryption steps, malicious participants cannot abuse this step removing one layer of decryption for attacks like exchanging the data to be decrypted, e.g. for the purpose of disclosure and stealing of keys used for ballot papers.

In the remainder of the paper where decryption order does not matter, we use again $P_1,...;P_N$ for the participants.

Safe Distribution of Private Key Boxes

As each private key box EB_k contains encrypted private keys $E_iB_k,...E_NB_k$, these encrypted keys have to be decrypted in order to use a private key box for certification. Here, the main challenge is guaranteeing that every participant will get a different key box, even in scenarios, where two cooperating malicious participants try to run a key theft attack (i.e., they let P_i decrypt synthetic data such that the attackers know any of the private keyboxes).

P_i can be sure that a key theft attack is not used, if he can reconstruct the original data by a sequence of verifiable decryption steps (as described in Section "Verifiable decryption"). This is why we use verifiable decryption for each decryption step when removing the decryption of all encrypted keys in a given private key box. As N decryption steps are done sequentially by all participants, the last decryption step, called *opening of the private key box* is done by exactly one participant per key box, which then *owns* the key box. When the private key box EB_k is opened, it contains the plain text of all N privates keys $E_1B_k,...E_NB_k$, and it is ready to use for certification. Each opened private key box legitimates one vote.

Key box decryption, i.e. removing one layer of encryption from key box keys and shuffling is done step-wise, and each decryption and shuffling step by a participant Pi includes the following aspects:

Given a private key box EBk containing N t times encrypted private keys K1,…,KN, Pi computes a new version of this private key box containing N (t-1) times encrypted private keys Di(K1),…,Di(KN). In other words, Pi applies its private key Di to all the N elements of the key box and thereby removes one layer of encryption from each of the N encrypted private keys in the box.

Pi adds a *decryption tuple* (Pi, K1,…,KN) to the history list mentioned in the previous section. All these history list entries are added to the private key box an can be used as a proof for each participant that he really decrypts one layer of an encrypted private key box and not something else. Furthermore, Pi uses the history list to avoid that he decrypts this partially decrypted key a second time by accident when shuffling is used.

Shuffling: Pi looks at the entries in the history list and randomly selects one partner that has not yet decrypted the key box and submits the box to this partner. To increase shuffling and to balance decryption, partners can randomly forward a key box without removing one decryption layer and without adding a decryption tuple to the history list of the box. Each participant watches the average decryption progress of the other participants and hesitates to encrypt much faster than the other participants.

Stepwise decryption and shuffling of each private key box proceeds until each participant has received a key box where he is the only one that has to remove a decryption layer. The participant that removes the last encryption layer becomes the *owner* of this private key box, and does not forward this private key box anymore.

No participant decrypts more than N private key boxes, more precisely, no participant decrypts more than N-1 private key boxes of which he is not an owner. The other key boxes received are randomly forwarded until they reach one of the other participants not having removed one level of decryption yet.

Altogether, validity of the key boxes and anonymity of the key box owners can be guaranteed at the same time by this safe distribution technique of private key boxes. This will be used for guaranteeing validity of the votes and anonymity of the voters at the same time.

PHASE 2: VERIFICATION

Local Verification of Own Private Key Boxes

Our third goal is to detect stealing of private key boxes or copying and suppressing private key boxes, i.e. exchanging key box keys with junk. In order to guarantee that each participant who opens its private key box can use it as a ballot paper after the last decryption step, each participant verifies its opened private key box within the procedure localyVerify (line (5) of Algorithm 1).

Our idea for local verification combines two steps: a first local verification step of the own private key box, and a second broadcast-based check for duplicate private key boxes.

In the local verification step, each participant Pi selects an individual string value xi of a fixed length L, constructs a unique dummy value 'verify'+xi, and signs this value 'verify'+xi with its private key box, i.e. Pi computes

signedVerify_i = E1Bk(…
(ENBk(h('verify'+xi))…).

Then, participant Pi checks that a corresponding public key box exists for this private key box. For this purpose, Pi uses the public keys D1Bx,…,DNBx of each public key box DBx for checking whether

h('verify'+xi) = D1Bx(…
(DNBx(signedVerify_i)…).

If this is the case for the public key box DBx, then Pi knows that DBx corresponds to his private key box. Only if this is the case for one key box, i.e., only if a public key box corresponding to its private key box exists, local verification returns *valid* for the participant Pi (in line (4) of Algorithm 1).

As Pi does this check locally, no other participant can learn that DBx corresponds to Pi's private key box.

Checking that all Private Key Boxes are Valid and Unique

The goal of this step is to ensure that each of the other participants has got a valid private key box too, and that the private key boxes are pairwise different. For this purpose, all participants anonymously broadcast their signature of the same message ('verify unique'). Thereafter, each participant checks whether it has received N different valid signatures of the message ('verify unique') that match to all N public key boxes, by checking whether all N public key boxes are needed to validated the N signatures that he has received. Furthermore, each participant checks that each public key box has been used only once for verification, i.e. for checking only one signature of ('verify unique'), and not for two different signatures of ('verify unique'). If N different private key boxes have been used for signing ('verify unique'), each participant has used a different private key box to sign ('verify unique'). As a consequence, no private key box can have been used by two different participants. Thus, the participant can be sure that everybody has received a unique key box, i.e. voting can start. Note that anonymous broadcast of signed ('verify unique') messages is crucial because otherwise, the public key box used for checking the signed ('verify unique') message could be used to associate the voter with a public key box, which after voting could be used identify the voter.

PHASE 3: VOTING PROCESS

Prevent Uncovering Votes Until it is Sure that all Ballot Papers are Valid

A fourth goal is to prevent uncovering of votes if ballot papers are invalid. That is why up to this point (between line (5) and line (6) of Algorithm 1) no vote has been exchanged. Up to this point, the voting process can be stopped by any participant that detects cheating – and it can be stopped by a malicious participant that pretends to have detected cheating. Note that no votes have been disclosed when the voting process is stopped at this point.

Voting is delayed until each participant has given his agreement. However, if all participants agreed by signing a message "localOK is valid" and sending it to at least one non-malicious participant that broadcasts this message, the voting process starts. Thus, the *Commit* decision (cf. Algorithm 1) can only be achieved if every participant has agreed by sending the signed message "localOK is valid". And as participants behave opportunistically, a participant only sends such a message if he has received a valid private key box and the test that his private key box is unique was successful.

We have to handle the problem that a malicious participant could try to send a message "localOK is valid" to some, but not to all participants. In this case, it has to be avoided that some participants compute *Commit = true* and start disclosing votes, but others do not vote.

In order to avoid this kind of malicious behavior, non-malicious participants that receive the signed message "localOK is valid" broadcast this signed messages to all other participants. Furthermore, by checking the signatures of all the received messages "localOK is valid", each receiving participant can verify the origin of the message, i.e. these messages cannot be invented by other participants.

- **Definition 2:** The decision is to vote, if and only if all participants have sent a signed message "localOK is valid" to at least one non-malicious participant.
- **Lemma 2:** As long as all non-malicious participants behave opportunistically and broadcast all signed messages "localOK is valid", they disclose their votes only if the decision is to vote.
- **Proof Sketch:** If no non-malicious participant receives a signed message "localOK is valid", no non-malicious participant discloses his vote as the participants behave opportunistically. Otherwise, there is at least one non-malicious participant that receives a signed message "localOK is valid". Since these messages are signed with the private key of the sender, its validity can be verified. And if the signature is valid, the message is broadcasted by the non-malicious participant to all other participants. Therefore, this message "localOK is valid" reaches all participants. The same is true for all other messages "localOK is valid" that reach any non-malicious participant. Therefore, non-malicious participants can be sure, that these messages reach all participants, which guarantees the requirement of Correct Termination. As they disclose their votes only in the case that all participants have signed and sent a "localOK is valid" message, they disclose their votes only if the common decision is to vote.

Global Verification of Private Key Boxes

Within the procedure globally Verify (line(6) of Alg. 1) participants send their result localOK of their local verification by using a message "localOK is valid" and non-malicious participants further broadcast these messages. If Pi has received a signed message "localOK is valid" of each participant, Pi knows the common decision to vote, and the procedure globallyVerify will return *true* to the variable *Commit* – otherwise it will wait or it returns the result *false*.

PHASE 4: VOTING AND COUNTING VOTES

Using Anonymous Broadcast for Votes

A sixth goal is that after the common decision was to go to vote, voting cannot be stopped anymore by a malicious participant. This is why votes have to be broadcasted to all participants. We use anonymous broadcasting to guarantee anonymity of the sender of the vote. Therefore, anonymous broadcast is being used for decrypting the votes and for sending all the decrypted votes to all participants. Anonymity of the voter requires also that the private key box used for decoding the voter's vote cannot be linked to the sender of a vote. This demands that also the public key box cannot be associated with the voter. And this is why we require that the signed 'verify unique' message that we have used for local verification cannot be associated with the voter, i.e. this is why the 'verify unique' signatures are also sent by anonymous broadcast.

Note that it does not matter if a vote arrives multiple times at a receiver because multiple votes signed with the same signature are counted as being one vote only.

Checking for Duplicate Votes

A fifth goal is to avoid multiple votes using the same private key box and the invention of ballot papers. The solution to this requirement combines the following ideas:

A vote is considered being *invalid*, if the private key box used for signing the vote has also been used for signing a different vote. As each signature generated with a private key box is checked by applying the corresponding public key box, the public key box can be used for checking for duplicate votes and for avoiding the counting of multiple votes signed with the same private key box. Note that multiple identical votes with identical signature are considered as being the same vote, i.e. they are counted as exactly one legal vote.

Provable Counting of Votes

Each vote that has been anonymously broadcasted is counted as a vote according to the following rules. A vote is considered valid if its signature corresponds to a public key box, i.e. every anonymously broadcasted vote message that has been signed by using a private key box is counted as a vote if it can be verified by a public key box.

Multiple vote messages signed with the same private key box of the same participant are considered being one vote only. However, if there are two or more different votes that have been signed with the same private key box, the votes are counted as being invalid.

Within the last decryption level of anonymous broadcasting of the signed votes, the signed votes reach every participant. Therefore, each participant receives the result of the voting process. And even more, each participant can use the received signed votes plus the corresponding public key boxes as a proof of the voting result. All the steps of the voting algorithm are summarized in the Algorithm given in Section "Overall voting protocol".

RESILIENCE AGAINST ATTACK VECTORS

Considering the previously mentioned attack vectors, the proposed protocol mitigates the attack vectors for the following reasons:

1. **Invention of Ballot Papers:** Someone tries to invent a ballot paper.

   a. **Attack Mitigation:** A ballot paper in our protocol consists of a pair of an encrypted public key box and a corresponding encrypted private key box. This pair is in fact invented by combining invented key pairs of each participant in the first phase of the protocol. Then, each public key box is distributed to all other participants, while the corresponding private key box is sent via an anonymous broadcast chain to a single participant. Due to the layered encryption of the private key box, only the last participant in the chain is able to retrieve the private keys in the box. After this, a commit phase follows in which all participants verify that all of them have received the same set of public key boxes, and that their own private key box matches exactly one public key box. This means, all participants agree on the set of valid ballot papers by digitally signing them, and thus the set of ballot papers is fixed. Any invention of a pair of key boxes before the agreement will result in the agreement to fail, while any invention of a key box after or during the agreement will not be considered.

2. **Copying Valid Ballot Papers:** Someone tries to copy a valid ballot paper.

   a. **Attack Mitigation:** As a ballot paper becomes valid after the last decryption step of a private key box, only the participant that is last to decrypt the private key box can cast a vote by using his private key box as a ballot paper. Copying the ballot paper or duplicating votes using the same ballot paper are prevented by the verification and the counting mechanism for submitted votes. As only verified votes are accepted and the generation of a verified vote requires using the private key box, only verified ballot papers are counted. Furthermore, counting guarantees that only one vote per private key box is counted, i.e., if there are different verifiable votes generated by using the same private key box, this is detected by using the corresponding public key box, and the copied ballot papers are provably invalid. Therefore, these ballot papers can be regarded as an abstention, not affecting the counting process of other votes.

3. **Stealing Unused Valid Ballot Papers:** Someone tries to steal or to copy an unused valid ballot paper of a different participant.

    a. **Attack Mitigation:** Copying a valid ballot paper of a different participant is mitigated by the fact that only the participant doing the last decryption step gets to know the public key box. When an unused valid ballot paper, i.e., a public key box is stolen, the victim would not be able to participate in the global verification phase, in which the reception of a valid ballot paper is acknowledged by each participant. In this case, the protocol would be stopped, as a positive response from the victim is not received. As this positive response is also digitally signed, the attacker could not forge this message.

4. **Exchanging Used Valid Ballot Papers with Another Vote:** Someone tries to steal a used valid ballot paper of a different participant and to replace it with an own vote

    a. **Attack Mitigation:** An attacker cannot exchange a used valid ballot paper by replacing the vote of a different partner with his own vote for the following reason. Without knowing the private key box of the victim, replacing a vote

is useless for other participants as only votes signed by public key boxes are counted.

5. **Suppressing Valid Ballot Papers:** Someone tries to suppress the reception of a ballot paper for a certain participant. Suppressing valid ballot papers is detected as follows. If one participant has not received a valid ballot paper, the same argumentation as for the previous attack holds: The participant could not acknowledge the reception in the global verification phase.

6. **Suppressing Casted Votes:** Someone tries to suppress the casted vote of another participant. If a malicious partner suppresses a casted vote of another participant during the vote process, this has no effect, as each vote is sent on multiple paths to each participant. Therefore, as long as there are enough paths being used to send the votes, no vote will be lost by this attack.

7. **De-Anonymization of Votes:** Someone tries to read the content of a casted vote of a different participant.

    a. **Attack Mitigation:** As each casted vote is encrypted multiple times, the attacker could not simply read the content of the vote if he intercepts the vote. Only the last participant of each vote decryption path has fully decrypted the vote and can see the contents. Furthermore, anonymity is preserved as each participant sends his vote on anonymous paths to each other participant, such that decryption and forwarding of partially decrypted votes hides the previous participants of the decryption path from the final participant reading the vote.

8. **De-Anonymization of Ballot Papers:** Someone tries to identify which ballot paper or which public key box belongs to which participant.

a. **Attack Mitigation:** The distribution of private key boxes is done via a random distribution of partially decrypted private key boxes such that no participant is able to associate a certain private key box or a certain public key box with a certain participant. Furthermore, the verification process of ballot papers uses anonymous paths in order to prevent an association of public key boxes with the participant owning the corresponding private key box.

9. **Spying Votes:** Someone spies the vote done on the ballot paper of a participant and later announces the vote to the public.

a. **Attack Mitigation:** Even if the announcement would include the encrypted and decrypted vote of the participant, there is no evidence that this vote is the correct vote and originated from a concrete participant. In order to prove that a participant has casted a certain vote, the complete voting trail that involves the decryption steps of each participant and the communication traffic of all participants would have to be revealed. This means, each participant's computer would have to be compromised in order to be able to verifiably announce the vote.

10. **Cooperative Attacks Against the Safety of the Protocol:** Two participants cooperate for undermining the correct voting process of the protocol.

a. **Attack Mitigation:** Two participants that cooperate are not sufficient to harm the safety of the voting protocol for the following reasons. In order to ensure the safety of the voting process, all participants are involved in each verification step, and the digital signature of all participants is required to agree on the set of ballot papers.

11. **Cooperative Attacks Against the Anonymity of the Protocol:** Two participants cooperate for undermining the anonymity of the protocol.

a. **Attack Mitigation:** Two participants that cooperate are not sufficient to harm the anonymity of the protocol for the following reason. To ensure anonymity, anonymous decryption paths have to include more than two decryption steps. In general, anonymous decryption paths have to be longer than the number of malicious partners planning a cooperative attack.

SUMMARY AND CONCLUSION

We have presented an anonymous and secure voting protocol which implements ballot papers by using the concept of private key boxes that contain one private key per participant. Keys in the key box are first encrypted to prevent spying or copying of keys, and are then step-wise decrypted and shuffled in order to save anonymity of the key box owner. Votes signed with private key boxes are distributed by anonymous broadcast to every participant. Thus every participant can use the public key boxes not only to count the votes, but also to prove the voting result to anybody.

Our protocol does not only allow detecting malicious behavior, e.g. message suppression, message altering, double voting or even faking votes, but also avoids that votes are uncovered, if this malicious behavior occurs before an agreement has been made to vote. Furthermore, it guarantees that after an agreement to vote, malicious behavior of single participants cannot stop or prevent the other participants from completing a fair voting. We have presented several attack vectors and have discussed how the protocol mitigates the attack vectors.

Although the desired voting protocol properties, i.e. Accuracy, Democracy, Anonymity,

Verifiability, Protocol Distribution, Provability, Fairness, and Correct Termination, seemed to be partially contradictory, our protocol combines all these desired properties. Therefore, we consider our voting protocol to be a good choice for all the voting scenarios where a trusted third party is not available and where the mentioned kinds of malicious behavior must be considered.

REFERENCES

Abe, M. (1997). Mix-networks on permutation networks. In *Proceedings of the International Conference on the Theory and Applications of Cryptology and Information Security* (pp. 258-273).

Benaloh, J., & Tuinstra, D. (1994). Receipt-free secret-ballot elections (extended abstract). In *Proceedings of the 26th Annual ACM Symposium on Theory of Computing* (pp. 544-553).

Benaloh, J., & Yung, M. (1986). Distributing the power of a government to enhance the privacy of voters. In *Proceedings of the Fifth Annual ACM Symposium on Principles of Distributed Computing* (pp. 52-62).

Boyd, C. (1989). A new multiple key cipher and an improved voting scheme. In J.-J. Quisquater & J. Vandewalle (Eds.), *Proceedings of the Workshop on the Theory and Application of Cryptographic Techniques on Advances in Cryptology* (LNCS 434, pp. 617-625).

Chaum, D. (1982). Blind signatures for untraceable payments. In *Proceedings of the International Cryptology Conference*, Santa Barbara, CA.

Chaum, D. (1988). Elections with unconditionally-secret ballots and disruption equivalent to breaking RSA. In D. Barstow, W. Brauer, P. Brinch Hansen, D. Gries, D. Luckham, C. Moler et al. (Eds.), *Proceedings of the Workshop on the Theory and Application of Cryptographic Techniques on Advances in Cryptology* (LNCS 330, pp. 177-182).

Cohen, J. (1986). *Improving privacy in cryptographic elections* (Tech. Rep. No. YALEU/DCS/TR-454). New Haven, CT: Yale University.

Cohen, J. D., & Fischer, M. J. (1985). A robust and verifiable cryptographically secure election scheme. In *Proceedings of the 26th Annual Symposium on Foundations of Computer Science* (pp. 372-382).

Cramer, R., Gennaro, R., & Schoenmakers, B. (1997). A secure and optimally efficient multi-authority election scheme. In W. Fumy (Ed.), *Proceedings of the Workshop on the Theory and Application of Cryptographic Techniques on Advances in Cryptology* (LNCS 1233, pp. 103-118).

Cranor, L., & Cytron, R. (1996). *Design and implementation of a practical security-conscious electronic polling system* (Tech. Rep. No. WUCS-96-02). St. Louis, MO: Washington University.

Desmedt, Y., & Kurosawa, K. (2000). How to break a practical MIX and design a new one. In B. Preneel (Ed.), *Proceedings of the Workshop on the Theory and Application of Cryptographic Techniques on Advances in Cryptology* (LNCS 1807, pp. 557-572).

Groth, J. (2004). Efficient maximal privacy in boardroom voting and anonymous broadcast. In A. Juels (Ed.), *Proceedings of the 8th International Conference on Financial Cryptography* (LNCS 3110, pp. 90-104).

Hirt, M., & Sako, K. (2000). Efficient receipt-free voting based on homomorphic encryption. In *Proceedings of the Workshop on the Theory and Application of Cryptographic Techniques on Advances in Cryptology* (LNCS 1807, pp. 539-556).

Kiayias, A., & Yung, M. (2002). Self-tallying elections and perfect ballot secrecy. In *Proceedings of the 5th International Workshop on Practice and Theory in Public Key Cryptosystems* (pp. 141-158).

Obermeier, S., & Böttcher, S. (2010, July). Distributed, anonymous, and secure voting among malicious partners without a trusted third party. In *Proceedings of the 4th International Conference on Methodologies, Technologies and Tools enabling e-Government*, Olten, Switzerland.

Okamoto, T. (1997). Receipt-free electronic voting schemes for large scale elections. In *Proceedings of the Security Protocols Workshop*, Paris, France (pp. 25-35).

Ray, I., Ray, I., & Narasimhamurthi, N. (2001). An anonymous electronic voting protocol for voting over the Internet. In *Proceedings of the Third International Workshop on Advanced Issues of E-Commerce and Web-based Information Systems*.

Rivest, R. (2006). *The three ballot voting system.* Retrieved from http://theory.csail.mit.edu/~rivest/Rivest-TheThreeBallotVotingSystem.pdf

Santin, A. O., Costa, R. G., & Maziero, C. A. (2008). A three-ballot-based secure electronic voting system. *IEEE Security and Privacy, 6*(3).

This work was previously published in the International Journal of E-Services and Mobile Applications, Volume 3, Issue 3, edited by Ada Scupola, pp. 31-49, copyright 2011 by IGI Publishing (an imprint of IGI Global).

Chapter 17
Applications of Intelligent Agents in Hospital Search and Appointment System

Tyrone Edwards
University of Technology, Jamaica

Suresh Sankaranarayanan
University of West Indies, Jamaica

ABSTRACT

Access to the correct healthcare facility is a major concern for most people, many of whom gather information about the existing hospitals and healthcare facilities in their locality. After gathering such information, people must do a comparison of the information, make a selection, and then make an appointment with the concerned doctor. The time spent for this purpose would be a major constraint for many individuals. Research is currently underway in this area on incorporating Information and Communication Technology (ICT) to improve the services available in the health industry. This paper proposes an agent based approach to replicate the same search operations as the individual would otherwise do, by employing an intelligent agent. The proposed agent based system has been simulated and also validated through implementation on an individual's smart phone or a PDA using JADE-LEAP agent development kit.

1. INTRODUCTION

In the current environment, diseases are easily passed on from one person to another person, and also other events such as accidents at home, school, road, etc require the assistance of a health-care professional. When an average person is in need of healthcare services, the concerned person goes through a process of identifying hospitals or healthcare facilities available within the locality. The individual then gathers data such as the proficiency/ expertise or skill levels of the doctors

DOI: 10.4018/978-1-4666-2654-6.ch017

available, the type of facilities, the civic nature of the environment and the cost estimate for services that would be offered. Based on the information gathered a comparison is then carried out and then an appropriate facility is selected. Consequently an appointment is fixed with an appropriate doctor in that facility. The task of carrying out such data gathering and subsequent analysis is not always a simple task; since the individual would have to first contact the various facilities during normal working hours which would facilitate speaking with a knowledgeable representatives available in that facility. Also gathering information from other persons who have already used that service at that facility would be a tedious job. After this only the individual can then make a good comparison of the information gathered and then select the appropriate health care facility. Usually, once an individual selects a facility he/she would normally use that facility whenever a healthcare need again arises. However, the need to access healthcare services can occur at anytime and at anyplace.

There are many possible ways to increase the quality of health care services and one such is through the application of Information and communication Technology (WHO, 2005). Intelligent agents (Bellavista, 1999; Serenko & Deltor, 2002; Suresh, 2006; Bailey & Suresh, 2009; Henry & Suresh, 2009; Miller & Suresh, 2009; Ryan & Suresh, 2009). Over the years, ICT has been successfully applied with benefits in areas such as m-commerce, e-commerce and Telemedicine. In the context, mobile devices have also gained popularity in usage over the years and many persons now own at least one mobile phone and also it is becoming cost effective. With these in mind, an intelligent agent based hospital search & appointment system has been proposed for Healthcare service search in this paper with Jamaica taken as an example to start with. This system proposed basically allows the user to set some preferences related to the attributes of the healthcare facility they would be looking for and would like to search and select the appropriate facility that matches the preference.

The system proposed also allows the user to make an appointment with an appropriate doctor available in the selected facility. The system has been implemented with mobile phones/PDA. The reminder of the paper has been organized as follows. Section 2 discusses about Health Care in Jamaica. Section 3 talks about ICT usage in Health care. Section 4 discusses Agent Technology application in Health care sector. Section 5 talks about the intelligent agent based Hospital Search & Appointment system. Section 6 provides the details of the implementation of an agent based hospital search using JADE-LEAP combination. Section 7 is concluding section.

2. HEALTH CARE IN JAMAICA

Health care in respect of the citizens of any country is very important. This is true for Jamaica also. It is seen by our experience and also as reported on web, that the quality of medical care in Jamaica is one of the main concerns. Presently health care in Jamaica (Sheila, 2000; Brice, 2001) is provided free for citizens and legal residents, at Government hospitals and Clinics, which also includes prescription for drugs. But the main drawback to the present healthcare system is towards waiting in long line with no prior appointment accepted by Physicians. There have always been concerns by people about going early in the morning and leaving the clinic late in the day and also not having seen a doctor. Also prescriptions for drugs are not easy to obtain. Although private doctors and clinics are widely available, one has to have enough money to pay to these doctors or depend on health insurance to cover the cost. In respect of the hospital pharmacies, more people congregate to receive the medicine they need. It is heard that some of these folks are even turned away because of either a lack of supply or the particular drug the pharmacy doesn't stock. If the person really needs an unavailable drug they must go to a public drugstore and pay for it or stay without that (Health Care, 2009).

In the context it is also felt that facilities towards health care in Jamaica are not that good enough when compared with those available in other developed countries like USA, UK and Europe. It is also seen that the use of ICT Technology in health care business is also not that well developed and they still rely on paper based system. If we can reduce or schedule appropriately the number of patients that need to travel to hospitals by improving access to healthcare via ICT in clinics, especially in rural and inner-city areas, it is felt that we would have fewer patients per doctor at the hospitals and as such congestion could be avoided. Prevention is always cheaper than the cost of the cure, and improving access to regular checkups and appropriate scheduling, may thus increase the chances of catching medical practitioners earlier thereby giving more people access to the limited number of specialists, through technology, instead of forcing them to travel miles thereby reducing patient load and also save money (Mullings, 2011).

All these factors have motivated us towards developing an agent based system which would allow citizens of Jamaica primarily, to search for the best hospitals and clinics for utilizing the medical facilities, by taking into consideration the cost and hospital rating into consideration. These factors are very important for people while choosing the hospital for getting the treatment and also while fixing the appointment with the Physician without needing to wait in long lines. All these and about the ICT in Health care would be discussed in the next section.

3. ICT IN HEALTH CARE

Information and Communication Technologies (ICTs) are being employed in many industries today, mainly for the significant benefits to be gained through the use of ICT (WHO, 2005) Benefits include improvements in process efficiency and information dissemination. The use of Information and Communication Technology (ICT) in the Health Care Industry (HCI) has also

been on the rise. The Health Care Industry (HCI) has been slow in accepting the use of ICT and for many years relied on paper based systems, for task such as patient data management and appointment scheduling.

As the health care industry moves toward the promotion of a personalized health experience, it has become important that, methods to improve the access, efficiency, effectiveness and quality of the processes related to both the clinical and business aspects of health care be employed (WHO, 2005) Health Care Facilities (HCFs) are able to use ICTs such as Internet based applications to provide current and potential clients with the information they need when they need it. Some facilities even allow clients to setup appointments to see a doctor using these internet based applications. The use of ICTs in health care led to a new term called Health Telematics (this term is used in Europe), and other now popular terms of e-Health and Telemedicine.

Electronic Health or E-Health is the use of Information and communication technology for storing and accessing the data stored electronically, sending the data digitally for clinical, educational and administrative purpose both locally and externally (WHO, 2005). Through the use of ICTs such as the Internet, caregivers and family member have easy access to health care information, which allows them better deal with persons within their care. Clinicians are also able to continue enhancing their knowledge by participating in online courses and research in new areas, and improvements in other treatment areas.

4. INTELLIGENT AGENT TECHNOLOGY IN HEALTH CARE

4.1. Intelligent Agents

In order to better understand the Intelligent Agent Technology, it is necessary to first understand what Agent Technology is. Agent Technology has been around for more than two decades, this area of research is still generating as much interest as it did when it was first introduced. Franklin and Graesser in 1996 gave the following definition "an autonomous agent situated within and a part of an environment that senses the environment and acts on it, over time, in pursuit of its own agenda and so as to effect what it senses in the future."

So an agent is a software that has the ability to sense an environment that is located in, then carries out some action, based on the information / data that it gathers from that environment. The agent is defined (Franklin & Graesser, 1996) as the feature of being autonomous, meaning independent or self-directed. So an agent being self-directed should have control over its own actions and not have to rely on the intervention of other agents or even its human creator.

Agents also have other features such as being mobile, able to move from one system to another within a networked environment. Another feature is that of intelligence, an intelligent agent is a computer system that is capable of flexible autonomous action in order to meet its design objectives (Jennings & Wooldridge, 1998). There are a number of toolkits available on the market ranging from general agent development platforms, like AgentBuilder developed by Reticular Systems, to highly specialized tools, like Excalibur developed by the Technical University of Berlin (Serenko & Detlor, 2002). The four major categories of agent toolkits are mobile agent toolkits, multi-agent toolkits, general purpose toolkits, and internet agent toolkits (Serenko & Detlor, 2002). Now for the purpose of this research a multi-agent toolkit would be ideal. Examples of multi-agent toolkits are Concordia, Gossip, FarGo, IBM Aglets and JADE (with the LEAP add-on).

The most suitable of the five toolkits above would be the JADE powered by the LEAP add-on (JADE-LEAP). JADE-LEAP is a free agent toolkit, the agent developer needs only to download and install the latest stable release (last stable version - 3.4.). According to the literature JADE-LEAP is the only multi-agent toolkit capable

of creating agents that can execute on a mobile device with limited resources (Bellifemine, 2003, 2007) However, it requires the agent developer to be very familiar with Java programming (i.e., Java Standard Edition and Java Micro Edition).

4.2. Agents in Health Care

Agent based technology has been used and proposed for use in many health care system applications (Foster, 2005; Kirn, 2002; Mazzi, 2002; Isern, 2003; Nealon & Moreno, 2002, 2003; Moreno, 2003; Vazquez-Salceda; 2004; Zhu, 2007; Miller & Suresh, 2009). The AgentCities Working Group on Health Care have proposed and created prototypes of agent applications for various areas in health care. These areas include patient data management, organ transplant and patient monitoring (Nealon & Moreno, 2002, 2003; Moreno, 2003). In the available literature, there are many ideas about using agents within the health care industry (Foster, 2005; Kirn, 2002; Mazzi, 2002; Isern, 2003; Nealon & Moreno, 2002, 2003; Moreno, 2003; Vazquez-Salceda; 2004; Zhu, 2007; Miller & Suresh, 2009; De Renzi, 2008; Jung, 2007) but to date there has been none which addresses the issue of hospital search and appointment scheduling via intelligent agents on mobile devices such as the cell phone.

From the literature it is clear that there is significant research being conducted in which single and multi agent systems are being proposed for solving problems within the health care domain. However there seems to be very limited work carried out in applying agent based technology in mobile devices for hospital search and appointment scheduling. With this in mind, we have proposed an Intelligent Agent based Hospital Search and Appointment System (IAHSAS) (Edwards & Suresh, 2009). It may be mentioned that IAHSAS uses agent technology that is compliant with the FIPA (Foundation for Intelligent Physical Agents, 2009 - FIPA, 2009) standard. The details of these will be discussed in the forthcoming sections.

5. INTELLIGENT AGENT BASED SEARCH AND APPOINTMENT SYSTEM

Before going into the details of intelligent agent based hospital search & Appointment system developed for Healthcare system (Edwards & Suresh, 2009) we need to see how the same is being carried out manually. The process may start with the recommendation of a General Practitioner (GP or Family Doctor), that the individual seek medical attention at a Health Care Facility (Hospital). The GP may suggest a Health Care Facility the individual may consider, but leaving the option to the individual.

The search process starts with the person trying to find all the Health Care Facilities within their locality that may offer the treatment services needed, by contacting each facility and requesting information from the customer representatives or from the website of the Health Care Facility. The information requested may include cost, treatment facility available, treatment success data, expertise of doctors, appointment opportunities, etc. Once this information is gathered, the individual would do some analysis of the data, to make their decision. The data gathering process is not without its challenges, due to mainly the availability of the information sources, i.e., the Health Care Facility Representatives. These persons are normally accessible during the normal working hours which makes the search process time restrictive.

The decision making process is not usually easy as in most cases as all the data needed is not available, for example, data on success of treatment and past patient feedback. This decision is important, because once a selection is made and an appointment setup, it is expected that the appointment will be kept or cancelled. Once the treatment process has started and if the individual has any problems it would be too late. Therefore it is important that this search process be as thorough as possible. Figure 1 shows the traditional search and appointment process.

One of the problems that the individual faces in the current scenario is the availability of the healthcare facility representative at the time one would like gathering data and fixing the appointment. We particularly seek to solve this problem by developing the proposed system and also reducing the time taken in identifying the appropriate facility and fixing an appointment. With this in view, the system developed uses software based intelligent agents to carry out the searching, selection of the healthcare facility and also allowing the user to fix an appointment at the selected facility. The use of artificial intelligence in making decisions is similar to what the human agents would do. The system will also allow remove the time restrictive issue, while also allowing the human agent to conduct a search anytime, anywhere, using their mobile phones.

5.1. Architectural Details of IAHSAS

The IAHSAS (Edwards & Suresh, 2009) supports four types of intelligent agents-Hospital Search Agent (HSA), Hospital Appointment Agent (HAA), Central Database Agent (CDBA) and the Hospital Agent (HA) as shown in Figure 2. Figure 3 shows the process flow of intelligent agent based Hospital Search and Appointment system. The Hospital Search Agent (HSA) and Hospital Appointment Agent (HAA) are both hosted on the mobile device, where the HSA is responsible for:

- *Collecting the user details on the Health Care Facility required*
- *Querying the Central Database Agent (CDBA) for facilities matching the user details*
- *Presenting the results of the query to the user in a suitable format, for selection*

The Hospital Appointment Agent (HAA) is also hosted on the mobile phone and has the following functionalities

Figure 1. Human agent search and appointment process

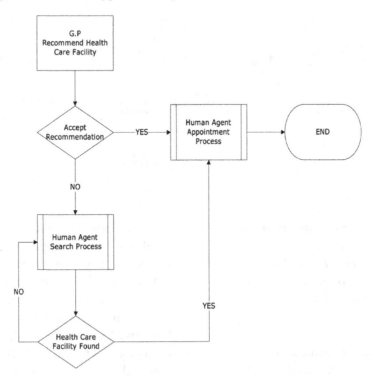

Figure 2. Agent architecture of IAHSAS

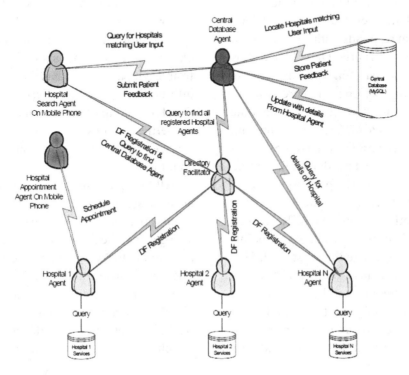

Figure 3. Process flow of IAHSAS

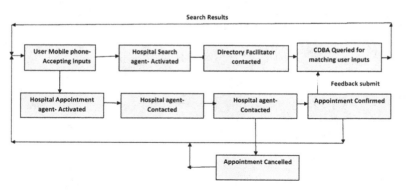

- *Collecting the user details on the desired appointment date, time and patient details*
- *Negotiating with the Hospital Agent based on the user's desired appointment details*
- *Presenting the user with the results of the negotiation*

To interact with the system the user is provided with a graphical user interface (GUI) - Hospital Search Agent. This interface is used to enter the details for locating a Health Care Facility. The Hospital Search Agent would contact the Directory Facilitator (DF) to query the Central Database Agent (CDBA) that responds with a list of facilities matching the user criteria. DF is like Yellow pages which give the list of agents registered in the system. This information is then presented to the user's mobile handset in a format so that the user may select a facility to initiate an appointment.

Another user interface is Hospital Appointment Agent that is used to accept the details for setting the appointment with the Hospital Agent (HA) of the Health Care Facility. The details include personal identification number i.e., TRN which is unique, name, date of birth, illness, appointment date and time, and date illness started. This information is submitted to HA and the HA checks for existing appointments for the same patient and availability of appointment date and time. The user is notified of the appointment request outcome and whatever needs to be done next. The

Hospital Agent (HA) represents the Health Care Facility (Hospital), and has the necessary intelligence to pursue the interests of the facility. The HA knows all the details of the facility and uses this knowledge in negotiating with the Hospital Appointment Agent. It also provides the Central Database Agent with these details too. The Central Database Agent (CDBA) is the regulator agent within the system; it has details on all Hospital Agents in the system. This information is used to match against the user details when search for a heath care facility is initiated. The CDBA is not under the control of the Health Care Facilities and therefore it maintains unbiased information on each facility, and also maintains feedback from each patient that has used the facility. This information is used for presenting the user with a ranking/rating of the various facilities based on properties such as patient experience, customer service, treatment success and environmental conditions.

All the agents discussed above communicate using messages; the messages passed between the agents are FIPA ACL Messages (FIPA, 2009). All the agents in the system are implemented to conform to the FIPA Agent Standards (FIPA, 2009). The conformance of the agents to these standards ensures and guarantees agent interoperability. The fact that this system has more than one agent working together to solve the general problem of searching and appointment scheduling, means the system is a multi-agent system (Jennings & Wooldridge, 1998). In a multi-agent system there is normally the issue of trusting the different agents in the system. In this system however there is no issue with trust as it is a closed system, which means all the components of the system can trust the information and services provided by other components.

5.2. Algorithm

The algorithm presented was used for the Intelligent Agent based Hospital Search and Appointment system in the initial stage of development:

- *Input the details for hospital search like type of facilities: General/Specialized, Price, Country, City, Illness*
- *Once the details are given, the hospital search Agent in the mobile phone would be started that queries the Central Database Agent*
- *The Central database Agent, based on the request from the Hospital search Agent, matches the user specifications with the hospitals available as follows:*
  - *If a hospital is available for lower price range in the same or different city*
  - *If a hospital is available for the price range specified in the same city or different city*
  - *If a hospital is not available within the price range, it finds a hospital which is 10-20% more than the maximum price in same or different city*
- *The Hospital Search Agent would then present the mobile phone user with the results of the central database search, along with ranking details on each result found.*
- *Now based on the hospital selected by the user, the Hospital Appointment Agent would then make an appointment request*

for the date chosen by the user with the appropriate hospital Agent.
- *Hospital Agent at the hospital site would send a confirmation for appointment requested or send other available dates for appointment.*

In the algorithm presented above, the intelligence that is built into the agent based hospital search and appointment system is shown in the actions of the Central database agent. The Central database agent matches available hospitals based on user input i.e., it finds hospitals within, below and $10-20$ percent above the price range. It also matches based on whether the hospital is located in the city specified by the user or in a different city. The Central database agent also provides a rating of some features of the hospital. These ratings are based on feedback stored in the central database by patients. This is the average rating calculated based on all feedback received. The idea of increasing the price by between $10-20$ percent and even searching in a different city, will increase the chance of finding a hospital, but there will be occasions when even these changes will not lead to a positive search result. In the event that the search does not result in a match being found the system will show no hospital found. To deal with the possibility that the price increase still did not lead to a match, the program was modified to allow the user to do a price insensitive search (i.e., a search based on any price). The implementation of our system carried out using JADE-LEAP been given in the forthcoming sections.

6. IMPLEMENTATION USING JADE -LEAP

The IAHSAS was implemented using technologies like Java Micro Edition (JME) profile for Mobile Information Device Profile (MIDP), Java Standard Edition (JSE), JADE with the LEAP add-on and MySQL Database Server Community Edition and

PHP. Before going into the results we will give a brief overview of technologies used (FIPA, 2009; Sun Microsystems, 2009a; Bellifemine, 2003, 2007; Sun Microsystems, 2009b).

JME is a specialized version of the popular java programming language, aimed at writing programs for machines with limited hardware resources, such as mobile phones, PDAs and other embedded and consumer electronics (Sun Microsystems, 2009a, 2009b). JME comes with a set of profiles each geared towards a particular type of device, for mobile phones the Mobile Information Device Profile is used for developing applications. Unlike the JSE version of java that is geared towards desktops and servers, which uses the Java Virtual Machine (JVM) (Sun Microsystems, 2009a). JME uses a stripped down version of the JVM called a KVM.

JME is delivered with a minimum set of class libraries required for each profile supported, and manufacturers of mobile phones may add proprietary classes to these devices. The MIDP profile allows the programmer to create applications that can provide the user with GUIs to interact with the applications (Sun Microsystems, 2009a; Moreno, 2003).

JADE (FIPA, 2009; Bellifemine, 2003, 2007) is the Java Agent Development toolkit, used to develop agents for desktop and server systems running JSE. JADE provides and an environment that can be used to create, execute and monitor agents. It provides two agents the Agent Management System (AMS) and the Directory Facilitator (DF) which monitors the agent platform and provides yellow page services. LEAP is the Lightweight Extensible Agent Platform, an add-on to JADE which replaces the JADE Kernel with a new kernel capable of creating and managing agents on devices with limited resources.

JADE with the LEAP add-on is called JADE-LEAP. MySQL (Sun Microsystems, 2009b) Database Server Community Edition is a free open source database server owned by Sun MicroSystems Inc, and has a large community, that provides technical and other troubleshoot support. PHP is a server side scripting language that is a widely accepted scripting language in website development. The combination of MySQL as a database server and PHP as a server side script leads to the development of feature rich websites and applications.

For our research purpose, we created five (5) hospitals. A sample details for one such hospital is shown in Figure 4. The list of five hospitals created in the JADE environment is shown in Figure 5. Figure 6 shows the user entering the details/requirements of the Health Care Facility given to perform the hospital search.

- Country
- City
- Illness
- Treatment Type
- Cost
- Rating Period

The user is able to select a country from a drop down list (currently available is Jamaica), the City is dependent on the country selected. The city option is another drop down list displaying the cities in the selected country. The user is able to select the medical condition from the Illness drop down control. The Treatment Type allows the user to select whether the treatment should be general or specialized (i.e., common or advanced – cutting edge). The Cost option allows a user to specify a price range for the treatment services offered. The Cost ranges from as low as $1000 to $10000, and the option of any price, which is from $0 to the highest listed price. The Rating Period option allows the user to specify a range to be used for considering patient feedback information. Past patients are allowed to submit a rating of a health care facility such as customer service, facilities, environment, their general experience and treatment success. This information is then used to rate a facility based on rating periods such as three months, six months and one year.

Figure 4. Web page of hospital

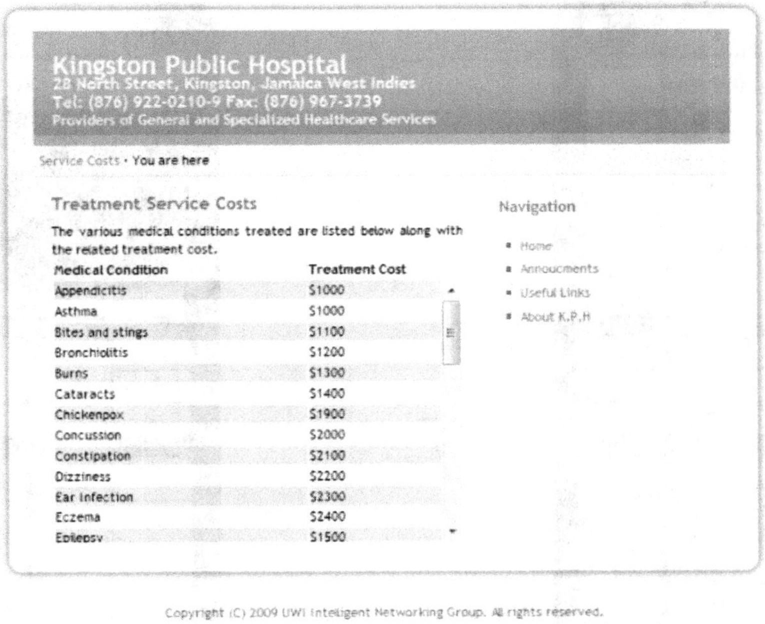

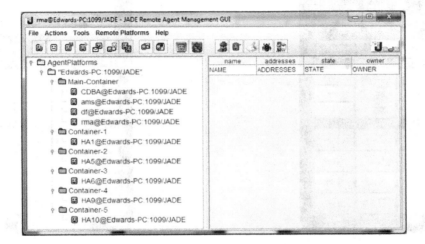

Now consider a scenario where the agent is been sent for searching for general hospitals within the price range of $1000 - $1500 for Bites and Stings illness in the location, Kingston. Based on the details entered Hospital Search Agent been activated as shown in Figure 5 for the search process. The results of the exact match i.e., hospitals falling within the range of $1000 to $1500 in the same location have been found which are shown in Figure 7 and 8. The results displayed, shows two facilities matching the user's requirements.

Let us now consider another scenario where the user enters the following details as shown in Figure 9:

Figure 5. Hospitals in JADE environment

Figure 6. User requirement search 1

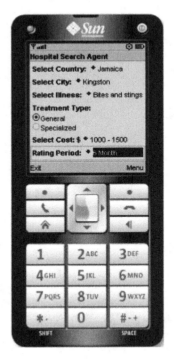

Figure 8. Search agent results-2

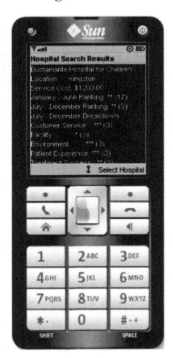

Figure 7. Search agent results-1

Figure 9. User requirement search-2

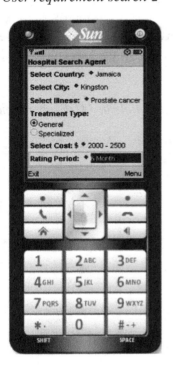

- **Country:** Jamaica
- **City:** Kingston
- **Illness:** Prostate cancer
- **Treatment Type**: General
- **Cost:** $2000 - $2500
- **Rating Period:** 6 Month

In here the search agent possess the intelligence to perform approximate match for the price $2000 - $2500 for prostate cancer in the location Kingston. This mean search agent couldn't find a hospital that treats prostate cancer illness for the price range of $2000 - $2500 in the location Kingston. So it finds a hospital which is more than the maximum price range of $2500 by 10-20% which is approximate match in Kingston as shown in Figure 10. The result displayed, shows one facility matching the user's requirements.

Let us consider another scenario where the user enters the following requirements as shown in Figure 11:

- **Country:** Jamaica
- **City:** St. Catherine
- **Illness:** Epilepsy
- **Treatment Type:** General
- **Cost:** $1000 - $1500
- **Rating Period:** 6 Month

For the details entered by the user in Figure 11, Search Agent possesses the intelligence to get hospitals that treat Epilepsy illness in St. Catherine location for price range of $1000 -$1500. It is seen that the search agent couldn't find exact or approximate match in St. Catherine but found a hospital that treats the illness in the price range of $1000 - $1500 in different location i.e., Kingston. This is shown in Figure 12. The result displayed, shows one facility matching the user's requirements.

Now consider another scenario where the user enters the following requirements as shown in Figure 13:

- **Country:** Jamaica
- **City:** Kingston
- **Illness:** Chickenpox
- **Treatment Type:** General
- **Cost:** $1000 - $1500
- **Rating Period:** 6 Month

Details entered by the user in Figure 13 show the result of an approximate match in a different location as shown in Figure 14. This means the search Agent possess the intelligence to perform the exact match and approximate match in location Kingston for the chickenpox illness for a cost of $1000 - $1500. It is found that no hospitals are found. So search agent now performs the same search in different location and it could find a hospital that treats Chickenpox illness for a price range more than 10-20% in different location and not in Kingston. The result displayed, shows one facility matching the user's requirements.

Now let us consider another scenario where the user enters the following requirement with any price as shown in Figure 15:

- **City:** St. Catherine
- **Illness:** Burns
- **Treatment Type:** Specialized
- **Cost:** Any Cost
- **Rating Period:** 6 Month

For the details entered by the user in Figure 15 Search Agent possess the intelligence to show the result of a match in the same location for any price in Figure 16. The result displayed, shows one facility matching the user's requirements for any price.

The same search Agent also possess intelligence to search for hospitals for any price for illness say Burns in different location say St. Catherine or so if no hospital with that particular specialty is not found in location asked for .say Kingston or so. Till now we have seen the intelligence possess by the agent in searching hospitals with and without price specified in same and different

313

Figure 10. Search agent results-3

Figure 12. Search agent results-3

Figure 11. User requirement search 3

Figure 13. User requirement search 4

Figure 14. Search agent results-4

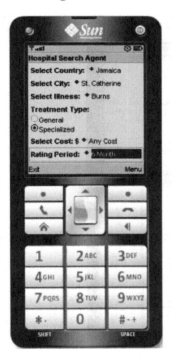

Figure 16. Search agent results-5

Figure 15. User requirement search 5

location by using their decision. Now we will see how once the hospital is found the user makes an appointment. Figures 17(a) and 17(b) show the user interface for specifying the details of the appointment to be scheduled. The information captured for this purpose includes:

- Personal Identification number (TRN)
- First and Last names
- Date of birth
- Contact No.
- Date Illness started
- Appointment Date
- Appointment Time

Once details for appointment are entered the user activates the Hospital Appointment Agent, to make the appointment; Figure 18 shows the result of the appointment request. It shows a confirmation message that the appointment request was success-fully completed, with a reminder of the Hospital, date and time of the appointment. If the user had

Figure 17. (a) User appointment details –screen 1. (b) User appointment details –screen 2

a pervious appointment still marked as new then an error message indicating the existence of such an appointment would be displayed. If there was also a clash based on the date and time selected an appropriate message would be displayed. Once the user has successfully completed setting up an appointment request, two additional actions become available to the user, the option to Cancel Appointment and the option to give Feedback. The option to cancel an appointment will do just that, contact is made with the Hospital Agent of the respective Hospital and a request sent to have the appointment cancelled.

Figure 19 shows a confirmation message to indicate the appointment was successfully cancelled. Though the Feedback option is available to the user, a user is only able to submit Feedback data once the user keeps the scheduled appointment. If not then any attempt to submit Feedback data will be rejected, with an appropriate message. These are shown in Figure 20, Figure 21, and Figure 22.

Till now we saw how hospital search is done for normal medical problems which include General and Specialized treatment. Now let us consider a situation where the user needs emergency appointment in case of situations like massive heart attack, Burns, Accident or so. In this case we cannot search based on price or so. In this section we present the results how search is made in emergency situation and appointment done. Figures 23 and 24 show the user selecting the option to carry out an Emergency Search and then completing a very short form that captures the following data:

- **Country:** Jamaica
- **City:** Kingston
- **Illness:** Asthma

Figure 25 shows the results of the search, which gives details on all Hospitals within the selected City – Kingston, the location, address, and contact number of the hospital. The user now completes

Figure 18. Appointment confirmation message

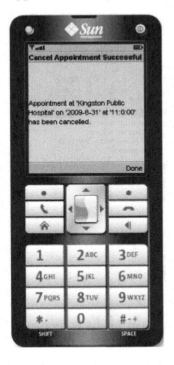

Figure 20. User feedback option

Figure 19. Appointment cancellation confirmation message

Figure 21. Feedback submission

Figure 22. Feedback confirmation

Figure 24. Activating search agent

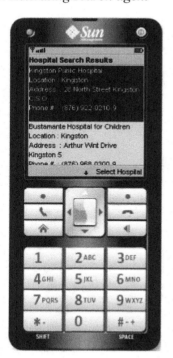

Figure 23. User requirement screen

Figure 25. Emergency search results

Figure 26. Appointment confirmation message

a short form as shown in Figures 17(a) and (b) and activating the Hospital Appointment Agent to setup the appointment/ notify, the Hospital of the Emergency case on route. Figure 26 shows the confirmation message received.

7. CONCLUSION AND FUTURE WORK

For such health applications, we normally employ a human agent to get the details for choosing the appropriate health care facility. But in such a mobile environment we can employ a mobile agent which can replicate the job of the human being. Also, quite recently mobile agents have gained considerable attention in the area of computer science which is from literature. With all these in mind we have attempted an Intelligent Agent Based hospital search and appointment system that searches the hospitals by getting necessary details, much like a human agent. The hospital search example attempted here is mainly concentrated

towards Jamaica. In this case, the Agent gathers the information about the hospital and compares them with the user preferences. The Agent here possesses adequate intelligence to search and select a hospital based on what the user needs. Also based on the hospital selected the appointment is been made. The results of our research have been shown as screenshots. In future the system can also be extended to include an ontology which can lead to additional information being captured, which may allow a doctor to start the diagnostic process early. This ontology can also lead to a more intuitive and proactive appointment scheduling process. The system can also be extended to include searching healthcare facilities offering general and specialized treatment and also appointment involving other facilities such as private practices. In addition the system can be extended for the entire globe using Google Map on Android 2.2. The system should implement a good security mechanism for accessing the system. This prevents malicious users to access patient personnel and health related data stored in hospital appointment agent residing in the mobile handset. Last but not the least whether the individual possesses health insurance, and the insurer, as this information can be used to determine the final cost of services.

REFERENCES

Bailey, L. S., & Suresh, S. (2009, March 16-17). Intelligent agent based mobile tour planner. In *Proceedings of the First Caribbean Conference on Information and Communication Technology*, Kingston, Jamaica (pp. 8-13).

Bellavista, P., Corradi, A., Tarantino, F., & Stefanelli, C. (1999). Mobile agents for web-based systems management. *Internet Research*, 9(5), 360–371. doi:10.1108/10662249910297769

Bellifemine, F., Caire, G., & Greenwood, D. (2007). *Developing multi-agent systems with JADE*. Chichester, UK: John Wiley & Sons. doi:10.1002/9780470058411

Bellifemine, F., Caire, G., Poggi, A., & Rimassa, G. (2003). JADE a white paper. *EXP in Search of Innovation, 3*(3), 6-19.

Brice, G. (2001). Implementing a continuous quality improvement programme in healthcare. *Caribbean Health Journal, 4*(3), 27–28.

DeRenzi, B., Gajos, K. Z., Parikh, T. S., Lesh, N., Mitchell, M., & Borriello, B. (2008). *Opportunities for intelligent interfaces aiding healthcare in low-income countries.* Retrieved from http://www.eecs.harvard.edu/~kgajos/papers/2008/derenzi08opportunities.shtml

Edwards, T., & Suresh, S. (2009, November 24-26). Intelligent agent based hospital search and appointment system. In *Proceedings of the 2nd ACM International conference on Interaction Sciences: Information Technology, Culture and Human*, Seoul, Korea (pp. 561-568).

FIPA. (2009). *The foundation for intelligent physical agents.* Retrieved from http://fipa.org/

Foster, D., McGregor, C., & El-Masri, S. (2005). *A survey of agent-based intelligent decision support systems to support clinical management and research.* Retrieved from http://www.diee.unica.it/biomed05/pdf/W22-102.pdf

Franklin, S., & Graesser, A. (1996). *Is it an agent, or just a program?: A taxonomy for autonomous agents.* Retrieved from http://www.msci.memphis.edu/~franklin/AgentProg.html

Health Care. (2009). *Health care in Jamaica.* Retrieved from http://www.jamaicans.com/articles/primearticles/health-care-in-jamaica.shtml

Henry, L., & Suresh, S. (2009, November 24-26). Applications of intelligent agent for mobile tutoring. In *Proceedings of the 2nd ACM International Conference on Interaction Sciences: Information Technology, Culture and Human*, Seoul, Korea (pp. 963-969).

Isern, D., Sánchez, D., Moreno, A., & Valls, A. (2003). *HeCaSe: An agent-based system to provide personalised medical services.* Retrieved from http://deim.urv.cat/~itaka/Publicacions/caepia03.pdf

Jennings, N. R., & Wooldridge, M. (1998). *Applications of intelligent agents.* Retrieved from http://eprints.ecs.soton.ac.uk/2188/2/agt-technology.pdf.gz

Jung, I., Thapa, D., & Wang, G.-N. (2007). *Intelligent agent based graphic user interface (GUI) for e-physician.* Retrieved from http://www.waset.org/journals/waset/v36/v36-35.pdf

Kirn, S. (2002). Ubiquitous healthcare: The OnkoNet mobile agents architecture. In M. Aksit, M. Mezini, & R. Unland (Eds.), *Proceedings of the International Conference on Objects, Components, Architectures, Services, and Applications for a Networked World* (LNCS 2591, pp. 265-277).

Lau, C., Churchill, R. S., Kim, J., Masten, F. A. III, & Kim, T. (2002). Asynchronous web-based patient-centered home telemedicine system. *IEEE Transactions on Bio-Medical Engineering, 49*(12), 1452–1462. doi:10.1109/TBME.2002.805456

Miller, K., & Suresh, S. (2009). Policy based agents in wireless body sensor mesh networks for patient health monitoring. *International Journal of u and e-Services. Science and Technology, 4*, 37–50.

Moreno, A. (2002). *Medical applications of multi-agent systems.* Retrieved from http://deim.urv.cat/~itaka/Xerrades/EUNITEworkshop.pdf

Moreno, A., Valls, A., & Viejo, A. (2003). *Using JADE-LEAP to implement agents in mobile devices.* Retrieved from http://jade.tilab.com/papers/EXP/02Moreno.pdf

Mullings, D. (2011). *Technological opportunities in telemedicine and telehealth.* Retrieved from http://www.jamaicaobserver.com/columns/Technology-opportunities--Telemedicine-and-Telehealth_8428572

Nealon, J. L., & Moreno, A. (2002). *The application of agent technology to health care.* Retrieved from http://deim.urv.cat/~itaka/Publicacions/aamas02ws.pdf

Nealon, J. L., & Moreno, A. (2003). *Agent-based applications in health care.* Retrieved from http://citeseerx.ist.psu.edu/viewdoc/summary?

Ryan, A. B., & Suresh, S. (2009, April 24-28). Intelligent agent based mobile shopper. In *Proceedings of the Sixth IFIP/IEEE International Conference on Wireless and Optical Communication Networks*, Cairo, Egypt.

Serenko, A., & Detlor, B. (2002). *Agent toolkits: A general overview of the market and an assessment of instructor satisfaction with utilizing toolkits in the classroom.* Retrieved from http://foba.lakeheadu.ca/serenko/Agent_Toolkits_Working_Paper.pdf

Sheila, S. (2000). Health promotion and health for all. *Caribbean Health Journal*, *2*(4), 10–11.

Sun Microsystems. (2009a). *Java ME technology.* Retrieved from http://java.sun.com/javame/technology/ index.jsp

Sun Microsystems. (2009b). *MySQL*. Retrieved from http://www.mysql.com

Suresh, S. (2006). *Studies in agent based IP traffic congestion management in DiffServ networks.* Unpublished doctoral dissertation, University of South Australia, Adelaide, Australia.

Vazquez-Salceda, J. (2004). Normative agents in health care: Uses and challenges. *AI Communications-Agents Applied in Health Care*, *18*(3), 175–189.

WHO. (2005). *Information technology in support of health care.* Retrieved from http://www.who.int/eht/en/InformationTech.pdf

Zhu, S., Abraham, J., Paul, S. A., Reddy, M., Yen, J., Pfaff, M., & DeFlitch, C. (2007). *R-CAST-MED: Applying intelligent agents to support emergency medical decision-making teams.* Retrieved from http://agentlab.psu.edu/lab/publications/Zhu_AIME07.pdf

This work was previously published in the International Journal of E-Services and Mobile Applications, Volume 3, Issue 4, edited by Ada Scupola, pp. 57-81, copyright 2011 by IGI Publishing (an imprint of IGI Global).

Chapter 18
Intelligent Store Agent for Mobile Shopping

Ryan Anthony Brown
University of West Indies, Jamaica

Suresh Sankaranarayanan
University of West Indies, Jamaica

ABSTRACT

The conventional shopping process involves a human being visiting a designated store and perusing first the items available. A purchase decision is then made based on the information so gathered. However, a number of unique challenges a human shopper would face, if he/she prefers to execute this process using a mobile device, such as a phone. Taking this aspect into consideration, the authors propose the use of an Intelligent Agent for performing the Mobile Shopping on behalf of customers. In this situation, the agents gather information about the products through the use of 'Store Coordinator Agents' and then use them for comparing with the user preferences. The proposed agent based system is composed of two agents, viz., a User Agent and Store Coordinator Agent. The implementation of the scheme so proposed has been done using JADE-LEAP development kit and the performance results are discussed in the paper.

INTRODUCTION

There has been an exponential growth in the use of digital mobile devices in various fields these days. This has resulted in an increased effort to develop various commercial applications that would provide leverage to this extensive use of these digital mobile devices rather than desktop PCs. One such area is the evolution of e-commerce having

application in mobile commerce (m-commerce). There is no precise meaning for m-commerce or mobile e-commerce as such, however, the core of mobile e-commerce (Abbott, 2001) uses a terminal such as a telephone, PDA, or custom terminal and the public mobile network to access information and conduct business transactions that result in the transfer of value in exchange for information, services or goods. Some examples of m-commerce

DOI: 10.4018/978-1-4666-2654-6.ch018

(Abbott, 2001) include the purchasing of airline tickets, purchasing of movie tickets, restaurant bookings and reservations, mobile banking and so on. We have taken the Mobile Shopping Activity as our research initiative and the results of this effort in using an intelligent agent to replicate the role of a human agent, are presented in this paper.

Normally when we think of buying a particular product (Thomas & Harold, 2003), things that normally come to our mind are the price, the quality, the brand, etc., of the desired product. To get this information, we often do window shopping in the conventional shopping method before we decide on buying the product. In electronic shopping we put an appropriate query, taking into consideration factors like the cost, the quality of product, etc. We also at times, compromise on the selection of the item, if we do not get an item suiting to our preconceived specifications. We human beings, under such circumstances, interpret various aspects depending on several considerations and make a balanced compromise before taking a decision on the deal.

In the mobile environment which we propose in this paper, the same job will be replicated by an intelligent Agent (Baldi, 1997; Carzniga, 1997; Ghezzi & Vigna, 1997; Puliafito & Tomarachio, 1999; Suresh, 2001; Suresh, 2006) for getting the details on the specifications of the customer desired item by performing the search operation – a replication of the job done by a human agent in window or electronic shopping. It may be mentioned here that considerable research attention is being paid to the application of agents in various areas, these days. Taking these aspects into consideration we propose in this paper an intelligent store agent based mobile shopping technique in which an intelligent store agent performs the job of getting the price, quality of product, brand etc., based on the request of the client/user, which amounts to replicating the functioning of a human agent. The intelligence possessed by the store agent is good enough to retrieve the details of the products available based on the exact specification given by the client but

in addition try to match the price specified with the other available likely products in the store. This paper however does not focus on payment, user authentication and such other things, as it is beyond the scope of this research. The reminder of the paper is organized as follows. E-commerce and m-commerce technology are discussed, and then the next section talks about the functioning of the intelligent store agent as a mobile shopper. Next, the details on the architecture of the proposed intelligent store agent are discussed, the details on the implementation of the intelligent store agent for mobile shopping activity, using J2ME and JADE-LEAP combination, are provided. Finally, results and a conclusion are discussed.

E- COMMERCE AND M-COMMERCE

Electronic commerce (e-Commerce) (Abbott, 2001) refers to the buying and selling of products or services over electronic systems such as the Internet and other computer networks. The definition has evolved since its inception; in as early as in 1970's, the technique of e-commerce was merely the sending and receiving of invoices electronically using varying technologies including EDI (Electronic Data Interchange). The period during 1980's saw the emergence of other forms of e-commerce, which included telephone banking, credit cards and the ATM (Automated Teller Machine). From 1990's onwards, e-commerce technique included enterprise resource planning systems (ERP), data mining and data warehousing. Most recently, since 2000, a significant amount of businesses have offered their services on the World Wide Web and persons have started to associate e-commerce with purchasing goods through the Internet via secure protocols and electronic payment services. Consequently, among the most widely used security technologies is the Secure Sockets Layer (SSL), which is built into both the leading Web browsers.

M-commerce (Abbott, 2001) is often represented as a derivative of e-commerce, implying that any e-commerce site should be made available from a mobile device, this however seems to be a misrepresentation. There are similarities in terms of being able to purchase a product or service in a virtual environment but there are however, unique characteristics and functions which distinguish both. Let us consider an example of a user wanting to buy a mobile phone through internet. Here the user performs a query search using a search engine like Google, to get the list of mobile phones say, with Wi-Fi technology in a certain price range. The search engine connects the Shop server and gets the details of mobile phones with the Wi-FI technology but nothing found in the required price range as desired by the user. In this case, since the match is not found, the user would have to refine the query in a different fashion to select an item that he could finally purchase. This type of querying by the human agent is however, tedious and time consuming.

Recently intelligent agents have gained considerable research attention in the field of Computer Science and also it is proved better (Baldi, 1997; Carzniga, 1997; Ghezzi & Vigna, 1997; Puliafito & Tomarachio, 1999; Suresh, 2001; Suresh, 2006) than the traditional client/server as it saves bandwidth and latency too. Taking these aspects into consideration, we have developed an intelligent agent based mobile shopper (Ryan and Suresh, 2009) which would replace the human agent in performing the shopping. The agent here possesses the necessary intelligence to perform the shopping like the human agent. The intelligence is good for it to get other products with different specification within the same price range.

AGENT BASED MOBILE SHOPPER - OUR PROPOSAL

Human shoppers in their quest to find goods and services at the best prices execute a shopping process which involves the following steps (Kotler & Armstrong, 2004):

- Determine the characteristics of the desired product or service
- Search stores that offer the product or service
- Reason about the price of the product or service
- Refine search characteristics if the search results are undesirable
- Select a product or service from the search results

The human shopper searches both physical stores and electronic stores (via a web-browser) for a product or service. A shopper can also search electronic stores using their mobile phones if enough support for m-commerce exists. In the mobile scenario there are a number of challenges faced by such a shopper. Some of which include:

- Entering information is time consuming
- Prices are difficult to compare
- Scrolling through lists is difficult
- Unreliable network connections

There has been some research work carried out on employing agents for mobile shopping (Guan, 2002; Li, 2007). The major drawback in the existing systems has been on the reasoning on the price of the product. It does not accept inputs from the user. Also it has been noted that adoption of Agent technology in real world applications is very slow though it has many advantages such as reduced network load, overcoming network latency etc., (Jha & Iyer, 2001; Nwana, 1997; Wooldridge & Jennings, 1995). Also it has been shown that mobile Agent technology is a better alternative to Client/server architecture. However, it may be noted that the idea of using standards and standard technologies in the implementation of agent based m-commerce has not yet been explored. Taking all these aspects into consideration, we have developed an Agent based mobile shopper (Ryan & Suresh, 2009) which gets the necessary input from the user (such as brand, price, quality etc.), which he desires and then would augment the agent based shopper developed - Handy Broker Agent. The agent developed here could also be fuzzy in nature as it might have to take a fuzzy decision in nature.

ARCHITECTURAL DETAILS OF AGENT BASED

Mobile Shopper

Our agent based mobile shopper (Ryan and Suresh, 2009) employs an event driven style of architecture. This event driven architecture has the following seven major components as shown in Figure 1.

1. Application GUI
2. Agent ControllerMidlet
3. EventManager
4. ShoppingRequestManager
5. DataObjectManger
6. MobileShopper and
7. Store Coordinator

The Application GUI consists of MIDP GUI components through which the user interacts. Its use has three main purposes: (i) collecting phone specifications (ii) informing the user of shopping progress and any errors encountered and (iii) displaying shopping results.

The AgentControllerMidlet acts as the main router, linking the other core components with the Application GUI. The user's inputs that are specified through the GUI are sent to this component via MIDP events. These events are transferred to the EventManager for further processing. AgentControllerMidlet has event listeners for a number of application specific events such as:

Figure 1. Agent based mobile shopper architecture

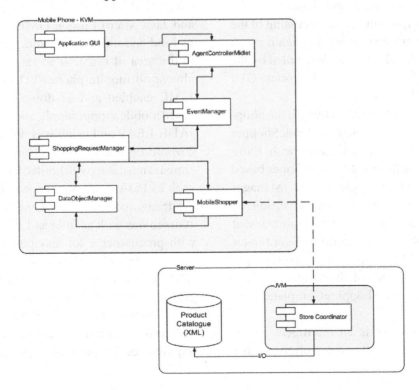

- ApplicationInitiationEvent,
- ApplicationClosingEvent,
- NextFormEvent,
- ShoppingCompleteEvent and
- ShoppingRequestException.

It also generates and displays appropriate Application GUI with the information provided by each event.

The EventManager generates the application specific events mentioned previously. These events are handled by event listeners registered with the EventManager. A user for example, interacts with the Application GUI to specify the details of the phone he/she is interested in then hits a "Shop!" button. This triggers a GUI event that is sent to AgentControllerMidlet who listens to all GUI events. AgentControllerMidlet component extracts the information from the GUI and asks the EventManager to raise a ShoppingRequestEvent. The EventManager sends this shopping request to the ShoppingRequestManager who coordinates the processing of the shopping request and informs the EventManager of the results. Upon completion of the processing of the shopping request the EventManager again raises a ShoppingCompleteEvent that is handled by the AgentControllerMidlet and the appropriate GUI is displayed for the user.

At the core of the shopping activity is the ShoppingRequestManager who creates MobileShopper agents that actually communicates with Store Coordinator agents to shop for user phones based on specification. The ShoppingRequestManager has the very important function of converting the user's specification into a shopping grammar that both the MobileShopper and the StoreCoordinator can understand. The ShoppingRequestManager also translates the result of the shopping into a form that listeners of ShoppingCompleteEvent can understand.

The MobileShopper is an intelligent agent, implemented using LEAP agent framework that goes to each store, interacting with the store's StoreCoordinator in search of a phone that meets the user's specification.

The StoreCoordinator is an agent that represents the store's interest in this shopping process. It is aware of the inventory that is available and the details of each item. It accepts request/inquiry from interested buyers. A store can have several StoreCoordinators but in our implementation we have a single StoreCoordinator per store.

The last component in our architecture is the DataObjectManger. This component manages the storage and retrieval of information on the mobile device. One of the biggest constraints of mobile devices is storage space and this has to be carefully managed to preserve the performance of the application.

IMPLEMENTATION USING JADE-LEAP

Our agent based mobile shopper was implemented using JADE-LEAP (Fabio, 2003; Fabio, 2007) and Java Micro Edition (JME) which is first of its kind and unique. No work has been reported in the area of research so far. As it is seen that almost all mobile phones/PDAs these days are J2ME enabled and so downloading the Agent based mobile shopper application developed using JADE-LEAP and using it is not a problem. Java 2 Micro Edition (J2ME) is Sun's version of Java aimed at machines with limited hardware resources such as PDAs, cell phones, and other consumer electronic and embedded devices. J2ME is aimed at machines with as little as 128KB of RAM and with processors a lot less powerful than those used on typical desktops and server machines. J2ME actually consists of a set of profiles. Each profile is defined for a particular type of device -- cell phones, PDAs, microwave ovens, etc. -- and consists of a minimum set of class libraries required for the particular type of device and a specification of a Java virtual machine required to support the device. The virtual machine specified in any profile is not necessarily the same as the virtual machine used in Java 2 Standard Edition (J2SE) and Java 2 Enterprise Edition (J2EE). It may be seen that the profile that we have used to develop a Palm OS device application is a subset of the Java Virtual Machine. Sun systems have released the following profiles: (i) The Foundation Profile - A profile for next generation consumer electronic devices and (ii) The Mobile Information Device Profile (MIDP) - A profile for mobile information devices, such as cellular phones and two-way pagers, and PDAs.

JADE (Java Agent Development Environment) is a full agent middleware platform. JADE-LEAP is an extension of JADE created specifically for J2ME-CLDC and CDC platforms. JADE-LEAP is the only agent development kit available for mobile devices. The JADE platform is composed

of agents and containers that they "live" in. The agent platform contains two default agents, Agent Management System (AMS) and directory facilitator (DF). The prior is responsible of the management of agents and the latter used to discover and mange the services an agent offers through its yellow page functionality. The agent containers can be distributed over the network. A special main container coordinates acts as the bootstrap for the JADE-LEAP platform and all other containers must register with this main container. In our implementation we have a main container that houses a number of Store Coordinator agents. We also have Store Coordinator agents in containers in different location that are linked back to the main container this is depicted in Figure 2 (Ryan & Suresh, 2009).

A Store Coordinator agent is created for each store and upon creation registers it the main container's DF. A Mobile Shopper is created on the

Figure 2. JADE-LEAP physical architecture

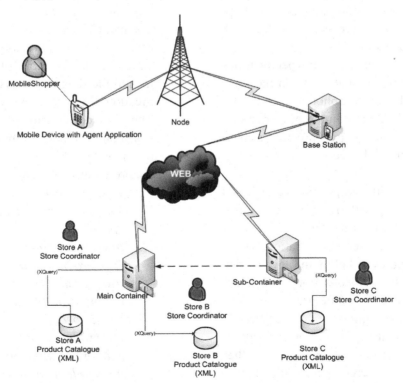

mobile device LEAP runtime that is linked back to the main container. It uses the DF of the main container to discover all stores that are available for shopping

Algorithm

The following is the algorithm (Ryan & Suresh, 2009) used in the development

- Input the user specification like the price range, EMS support, MMS support, Bluetooth, Wi-Fi etc for the mobile phones shopping through handheld device
- Start mobile shopper agent from the handheld device to shop
- The mobile agent gets all the details of the available stores from the Directory of Stores called the Directory Facilitator
- Based on the directory of stores available the mobile shopper agent sends the product specification to each and every store.
- The Store coordinator Agent of each shop, based on the request from the mobile shopper, matches the user specifications with the products available.
- If a product with the user specifications requested is not available within the price range, it finds a mobile phone which is 10% more than his/her maximum price range.
- Results of Agent shopping are produced on his/her handheld device.

The drawback with the present agent based mobile shopping algorithm (Ryan & Suresh, 2009) is that the store coordinator agent retrieves the product with the user specification within the price range and even below the lower price range. If not available, the store coordinator agent would then search for the same product by adding extra 10% of maximum price specified. If nothing is available it is going to return with no results. This is not true as we can get some other quality product with different specification for the

price range specified. So we have now modified this algorithm and propose a new one. This new algorithm, for the same user specification and price range specified for the store coordinator agent would not only possess the intelligence of retrieving the products with same user specification below the lower price range, within the price range specified and also above the maximum price range by 10%. It also possesses the intelligence of retrieving the products with different specification below the lower price range, within price range specified and also above the maximum price range specified by 10%. So we here have improved the agent based mobile shopper algorithm for achieving this intelligence as follows:

- Input the user specification like the price range, EMS support, MMS support, Camera Pixel, Bluetooth, Wi-Fi etc for the mobile phones shopping through handheld device
- Start mobile shopper from the handheld device to shop
- The mobile agent gets all the details of the available stores from the Directory of Stores called the Directory Facilitator
- Based on the directory of stores available the mobile shopper sends the product specification to each and every store.
- The Store coordinator Agent of each shop, based on the request from the mobile shopper, matches the user specifications with the products available as follows:
  ○ *If a product is available with the same user specification or different specifications available in store for lower price range*
  ○ *If a product is available with same user specification or different specification available in store for the price range specified*
  ○ *If a product with the same user specification or different specifications is not available in store within the price*

range, it finds a mobile phone which is 10% more than his/her maximum price range

- Results of Agent shopping are produced on his/her handheld device.

Agent Communication Language

The agents, in our shopping environment, uses FIPA (Fabio, 2003; Fabio, 2007) standard and ACL message standard, to communicate. All shops in the environment are registered with the DF agent. The shopping agent on the mobile phone queries the DF to find all available stores. The shopping agent then sends an ACL message containing the XPath (Farb, 2003) specification as message content to each stores message inbox. The store coordinator agent reads the messages from the inbox, executes the XPath (Farb, 2003) query against the catalogue and returns the results.

RESULTS

The implementation of Agent based shopper is done using JADE-LEAP toolkit. The Graphical user interface for the mobile phone is done using Java 2 Micro edition (J2ME) for the user to give his specification for the agent to shop.

Here we have created five (5) stores for our research, each holding a few mobile phones. A sample webpage of Store is shown in Figure 3. The list of five stores in the JADE environment is shown in Figure 4. The agent starts by getting the user specification to shop, as shown in Figure 5(a) and 5(b).

In Figure 6, the agent being sent for shopping is shown. The agent here possesses the intelligence to shop products below the price range specified, if it matches the user specifications. Once the agent is sent shopping with the price range (say $3000 to $5000) specified by the user it matches the criteria what he asks for as shown in Figure 7. Also if the exact user specification is not available in the price range ($3000-$5000) specified it has the intelligence to shop and get phones with the user specification 10% above the maximum price range by and produce it to the user as shown in Figure 8. We believe that normally users do not mind spending extra 10%, if he gets a mobile phone within specification.

In certain cases the customer might be asking for a phone for a some specifications like say e-mail and other facilities for a lower price range but these are available in higher price range only.

Figure 3. XML page of shop1 with product details

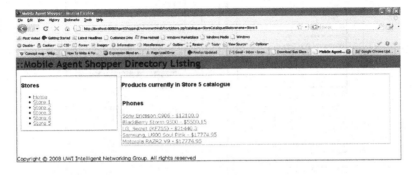

Figure 4. Five stores in JADE environment

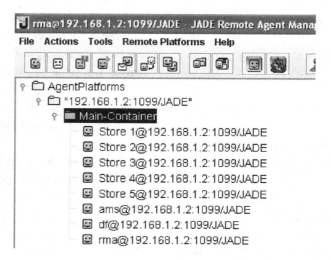

Our store agent here would retrieve phones with these specifications asked available in other price range and produce it to user on his mobile phone in rather than returning no products available.

Consider a scenario where the user request for a phone that is less than 3000 with e-mail facility support. It shows phones with the facilities we ask for in the price range and also what phone is available for the price range we have requested too. It is seen phones in range of 5000 and above supports facilities like e-mail, camera and so. Now it is for the user to decided whether to opt for

Figure 5. (a) User specifications – screen 1; (b) User specifications – screen 1

(a)   (b)

Figure 6. Agent being sent for shopping

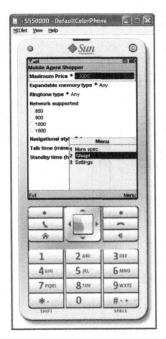

Figure 8. Approximate matching of mobile phone

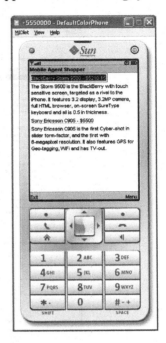

Figure 7. Results of exact matching of mobile phone

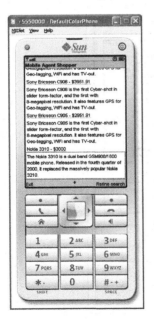

higher price range phone or buy a simple phone in the budget he has requested for. This is shown in Figure 9(a) and Figure 9(b).

Now say the customer wants to refine the above search with phones supporting HCSD see for the same $5000. It is found that no phones with that facility is available in that price range or even more than price range by 10%.But phone with the feature asked is available in other price range and is produce to the user on mobile phone in addition to phone available for the price he requested for too. These are shown in Figure 10(a) and Figure 10(b) respectively.

Now say customer is requesting for a phone with same HCSD facility for price range above 40,000 dollar. Agent shopper finds that phone with that facility is available for a lower price range and phones with price range requested for provides even more advanced features. These are shown in Figure 11 (a) and Figure 11 (b).

There would be some extreme cases where no shops are seen registered with the Agent environ-

Figure 9. (a) Result of shopping - screen 1; (b) result of shopping - screen 2

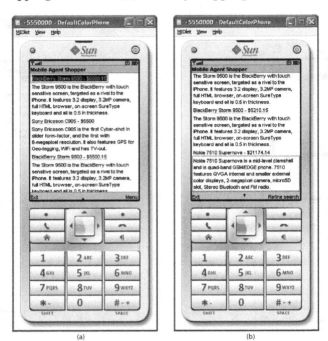

Figure 10. (a) Result of shopping - screen 3; (b) result of shopping - screen 3

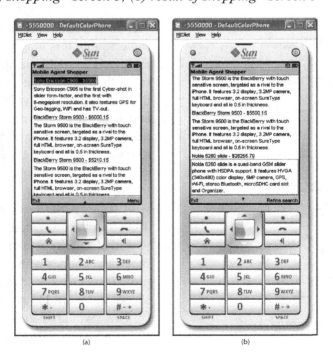

Figure 11. (a) Result of shopping - screen 3; (b) result of shopping - screen 3

(a)   (b)

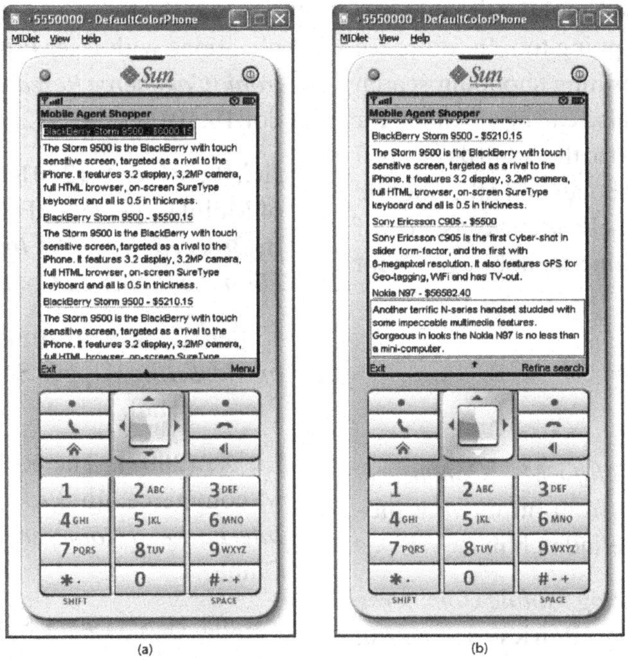

ment due to some network problem or so and user tries shopping. This would results in no shops found. Another extreme case is when agent based shopper is shut down completely due to some problem and user tries to shop. It would result in no agent been created for shopping.

CONCLUSION

We normally employ a human agent to get the details for our business transactions. But in a mobile environment we have shown that we can employ a mobile agent which can replicate the job of the human being. Also, quite recently mobile agents have gained considerable attention in the area of computer science which is evident from available literature. With all these in mind, we have attempted a new technology, viz. Intelligent Agent Based Mobile Shopper that performs shopping for customers by getting necessary details, much like a human agent. The shopping example

attempted here to show the feasibility of the proposed technology, is mainly concentrated towards mobile phones. In this case, the Agent gathers the information about the mobile phones from several Store Server agents and compares them with the user preferences of products. The Agent possess the intelligence to shop for products requested with the specification requested within the price range specified and even below and above the price range. In addition it also possess the intelligence to retrieve the products available for that price range and also show the price range of product for the specification requested. This here gives a complete agent based shopper which is been done for mobile phone shopping. The results of our research have been shown as screenshots. The limitations of agent based shopper so developed is that the price mark up is fixed, i.e., 10% and also selection criteria cannot be given by the user but got be chosen from what is been provided. In addition the review result of shops and products to be shopped cannot be provided in the software developed which can be taken as future work. Also the same agent based mobile shopper can be extended for other shopping too. In future we propose to work on performing shopping search using model number and also can be extended for other product search too. Also the agent developed would be tested on smart phones and other mobile devices too in real time. Finally security aspects of mobile commerce, pertaining to mobile payment would also be taken.

REFERENCES

Abbott, L. (2001). *MobileInfo M – Commerce*. Retrieved June 29, 2009, from http://www.mobileinfo.com/ mcommerce/index.htm

Abbott, L. (2001). *Separating Mobile Commerce from Electronic Commerce*. Retrieved June 30, 2009, from http://www.mobileinfo.com/mcommerce/differences.htm

Baldi, M., et al. (1997). Exploiting Code Mobility in Decentralized and Flexible Network Management. In *Proceedings of the First International Workshop on Mobile Agents*, Berlin.

Carzaniga, A., et al. (1997). Designing distributed applications with mobile code paradigms. In *Proceedings of the 19th International Conference on Software Engineering (ICSE'97)*. Washington, DC: IEEE.

Fabio, G., et al. (2003). JADE White Paper. *Telecom Italia Lab, 3*(3).

Fabio, G. (2007). *Developing MultiAgent System with JADE*. London: John Wiley & Sons.

Ghezzi, C., & Vigna, G. (1997). Mobile Code Paradigms and Technologies: A Case Study. In *Proceedings of the First International Workshop on Mobile Agents*, Berlin.

GoldFarb. C. F. (2003). XML handbook. Upper Saddle River, NJ: Prentice Hall.

Guan, S. (2002). Handy Broker - An Intelligent Product-Brokering Agent for M-Commerce Applications with User Preference Tracking. *Electronic Commerce Research and Applications, 1*(4). doi:10.1016/S1567-4223(02)00023-6

Jha, R., & Iyer, S. (2001). Performance Evaluation of Mobile Agents for E-Commerce Applications. In *Proceedings of 8th Internation Conference on High Perfromance Computing*, Hyderbad, India.

Kotler, P., & Armstrong, G. (2004). *Principles of Marketing*. Upper Saddle River, NJ: Pearson Prentice Hall.

Li, X. (2007). The Role of Mobile Agents in M-commerce. In *Proceedings of Sixth Wuhan International Conference on E-Business*, Wuhan, China.

Nwana, H. S. (1997). Software Agents: An Overview. *The Knowledge Engineering Review, 11*(3).

Puliafito, A., & Tomarchio, O. (1999). Advanced Network Management Functionalities through the use of Mobile Software Agents. In *Proceedings of 3rd International Workshop on Intelligent Agents for Telecommunication Applications (IATA'99)*, Stockholm, Sweden.

Ryan, A. B., & Suresh, S. (2009). Intelligent Agent based Mobile Shopper. In *Proceedings of sixth IFIP/IEEE International conference in Wireless Optical Communications and Networks*, Cairo, Egypt.

Suresh, S. (2001). *Applications of Policy and Mobile Agents in the Management of Computer Networks*. Unpublished Master's thesis, Anna University, Chennai, India.

Suresh, S. (2006). *Studies in Agent based IP Traffic Congestion Management in DiffServ Networks*. Unpublished doctoral dissertation, University of South Australia, Adelaide, Australia.

Thomas, R., & Harold, K. (2003). *Handbook of Consumer Behavior*. Upper Saddle River, NJ: Prentice-Hall.

Wooldridge, M., & Jennings, N. (1995). Intelligent Agents: Theory and Practice. *The Knowledge Engineering Review, 10*(2). doi:10.1017/S0269888900008122

This work was previously published in the International Journal of E-Services and Mobile Applications, Volume 3, Issue 1, edited by Ada Scupola, pp. 57-72, copyright 2011 by IGI Publishing (an imprint of IGI Global).

Compilation of References

Abbott, L. (2001). *MobileInfo M – Commerce*. Retrieved June 29, 2009, from http://www.mobileinfo.com/ mcommerce/index.htm

Abbott, L. (2001). *Separating Mobile Commerce from Electronic Commerce*. Retrieved June 30, 2009, from http://www.mobileinfo.com/mcommerce/differences.htm

Abe, M. (1997). Mix-networks on permutation networks. In *Proceedings of the International Conference on the Theory and Applications of Cryptology and Information Security* (pp. 258-273).

Acar, M., & Kumaş, E. (2008). Türkiye'nin Dönüşüm Sürecinde Anahtar bir Mekanizma Olarak E-Devlet, E-Dönüşüm ve Entegrasyon Standartları. In *İAS2008*. Turkey: TÜİK.

Achrol, R. S., & Gundlach, G. T. (1999). Legal and social safeguards against opportunism in exchange. *Journal of Retailing, 75*(1), 107–124. doi:10.1016/S0022-4359(99)80006-2

ADAE. (2011). *Association des Dirigeants et Administrateurs d'Entreprise*. Retrieved from http://www.adae.asso.fr

Adams, D. A., Nelson, R. R., & Todd, P. A. (1992). Perceived usefulness, ease of use, and usage of information technology: A replication. *Management Information Systems Quarterly, 16*(2), 227–247. doi:10.2307/249577

Adler, M., & Ziglio, E. (1996). *Gazing into the oracle: The Delphi Method and its application to social policy and public health*. London, UK: Jessica Kingsley Publishers.

Agarwal, R., & Karahanna, E. (2000). Time flies when you're having fun: Cognitive absorption and beliefs about information technology usage. *Management Information Systems Quarterly, 24*(4), 665–694. doi:10.2307/3250951

Agarwal, R., & Prasad, J. (1998). A conceptual and operational definition of personal innovativeness in the domain of information technology. *Information Systems Research, 9*(2), 204–215. doi:10.1287/isre.9.2.204

Agarwal, R., & Prasad, J. (1999). Are individual differences germane to the acceptance of new information technologies. *Decision Sciences, 30*(2), 361–391. doi:10.1111/j.1540-5915.1999.tb01614.x

Aier, S. (2007). *Integrationstechnologien als basis einer nachhaltigen enterprise architecture – abhängigkeiten zwischen organisation und informationstechnologie*. Berlin, Germany: Gito Verlag.

Aier, S., & Schönherr, M. (2004). Flexibilisierung von organisations- und IT-architekturen durch EAI. In Aier, S., & Schönherr, M. (Eds.), *Enterprise application integration [Flexibilisierung komplexer enterprise architekturen]*. (pp. 1–60). Berlin, Germany: Gito Verlag.

AIS - Association for Information Systems. (n.d.). *Senior Scholars' Basket of Journals*. Retrieved from http://home.aisnet.org/displaycommon.cfm?an=1&subarticlenbr=346

Ajzen, I. (1985). From intentions to actions: A theory of planned behavior. In Kuhl, J., & Beckmann, J. (Eds.), *Action control: From cognition to behavior*. Berlin: Springer Verlag.

Ajzen, I. (1991). The theory of planned behavior. *Organizational Behavior and Human Decision Processes, 50*(2), 179–211. doi:10.1016/0749-5978(91)90020-T

Ajzen, I. (1991). Theory of planned behavior. *Organizational Behavior and Human Decision Processes, 50*(2), 179–211. doi:10.1016/0749-5978(91)90020-T

Ajzen, I., & Fishbein, M. (1980). *Understanding attitudes and predicting social behavior*. Upper Saddle River, NJ: Prentice- Hall.

Aktaş, Z. (1987). *Structured Analysis and Design of Information Systems*. Upper Saddle River, NJ: Prentice-Hall.

AlAwadhi, S., & Morris, A. (2008). The use of the UTAUT model in the adoption of e-government services in Kuwait. In *Proceedings of the 41st Hawaii International Conference on System Sciences* (pp. 1-5).

Aldas-Manzano, J., Ruiz-Mafe, C., & Sanz-Blas, S. (2009). Exploring individual personality factors as drivers of M-shopping acceptance. *Industrial Management & Data Systems, 109*(6), 739–757. doi:10.1108/02635570910968018

Al-Qirim, N. (2005). An empirical investigation of an e-commerce adoption-capability model in small businesses in New Zealand. *Electronic Markets, 15*(4), 418–437. doi:10.1080/10196780500303136

Amar, A. D. (2002). *Managing Knowledge Workers. Unleashing Innovation and Productivity*. London: Quorum Books.

Amin, H., Rizal, M., Hamid, A., Lada, S., & Anis, Z. (2008). The adoption of mobile banking in Malaysia: The case of Bank Islam Malaysia Berhad (BIMB). *International Journal of Business and Society, 9*(2), 43–53.

Anastasios, E., & Vasileios, T. (2008). Evaluating tax sites: an evaluation framework and its application. *Electronic Government, 5*(3), 321–343. doi:10.1504/EG.2008.018878

Andersen, K. V., & Medaglia, R. (2008). e-Government Front-End Services: Administrative and Citizen Cost-Benefits. In *Proceedings of the 7th International Conference (EGOV 2008, TORINO, ITALY)* (Vol. 5184).

Andersen, M. G., & Katz, R. B. (1998). Strategic sourcing. *International Journal of Logistics Management, 9*(1), 1–13. doi:10.1108/09574099810805708

Anderson, J., & Schwager, P. (2003). SME adoption of wireless LAN technology: Applying the UTAUT model. In *Proceedings of the 7th Annual Conference of the Southern Association for Information Systems*, Savannah, GA (pp. 39-43).

Anderson, J. C., & Gerbing, D. W. (1988). Structural equation modeling in practice: A review and recommended two-step approach. *Psychological Bulletin, 103*(3), 411–423. doi:10.1037/0033-2909.103.3.411

Anderson, J. E., Schwager, P. H., & Kerns, R. L. (2006). The drivers for acceptance of Tablet PCs by faculty in a college of business. *Journal of Information Systems Education, 17*(4), 429–440.

Anderson, W. T. Jr, & Cunningham, W. L. (1972). The socially conscious consumer. *Journal of Marketing, 36*(3), 23. doi:10.2307/1251036

Andraski, J. C. (1998). Leadership and the realization of supply chain collaboration. *Journal of Business Logistics, 19*(2), 9–11.

Androulidakis, N., & Androulidakis, I. (2005). Perspectives of mobile advertising in Greece. In *Proceedings of the International Conference on Mobile Business*, Sydney, Australia (pp. 441-444).

Angeli, A. D., & Kyriakoullis, L. (2006). Globalization vs. localization in e-commerce: Cultural-aware interaction design. In *Proceedings of the Working Conference on Advanced Visual Interfaces*.

Anthopoulos, L., Siozos, P., Nanopoulos, A., & Tsoukalas, I. A. (2006). The Bottom-up Design of e-Government: A Development Methodology based on a Collaboration Environment. *e-Service Journal, 4*(3).

Anthopoulos, L. G., Siozosa, P., & Tsoukalas, I. A. (2007). Applying participatory design and collaboration in digital public services for discovering and re-designing e-Government services. *Government Information Quarterly, 24*(2), 353–376. doi:10.1016/j.giq.2006.07.018

Antil, J. (1984). Socially responsible consumers: Profile and implications for public policy. *Journal of Macromarketing, 4*, 19–32. doi:10.1177/027614678400400203

Anttiroiko, A. V. (2003). Building strong e-democracy: the role of technology in developing democracy for the information age. *Communications of the ACM, 46*(9), 121–128. doi:10.1145/903893.903926

Arendsen, R., & Hedde, M. Jt. (2009). On the Origin of Intermediary E-Government Services. In *Proceedings of the 8th International Conference (EGOV 2009)*, Linz, Austria (Vol. 5693).

Arendsen, R., Engers, T. M. V., & Schurink, W. (2008). Adoption of High Impact Governmental eServices: Seduce or Enforce? In *Proceedings of the 7th International Conference (EGOV 2008)*, Torino, Italy (Vol. 5184).

Ariadne Training Limited. (2001). *Engineering Software - Applied Object Oriented Analysis And Design Using The Uml*. Ariadne Training Limited.

Armen, C., & Morelli, R. (2005). E-voting and computer science: Teaching about the risks of electronic voting technology. In *Proceedings of the Tenth Annual Conference on Innovation and Technology in Computer Science Education*, Bologna, Italy (pp. 227-231).

Asgarkhani, M. (2005). The Effectiveness of e-Service in Local Government: A Case Study. *Electronic. Journal of E-Government*, *3*(4), 157–166.

Ask, A., Hatakka, M., & Grönlund, Å. (2008). The Örebro City Citizen-Oriented E-Government Strategy. [IJEGR]. *International Journal of Electronic Government Research*, *4*(4), 69–88.

ATHENA. (n. d.). *Advanced technologies for interoperability of heterogeneous enterprise networks and their application*. Retrieved from http://www.athena-ip.org/

Athens University of Economics and Business & ICAP GROUP. (2008). *Social-financial overview of the mobile phone industry in Greece*. Retrieved from http://www.sepe.gr/files/pdf/Executive%20Summary.pdf

Autio, M., Heiskanen, E., & Heinonen, V. (2009). Narratives of green consumers – the antihero, the environmental hero and the anarchist. *Journal of Consumer Behaviour*, *8*(1), 40–53. doi:10.1002/cb.272

Autry, C. W., Griffis, S. E., Goldsby, T. J., & Bobbitt, L. M. (2005). Warehouse management systems: Resource commitment, capabilities, and organizational performance. *Journal of Business Logistics*, *26*(2), 165–182. doi:10.1002/j.2158-1592.2005.tb00210.x

Avdagić, M., Šabić, Z., Zaimović, T., & Nazečić, N. (2008, September). eGovernment and mGovernment Integration: Role in Public Administration Reform. In *Proceedings of the Third International Conference & Exhibitions on Mobile Government (mLife 2008), Mobile Government Consortium International (mGCI) UK*, Turkey.

Axelsson, K., & Melin, U. (2007). Talking to, Not About, Citizens – Experiences of Focus Groups in Public E-Service Development. In *Proceedings of the 6th International Conference (EGOV 2007)*, Regensburg, Germany (Vol. 4656).

Axelsson, K., & Melin, U. (2008). Citizen Participation and Involvement in eGovernment Projects: An Emergent Framework. In *Proceedings of the 7th International Conference (EGOV 2008)*, Torino, Italy (Vol. 5184).

Axelsson, K., Melin, U., & Persson, A. (2007). Communication Analysis of Public Forms - Discovering Multi-Functional Purposes in Citizen and Government Communication. *International Journal of Public Information Systems*, *3*, 161–181.

Bachmann, G. R., John, D. R., & Rao, A. R. (1993). Children's susceptibility to peer group purchase influence: An exploratory investigation. *Advances in Consumer Research. Association for Consumer Research (U. S.)*, *20*, 463–468.

Bailey, L. S., & Suresh, S. (2009, March 16-17). Intelligent agent based mobile tour planner. In *Proceedings of the First Caribbean Conference on Information and Communication Technology*, Kingston, Jamaica (pp. 8-13).

Baker, J., Grewal, D., & Parasuraman, A. (1994). The influence of store environment on quality inferences and store image. *Journal of the Academy of Marketing Science*, *22*, 328–339. doi:10.1177/0092070394224002

Balasubramanian, S., Peterson, R. A., & Jarvenpaa, S. L. (2002). Exploring the implications of m-commerce for markets and marketing. *Journal of the Academy of Marketing Science*, *30*, 348–361. doi:10.1177/009207002236910

Balci, A., Kumaş, E., Taşdelen, H., Süngü, E., Medeni, T., & Medeni, T. D. (2008). Development and implementation of e-government services in Turkey: issues of standardization, inclusion, citizen and satisfaction. In *Proceedings of the ICEEGOV 2008* (Vol. 351, pp. 337-342).

Balcı, A., & Kirilmaz, H. (2009). Kamu Yönetiminde Yeniden Yapilanma Kapsaminda E-Devlet Uygulamalari. [Reorganization of Public Administration and e-Government Applications]. *Türk İdare Dergisi*, *81*, 463–464.

Balderjahn, I. (1988). Personality variables and environmental attitudes as predictors of ecologically responsible consumption patterns. *Journal of Business Research*, *17*(1), 51–56. doi:10.1016/0148-2963(88)90022-7

Baldi, M., et al. (1997). Exploiting Code Mobility in Decentralized and Flexible Network Management. In *Proceedings of the First International Workshop on Mobile Agents*, Berlin.

Ball, A. D., & Tasaki, L. (1992). The role and measurement of attachment in consumer behavior. *Journal of Consumer Psychology*, *1*, 155–172. doi:10.1207/s15327663jcp0102_04

Bandura, A. (1982). Self-efficacy mechanism in human agency. *The American Psychologist*, *37*, 122–147. doi:10.1037/0003-066X.37.2.122

Bandura, A. (1986). *Social foundations of thought and action: a social cognitive theory*. Upper Saddle River, NJ: Prentice-Hall.

Bandura, A. (1997). *Self-efficacy: The exercise of control*. New York, NY: W.H. Freeman.

Barbonis, P. A., & Laspita, S. (2005). Some factors influencing adoption of e-commerce in Greece. In *Proceedings of the IEEE International Conference on Engineering Management, 1*, 31–35. doi:10.1109/IEMC.2005.1559082

Barnes, S. J. (2002). The mobile commerce value chain: Analysis and future developments. *International Journal of Information Management*, *22*, 91–108. doi:10.1016/S0268-4012(01)00047-0

Barnes, S. J., & Scornavacca, E. (2004). Mobile marketing: The role of permission and acceptance. *International Journal of Mobile Communications*, *2*, 128–139. doi:10.1504/IJMC.2004.004663

Barney, J. B. (1991). Firm resources and sustained competitive advantage. *Journal of Management*, *17*(2), 99–120. doi:10.1177/014920639101700108

Barrat, J. (2008). Electronic voting certification procedures. Who should carry out the technical analysis? In Barrat, J. (Ed.), *E-voting: The last electoral revolution* (1st ed.). Ann Arbor, MI: ICPSR.

Barwise, P., & Strong, C. (2001). Permission-based mobile advertising. *Journal of Interactive Marketing*, *16*(1), 14–24. doi:10.1002/dir.10000

Bass, L., Clements, P., & Kazman, R. (2003). *Software architecture in practice* (2nd ed.). Reading, MA: Addison-Wesley.

Bauer, H. H., Reichardt, T., Barnes, S. J., & Neumann, M. M. (2005). Driving consumer acceptance of mobile marketing: A theoretical framework and empirical study. *Journal of Electronic Commerce Research*, *6*, 181–192.

Baumgartner, H. (2002). Toward a personology of the consumer. *The Journal of Consumer Research*, *29*, 286–292. doi:10.1086/341578

Bearden, W. O., Netemeyer, R. G., & Teel, R. J. (1989). Measurement of consumer susceptibility to interpersonal influence. *The Journal of Consumer Research*, *15*(4), 473–481. doi:10.1086/209186

Becerra, E., & Stutts, M. (2008). Ugly duckling by day, super model by night: The influence of body image on the use of virtual worlds. *Journal of Virtual Worlds Research*, *1*(2), 1–19.

Belanger, F., & Carter, L. (2006). The Effects of the Digital Divide on E-Government: An Empirical Evaluation. In *Proceedings of the 39th Annual Hawaii International Conference on System Sciences (HICSS'06)*.

Belanger, F., & Carter, L. (2008). Trust and risk in e-government adoption. *The Journal of Strategic Information Systems*, *17*(2). doi:10.1016/j.jsis.2007.12.002

Belch, M. A. (1979). Identifying the socially and ecologically concerned segment through lifestyle research: Initial findings. In Henion, K. E. II, & Kinnear, T. C. (Eds.), *The conserver society* (pp. 69–81). Chicago, IL: American Marketing Association.

Belch, M. A. (1982). A segmentation strategy for the 1980's: Profiling the socially-concerned market through life-style analysis. *Journal of the Academy of Marketing Science*, *10*(4), 345–358. doi:10.1007/BF02729340

Bellavista, P., Corradi, A., Tarantino, F., & Stefanelli, C. (1999). Mobile agents for web-based systems management. *Internet Research*, *9*(5), 360–371. doi:10.1108/10662249910297769

Bellifemine, F., Caire, G., Poggi, A., & Rimassa, G. (2003). JADE a white paper. *EXP in Search of Innovation*, *3*(3), 6-19.

Bellifemine, F., Caire, G., & Greenwood, D. (2007). *Developing multi-agent systems with JADE*. Chichester, UK: John Wiley & Sons. doi:10.1002/9780470058411

Benaloh, J., & Tuinstra, D. (1994). Receipt-free secret-ballot elections (extended abstract). In *Proceedings of the 26th Annual ACM Symposium on Theory of Computing* (pp. 544-553).

Benaloh, J., & Yung, M. (1986). Distributing the power of a government to enhance the privacy of voters. In *Proceedings of the Fifth Annual ACM Symposium on Principles of Distributed Computing* (pp. 52-62).

Benjamin, M., & Whitley, E. A. (2009). Critically classifying: UK e-government website benchmarking and the recasting of the citizen as customer. *Information Systems Journal, 19*(2).

Berque, A. (1982). *Vivre L'espace au Japon* (Miyahara, M., Trans.). Tokyo, Japan: Presses Universitaires de France.

Bharadwaj, A. S. (2000). A resource-based perspective on information technology capability and firm performance: An empirical investigation. *Management Information Systems Quarterly, 24*(1), 169–196. doi:10.2307/3250983

Bharat Electronics. (2009). *Electronic voting machines.* Retrieved from http://www.bel-india.com/index.aspx?q=&sectionid=237

Bhattacherjee, A. (2000). Acceptance of E-commerce services: The case of electronic brokerages. *IEEE Transactions on Systems Man and Cybernetics and Humans, 30*(4), 411–420. doi:10.1109/3468.852435

Bhattacherjee, A. (2001). Understanding information systems continuance: An expectation-confirmation model. *Management Information Systems Quarterly, 25*(3), 351–370. doi:10.2307/3250921

Bhattacherjee, A. (2002). Individual trust in online firms: Scale development and initial test. *Journal of Management Information Systems, 19*(1), 211–241.

Bhatti, T. (2007). Exploring factors influencing the adoption of mobile commerce. *Journal of Internet Banking and Commerce, 12*, 2–13.

Bianchi, A., & Phillips, J. G. (2005). Psychological predictors of problem mobile phone use. *Cyberpsychology & Behavior, 8*, 39–51. doi:10.1089/cpb.2005.8.39

Bigne, E., Ruiz, C., & Sanz, S. (2005). The impact of Internet user shopping patterns and demographics on consumer mobile buying behavior. *Journal of Electronic Commerce Research, 6*(3), 193–209.

Bisaillon, V. (2005). Le consumérisme politique comme nouveau mouvement social économique, in Consumérisme politique I: Du boycott au boycott. Chaire de responsabilité sociale et de développement durable ESG-UQAM, Receuil de textes CEH/RT-30-2005. In *Proceedings of the 8ème Séminaire de la Série Annuelle sur les Nouveaux Mouvements Sociaux Économiques* (pp. 6-17).

Blechar, J., Constantiou, I. D., & Damsgaard, J. (2006). Exploring the influence of reference situations and reference pricing on mobile service user behaviour. *European Journal of IS, 15*(3), 285–291.

Blythe, J. (1999). Innovativeness and newness in high-tech consumer durables. *Journal of Product and Brand Management, 8*(5), 415–429. doi:10.1108/10610429910296028

Bolton, R. N. (1998). A dynamic model of the duration of the customer's relationships with a continuous service provider: the role of satisfaction. *Marketing Science, 17*(1), 45–65. doi:10.1287/mksc.17.1.45

Boostrom, R. (2008). The social construction of virtual reality and the stigmatized identity of the newbie. *Journal of Virtual Worlds Research, 1*(2), 1–19.

Bosch, J. (2000). *Design and use of software architecture: Adopting and evolving a product-line approach.* Reading, MA: Addison-Wesley.

Bowersox, D. J., & Closs, D. C. (1996). *Logistical management: The integrated supply chain process.* New York, NY: McGraw-Hill.

Boyd, C. (1989). A new multiple key cipher and an improved voting scheme. In J.-J. Quisquater & J. Vandewalle (Eds.), *Proceedings of the Workshop on the Theory and Application of Cryptographic Techniques on Advances in Cryptology* (LNCS 434, pp. 617-625).

Boyer-Wright, K. M., & Kottemann, J. E. (2008). High-Level Factors Affecting Global Availability of Online Government Services. In *Proceedings of the 41st Annual Hawaii International Conference on System Sciences (HICSS'08)*.

Boynton, A. C., Zmud, R. W., & Jacobs, G. C. (1994). The influence of IT management practice on IT use in large organizations. *Management Information Systems Quarterly, 18*(3), 299–318. doi:10.2307/249620

BPEL. (2007). *Business process execution language for web services*. Retrieved from http://www.ibm.com/developerworks/library/specification/ws-bpel/

Brandach, J. L., & Eccles, R. G. (1989). Market versus hierarchies: From ideal types to plural forms. *Annual Review of Sociology, 15*(1), 97–118. doi:10.1146/annurev.so.15.080189.000525

Breuil, H., Burette, D., & Flüry-Hérard, B. (2008, Décembre). *TIC et Développement durable, (Rapport), Ministère de l'Ecologie, de l'Energie, du Développement Durable et de l'Aménagement du Territoire, Conseil général de l'environnement et du développement durable, N° 005815-01*. Retrieved from http://www.telecom.gouv.fr/fonds_documentaire/rapports/09/090311rapport-ticdd.pdf

Brice, G. (2001). Implementing a continuous quality improvement programme in healthcare. *Caribbean Health Journal, 4*(3), 27–28.

Brooker, G. (1976). The self-actualizing socially conscious consumer. *The Journal of Consumer Research, 3*(2), 107–112. doi:10.1086/208658

Brown, I., & Jayakody, R. (2008). B2C e-commerce success: A test and validation of a revised conceptual model. *The Electronic Journal Information Systems Evaluation, 11*(3), 167–184.

Bruner, G. C., & Kumar, A. (2005). Explaining consumer acceptance of handheld Internet devices. *Journal of Business Research, 58*, 553–558. doi:10.1016/j.jbusres.2003.08.002

Brusa, G., Caliusco, M. L., & Chiotti, O. (2007). Enabling Knowledge Sharing within e-Government Back-Office Through Ontological Engineering. *Journal of Theoretical and Applied Electronic Commerce Research, 2*(1), 33–48.

Buccella, A., & Cechich, A. (2009). A semantic-based architecture for supporting geographic e-services. In *Proceedings of the 3rd International Conference on Theory and Practice of Electronic Governance (ICEGOV2009)* (Vol. 322, pp. 27-35).

Buckley, M. R., Cote, J. A., & Comstock, S. M. (1990). Measurement errors in behavioral sciences: The case of personality/attitude research. *Educational and Psychological Measurement, 50*(3), 447–474. doi:10.1177/0013164490503001

Budner, S. (1962). Intolerance of ambiguity as a personality variable. *Journal of Personality, 30*, 29–50. doi:10.1111/j.1467-6494.1962.tb02303.x

Buhalis, D., & Deimezi, O. (2003). Information technology penetration and ecommerce developments in Greece, with a focus on small to medium-sized enterprises. *Tourism Research, 13*(4), 309–324.

Burke, R. R. (1996). Virtual shopping: Breakthrough in marketing research. *Harvard Business Review, 74*(2), 120–131.

Burke, R. R. (2002). Technology and the customer interface: What consumers want in the physical and virtual store. *Journal of the Academy of Marketing Science, 30*(4), 411–432. doi:10.1177/009207002236914

Butt, S., & Phillips, J. G. (2008). Personality and self reported mobile phone use. *Computers in Human Behavior, 24*, 346–360. doi:10.1016/j.chb.2007.01.019

Büyüközkan, G. (2009). Determining the mobile commerce user requirements using an analytic approach. *Computer Standards & Interfaces, 31*, 144–152. doi:10.1016/j.csi.2007.11.006

Cabinet Office. (2010). *GovTalk*. Retrieved from http://www.govtalk.gov.uk

Cachon, G. P. (2004). The allocation of inventory risk in a supply chain: Push, pull, and advance-purchase discount contracts. *Management Science, 50*(2), 222–238. doi:10.1287/mnsc.1030.0190

Cacioppo, J. T., & Petty, R. E. (1982). The need for cognition. *Journal of Personality and Social Psychology, 42*, 116–131. doi:10.1037/0022-3514.42.1.116

Cannon, J. P., Achrol, R. S., & Gundlach, G. T. (2000). Contracts, norms, and plural form governance. *Journal of the Academy of Marketing Science, 28*(2), 180–195. doi:10.1177/0092070300282001

Capgemini, R. E. IDC, Sogeti, & DTi. (2009). *Smarter, faster, better egovernment* (8th benchmark measurement). Retrieved from http://ec.europa.eu/information_society/eeurope/i2010/docs/benchmarking/egov_benchmark_2009.pdf

Capgemini. (2007). *The user challenge: Benchmarking the supply of online public services* (7th measurement). Retrieved from http://ec.europa.eu/information_society/eeurope/i2010/docs/benchmarking/egov_benchmark_2007.pdf

Carr, A. S., & Pearson, J. N. (1999). Strategically managed buyer-seller relationships and performance outcomes. *Journal of Operations Management, 17*(5), 497–519. doi:10.1016/S0272-6963(99)00007-8

Carratta, T., Dadayan, L., & Ferro, E. (2006). ROI Analysis in e-Government Assessment Trials: The Case of Sistema Piemonte'. In *Proceedings of the 5th International Conference (EGOV 2006),* Krakow, Poland (Vol. 4084).

Carroll, A., Barnes, S. J., Scornavacca, E., & Fletcher, K. (2007). Consumer perceptions and attitudes towards SMS advertising: recent evidence from New Zealand. *International Journal of Advertising, 26,* 79–98.

Carter, L., & Belanger, F. (2005). The utilization of e-government services: citizen trust, innovation and acceptance factors. *Information Systems Journal, 15*(1). doi:10.1111/j.1365-2575.2005.00183.x

Carter, L., & Schaupp, L. C. (2009). Relating Acceptance and Optimism to E-File Adoption. *International Journal of Electronic Government Research, 5*(3), 62–74.

Carzaniga, A., et al. (1997). Designing distributed applications with mobile code paradigms. In *Proceedings of the 19th International Conference on Software Engineering (ICSE'97).* Washington, DC: IEEE.

Cay, S., & Yang, Z. (2008). Development of cooperative norms in the supply-supplier relationship: The Chinese experience. *Journal of Supply Chain Management, 44*(1), 60.

Chae, M., Kim, J., Kim, H., & Ryu, H. (2002). Information quality for mobile internet services: a theoretical model with empirical validation. *Electronic Markets, 12,* 38–46. doi:10.1080/101967802753433254

Chakrabarty, K. C. (2010). Mobile commerce, mobile banking: The emerging paradigm reserve bank of India. *Monthly Bulletin,* 23-30.

Chan, C. M. L., & Pan, S. L. (2008). User engagement in e-government systems implementation: A comparative case study of two Singaporean e-government initiatives. *The Journal of Strategic Information Systems, 17*(2). doi:10.1016/j.jsis.2007.12.003

Charalabidis, Y., Askounis, D., Gionis, G., & Lampathaki, F. (2006). Organizing Municipal e-Government Systems: A Multi-facet Taxonomy of e-Services for Citizens and Businesses. In *Proceedings of the 5th International Conference (EGOV 2006),* Krakow, Poland (Vol. 4084).

Charalabidis, Y., Askounis, D., & Gionis, G. (2007). A model for assessing the impact of enterprise application interoperability in the typical European enterprise. In Doumeingts, G., Müller, J., Morel, G., & Vallespir, B. (Eds.), *Enterprise interoperability: New challenges and approaches* (pp. 287–296). London, UK: Springer.

Charalabidis, Y., Lampathaki, F., Kavalaki, A., & Askounis, D. (2010). A review of electronic government interoperability frameworks: Patterns and challenges. *International Journal of Electronic Government, 3*(2), 189–221. doi:doi:10.1504/IJEG.2010.034095

Charalabidis, Y., Panetto, H., Loukis, E., & Mertins, K. (2008). *Interoperability approaches for enterprises and administrations worldwide.* Electronic Journal for e-Commerce Tools and Applications.

Chatterjee, D., Grewal, R., & Sambamurthy, V. (2002). Shaping up for e-commerce: Institutional enablers of the organizational assimilation of web technologies. *Management Information Systems Quarterly, 26*(2), 65–89. doi:10.2307/4132321

Chaudhuri, A., & Holbrook, M. B. (2001). The chain of effects from brand trust and brand affect to brand performance: The role of brand loyalty. *Journal of Marketing, 65*(2), 81–93. doi:10.1509/jmkg.65.2.81.18255

Chaum, D. (1982). Blind signatures for untraceable payments. In *Proceedings of the International Cryptology Conference,* Santa Barbara, CA.

Chaum, D. (1988). Elections with unconditionally-secret ballots and disruption equivalent to breaking RSA. In D. Barstow, W. Brauer, P. Brinch Hansen, D. Gries, D. Luckham, C. Moler et al. (Eds.), *Proceedings of the Workshop on the Theory and Application of Cryptographic Techniques on Advances in Cryptology* (LNCS 330, pp. 177-182).

Chau, P. Y. K. (1996). An empirical assessment of a modified technology acceptance model. *Journal of Management Information Systems, 13*(2), 185–204.

Chee-Wee, T., Benbasat, I., & Cenfetelli, R. T. (2008). Building Citizen Trust towards E-Government Services: Do High Quality Websites Matter? In *Proceedings of the 41st Annual Hawaii International Conference on System Sciences (HICSS'08).*

Chen, A. J., Pan, S. L., Zhang, J., Huang, W. W., & Zhu, S. (2009). Managing e-government implementation in China: A process perspective. *Information & Management, 46*(4). doi:10.1016/j.im.2009.02.002

Chen, C. C., Wu, C. S., & Wu, R. C. F. (2006). e-Service enhancement priority matrix: The case of an IC foundry company. *Information & Management, 43*(5). doi:10.1016/j.im.2006.01.002

Chen, C., Czerwinski, M., & Macredie, R. (2000). Individual differences in virtual environments - Introduction and overview. *Journal of the American Society for Information Science American Society for Information Science, 51*(6), 499–507. doi:10.1002/(SICI)1097-4571(2000)51:6<499::AID-ASI2>3.0.CO;2-K

Cheng, L., & Grimm, C. M. (2006). The application of empirical strategic management research to supply chain management. *Journal of Business Logistics, 27*(1), 1–57. doi:10.1002/j.2158-1592.2006.tb00240.x

Chen, I. J., & Paulraj, A. (2004). Understanding supply chain management: Critical research and a theoretical framework. *International Journal of Production Research, 42*(1), 131–163. doi:10.1080/00207540310001602865

Chen, J., & Tong, L. (2003). Analysis of mobile phone's innovative will and leading customers. *Science Management, 24*(3), 25–31.

Chen, L. (2008). A model of consumer acceptance of mobile payment. *International Journal of Mobile Communications, 6*, 32–52. doi:10.1504/IJMC.2008.015997

Cheong, J. H., & Park, M. C. (2005). Mobile internet acceptance in Korea. *Internet Research, 15*, 125–140. doi:10.1108/10662240510590324

Cherry, S. (2005). South Korea pushes mobile broadband-The WiBro scheme advances. *IEEE Spectrum*, 14–16. doi:10.1109/MSPEC.2005.1502522

Childers, T. L. (1986). Assessment of the psychometric properties of an opinion leadership scale. *JMR, Journal of Marketing Research, 23*, 184–188. doi:10.2307/3151666

Child, J., & David, F. (1998). *Strategies of cooperation.* Oxford, UK: Oxford University Press.

Chiu, C. M., Lin, H. Y., Sun, S. Y., & Hsu, M. H. (2009). Understanding customers' loyalty intentions towards online shopping: An integration of technology acceptance model and fairness theory. *Behaviour & Information Technology, 28*(4), 347–360. doi:10.1080/01449290801892492

Chiu, Y. B., Lin, C. P., & Tang, L. L. (2005). Gender differs: Assessing a model of online purchase intentions in e-tail service. *International Journal of Service Industry Management, 16*, 416–435. doi:10.1108/09564230510625741

Cho, D. Y., Kwon, H. J., & Lee, H. Y. (2007). Analysis of trust in internet and mobile commerce adoption. In *Proceedings of the 40th Hawaii International Conference on System Science* (p. 50).

Christiaanse, E., & Kumar, K. (2000). ICT-enabled coordination of dynamic supply webs. *International Journal of Physical Distribution and Logistics Management, 30*(3-4), 268–286.

CIMOSA. (n. d.). *CIM open system architecture.* Retrieved from http://www.pera.net/Methodologies/Cimosa/CIMOSA.html

Clarke, I. (2001). Emerging value propositions for m-commerce. *The Journal of Business Strategy, 18*, 133–149.

Claycomb, C., Iyer, K., & Germain, R. (2005). Predicting the level of B2B e-commerce in industrial organizations. *Industrial Marketing Management, 34*(3), 221–234. doi:10.1016/j.indmarman.2004.01.009

Clements, P., Bachmann, F., Bass, L., Gralan, D., Ivers, J., & Litle, R. (2002). *Documenting software architectures: Views and beyond.* Reading, MA: Addison-Wesley.

Closs, D. J., Goldsby, T. J., & Clinton, S. R. (1997). Information technology influences on world class logistics capability. *International Journal of Physical Distribution and Logistics Management, 27*(1), 4–17. doi:10.1108/09600039710162259

Coakes, S., Steed, L., & Ong, C. (2009). SPSS: *Vol. 16. Analysis without anguish.* Chichester, UK: John Wiley & Sons.

Codagnone, C., & Wimmer, M. A. (Eds.). (2007). *Road-mapping egovernment research. Visions and measures towards innovative governments in 2020: Results from the EC-funded project eGovRTD2020.* Retrieved from http://www.egovrtd2020.org/EGOVRTD2020/FinalBook.pdf

Cohen, J. (1986). *Improving privacy in cryptographic elections* (Tech. Rep. No. YALEU/DCS/TR-454). New Haven, CT: Yale University.

Cohen, J. D., & Fischer, M. J. (1985). A robust and verifiable cryptographically secure election scheme. In *Proceedings of the 26th Annual Symposium on Foundations of Computer Science* (pp. 372-382).

Cohen, S. (2005). *Auditing technology for electronic voting machines.* Cambridge, MA: MIT Press.

Coleman, S., Macintosh, A., & Schneeberger, A. (2007). *D12.3: eParticipation research direction based on barriers, challenges and needs* (Tech. Rep. No. FP6-2004-ist-4-027219). Leeds, UK: University of Leeds.

Commission of the European Communities (CEC). (2003). *Linking up Europe: The importance of interoperability for egovernment services.* Retrieved from http://www.csi.map.es/csi/pdf/interoperabilidad_1675.pdf

Commission on Electronic Voting. (2010). *Independent commission on electronic voting and counting at elections.* Retrieved from http://www.cev.ie/

Compeau, D. R., & Higgins, C. A. (1995). Computer self-efficacy: Development of a measure and initial test. *Management Information Systems Quarterly, 19*(2), 189–211. doi:10.2307/249688

Compeau, D. R., Higgins, C. A., & Huff, S. (1999). Social cognitive theory and individual reaction to computing technology: A longitudinal study. *Management Information Systems Quarterly, 23*(2), 145–158. doi:10.2307/249749

Condos, C., James, A., Every, P., & Simpson, T. (2002). Ten usability principles for the development of effective WAP and m-commerce services. *Aslib Proceedings, 54*(6), 345–355. doi:10.1108/00012530210452546

Connolly, R. (2007). Trust and the Taxman: A Study of the Irish Revenue's Website Service Quality. *Electronic. Journal of E-Government, 5*(2), 127–134.

Constantinides, E. (2002). The 4S web-marketing mix model. *Electronic Commerce Research and Applications, 1*(1), 57–76. doi:10.1016/S1567-4223(02)00006-6

Consulting, D. (2009). *Supporting the European interoperability strategy elaboration.* Retrieved from http://ec.europa.eu/idabc/servlets/Doc9cff.pdf?id=32455

Cooper, M. C., Ellram, L. M., Gardner, J. T., & Hanks, A. M. (1997). Meshing multiple alliances. *Journal of Business Logistics, 18*(1), 67–89.

Cooper, M. C., Lambert, D. M., & Pagh, J. D. (1997). Supply chain management: More than a new name for logistics. *International Journal of Logistics Management, 8*(1), 1–14. doi:10.1108/09574099710805556

Cooper, R. (1998). Assemblage Notes. In Chia, R. C. H. (Ed.), *Organization Worlds: Exploring in Technology and Organization with Robert Cooper* (pp. 108–129). London: Routledge.

Cooper, V., Lichtenstein, S., & Smith, R. (2009). Successful Web-Based IT Support Services: Service Provider Perceptions of Stakeholder-Oriented Challenges. *International Journal of E-Services and Mobile Applications, 1*(1), 1–20. doi:10.4018/jesma.2009092201

CORDIS. (n. d.). *European Commission homepage.* Retrieved from http://cordis.europa.eu/home_en.html

Corradini, F., Angelis, F. D., Polini, A., & Polzonetti, A. (2008). Improving Trust in Composite eServices Via Run-Time Participants Testing. In *Proceedings of the 7th International Conference (EGOV 2008)*, Torino, Italy (Vol. 5184).

Costa, P. T., & McCrae, R. R. (1990). Primary traits of Eysenck's P-E-N system: Three- and five-factor solutions. *Journal of Personality and Social Psychology, 69*, 308–317. doi:10.1037/0022-3514.69.2.308

Cox, C., & Rubin, A. (2004). *Is the U.S. ready for electronic voting?* Retrieved from http://teacher.scholastic.com/scholasticnews/indepth/upfront

Cox, A. D., Chicksand, P. I., & Davies, T. (2005). Sourcing indirect spend: A survey of current internal and external strategies for non-revenue-generating goods and services. *Journal of Supply Chain Management, 41*(2), 39–51. doi:10.1111/j.1055-6001.2005.04102004.x

Cox, D. F. (1967). The influence of cognitive needs and styles on information handling in making product evaluations. In Cox, D. F. (Ed.), *Risk taking and information handling in consumer behavior*. Boston, MA: Harvard University Press.

Crabbe, M., Standing, C., & Standing, S. (2009). An adoption model for mobile banking in Ghana. *International Journal of Mobile Communications, 7*(5), 515–543. doi:10.1504/IJMC.2009.024391

Cramer, R., Gennaro, R., & Schoenmakers, B. (1997). A secure and optimally efficient multi-authority election scheme. In W. Fumy (Ed.), *Proceedings of the Workshop on the Theory and Application of Cryptographic Techniques on Advances in Cryptology* (LNCS 1233, pp. 103-118).

Cranor, L., & Cytron, R. (1996). *Design and implementation of a practical security-conscious electronic polling system* (Tech. Rep. No. WUCS-96-02). St. Louis, MO: Washington University.

Crowley, A. G. (1998). Virtual logistics: Transport in the marketspace. *International Journal of Physical Distribution & Logistics Management, 28*(7), 547–574. doi:10.1108/09600039810247470

Cullen, R., & Reilly, P. (2007). Information Privacy and Trust in Government: a citizen-based perspective from New Zealand'. In *Proceedings of the 40th Annual Hawaii International Conference on System Sciences (HICSS'07)*.

Curran, J. M., & Meuter, M. L. (2005). Self-service technology adoption: Comparing three technologies. *Journal of Services Marketing, 19*(2), 103–113. doi:10.1108/08876040510591411

Curran, J. M., Meuter, M. L., & Surprenant, C. F. (2003). Intentions to use self-service technologies: A confluence of multiple attitudes. *Journal of Service Research, 5*(3), 209–224. doi:10.1177/1094670502238916

Cusumano, M. A., & Yoffie, D. B. (1998). *Competing on Internet time: Lessons from Netscape and its battle with Microsoft*. New York, NY: Free Press.

Dabholkar, P. (1996). Consumer evaluations of new technology-based self-service options: An investigation of alternative models of service quality. *International Journal of Research in Marketing, 13*, 29–51. doi:10.1016/0167-8116(95)00027-5

Dabholkar, P., & Bagozzi, R. P. (2002). An attitudinal model of technology-based self-service: Moderating effects of consumer traits and situational factors. *Journal of the Academy of Marketing Science, 30*, 184–201.

Dahlstrom, R., McNeilly, K., & Speh, T. (1996). Buyer-seller relationships in the procurement of logistical services. *Journal of the Academy of Marketing Science, 24*(2), 110–124. doi:10.1177/0092070396242002

Dailey, L. (1999). Designing the world we surf in: A conceptual model of web atmospherics. In *Proceedings of the AMA Summer Educator's Conference*, Chicago, IL.

Daugherty, P. J., Autry, C. W., & Ellinger, A. E. (2001). Reversel: The relationship between resource commitment and program performance. *Journal of Business Logistics, 22*(1), 107–112. doi:10.1002/j.2158-1592.2001.tb00162.x

Daugherty, P. J., Ellinger, A. E., & Dale, S. R. (1995). Information accessibility: Customer responsiveness and enhanced performance. *International Journal of Physical Distribution and Logistics Management, 25*(1).

Davis, F. D. (1989). Perceived usefulness, perceived ease of use, and user acceptance of information technology. *Management Information Systems Quarterly, 13*(3), 319–340. doi:10.2307/249008

Davis, F. D., Bagozzi, R. P., & Warshaw, P. R. (1989). User acceptance of computer technology: A comparison of two theoretical models. *Management Science, 35*(8), 982–1003. doi:10.1287/mnsc.35.8.982

Davis, R., & Sajtos, L. (2009). Anytime, anywhere: Measuring the ubiquitous consumer's impulse purchase behavior. *International Journal of Mobile Marketing, 4*, 15–22.

Dawes, S. (2002). *The future of e-government*. Albany, NY: University at Albany/SUNY. Retrieved from www.ctg.albany.edu/publications/reports/future_of_egov/future_of_egov.pdf

De Biji, P. W. J., & Peitz, M. (2005). Local loop unbundling in Europe: Experience, prospects, and policy challenges. *Communications and Strategies, 57*(1), 33–57.

De Moor, A., Park, J., & Croitoru, M. (2009). Argumentation map generation with conceptual graphs: The case for ESSENCE. In *Proceedings of the Fourth Conceptual Structures Tool Interoperability Workshop collocated with the 17th International Conference on Conceptual Structures.*

De Soria, I. M., Alonso, J., Orue-Echevarria, L., & Vergara, M. (2009). Developing an enterprise collaboration maturity model: research challenges and future directions. In *Proceedings of the 15th International Conference on Concurrent Enterprising*, Leiden, The Netherlands. ePractice (n. d.). *Meet, share, learn.* Retrieved from http://www.epractice.eu/en/factsheets/

de Vos, H., Haaker, T., Teerling, M., & Kleijnen, M. (2008). Consumer Value of Context Aware and Location Based Mobile Services. *International Journal of E-Services and Mobile Applications*, 36–50.

Dechow, D. (2008). Surveillance, consumers, and virtual worlds. *Journal of Virtual Worlds Research, 1*(2), 1–4.

Dembkowsky, S., & Hammer-Lloyd, S. (1994). The environmental value-attitude system model: A framework to guide the understanding of environmentally-conscious consumer behavior. *Journal of Marketing Management, 10*(4), 593–603. doi:10.1080/0267257X.1994.9964307

Derballa, V., & Pousttchi, K. (2006). Mobile Knowledge Management. In Schwartz, D. G. (Ed.), *Encyclopedia of Knowledge Management*. Hershey, PA: IGI Global.

DeRenzi, B., Gajos, K. Z., Parikh, T. S., Lesh, N., Mitchell, M., & Borriello, B. (2008). *Opportunities for intelligent interfaces aiding healthcare in low-income countries.* Retrieved from http://www.eecs.harvard.edu/~kgajos/papers/2008/derenzi08opportunities.shtml

Desmedt, Y., & Kurosawa, K. (2000). How to break a practical MIX and design a new one. In B. Preneel (Ed.), *Proceedings of the Workshop on the Theory and Application of Cryptographic Techniques on Advances in Cryptology* (LNCS 1807, pp. 557-572).

Deursen, A. V. (2007). Where to Go in the Near Future: Diverging Perspectives on Online Public Service Delivery. In *Proceedings of the 6th International Conference (EGOV 2007)*, Regensburg, Germany (Vol. 4656).

Deutsch, M., & Gerard, H. (1995). A study of normative and informational social influences upon individual judgment. *Journal of Abnormal and Social Psychology, 51*, 624–636.

Devaraj, S., Fan, M., & Kohli, R. (2002). Antecedents of B2C channel satisfaction and preference: Validating e-commerce metrics. *Information Systems Research, 13*(3), 316–333. doi:10.1287/isre.13.3.316.77

Di Franco, A., Petro, A., Shear, E., & Vladimirov, V. (2004). Small vote manipulations can swing elections. *Communications of the ACM, 47*(10), 43–45. doi:10.1145/1022594.1022621

Dick, A., & Basu, K. (1994). Customer loyalty: Toward an integrated conceptual framework. *Journal of the Academy of Marketing Science, 22*(2), 99–113. doi:10.1177/0092070394222001

Digitalisér.dk. (n. d.). *About Digitalisér.dk.* Retrieved from http://digitaliser.dk/resource/432461

Dikel, D. M., Kane, D., & Wilson, J. R. (2001). *Software architecture: Organizational principles and patterns.* Upper Saddle River, NJ: Prentice Hall.

Dimitriadis, S., & Kyrezis, N. (2008). Does trust in the bank build trust in its technology based channels? *Journal of Financial Services Marketing, 23*(1), 28–38. doi:10.1057/fsm.2008.3

Doll, W. J., Hendrickson, A., & Deng, X. (1998). Using Davis's perceived usefulness and ease of use instruments for decision making: A confirmatory and multi group invariance analysis. *Decision Sciences, 29*(4), 839–870. doi:10.1111/j.1540-5915.1998.tb00879.x

Doney, P. M., Cannon, J. P., & Mullen, M. R. (1998). Understanding the influence of national culture on the development of trust. *Academy of Management Review, 23*(3), 601–620. doi:10.2307/259297

Donthu, N., & Garcia, A. (1999). The Internet shopper. *Journal of Advertising Research, 39*, 52–58.

Doumeingts, G., Vallespir, B., & Chen, D. (1998). Decisional modelling using the GRAI grid. In Bernus, P., Meritns, K., & Schmidt, G. (Eds.), *Handbook on architectures of information systems* (pp. 313–338). Berlin, Germany: Springer-Verlag.

Doyle, S. (2001). Software review: Using short message services as a marketing tool. *Journal of Database Marketing, 8*(3), 273–277. doi:10.1057/palgrave.jdm.3240043

DPT. (2008). *Bilgi Toplumu Stratejisi ve Eylem Planı 1. Değerlendirme Raporu.* Devlet: Planlanma Teşkilatı.

Droge, C., Jayaraman, J., & Vickery, S. K. (2004). The effects of internal versus external integration practices on time-based performance and overall firm performance. *Journal of Operations Management, 22*(6), 557–573. doi:10.1016/j.jom.2004.08.001

Drucker, P. F., & Maciariello, J. A. (2004). T*he Daily Drucker*.

Dubois, A., & Gadde, L. E. (2002). Systematic combining: An abductive approach to case research. *Journal of Business Research, 55*.

Dutton, W. H. (2007). *Through the network (of networks) – the fifth estate*. Oxford, UK: Oxford Internet Institute. doi:10.2139/ssrn.1134502

Earl, M. J. (1993). Experiences in strategic information systems planning: Editor's comments. *Management Information Systems Quarterly, 17*(3), 2–3.

Eastin, M. S. (2002). Diffusion of e-commerce: An analysis of the adoption of four e-commerce activities. *Telematics and Informatics, 19*(3), 251–267. doi:10.1016/S0736-5853(01)00005-3

Edvardsson, B. (1988). Service quality in customer relationships: A study of critical incidents in mechanical engineering companies. *The Service Industries Journal,* [*8*(4), 427–445.

Edwards, T., & Suresh, S. (2009, November 24-26). Intelligent agent based hospital search and appointment system. In *Proceedings of the 2nd ACM International conference on Interaction Sciences: Information Technology, Culture and Human*, Seoul, Korea (pp. 561-568).

EIF. (2009). *European interoperability framework for pan-european egovernment services*. Retrieved from http://ec.europa.eu/idabc/en/document/2319/5644.html

Eisenhardt, K. M. (1989). Building theory from case study research. *Academy of Management Review, 14*(4).

Ekong, U. O., & Ekong, V. E. (2010). M-Voting: A Panacea for Enhanced E-Participation. *Asian Journal of Information Technology, 9*(2), 111–116. doi:10.3923/ajit.2010.111.116

Electoral Advisory Systems for Citizens. (2004). *Indra.* Retrieved from http://94.126.241.45/webelecta/electa_indra_EN.htm

Electronic Frontier Finland. (2008). *Incompatibility of the Finnish e-voting system with the council of Europe e-voting recommendations.* Retrieved from http://www.effi.org/

Elias, N. (1992). *Time: An Essay*. Oxford, UK: Blackwell Publishers.

El-Kasheir, D., Ashour, A., & Yacout, O. (2009). Factors affecting continued usage of internet banking among Egyptian customers. *Communications of the IBIMA, 9*, 252–263.

Ellinger, A. E., Lynch, D. F., Andzulis, J. K., & Smith, R. J. (2003). B-to-B e-commerce: A content analytical assessment of motor carrier websites. *Journal of Business Logistics, 24*(1), 119–220. doi:10.1002/j.2158-1592.2003.tb00037.x

Ellinger, A. E., Lynch, D. F., & Hansen, J. D. (2003). Firm size, web site content and financial performance in the transportation industry. *Industrial Marketing Management, 32*, 177–185. doi:10.1016/S0019-8501(02)00261-4

eMarketer. (2010). *Boom time in second life: No recession for virtual economy.* Retrieved from http://www.emarketer.com/Article.aspx?R=1007482

Engel, J. F., & Blackwell, R. D. (1982). *Consumer behaviour*. New York, NY: Oxford University Press.

Erikson, E. H. (1968). *Identity: Youth and Crisis*. New York: Norton.

Eroglu, S. A., Machleit, K. A., & Davis, L. M. (2000). Online retail atmospherics: Empirical test of a cue typology. In *Retailing 2000: Launching the new millennium: Proceedings of the Sixth Triennial National Retailing Conference* (pp. 144-150).

European Commission. (2009). *Report on cross-border e-commerce in the EU*. Brussels, Belgium: Commission Staff Working Document.

European Public Administration Network (EPAN). (2004). *eGovernment working group: Key principles of an interoperability architecture*. Retrieved from http://www.epractice.eu/document/2963

Eurostat. (2009). *i2010 Benchmarking indicators*. Retrieved from http://epp.eurostat.ec.europa.eu/portal/page/portal/statistics/search_database?_pir ef458_1209540_458_211810_211810.node_code=tin00115

Evangelista, P., & Sweeney, E. (2006). Technology usage in the supply chain: The case of small 3PLs. *International Journal of Logistics Management, 17*(1), 55–74. doi:10.1108/09574090610663437

Evans, P. B., & Wurster, T. (1997). Strategy and the new economies of information. *Harvard Business Review, 75*(5), 15–21.

Everett, S. (2003). The policy cycle – democratic process or rational paradigm revisited? *Australian Journal of Public Administration, 62*(2), 65–70. doi:10.1111/1467-8497.00325

Fabio, G., et al. (2003). JADE White Paper. *Telecom Italia Lab, 3*(3).

Fabio, G. (2007). *Developing MultiAgent System with JADE*. London: John Wiley & Sons.

Falcão, J., Cunha, M., Leitão, J., Faria, J., Pimenta, M., & Carravilla, A. (2006). A methodology for auditing e-voting processes and systems used at the elections for the Portuguese parliament. In R. Krimmer (Ed.), *Proceedings of the International Workshop on Electronic Voting in Europe: Technology, Law, Politics and Society* (LNI 86, pp. 145-154).

Fang, Z. (2002). E-Government in digital era: Concept, practice, and development. *International Journal of the Computer, the Internet and Management, 10*(2), 1-22.

Featherman, M. S., & Pavlou, P. A. (2003). Predicting e-services adoption: A perceived risk facets perspective. *International Journal of Human-Computer Studies, 59*, 451–474. doi:10.1016/S1071-5819(03)00111-3

Feick, L., Guskey, A., & Price, L. (1995). Everyday market helping behavior. *Journal of Public Policy & Marketing, 14*, 255–266.

Feick, L., & Price, L. (1987). The market maven: A diffuser of marketplace information. *Journal of Marketing, 51*, 83–87. doi:10.2307/1251146

Felder, R. M., & Silverman, L. K. (1988). Learning and teaching styles in engineering education. *English Education, 78*, 674–681.

Feldman, A., Halderman, J., & Felten, E. (2007). Security analysis of the Diebold AccuVote-TS voting machine. In *Proceedings of the USENIX Workshop on Accurate Electronic Voting Technology* (p. 2).

Fenigstein, A., Scheier, M. F., & Buss, A. H. (1975). Public and private self-consciousness: Assessment and theory. *Journal of Consulting and Clinical Psychology, 43*, 522–527. doi:10.1037/h0076760

FEVAD. (2010). *Fédération du e-commerce et de la vente à distance, Chiffres Clés disponibles sur le site web de cet organism*. Retrieved from http://www.fevad.com

Fiore, A. M., Jin, H. J., & Kim, J. (2005). For fun and profit: Hedonic value from image interactivity and responses toward an online store. *Psychology and Marketing, 22*(8), 669–694. doi:10.1002/mar.20079

FIPA. (2009). *The foundation for intelligent physical agents*. Retrieved from http://fipa.org/

Fischer, E., & Coleman, K. (2006). *The direct recording electronic voting machine (DRE)* (Tech. Rep. No. RL33190). Washington, DC: The Library of Congress.

Fishbein, M., & Ajzen, I. (1975). *Belief, attitude, intention and behavior: An introduction to theory and research*. Reading, MA: Addison-Wesley.

Fisher, R. J., Maltz, E., & Jaworski, B. J. (1997). Enhancing communication between marketing and engineering: The moderating role of relative functional identification. *Journal of Marketing, 61*(3), 54–70. doi:10.2307/1251789

Flavian, C., Gurrea, R., & Orus, C. (2009). The impact of online product presentation on consumers' perceptions: an experimental analysis. *International Journal of E-Services and Mobile Applications*, *1*(3), 17–37. doi:10.4018/jesma.2009070102

Flynn, L., & Goldsmith, R. (1993). A validation of the Goldsmith and Hofacker innovativeness scale. *Educational and Psychological Measurement*, *53*, 1105–1116. doi:10.1177/0013164493053004023

Fornell, C., & Larcker, D. (1981). Structural equation models with unobservable variables and measurement error. *JMR, Journal of Marketing Research*, *18*(1), 39–50. doi:10.2307/3151312

Fornell, C., & Wernerfelt, B. (1987). Defensive marketing strategy by customer complaint management. *JMR, Journal of Marketing Research*, *24*(4), 337–346. doi:10.2307/3151381

Forrester, J. W. (1958). Industrial dynamics: A major breakthrough for decision makers. *Harvard Business Review*, *38*, 37–66.

Foster, D., McGregor, C., & El-Masri, S. (2005). *A survey of agent-based intelligent decision support systems to support clinical management and research*. Retrieved from http://www.diee.unica.it/biomed05/pdf/W22-102.pdf

Fraj, E., & Martinez, E. (2006). Influence of personality on ecological consumer behaviour. *Journal of Consumer Behaviour*, *5*, 167–181. doi:10.1002/cb.169

Franklin, S., & Graesser, A. (1996). *Is it an agent, or just a program?: A taxonomy for autonomous agents.* Retrieved from http://www.msci.memphis.edu/~franklin/AgentProg.html

Fraser, J., Adams, N., Macintosh, A., McKay-Hubbard, A., Lobo, T. P., & Pardo, P. F. (2003). Knowledge Management Applied to e-Government Services: the Use of an Ontology. In *Knowledge Management in Electronic Government*. Berlin: Springer. doi:10.1007/3-540-44836-5_13

Frick, A., & Marre, R. (1995). *Der software-entwicklungsprozess*. Munich, Germany: Hanser.

Friedman, B., Kahn, P. H. Jr, & Howe, D. C. (2000). Trust online. *Communications of the ACM*, *43*(12), 34–40. doi:10.1145/355112.355120

Frost, J. H., Chance, Z., Norton, M. I., & Ariely, D. (2008). People are experience goods: Improving online dating with virtual dates. *Journal of Interactive Marketing*, *22*(1), 51–61. doi:10.1002/dir.20107

Fugate, B., Sahin, F., & Mentzer, J. T. (2006). Supply chain management coordination mechanism. *Journal of Business Logistics*, *27*(2), 129–134. doi:10.1002/j.2158-1592.2006.tb00220.x

Fu, J. R., Farn, C. K., & Chao, W. P. (2006). Acceptance of electronic tax filing: A study of taxpayer intentions. *Information & Management*, *43*(1). doi:10.1016/j.im.2005.04.001

Fujioka, A., Okamoto, T., & Ohta, K. (1992). A practical secret voting scheme for large scale elections. In *Proceedings of the Workshop on the Theory and Application of Cryptographic Techniques: Advances in Cryptology* (pp. 244-251).

Fukuyama, E. (1995). *Trust: The social virtues & the creation of prosperity*. New York, NY: Free Press.

Furuli, K., & Kongsrud, S. (2007). Mypage and Borger.dk - a Case Study of Two Government Service Web Portals. *Electronic. Journal of E-Government*, *5*(2), 165–176.

Gallant, L. M., Culnan, M. J., McLoughlin, P., Bentley, C., & Waltham, M. A. (2007). Why People e-File (or Don't e-File) Their Income Taxes. In *Proceedings of the 40th Annual Hawaii International Conference on System Sciences (HICSS'07)*.

Galliers, R. (1992). *Choosing information system research approaches, information systems research: Issues, methods and practical guidelines*. Oxford, UK: Blackwell Scientific.

Gartner. (2007, October 7-12). *Green IT: The new industry shockwave*. Paper presented at the ITXPO Symposium, Orlando, FL.

Gartner. (2009). *NIFO project – final report: A report for European commission directorate general for informatics* (Version 130). Retrieved from http://ec.europa.eu/idabc/servlets/Doc?id=32120

Gasmelseid, T. M. (2007). A Multiagent Service-oriented Modeling of E-Government Initiatives. *International Journal of Electronic Government Research*, *3*(3), 87–106.

Gefen, D. (2000). E-commerce: the role of familiarity and trust. *Omega, 28,* 725–777. doi:10.1016/S0305-0483(00)00021-9

Gefen, D. (2003). TAM or just plain habit: A look at experienced online shoppers. *Journal of End User Computing, 15,* 1–13. doi:10.4018/joeuc.2003070101

Gefen, D., Karahanna, E., & Straub, D. W. (2003). Trust and TAM in online shopping: An integrated model. *Management Information Systems Quarterly, 27,* 51–90.

Gefen, D., & Straub, D. (2000). The Relative Importance of Perceived Ease of Use in IS Adoption: A Study of E-Commerce Adoption. *Journal of the Association for Information Systems, 1*(8).

Gefen, D., & Straub, D. W. (1997). Gender differences in the perception and use of e-mail: An extension to the technology acceptance model. *Management Information Systems Quarterly, 21,* 389–400. doi:10.2307/249720

Gefen, D., & Straub, D. W. (2003). Managing user trust in B2C e-services. *E-Service Journal, 2*(2), 7–24. doi:10.2979/ESJ.2003.2.2.7

Geser, H. (2004). Towards a sociological theory of the mobile phone. In H. Geser (Ed.), *Sociology in Switzerland: Sociology of the mobile phone.* Zurich, Switzerland: University of Zurich. Retrieved from http://socio.ch/mobile/t_geser1.htm

Ghezzi, C., & Vigna, G. (1997). Mobile Code Paradigms and Technologies: A Case Study. In *Proceedings of the First International Workshop on Mobile Agents,* Berlin.

Ghinea, G., & Chen, S. Y. (2008). Measuring quality of perception in distributed multimedia: Verbalizers vs. imagers. *Computers in Human Behavior, 24,* 1317–1329. doi:10.1016/j.chb.2007.07.013

Gibson, A. N., Bertot, J. C., & McClure, C. R. (2009). Emerging Role of Public Librarians as E-Government Providers. In *Proceedings of the 41st Annual Hawaii International Conference on System Sciences (HICSS'09).*

Giese, J. L., & Cote, J. A. (2000). Defining consumer satisfaction. *Academy of Marketing Science Review.* Retrieved from http://www.amsreview.org/articles/giese01- 2000.pdf

Giunipero, L., Handfield, R. B., & Eltantawy, R. (2006). Supply management's evolution: Key skill sets for the supply manager of the future. *International Journal of Operations & Production Management, 26*(7), 822–844. doi:10.1108/01443570610672257

Global Logistics Research Team at Michigan State University. (1995). *World class logistics: The challenge of managing continuous change.* Oak Brook, IL: Council of Logistics Management.

Godin, S. (1999). *Permission marketing: Turning strangers into friends, and friends into customers.* New York, NY: Simon and Schuster.

Goldberg, L. R. (1990). An alternative "description of personality": The big-five factor structure. *Journal of Personality and Social Psychology, 59,* 1216–1229. doi:10.1037/0022-3514.59.6.1216

GoldFarb. C. F. (2003). XML handbook. Upper Saddle River, NJ: Prentice Hall.

Goldsmith, R. E., d'Hauteville, F., & Flynn, L. R. (1998). Theory and measurement of consumer innovativenss: a transactional evaluation. *European Journal of Marketing, 32*(3/4), 340–353. doi:10.1108/03090569810204634

Goldsmith, R. E., & Flynn, L. R. (1992). Identifying innovators in consumer product markets. *European Journal of Marketing, 26*(2), 42–55. doi:10.1108/03090569210022498

Golubeva, A., & Merkuryeva, I. (2006). Demand for online government services: Case studies from St. Petersburg. *Information Polity, 11*(3-4), 241–254.

Gomaa, H. (2000). *Designing concurrent, distributed and real-time applications with UML.* Reading, MA: Addison-Wesley.

Gottschalk, P. (2008). Maturity levels for interoperability in digital government. *Government Information Quarterly, 26,* 75–81. doi:doi:10.1016/j.giq.2008.03.003

Gottschalk, P., & Abrahamsen, A. F. (2002). Plans to utilize electronic marketplaces: The case of B2B procurement markets in Norway. *Industrial Management & Data Systems, 102*(6), 325–331. doi:10.1108/02635570210432028

Gøtze, J., Christiansen, P. E., Mortensen, R. K., & Paszkowski, S. (2009). Cross-national interoperability and enterprise architecture. *Informatica*, *20*(3), 369–396.

Gounaris, S., Dimitriadis, S., & Stathakopoulos, V. (2010). An examination of the effects of service quality and satisfaction on customers' behavioral intentions in e-shopping. *Journal of Services Marketing*, *24*(2), 142–156. doi:10.1108/08876041011031118

Gouscosa, D., Kalikakisa, M., Legalb, M., & Papadopouloub, S. (2007). A general model of performance and quality for one-stop e-Government service offerings. *Government Information Quarterly*, *24*(4), 860–885. doi:10.1016/j.giq.2006.07.016

Gravill, J., & Compeau, D. (2008). Self-regulated learning strategies and software training. *Information & Management*, *45*(5), 288–296. doi:10.1016/j.im.2008.03.001

Grazioli, S., & Jarvenpaa, S. L. (2000). Perils of Internet fraud: An empirical investigation of deception and trust with experienced Internet. *IEEE Transactions on Systems, Man, and Cybernetics. Part A, Systems and Humans*, *30*(4), 395–410. doi:10.1109/3468.852434

Greek Interoperability Centre. (2008). *Interoperability guide* (Version 1). Retrieved from http://www.iocenter.eu/

Grewal, D., & Baker, J. (1994). Do retail store environment cues affect consumer price perceptions? An empirical examination. *International Journal of Research in Marketing*, *11*(2), 107–115. doi:10.1016/0167-8116(94)90022-1

Grewal, R., Comer, J. M., & Mehta, R. (2001). An investigation into the antecedents of organizational participation in business-to-business electronic markets. *Journal of Marketing*, *65*(3), 17–33. doi:10.1509/jmkg.65.3.17.18331

Griffith, D. A. (2005). An examination of the influences of store layout in online retailing. *Journal of Business Research*, *58*(10), 1391–1396. doi:10.1016/j.jbusres.2002.08.001

Griffith, D. A., & Chen, Q. (2004). The influence of virtual direct experience (vde) on on-line ad message effectiveness. *Journal of Advertising*, *33*(1), 55–68.

Grimm, P. E., Agrawal, J., & Richardson, P. S. (1999). Product conspicuousness and buying motives as determinants of reference group influences. *European Advances in Consumer Research*, *4*, 97–103.

Grönlund, Å., & Andersson, A. (2006). e-Gov Research Quality Improvements Since 2003: More Rigor, but Research (Perhaps) Redefined. In *Proceedings of 5th International Conference (EGOV 2005)*, Krakow, Poland (LNCS 4084, pp. 1-13). Berlin: Springer.

Grönlund, Å., Hatakka, M., & Ask, A. (2007). Inclusion in the E-Service Society – Investigating Administrative Literacy Requirements for Using E-Services. In *Proceedings of the 6th International Conference (EGOV 2007)*, Regensburg, Germany (Vol. 4656).

Groth, J. (2004). Efficient maximal privacy in boardroom voting and anonymous broadcast. In A. Juels (Ed.), *Proceedings of the 8th International Conference on Financial Cryptography* (LNCS 3110, pp. 90-104).

Grover, V., & Malhotra, M. (1997). Business process re-engineering: A tutorial on the concept, evolution, method, technology and application. *Journal of Operations Management*, *15*, 192–213. doi:10.1016/S0272-6963(96)00104-0

Guan, S. (2002). Handy Broker - An Intelligent Product-Brokering Agent for M-Commerce Applications with User Preference Tracking. *Electronic Commerce Research and Applications*, *1*(4). doi:10.1016/S1567-4223(02)00023-6

Gu, J. C., Lee, S. C., & Suh, Y. H. (2009). Determinants of behavioral intention to mobile banking. *Expert Systems with Applications*, *36*, 11605–11616. doi:10.1016/j.eswa.2009.03.024

Gundlach, G. T., & Murphy, P. E. (1993). Ethical and legal foundations of relational marketing exchanges. *Journal of Marketing*, *57*(4), 35–46. doi:10.2307/1252217

Gundlach, G. T., Ravi, S. A., & Mentzer, J. T. (1995). The structure of commitment in exchange. *Journal of Marketing*, *59*(1), 78–92. doi:10.2307/1252016

Gustin, C. M., Daugherty, P. J., & Stank, T. P. (1995). The effects of information availability on logistics integration. *Journal of Business Logistics*, *16*(1), 1–21.

Gutiérrez-Rubí, A. (2009). *El voto electrónico llega a España con las elecciones europea.* Retrieved from http://www.gutierrez-rubi.es/2009/06/04/el-voto-electronico-llega-a-espana-con-las-elecciones-europeas/

Hach, H. (2005). *Evaluation und optimierung kommunaler e-government-prozesse.* Unpublished doctoral dissertation, Universität Flensburg, Flensburg, Germany.

Hague, B. N., & Loader, B. (1999). *Digital democracy: Discourse and decision making in the information age.* London, UK: Routledge.

Hair, J. F., Tatham, R. L., Anderson, R. E., & Black, W. (1998). *Multivariate data analysis.* Upper Saddle River, NJ: Prentice Hall.

Hair, J., Black, W., Babin, B., Anderson, R., & Tatham, R. (2005). *Multivariate data analysis* (6th ed.). Upper Saddle River, NJ: Prentice Hall.

Hair, J. Jr, Anderson, J., Norman, J., & Black, W. (1995). *Multivariate data analysis with readings* (4th ed.). Upper Saddle River, NJ: Prentice Hall.

Hall, D., & Mansfield, R. (1995). Relationships of age and seniority with career variables of engineers and scientists. *The Journal of Applied Psychology*, *60*(2), 201–210. doi:10.1037/h0076549

Hall, R. H., & Hanna, P. (2003). The impact of web page text-background color combinations on readability, retention, aesthetics, and behavioral intention. *Behaviour & Information Technology*, *23*(3), 183–195. doi:10.1080/01449290410001669932

Handfield, R. B. (1995). *Re-engineering for time-based competition.* Westport, CT: Quorum Books.

Harkiolakis, N., & Halkias, D. (2007). Online buyer behaviour and perceptions in Greece. *International Journal of Applied Systemic Studies*, *1*(3), 317–328. doi:10.1504/IJASS.2007.017714

Ha, S., & Stoel, L. (2009). Consumer e-shopping acceptance: Antecedents in a technology acceptance model. *Journal of Business Research*, *62*, 565–571. doi:10.1016/j.jbusres.2008.06.016

Hasan, B., & Mesbah, U. A. (2007). Effects of interface style on user perceptions and behavioral intention to use computer systems. *Computers in Human Behavior*, *23*, 3025–3037. doi:10.1016/j.chb.2006.08.016

Haven, B., Bernoff, J., Glass, S., & Feffer, K. A. (2007). *A second life for marketers?* Cambridge, MA: Forrester Research.

Hawes, J. M., Kenneth, E. M., & Swan, J. E. (1989). Trust earning perceptions of sellers and buyers. *Journal of Personal Selling & Sales Management*, *9*, 1–8.

Hayashi, T. (2004). Captured Nature and Japanese Way of Tolerance. *MAJA Estonian Architectural Review.* Retrieved from http://www.solness.ee/majaeng/index.php?gid=44&id=453

Hayes, R. H., & Wheelwright, S. C. (1979). Linking manufacturing process and product life cycles. *Harvard Business Review*, *5*(1), 133–140.

Hayes, R. H., & Wheelwright, S. C. (1984). *Restoring our competitive edge: Competing through manufacturing.* New York, NY: John Wiley & Sons.

Hayes, R. H., Wheelwright, S. C., & Clark, K. (1988). *Dynamic manufacturing.* New York, NY: Free Press.

Health Care. (2009). *Health care in Jamaica.* Retrieved from http://www.jamaicans.com/articles/primearticles/health-care-in-jamaica.shtml

Heeks, R. (2003). *Most eGovernment-for-development projects fail: How can risks be reduced?* Manchester, UK: Institute for Development Policy and Management.

Heide, J. B., & John, G. (1992). Do norms matter in marketing relationships? *Journal of Marketing*, *56*(2), 32–44. doi:10.2307/1252040

Heinonen, K., & Strandvik, T. (207). Consumer responsiveness to mobile marketing. *International Journal of Mobile Communications*, *5*, 603–617. doi:10.1504/IJMC.2007.014177

Helbach, J., & Schwenk, J. (2007). Secure Internet voting with code sheets. In A. Alkassar & M. Volkamer (Eds.), *Proceedings of the 1st International Conference on E-Voting and Identity* (LNCS 4896, pp. 166-177).

Hendaoui, A., Limayem, A., & Thompson, C. W. (2008). 3D social virtual world: Research issues and challenges. *IEEE Internet Computing*, 88–92. doi:10.1109/MIC.2008.1

Hendrickson, A. R., Massey, P. D., & Cronan, T. P. (1993). On the test–retest reliability of perceived ease of use scales. *Management Information Systems Quarterly*, *17*(2), 227–230. doi:10.2307/249803

Henry, L., & Suresh, S. (2009, November 24-26). Applications of intelligent agent for mobile tutoring. In *Proceedings of the 2nd ACM International Conference on Interaction Sciences: Information Technology, Culture and Human*, Seoul, Korea (pp. 963-969).

Henten. (2010). Services, E-Services, and Nonservices. *Electronic Services: Concepts, Methodologies, Tools and Applications* (Vol. 3, pp. 1-9). Hershey, PA: IGI Global.

Héritier, A. (1993). *Policy-analyse: Kritik und neuorientierung*. Opladen, Germany: Westdeutscher Verlag.

Hernández, B., Jiménez, J., & Martín, M. J. (2009). Customer behavior in electronic commerce: The moderating effect of e-purchasing experience. *Journal of Business Research*, *63*(9-10), 964–971. doi:10.1016/j.jbusres.2009.01.019

Hirschman, E. C. (1980). Innovativeness, novelty seeking and consumer creativity. *The Journal of Consumer Research*, *7*, 283–295. doi:10.1086/208816

Hirt, M., & Sako, K. (2000). Efficient receipt-free voting based on homomorphic encryption. In *Proceedings of the Workshop on the Theory and Application of Cryptographic Techniques on Advances in Cryptology* (LNCS 1807, pp. 539-556).

Hoch, S. J., & Deighton, J. (1989). Managing what consumers learn from experience. *Journal of Marketing*, *53*, 1–20. doi:10.2307/1251410

Hoffman, D. L., Novak, T. P., & Peralta, M. (1999). Building consumer trust online. *Communications of the ACM*, *42*(4), 80–85. doi:10.1145/299157.299175

Hofmeister, C., Kruchten, P., Nord, R., Obbink, H., Ran, A., & America, P. (2006). A general model of software architecture design derived from five industrial approaches. *Journal of Systems and Software*, *80*, 106–126. doi:10.1016/j.jss.2006.05.024

Hollands, J. G., Parker, H. A., McFadden, S., & Boothby, R. (2002). LCD versus CRT displays: A comparison of visual search performance for colored symbols. *Human Factors*, *44*, 210. doi:10.1518/0018720024497862

Holsapple, C. W., & Sasidharan, S. (2005). The dynamics of trust in online B2C e-commerce: A research model and agenda. *Information Systems and E-business Management*, *3*(4), 377–403. doi:10.1007/s10257-005-0022-5

Hong, S., Thong, J., & Tam, K. (2006). Understanding continued information technology usage behavior: A comparison of three models in the context of Mobile Internet. *Decision Support Systems*, *42*, 1819–1834. doi:10.1016/j.dss.2006.03.009

Hong, W., Thong, J. Y. L., Wong, W. M., & Tam, K. Y. (2001). Determinants of user acceptance of digital libraries: An empirical examination of individual differences and system characteristics. *Journal of Management Information Systems*, *18*(3), 97–124.

Horan, T. A., Abhichandani, T., & Rayalu, R. (2006). Assessing User Satisfaction of E-Government Services: Development and Testing of Quality-in-Use Satisfaction with Advanced Traveler Information Systems (ATIS). In *Proceedings of the 39th Annual Hawaii International Conference on System Sciences (HICSS'06)*.

Hourahine, B., & Howard, M. (2004). Money on the move: Opportunities for financial service providers in the third space. *Journal of Financial Services Marketing*, *9*, 57–67. doi:10.1057/palgrave.fsm.4770141

Houston, M. B., & Johnson, S. A. (2000). Buyer-supplier contracts versus joint ventures: Determinants and consequences of transaction structure. *JMR, Journal of Marketing Research*, *37*(1), 1–15. doi:10.1509/jmkr.37.1.1.18719

Howard, C. (2005). The policy cycle: A model of post-machiavellian policy making? *Australian Journal of Public Administration*, *64*(3), 3–13. doi:10.1111/j.1467-8500.2005.00447.x

Hsu, H.-H., & Lu, H.-P. (2008). Multimedia messaging service acceptance of pre- and post- adopters: A sociotechnical perspective. *International Journal of Mobile Communications*, *6*, 598–615. doi:10.1504/IJMC.2008.019324

Hsu, M., Yen, C., Chiu, C., & Chang, C. (2006). A longitudinal investigation of continued online shopping behavior: An extension of the theory of planned behavior. *International Journal of Human-Computer Studies, 64*(9), 889–904. doi:10.1016/j.ijhcs.2006.04.004

Hu, G., Zhong, W., & Mei, S. (2008). Electronic Public Service (EPS) and its implementation in Chinese local governments. *Int. J. of Electronic Governance, 1*(2), 118–138. doi:10.1504/IJEG.2008.017900

Hu, L. T., & Bentler, P. M. (1999). Cutoff criteria for fit indexes in covariance structure analysis: Conventional criteria versus new alternatives. *Structural Equation Modeling, 6*, 1–55. doi:10.1080/10705519909540118

Hung, S. Y., Changa, C. M., & Yu, T. J. (2006). Determinants of user acceptance of the e-Government services: next term the case of online tax filing and payment system. *Government Information Quarterly, 23*(1), 97–122. doi:10.1016/j.giq.2005.11.005

Hung, S. Y., Ku, C. Y., & Chang, C. M. (2003). Critical factors of WAP services adoption: An empirical study. *Electronic Commerce Research and Applications, 2*(1), 46–60. doi:10.1016/S1567-4223(03)00008-5

Hunt, S. D., & Chonko, L. B. (1984). Marketing and Machiavellianism. *Journal of Marketing, 48*, 30–42. doi:10.2307/1251327

Hunt, S. D., & Davis, D. F. (2008). Grounding supply chain management in resource-advantage theory. *Journal of Supply Chain Management, 44*(1), 10–21. doi:10.1111/j.1745-493X.2008.00042.x

Hu, P. J., Chau, P. Y. K., Liu Sheng, O. R., & Yan Tam, K. (1999). Examining the technology acceptance model using physician acceptance of telemedicine technology. *Journal of Management Information Systems, 16*(2), 91–112.

Hurley, R. F., & Hult, G. T. M. (1998). Innovation, market orientation, and organizational integration and empirical examination. *Journal of Marketing, 62*(3), 42–54. doi:10.2307/1251742

Hypponen, H., Salmivalli, L., & Suomi, R. (2005). Organizing for a National Infrastructure Project: The Case of the Finnish Electronic Prescription. In *Proceedings of the 38th Annual Hawaii International Conference on System Sciences (HICSS'05)*.

IDABC. (2009). *European interoperability framework for pan-European e-government services.* Retrieved from http://ec.europa.eu/idabc/en/document/2319/5644

IDABC. (2009). *National interoperability frameworks observatory (NIFO). Retrieved* from http://ec.europa.eu/idabc/en/document/7796

IDABC. (n. d.). *Interoperable delivery of European e-government services to public administrations, businesses and citizens.* Retrieved from http://europa.eu.int/idabc

IEEE. Computer Society. (1990). *IEEE standard glossary of software engineering terminology.* Washington, DC: IEEE Computer Society.

IEPC. (2008). *Memoria electoral: Electoral and voter participation institute of the State of Coahuila.* Retrieved from http://www.iepcc.org.mx/

Igarashi, T., Motoyoshi, T., Takai, J., & Yoshida, T. (2008). No mobile, no life: Self-perception and text-message dependency among Japanese high school students. *Computers in Human Behavior, 24*, 2311–2324. doi:10.1016/j.chb.2007.12.001

Igbaria, M., & Iivari, J. (1995). The effects of self-efficacy on computer usage. *Omega, 23*(6), 587–605. doi:10.1016/0305-0483(95)00035-6

Igbaria, M., & Parasuraman, S. (1989). A path analytic study of individual characteristics, computer anxiety and attitudes toward microcomputers. *Journal of Management, 15*, 373–388. doi:10.1177/014920638901500302

Im, I., Kim, Y., & Han, H. (2008). The effects of perceived risk and technology type on users' acceptance of technologies. *Information & Management, 45*(1), 1–9.

Im, S., Mason, C. H., & Houston, M. B. (2007). Does innate consumer innovativeness relate to new product/service adoption behavior? The intervening role of social learning via vicarious innovativeness. *Journal of the Academy of Marketing Science, 35*, 63–75. doi:10.1007/s11747-006-0007-z

Indra. (2009). *Indra realizará el escrutinio de las Elecciones al Parlamento Europeo del 7-J.* Retrieved from http://www.indra.es/servlet/ContentServer?pagename=IndraES/SalaPrensa_FA/DetalleEstructuraSalaPrensa&cid=12434823 16640&pid=1087577300456&Language=es_ES

Indrajit, R., Indrakshi, R., & Natarajan, N. (2001). An anonymous electronic voting protocol for voting over the Internet. In *Proceedings of the Third International Workshop on Advanced Issues of E-Commerce and Web-Based Information Systems* (p. 188).

Information Society Technologies. (2008). *Enterprise interoperability research roadmap* (Version 5.0). Retrieved from ftp://ftp.cordis.europa.eu/pub/fp7/ict/docs/enet/ei-research-roadmap-v5-final_en.pdf

Information Systems Technologies Laboratory (IST Lab). (2007). *Research about the tendency in the use of mobile data services in Greece (Comparison study 2006-2007)*. Athens, Greece: Athens University of Economics Wireless Research Center.

Infos-Industrielles. (2007). *Les TIC au service de l'efficacité énergétique.* Retrieved from http://www.infos-industrielles.com/dossiers/1156.asp

Irani, Z., Lee, H., Weerakkody, V., Kamal, M., Topham, S., & Simpson, G. (2010). Ubiquitous Participation Platform for POLicy Makings (UbiPOL): A Research Note. *International Journal of Electronic Government Research, 6*(1), 78–106. doi:10.4018/jegr.2010102006

Isern, D., Sánchez, D., Moreno, A., & Valls, A. (2003). *HeCaSe: An agent-based system to provide personalised medical services.* Retrieved from http://deim.urv.cat/~itaka/Publicacions/caepia03.pdf

Islam, M. S., & Grönlund, Å. (2007). Agriculture Market Information E-Service in Bangladesh: A Stakeholder-Oriented Case Analysis. In *Proceedings of the 6th International Conference (EGOV 2007)*, Regensburg, Germany (Vol. 4656).

Jackson, C. M., Chow, S., & Leitch, R. A. (1997). Towards an understanding of the behavioral intention to use an information system. *Decision Sciences, 28*(2), 357–389. doi:10.1111/j.1540-5915.1997.tb01315.x

Jahng, J. J., Jain, H., & Ramamurthy, K. (2002). Personality traits and effectiveness of presentation of product information in e-business systems. *European Journal of Information Systems, 11*, 181–195. doi:10.1057/palgrave.ejis.3000431

Janssen, M., & Feenstra, R. (2008). Socio-technical design of service compositions: a coordination view. In *Proceedings of the 2nd International Conference on Theory and Practice of Electronic Governance (ICEGOV 2008)* (Vol. 351, pp. 323-330).

Janssen, M., & Klievink, B. (2009). The Role of Intermediaries in Multi-Channel Service Delivery Strategies. [IJEGR]. *International Journal of Electronic Government Research, 5*(3), 36–46.

Janssen, M., & Kuk, G. (2007). E-Government Business Models for Public Service Networks. [IJEGR]. *International Journal of Electronic Government Research, 3*(3), 54–71.

Java Community Process. (2005). *JSR 208: JavaTM Business Integration (JBI).* Retrieved from http://www.jcp.org/en/jsr/detail?id=208

Jayaram, J., Vickery, S. K., & Droge, C. (1999). An empirical study of time-based competition in the north American automotive supplier industry. *International Journal of Operations & Production Management, 19*(10), 1010–1033. doi:10.1108/01443579910287055

Jennings, N. R., & Wooldridge, M. (1998). *Applications of intelligent agents.* Retrieved from http://eprints.ecs.soton.ac.uk/2188/2/agt-technology.pdf.gz

Jha, R., & Iyer, S. (2001). Performance Evaluation of Mobile Agents for E-Commerce Applications. In *Proceedings of 8th Internation Conference on High Perfromance Computing*, Hyderbad, India.

Jiun-Haw, L., Liu, D. N., & Wu, S.-T. (2008). *Introduction to flat panel displays.* New York, NY: John Wiley & Sons.

John, O. P., & Srivastava, S. (1999). The big five trait taxonomy: History, measurement, and theoretical perspectives. In Pervin, L. A., & John, O. P. (Eds.), *Handbook of personality* (pp. 102–138). New York, NY: Guilford Press.

Jones, J. W. (1989). Personality and epistemology: Cognitive social learning theory as a philosophy of science. *Zygon, 24*(1), 23–38. doi:10.1111/j.1467-9744.1989.tb00974.x

Joseph, B., & Vyas, S. J. (1984). Concurrent validity of a measure of innovative cognitive style. *Journal of the Academy of Marketing Science, 12*(2), 159–175. doi:10.1007/BF02729494

Jung, I., Thapa, D., & Wang, G.-N. (2007). *Intelligent agent based graphic user interface (GUI) for e-physician.* Retrieved from http://www.waset.org/journals/waset/v36/v36-35.pdf

Jung, Y., & Kang, H. (2010). User goals in social virtual worlds: A means-end chain approach. *Computers in Human Behavior, 26*(2), 218–225. doi:10.1016/j.chb.2009.10.002

Juniper Research. (2008). *Mobile advertising strategies and forecasts.* Retrieved from http://juniperresearch.com/reports/mobile_advertising

Kaaya, J. (2009). Determining Types of Services and Targeted Users of Emerging E-Government Strategies: The Case of Tanzania. [IJEGR]. *International Journal of Electronic Government Research, 5*(2), 16–36.

Kaliannan, M., Awang, H., & Raman, M. (2009). Electronic procurement: a case study of Malaysia's e-Perolehan (e-procurement) initiative. *Int. J. of Electronic Governance, 2*(2/3), 103–117. doi:10.1504/IJEG.2009.029124

Kanat, I. E., & Özkan, S. (2009). Exploring citizens' perception of government to citizen services: A model based on theory of planned behaviour (TBP). *Transforming Government: People, Process and Policy, 3*(4).

Kannan, P. K., Chang, A., & Whinston, A. B. (2001). Wireless commerce: Marketing issues and possibilities. In *Proceedings of the 34th Hawaii International Conference on System Sciences* (pp. 1-6).

Kao, G. Y., Lei, P., & Sun, C. T. (2008). Thinking style impacts on web search strategies. *Computers in Human Behavior, 24*, 1330–1341. doi:10.1016/j.chb.2007.07.009

Kariofillis-Christos, C., & Economides, A. A. (2009). A holistic evaluation of Greek municipalities' websites. *Electronic Government: An International Journal, 6*(2), 193–212. doi:10.1504/EG.2009.024442

Karjaluoto, H., Leppaniemi, M., Standing, C., Kajalo, S., Merisavo, M., Virtanen, V., & Salmenkivi, S. (2006). Individual differences in the use of mobile services among Finnish consumers. *International Journal of Mobile Marketing, 1*, 4–10.

Karoway, C. (1997). Superior supply chains pack plenty of byte. *Purchasing Technology, 8*(11), 32–35.

Kassarjian, H. (1965). Social character and differential preference for mass communication. *JMR, Journal of Marketing Research, 2*, 146–154. doi:10.2307/3149978

Kassarjian, H. (1971). Personality and consumer behavior: A review. *JMR, Journal of Marketing Research, 8*, 409–414. doi:10.2307/3150229

Kasunic, M., & Anderson, W. (2004). *Measuring systems interoperability: Challenges and opportunities: Software engineering measurement and analysis initiative* (Tech. Rep. No. CMU/SEI-2004-TN-003). Pittsburgh, PA: Carnegie Mellon University.

Kathuria, R. (2000). Competitive priorities and managerial performance: A taxonomy of small manufacturers. *Journal of Operations Management, 18*(6), 627–641. doi:10.1016/S0272-6963(00)00042-5

Kaufman, C. F., Lane, P. M., & Lindquist, J. D. (1991). Exploring more than 24 hours a day: A preliminary investigation of polychromic time-use. *The Journal of Consumer Research, 18*, 392–401. doi:10.1086/209268

Kawamoto, K., Koomey, J. G., Nordman, B., Brown, R. E., Piette, M. A., Ting, M., et al. (2001). Electricity used by office equipment and network equipment in the US (Tech. Rep. No. LBNL-45917) Berkeley, CA: National Laboratory.

Kaynar, O., & Amichai-Hamburger, Y. (2008). The effects of need for cognition on internet use revisited. *Computers in Human Behavior, 24*, 361–371. doi:10.1016/j.chb.2007.01.033

Kerkhove, D. D. (2003). *NextD Journal: ReRethinking Design Issue Two, Conversation 2.3.* Retrieved from http://www.nextd.org/02/pdf_download/NextD_2_3.pdf

Kiayias, A., & Yung, M. (2002). Self-tallying elections and perfect ballot secrecy. In *Proceedings of the 5th International Workshop on Practice and Theory in Public Key Cryptosystems* (pp. 141-158).

Kim, E., & Choi, J. (2006). An Ontology-Based Context Model in a Smart Home. In M. Gavrilova et al. (Eds.), *ICCSA 2006* (LNCS 3983, pp. 11-20). Berlin: Springer Verlag.

Kim, G., Shin, B. S., & Lee, H. G. (2009). Understanding dynamics between initial trust and usage intentions of mobile banking. *Information Systems Journal, 19*, 283–311. doi:10.1111/j.1365-2575.2007.00269.x

Kim, H., Chan, H. C., & Gupta, S. (2007). Value-based adoption of mobile Internet: An empirical investigation. *Decision Support Systems, 43*, 111–126. doi:10.1016/j.dss.2005.05.009

Kim, H., & Markus, H. R. (1999). Deviance or uniqueness, harmony or conformity? A ultural analysis. *Journal of Personality and Social Psychology, 77*(4), 785–800. doi:10.1037/0022-3514.77.4.785

Kim, J., Fiore, M. A., & Lee, H. H. (2007). Influences of online store perception, shopping enjoyment, and shopping involvement on consumer patronage behavior towards an online retailer. *Journal of Retailing and Consumer Services, 14*, 95–107. doi:10.1016/j.jretconser.2006.05.001

Kim, J., & Forsythe, S. (2008). Adoption of virtual try-on technology for online apparel shopping. *Journal of Interactive Marketing, 22*(2), 45–59. doi:10.1002/dir.20113

King, S. F. (2007). Citizens as customers: Exploring the future of CRM in UK local government. *Government Information Quarterly, 24*(1), 47–63. doi:10.1016/j.giq.2006.02.012

Kirn, S. (2002). Ubiquitous healthcare: The OnkoNet mobile agents architecture. In M. Aksit, M. Mezini, & R. Unland (Eds.), *Proceedings of the International Conference on Objects, Components, Architectures, Services, and Applications for a Networked World* (LNCS 2591, pp. 265-277).

Kleijnen, M., Ruyter, K., & Wetzels, M. (2004). Consumer adoption of wireless services: Discovering the rules, while playing the game. *Journal of Interactive Marketing, 18*, 51–61. doi:10.1002/dir.20002

Kleijnen, M., Ruyter, K., & Wetzels, M. (2007). An assessment of value creation in mobile service delivery and the moderating role of time consciousness. *Journal of Retailing, 83*, 33–46. doi:10.1016/j.jretai.2006.10.004

Kleijnen, M., Wetzels, M., & Ruyter, K. (2004). Consumer acceptance of wireless finance. *Journal of Financial Services Marketing, 8*, 206–217. doi:10.1057/palgrave.fsm.4770120

Klischewski, R., & Ukena, S. (2008). An Activity-Based Approach towards Development and Use of E-Government Service Ontologies. In *Proceedings of the 41st Annual Hawaii International Conference on System Sciences (HICSS'08)*.

Knemeyer, A. M., & Murphy, P. R. (2004). Evaluating the performance of third-party logistics arrangements: A relationship marketing perspective. *Journal of Supply Chain Management, 40*(1), 35–51. doi:10.1111/j.1745-493X.2004.tb00254.x

Knutsen, L., Constantiou, I. D., & Damsgaard, J. (2005). Acceptance and perceptions of advanced mobile services: Alterations during a field study. In *Proceedings of the International Conference on Mobile Business*, Sydney, Australia (pp. 326-331).

Kohno, T., Stubblefield, A., Wallacj, D., & Rubin, A. (2004). Analysis of an electronic voting system. In *Proceedings of the IEEE Symposium on Security and Privacy* (pp. 27-40).

Kotabe, M., Martin, X., & Domoto, H. (2003). Gaining from vertical partnerships: Knowledge transfer, relationship duration, and supplier performance improvement in the U.S. and Japanese automotive industries. *Strategic Management Journal, 24*(4), 293–316. doi:10.1002/smj.297

Kotler, P. (1973-4). Atmospherics as a marketing tool. *Journal of Retailing, 49*, 48–63.

Kotler, P., & Armstrong, G. (2004). *Principles of Marketing*. Upper Saddle River, NJ: Pearson Prentice Hall.

Koussouris, S., Lampathaki, F., Tsitsanis, A., Psarras, J., & Pateli, A. (2007). A methodology for developing local administration services portals. In P. Cunningham & M. Cunningham (Eds.), *Proceedings of the eChallenges conference: Expanding the knowledge economy: Issues, applications, case studies*. Amsterdam, The Netherlands: IOS Press.

KPMG. (2003). *Logistica integrata ed operatori di settore: Trende scenari evolutivi del mercato Italiano*. Milan, Italy. KPMG Business Advisory Services.

Krasonikolakis, I., & Vrechopoulos, A. (2009, September 25-27). Setting the research agenda for store atmosphere studies in virtual reality retailing. In *Proceedings of the 4th Mediterranean Conference on Information Systems*, Athens, Greece.

Kraussl, Z., Yao-Hua, T., & Gordijn, J. (2009). A Model-Based Approach to Aid the Development of E-Government Projects in Real-Life Setting Focusing on Stakeholder Value. In *Proceedings of the 41nd Annual Hawaii International Conference on System Sciences (HICSS'09)*.

Krcmar, H. (2005). *Informationsmanagement*. Berlin, Germany: Springer-Verlag.

Kunstelj, M., Jukić, T., & Vintar, M. (2007). Analysing the Demand Side of E-Government: What Can We Learn From Slovenian Users? In *Proceedings of the 6th International Conference (EGOV 2007)*, Regensburg, Germany (Vol. 4657).

Kuo, Y., & Yen, S. (2009). Towards an understanding of the behavioral intention to use 3G mobile value-added services. *Computers in Human Behavior, 25*(1), 103–110. doi:10.1016/j.chb.2008.07.007

Kuscu, M. H., Kushchu, İ., & Yu, B. (2007). Introducing Mobile Government. In Kuschu, I. (Ed.), *Mobile Government: An Emerging Direction in e-Government* (pp. 1–11). Hershey, PA: IGI Global. doi:10.4018/978-1-59140-884-0.ch001

Kwak, H., Fox, R., & Zinkhan, G. (2002). What products can be successfully promoted and sold via the Internet? *Journal of Advertising Research, 42*, 23–38.

Lacey, R. (2007). Relationship drivers of customer commitment. *Journal of Marketing Theory and Practice, 15*(4), 315–333. doi:10.2753/MTP1069-6679150403

Ladendorf, K. (2008). *Casting its lot with e-voting*. Retrieved from http://www.hartic.com/news/77

Lai, F., Li, D., Wang, Q., & Zhao, X. (2008). The information technology capability of third party logistics providers: A resource based view and empirical evidence from China. *Journal of Supply Chain Management, 44*(3), 22. doi:10.1111/j.1745-493X.2008.00064.x

Lambert, D. M., Stock, J. R., & Ellram, L. M. (1998). *Fundamentals of logistics*. New York, NY: McGraw-Hill.

Lamone, L. H. (2003). *State of Maryland Diebold AccuVote-TS voting system and processes*. Annapolis, MD: Maryland State Board of Elections.

Lam, S., Chiang, J., & Parasuraman, A. (2008). The effects of the dimensions of technology readiness on technology acceptance: An empirical analysis. *Journal of Interactive Marketing, 22*(4), 19–39. doi:10.1002/dir.20119

Landay, L. (2008). Having but not holding: Consumerism & commodification in second life. *Journal of Virtual Worlds Research, 1*(2), 1–5.

Lange, C., Bojārs, U., Groza, T., Breslin, J. G., & Handschuh, S. (2008). Expressing argumentative discussions in social media sites. In *Proceedings of the Conference on Social Data on the Web at the International Semantic Web*.

LaRose, R., & Eastin, M. (2002). Is online buying out of control? Electronic commerce and consumer self-regulation. *Journal of Broadcasting & Electronic Media, 46*, 549–564. doi:10.1207/s15506878jobem4604_4

Lascu, D. N., Bearden, W. O., & Rose, R. L. (1995). Norm extremity and interpersonal influences on consumer conformity. *Journal of Business Research, 32*(3), 201–212. doi:10.1016/0148-2963(94)00046-H

Lasswell, H. D. (1956). *The decision process: Seven categories of functional analysis*. College Park, MD: University of Maryland.

Lasswell, H. D. (1971). *A pre-view of policy sciences*. New York, NY: Elsevier.

Lau, C., Churchill, R. S., Kim, J., Masten, F. A. III, & Kim, T. (2002). Asynchronous web-based patient-centered home telemedicine system. *IEEE Transactions on Bio-Medical Engineering, 49*(12), 1452–1462. doi:10.1109/TBME.2002.805456

Laudi, A. (2010). The semantic interoperability centre Europe – reuse and the negotiation of meaning. In Charalabidis, Y. (Ed.), *Interoperability in digital public services and administration: Bridging e-government and e-business* (pp. 144–161). Hershey, PA: IGI Global.

Layne, K., & Lee, J. (2001). Developing fully functional E-government: A four stage model. *Government Information Quarterly, 18*(2), 122. doi:10.1016/S0740-624X(01)00066-1

Leben, A., Kunstelj, M., Bohanec, M., & Vintar, M. (2006). Evaluating public administration e-portals. *Information Polity, 11*(3-4), 207–225.

Lee, C. B. P., & Lei, U. L. E. (2007). Adoption of E-Government Services in Macao. In *Proceedings of the 1st International Conference on Theory and Practice of Electronic Governance (ICEGOV 2007)*, Macao, China (pp. 217-220).

Lee, D., Peristeras, V., Price, D., Kumar, S., Liotas, N., Ruston, S., et al. (2009). *D2.2: functional specification.* Retrieved from http://www.n4c.eu/Download/n4c-wp2-023-dtn-infrastructure-fs-12.pdf

Lee, H., Irani, Z., Osman, I. H., Balci, A., Ozkan, S., & Medeni, T. D. (2008). Toward a reference process model for citizen-oriented evaluation of e-Government services. *Transforming Government: People, Process and Policy, 2*(4).

Lee, Y., & Kwon, O. (2009). Can affective factors contribute to explain continuance intention of web-based services? In *Proceedings of the 11th International Conference on Electronic Commerce* (pp. 302-310).

Lee, G. G., & Lin, H. F. (2005). Customer perceptions of e-service quality in online shopping. *International Journal of Retail and Distribution Management, 33*(2), 161–175. doi:10.1108/09590550510581485

Lee, H. L., & Billington, C. (1992). Managing supply chain inventory: Pitfalls and opportunities. *Sloan Management Review*, 65–73.

Lee, H. L., Padmanabhan, V., & Whang, S. (1997). Information distortion in a supply chain: The bullwhip effect. *Management Science, 43*(4), 546–558. doi:10.1287/mnsc.43.4.546

Lee, J. (2002). A key to marketing financial services: The right mix of products, services, channels and customers. *Journal of Services Marketing, 16*(3), 238–258. doi:10.1108/08876040210427227

Lee, K. S., Lee, H. S., & Kim, S. Y. (2007). Factors influencing the adoption behavior of mobile banking: A South Korean perspective. *Journal of Internet Banking and Commerce, 12*(2).

Lee, M. K. O., & Turban, E. (2001). A trust model for consumer Internet shopping. *International Journal of Electronic Commerce, 6*(1), 75–91.

Legris, P., Ingham, J., & Collerette, P. (2003). Why do people use information technology? A critical review of the technology acceptance model. *Information & Management, 40*(2), 191–204. doi:10.1016/S0378-7206(01)00143-4

Lepouras, G., Vassilakis, C., Sotiropoulou, A., Theotokis, D., & Katifori, A. (2008). An active blackboard for service discovery, composition and execution. *Int. J. of Electronic Governance, 1*(3), 275–295. doi:10.1504/IJEG.2008.020450

Leppäniemi, M., Sinisalo, J., & Karjaluoto, H. (2006). A review of mobile marketing research. *International Journal of Mobile Marketing, 1*, 30–40.

Le, T., Rao, S., & Truong, D. (2004). Industry-sponsored marketplaces: A platform for supply chain integration or a vehicle for market aggregation? *Electronic Markets, 14*(4), 295–307. doi:10.1080/10196780412331311748

Lewis, M. W. (1998). Iterative triangulation: A theory development process using existing case studies. *Journal of Operations Management, 16*, 455–469. doi:10.1016/S0272-6963(98)00024-2

Lewison, M. (1994). *Retailing* (5th ed.). New York, NY: Macmillan.

Ley. (2007). *Ley 11/2007 de acceso electrónico de los ciudadanos a los Servicios Públicos.* Retrieved from http://www.boe.es/aeboe/consultas/bases_datos/doc.php?coleccion=iberlex&id=2007/12352

Li, X. (2007). The Role of Mobile Agents in M-commerce. In *Proceedings of Sixth Wuhan International Conference on E-Business*, Wuhan, China.

Lian, J. W., & Lin, T. M. (2008). Effects of consumer characteristics on their acceptance of online shopping: Comparisons among different product types. *Computers in Human Behavior, 24*, 48–65. doi:10.1016/j.chb.2007.01.002

Li, H. S., Edwards, M., & Lee, J. (2002). Measuring the intrusiveness of advertisements: Scale development and validation. *Journal of Advertising, 31*, 37–47.

Lin, C. S., Wu, S., & Tsai, R. J. (2005). Integrated perceived playfulness into expectation–confirmation model for web portal context. *Information & Management, 42*(5), 683–693. doi:10.1016/j.im.2004.04.003

Linder, W. (2005). *Schweizerische demokratie – institutionen, prozesse, perspektiven.* Bern, Germany: Haupt.

Lin, H. H., & Wang, Y. S. (2006). An examination of the determinants of customer loyalty in mobile commerce contexts. *Information & Management, 43*(3), 271–282. doi:10.1016/j.im.2005.08.001

Lin, H.-F. (2008). Determinants of successful virtual communities: Contributions from system characteristics and social factors. *Information & Management, 45*(8), 522–527. doi:10.1016/j.im.2008.08.002

Linstone, H., & Turloff, M. (1975). *The Delphi method: Techniques and applications.* London, UK: Addison-Wesley.

Li, S., Glass, R., & Records, H. (2008). The influence of gender on new technology adoption and use- mobile commerce. *Journal of Internet Commerce, 7*(2), 270–289. doi:10.1080/15332860802067748

Locander, W. B., & Hermann, P. W. (1979). The effect of self-confidence and anxiety on information seeking in consumer risk reduction. *JMR, Journal of Marketing Research, 16*, 268–274. doi:10.2307/3150690

Lohse, L. G., & Spiller, P. (1999). Internet retail store design: How the user interface influences traffic and sales. *Journal of Computer-Mediated Communication, 5*(2).

Lourdes, T., Vicente, P., & Basilio, A. (2005). E-government developments on delivering public services among EU cities. *Government Information Quarterly, 22*(2), 217–238. doi:10.1016/j.giq.2005.02.004

Luarn, P., & Lin, H. H. (2005). Toward an understanding of the behavioral intention to use mobile banking. *Computers in Human Behavior, 21*, 873–891. doi:10.1016/j.chb.2004.03.003

Luder, E. (1997). Active matrix addressing of LCDs: Merits and shortcomings. In MacDonald, L. W., & Lowe, A. C. (Eds.), *Display systems* (pp. 157–172). Chichester, UK: John Wiley & Sons.

Lu, H. P., & Gustafson, D. H. (1994). An empirical study of perceived usefulness and perceived ease of use on computerized support system use over time. *International Journal of Information Management, 14*(5), 317–329. doi:10.1016/0268-4012(94)90070-1

Lu, J., Yao, J., & Yu, C. (2005). Personal innovativeness, social influences and adoption of wireless internet services via mobile technology. *The Journal of Strategic Information Systems, 14*(3), 245–268. doi:10.1016/j.jsis.2005.07.003

Lu, J., Yu, C., Liu, C., & Yao, J. E. (2003). Technology acceptance model for wireless Internet. *Internet Research: Electronic Networking Applications and Policy, 13*, 206–222. doi:10.1108/10662240310478222

Luka, S. C. Y. (2009). The impact of leadership and stakeholders on the success/failure of e-government service using the case study of e-stamping service in Hong Kong. *Government Information Quarterly, 26*(4), 594–604. doi:10.1016/j.giq.2009.02.009

Luo, X., Li, H., Zhang, J., & Shim, J. P. (2010). Examining multi-dimensional trust and multi-faceted risk in initial acceptance of emerging technologies: An empirical study of mobile banking services. *Decision Support Systems, 49*, 222–234. doi:10.1016/j.dss.2010.02.008

Lusch, R. F., & Brown, J. R. (1996). Interdependency, contracting, and relational behavior in marketing channels. *Journal of Marketing, 60*(4), 19–39. doi:10.2307/1251899

Lynagh, P. M., Murphy, P. R., Poist, R. F., & Grazer, W. F. (2001). Web-based informational practices of logistics service providers: An empirical assessment. *Transportation Journal, 40*(4), 34–45.

Lynott, P. P., & McCandless, N. J. (2000). The impact of age vs. life experiences on gender role attitudes of women in different cohorts. *Journal of Women & Aging, 12*(2), 5–21. doi:10.1300/J074v12n01_02

Macintosh, A. (2004) Characterizing e-participation in policy-making. In *Proceedings of the 37th Annual Hawaii International Conference on System Sciences* (p. 10).

Macintosh, A. (2007). *Challenges and barriers of eParticipation in Europe?* Paper presented at the Forum for Future Democracy.

Maditinos, D., Sarigiannidis, L., & Dimitriadis, E. (2010). The role of perceived risk on Greek internet users' purchasing intention: An extended TAM approach. *International Journal of Trade and Global Markets, 3*(1), 99–114. doi:10.1504/IJTGM.2010.030411

Magoutas, B., & Mentzas, G. (2009). Refinement, Validation and Benchmarking of a Model for E-Government Service Quality. In *Proceedings of the 8th International Conference (EGOV 2009)*, Linz, Austria (Vol. 5693).

Magoutas, B., Halaris, C., & Mentzas, G. (2007). An Ontology for the Multi-perspective Evaluation of Quality in E-Government Services. In *Proceedings of the 6th International Conference (EGOV 2007)*, Regensburg, Germany (Vol. 4657).

Mahatanankoon, P. (2007). The effects of personality traits and optimum stimulation level on text-messaging activities and m-commerce intention. *International Journal of Electronic Commerce, 12*, 7–30. doi:10.2753/JEC1086-4415120101

Mahoney, J. T., & Pandain, J. R. (1992). The resource-based view within the conversation of strategic management. *Strategic Management Journal, 20*(10), 935–952.

Malhotra, N. K., & Birks, D. F. (2000). *Marketing research: An applied approach*. London, UK: Pearson.

Malhotra, N. K., Kim, S. S., & Agarwal, J. (2004). Internet users' information privacy concerns (iuipc): The construct, the scale, and a causal model. *Information Systems Research, 15*, 336–355. doi:10.1287/isre.1040.0032

Mallat, N., Rossi, M., & Tuunainen, V. K. (2004). Mobile banking services. *Communications of the ACM, 47*(5), 42–46. doi:10.1145/986213.986236

Maloney, M. P., & Ward, M. P. (1973). Ecology, let's hear it from the people. *The American Psychologist, 28*, 583–586. doi:10.1037/h0034936

Maloni, M. J., & Carter, C. R. (2006). Opportunities for research in third-party logistics. *Transportation Journal, 45*(2), 23–38.

Maloni, M., & Benton, W. C. (2000). Power influences in the supply chain. *Journal of Business Logistics, 21*(11), 49–73.

Marez, L., Vyncke, P., Berte, K., Schurman, D., & Moor, K. (2007). Adopter segments, adoption determinants and mobile marketing. *Journal of Targeting. Measurement and Analysis for Marketing, 16*, 78–96. doi:10.1057/palgrave.jt.5750057

Martin, J. (2008). Consuming code: Use-value, exchange-value, and the role of virtual goods in second life. *Journal of Virtual Worlds Research, 1*(2), 1–21.

Massey, D. (1994). *Space, Place, and Gender*. Open University.

Mathieson, K. (1991). Predicting user intentions: Comparing the technology acceptance model with the theory of planned behavior. *Information Systems Research, 84*(1), 123–136.

Mathieson, K., Peacock, E., & Chin, W. W. (2001). Extending the technology acceptance model: The influence of perceived user resources. *The Data Base for Advances in Information Systems, 32*(3), 86–112.

Mattila, M. (2003). Factors affecting the adoption of mobile banking services. *Journal of Internet Banking and Commerce, 8*(1).

Mayer, R. C., Davis, J. H., & Schoorman, F. D. (1995). An integration model of organizational trust. *Academy of Management Review, 20*(3), 709–734. doi:10.2307/258792

Mazursky, D., & Vinitzky, G. (2005). Modifying consumer search processes in enhanced on-line interfaces. *Journal of Business Research, 58*, 1299–1309. doi:10.1016/j.jbusres.2005.01.003

McCloskey, D. W. (2006). The importance of ease of use, usefulness, and trust to online consumers: An examination of the technology acceptance model with older consumers. *Journal of Organizational and End User Computing, 18*(4), 47–65. doi:10.4018/joeuc.2006070103

McCrae, R. R., & Costa, P. T. (2003). *Personality in adulthood, a five-factor theory perspective*. New York, NY: Guilford Press. doi:10.4324/9780203428412

McDuffie, J. M., West, S., Welsh, J., & Baker, B. (2001). Logistics transformed: The military enters a new age. *Supply Chain Management Review, 5*(3), 92–100.

McGaley, M., & McCarthy, J. (2004). Transparency and e-voting democratic vs. commercial interests. In R. Krimmer & Grimm, R. (Eds.), *Proceedings of the 1ˢᵗ International Workshop on Electronic Voting in Europe: Technology, Law, Politics and Society* (LNI 47, pp. 143-152).

McKnight, D. H., & Chervany, N. L. (2001-2002). What trust means in e-commerce customer relationships: An interdisciplinary conceptual typology. *International Journal of Electronic Commerce, 6*(2), 35–59.

McKnight, D. H., Choudhury, V., & Kacmar, C. (2002). Developing and validating trust measures for e-commerce: An integrated typology. *Information Systems Research, 13*(3), 334–359. doi:10.1287/isre.13.3.334.81

McKnight, D. H., Choudhury, V., & Kacmar, C. (2002). The impact of initial consumer trust on intentions to transact with a web site: A trust building model. *The Journal of Strategic Information Systems, 11*, 297–323. doi:10.1016/S0963-8687(02)00020-3

McKnight, D. H., Cummings, L. L., & Chervany, N. L. (1998). Initial trust formation in new organizational relationships. *Academy of Management Review, 23*, 473–490.

McLeod, A. J., & Pippin, S. E. (2009). Security and Privacy Trust in E-Government: Understanding System and Relationship Trust Antecedents. In *Proceedings of the 41nd Annual Hawaii International Conference on System Sciences (HICSS'09).*

Medeni, T., Tutkun, C., Medeni, İ. T., Kumas, E., & Balci, A. (2008). Proposing a Modeling of Ubiquitous Knowledge Amphora for the Transition from e-Government to m-Government in Turkey. In *Proceedings of the E-Government Gateway Project mLife Events 2008,* Turkey.

Medeni, T., Elwell, M., & Cook, S. (2007). "Digitally Deaf" into Games for Learning: Towards a Theory of Reflective and Refractive Space-Time for Knowledge Management. In *BEYKON 2007.* Turkey: Immersing.

Medeni, T., Iwatsuki, S., & Cook, S. (2008). Reflective *Ba* and Refractive *Ma* in Cross-Cultural Learning. In Putnik, G. D., & Cunha, M. M. (Eds.), *Encyclopedia of Networked and Virtual Organizations.* Hershey, PA: IGI Global. doi:10.4018/978-1-59904-885-7.ch178

Medeni, T., Medeni, İ. T., Balci, A., & Dalbay, Ö. (2009). *Suggesting a Framework for Transition towards more Interoperable e-Government in Turkey: A Nautilus Model of Cross-Cultural Knowledge Creation and Organizational Learning.* Ankara, Turkey: ICEGOV.

Medeni, T., & Umemoto, K. (2008). An Action Research into International Masters Program in Practicing Management (IMPM): Suggesting Refraction to Complement Reflection for Management Learning in the Global Knowledge Economy. *Eurasian Journal of Business and Economics, 1*(1), 99–136.

Medeni, T., & Umemoto, K. (2010). *Educating Managers for the Global Knowledge Economy.* VDM Publications.

Mehrabian, A., & Russell, J. A. (1974). *An approach to environmental psychology.* Cambridge, MA: MIT Press.

Melin, U., & Axelsson, K. (2009). Managing e-service development – comparing two e-government case studies. *Transforming Government: People, Process and Policy, 3*(3).

Melquiot, P. (2009). *Technologies de l'information et de la communication (TIC), impacts sur l'environnement et le climat.* Retrieved from http://www.actualites-news-environnement.com/19871-Technologies-information-communication-TIC-environnement-climat.html

Menon, M. K., McGinnis, M. A., & Ackerman, K. B. (1998). Selection criteria for providers of third party logistics services: An exploratory study. *Journal of Business Logistics, 19*(1), 21–37.

Menon, S., & Kahn, B. (1995). The impact of context on variety seeking in product choices. *The Journal of Consumer Research, 22*, 285–295. doi:10.1086/209450

Menozzi, M., Napflin, U., & Krueger, H. (1999). CRT versus LCD: A pilot study on visual performance and suitability of two display technologies for use in office work. *Displays, 20*, 3–10. doi:10.1016/S0141-9382(98)00051-1

Mentzer, J. T., DeWitt, W., Keebler, J. S., Min, S., Nix, N. W., & Smith, C. D. (2001). Defining supply chain management. *Journal of Business Logistics, 22*(2), 1–25. doi:10.1002/j.2158-1592.2001.tb00001.x

Mercury, R. (2007). *Mercury's statement on electronic voting.* Retrieved from http://www.notablesoftware.com/RMstatement.html

Merlino, M., & Testa, S. (1998, May 21-22). L'adozione delle tecnologie dell'informazione nelle aziende fornitrici di servizi logistici dell'area genovese-savonese: i risultati di un'indagine empirica. In *Proceedings of the 2nd Workshop I Processi Innovativi Nella Piccola Impresa*, Urbino, Italy.

Merrilees, B., & Miller, D. (2001). Superstore interactivity: A new self-service paradigm of retail service. *International Journal of Retail & Distribution Management*, *29*(8), 379–389. doi:10.1108/09590550110396953

Messinger, P. R., Ge, X., Stroulia, E., Lyons, K., Smirnov, K., & Bone, M. (2008). On the relationship between my avatar and myself. *Journal of Virtual Worlds Research*, *1*(2), 1–17.

Messinger, P. R., Stroulia, E., Lyons, K., Bone, M., Niu, H., & Smirnov, K. (2009). Virtual worlds — past, present, and future: New directions in social computing. *Decision Support Systems*, *47*, 204–228. doi:10.1016/j.dss.2009.02.014

Meuter, M. L., Ostrom, A. L., Roundtree, R. I., & Bitner, M. J. (2000). Self-service technologies: Understanding customer satisfaction with technology-based service. *Journal of Marketing*, *64*(3), 50–64. doi:10.1509/jmkg.64.3.50.18024

Microvote General Corporation. (2008). *Election solutions*. Retrieved from http://www.microvote.com/products.htm

Midgley, D. F., & Dowling, G. R. (1978). Innovativeness: The concept and its measurement. *The Journal of Consumer Research*, *4*, 229–241. doi:10.1086/208701

Mike, G., & Anthony, M. (2007). e-Government information systems: Evaluation-led design for public value and client trust. *European Journal of Information Systems*, *16*(2).

Miller, C., Mukerji, J., Burt, C., Dsouza, D., Duddy, K., & El Kaim, W. (2001). *Model Driven Architecture (MDA) (Document No. ormsc/2001-07-01)*. Architecture Board ORMSC (Miller, C., & Mukerji, J., Eds.). OMG.

Miller, K., & Suresh, S. (2009). Policy based agents in wireless body sensor mesh networks for patient health monitoring. *International Journal of u and e-Services. Science and Technology*, *4*, 37–50.

Minguzzi, A., & Morvillo, A. (1999, June 20-23). Entrepreneurial culture and the spread of information technology in transport firms: First results on a Southern Italy sample. In *Proceedings of the 44th ICSB World Conference Innovation and Economic Development: The Role of Entrepreneurship and Small and Medium Enterprises*, Naples, Italy.

Min, Q., Ji, S., & Qu, G. (2008). Mobile commerce user acceptance study in China: A revised UTAUT model. *Tsinghua Science and Technology*, *13*(3), 257–264. doi:10.1016/S1007-0214(08)70042-7

Mitchell, V., & Boustani, P. (1993). Market development using new products and new customers: A role for perceived risk. *European Journal of Marketing*, *27*, 17–32. doi:10.1108/03090569310026385

Mitra, A. (2005). Direction of electronic governance initiatives within two worlds: case for a shift in emphasis. *Electronic Government, an Int. J.*, *2*(1), 26-40.

MODINIS. (2007). *Study on interoperability at local and regional level* (Final version). Retrieved from http://www.epractice.eu/files/media/media1309.pdf

Moon, J.-W., & Kim, Y.-G. (2001). Extending the TAM for the World-Wide-Web context. *Information & Management*, *38*, 217–230. doi:10.1016/S0378-7206(00)00061-6

Mooradian, T. A. (1996). Personality and ad-evoked feelings: The case for extraversion and neuroticism. *Journal of the Academy of Marketing Science*, *24*, 99–110. doi:10.1177/0092070396242001

Mooradian, T. A., & Olver, J. M. (1997). I can't get no satisfaction: The impact of personality and emotion on post purchase processes. *Psychology and Marketing*, *14*, 379–393. doi:10.1002/(SICI)1520-6793(199707)14:4<379::AID-MAR5>3.0.CO;2-6

Morales, V. M. (2009). *Seguridad en los procesos de voto electrónico remoto: Registro, votación, consolidación de resultados y auditoria*. Unpublished doctoral dissertation, Universitat Politecnica de Catalunya, Barcelona, Spain.

Moreno, A. (2002). *Medical applications of multi-agent systems*. Retrieved from http://deim.urv.cat/~itaka/Xerrades/EUNITEworkshop.pdf

Moreno, A., Valls, A., & Viejo, A. (2003). *Using JADE-LEAP to implement agents in mobile devices.* Retrieved from http://jade.tilab.com/papers/EXP/02Moreno.pdf

Morgan, R. M. (2000). Relationship marketing and marketing strategy: The evolution of relationship marketing within the organization. In Sheth, J. N., & Parvatiyar, A. (Eds.), *Handbook of relationship marketing* (pp. 481–505). Thousand Oaks, CA: Sage.

Morrison, P. D., & Roberts, J. H. (1998). Matching electronic distribution channels to product characteristics: Role of congruence in consideration set formation. *Journal of Business Research, 41*(3), 223–229. doi:10.1016/S0148-2963(97)00065-9

Mort, G. S., & Drennan, J. (2005). Marketing m-services: Establishing a usage benefit typology related to mobile user characteristics. *Journal of Database Marketing and Customer Strategy Management, 12*, 327–342. doi:10.1057/palgrave.dbm.3240269

Mort, G., & Drennan, J. (2005). Marketing m-services: Establishing a usage benefit typology related to mobile user characteristics. *Database Marketing & Customer Strategy Management, 12*(4), 327–341. doi:10.1057/palgrave.dbm.3240269

Mowen, J. (2000). *The 3M model of motivation and personality.* Norwell, MA: Kluwer Academic.

Mucchielli, A. (1991). *Les Méthodes de Contenus, Que sais-je?* Paris, France: Presses Universitaires de France.

Mullings, D. (2011). *Technological opportunities in telemedicine and telehealth.* Retrieved from http://www.jamaicaobserver.com/columns/Technology-opportunities--Telemedicine-and-Telehealth_8428572

Murugesan, S. (2008). Harnessing green IT: Principles and practices. *IEEE IT Professional,* 24-33.

Mylonakis, J. (2004). Can mobile services facilitate commerce? Findings from the Greek telecommunication market. *International Journal of Mobile Communications, 2*(2), 188–198. doi:10.1504/IJMC.2004.004667

Nahm, A. Y., Vonderembse, M. A., Rao, S. S., & Ragu-Nathan, T. S. (2006). Time-based manufacturing improves business performance - results from a survey. *International Journal of Production Economics, 101*(2), 213–229. doi:10.1016/j.ijpe.2005.01.004

Nealon, J. L., & Moreno, A. (2002). *The application of agent technology to health care.* Retrieved from http://deim.urv.cat/~itaka/Publicacions/aamas02ws.pdf

Nealon, J. L., & Moreno, A. (2003). *Agent-based applications in health care.* Retrieved from http://citeseerx.ist.psu.edu/viewdoc/summary?

Niemann, K. D. (2005). *Von der enterprise architecture zur IT-governance.* Braunschweig, Germany: Vieweg.

Nikolaou, I., Bettany, S., & Larsen, G. (2010). Brands and consumption in virtual worlds. *Journal of Virtual Worlds Research, 2*(5), 1–15.

Nonaka, I., Toyama, R., & Scharmer, O. (2001). *Building Ba to Enhance Knowledge Creation and Innovation at Large Firms.* Retrieved from http://www.dialogonleadership.org/Nonaka_et_al.html

Nonaka, I., & Takeuchi, H. (1995). *The Knowledge-creating Company: How Japanese Companies Create the Dynamics of Innovation.* Oxford, UK: Oxford University Press. doi:10.1016/0024-6301(96)81509-3

Novack, R. A., Langley, C. J., & Rinehart, L. M. (1995). *Creating logistics value: Themes for the future.* Oak Brooks, IL: Council of Logistics Management.

Nwana, H. S. (1997). Software Agents: An Overview. *The Knowledge Engineering Review, 11*(3).

Nysveen, H., Pedersen, P. E., & Thorbjørnsen, H. (2005). Intentions to use mobile services: Antecedents and cross-service comparisons. *Journal of the Academy of Marketing Science, 33*, 330–347. doi:10.1177/0092070305276149

Nysveen, H., Pedersen, P., Thorbjornsen, H., & Berthon, P. (2005). Mobilizing the brand: The effects of mobile services on brand relationships and main channel use. *Journal of Service Research, 7*(3), 257–276. doi:10.1177/1094670504271151

O'Reilly, T. (2006). *What is Web 2.0: Design patterns and business models for the next generation of software.* Retrieved from http://www.oreillynet.com/pub/a/oreilly/tim/news/2005/09/30/what-is-Web-20.html

OASIS SOA. (2006). *OASIS reference model for service oriented architecture 1.0.* Retrieved from http://www.oasis-open.org/committees/download.php/19679/soa-rm-cs.pdf

OASIS. (2006). *Oasis web services security (WSS).* Retrieved from http://www.oasis-open.org/committees/tc_home.php?wg_abbrev=wss

Obermeier, S., & Böttcher, S. (2010, July). Distributed, anonymous, and secure voting among malicious partners without a trusted third party. In *Proceedings of the 4th International Conference on Methodologies, Technologies and Tools enabling e-Government*, Olten, Switzerland.

Observatorio Voto Electrónico (OVE). (2005). *Electronic voting observation unit of the University of Leon.* Retrieved from http://www.votobit.org/ove/index.html

Observatory of the Greek Information Society. (2010). *The attitude of Greeks against online shopping, indicators of consumer's behavior.* Retrieved from http://www.observatory.gr/page/default.asp?la=1&id=2101&pk=439&return=183

OECD. (2006). *OECD e-Government Studies.* Turkey: OECD.

Okamoto, T. (1997). Receipt-free electronic voting schemes for large scale elections. In *Proceedings of the Security Protocols Workshop*, Paris, France (pp. 25-35).

Okazaki, S. (2004). How do Japanese consumers perceive wireless ads? A multivariate analysis. *International Journal of Advertising, 23*, 429–454.

Okazaki, S. (2005). Mobile advertising adoption by multinationals: Senior executives initial responses. *Internet Research, 15*(2), 160–180. doi:10.1108/10662240510590342

Okazaki, S. (2009). Social influence model and electronic word of mouth PC versus mobile internet. *International Journal of Advertising, 28*, 439–472. doi:10.2501/S0265048709200692

Oliver, R. L. (1993). Cognitive, affective, and attribute bases of the satisfaction response. *The Journal of Consumer Research, 20*(3), 418–430. doi:10.1086/209358

Oliver, R. L., & DeSarbo, W. S. (1988). Response determinants in satisfaction judgments. *The Journal of Consumer Research, 14*(4), 495–507. doi:10.1086/209131

OLPC. (2010). *One laptop per child.* Retrieved from http://laptop.org/en/

Ong, C. S., Laia, J. Y., & Wang, Y. S. (2004). Factors affecting engineers' acceptance of asynchronous e-learning systems in high-tech companies. *Information & Management, 41*(6), 795–804. doi:10.1016/j.im.2003.08.012

Oracle. (n. d.). *Oracle tuxedo application runtime for CICS and Batch.* Retrieved from http://www.oracle.com/products/middleware/tuxedo/tuxedo.html

Osimo, D. (2008). *Web 2.0 in government: Why and how? (Tech. Rep. No. EUR 23358 EN).* Brussels, Belgium: European Commission.

Ozdemir, S., Trott, P., & Hoecht, A. (2007). New service development: Insight from an explorative study into the Turkish retail banking sector. *Innovation: Management. Policy & Practice, 9*(3-4), 276–289. doi:10.5172/impp.2007.9.3-4.276

Pagani, M. (2004). Determinants of adoption of third generation mobile multimedia services. *Journal of Interactive Marketing, 18*(3), 46–59. doi:10.1002/dir.20011

Page, E. C., & Jenkins, B. (2005). *Policy bureaucracy: Government with a cast of thousands.* Oxford, UK: Oxford University Press.

Palvia, P. (2009). The role of trust in e-commerce relational exchange: A unified model. *Information & Management, 46*(4), 213–220. doi:10.1016/j.im.2009.02.003

Panetto, H., Scannapieco, M., & Zelm, M. (2004). INTEROP NoE: Interoperability research for networked enterprises applications and software. In *Proceedings of the Workshops of On the Move to Meaningful Internet Systems* (pp. 866-882).

Panopoulou, E., Tambouris, E., & Tarabanis, K. (2009). *D4.2b: eParticipation Good practice cases.* Retrieved from http://islab.uom.gr/eP/index.php?searchword=practice&option=com_search

Papadomichelaki, X., & Mentzas, G. (2009). A Multiple-Item Scale for Assessing E-Government Service Quality. In *Proceedings of the 8th International Conference (EGOV 2009),* Linz, Austria (Vol. 5693).

Papadomichelaki, X., Magoutas, B., Halaris, C., Apostolou, D., & Mentzas, G. (2006). A Review of Quality Dimensions in e-Government Services. In *Proceedings of the 5th International Conference (EGOV 2006)*, Krakow, Poland (Vol. 4084).

Papazafeiropoulou, A., Pouloudi, A., & Doukidis, G. (2001). *Electronic commerce policy making in Greece.* Melbourne, Australia: Center for Strategic Information Systems.

Pappas, F. C., & Volk, F. (2007). Audience counts and reporting system: Establishing a cyber-infrastructure for museum educators. *Journal of Computer-Mediated Communication, 12*(2), 752–768. doi:10.1111/j.1083-6101.2007.00348.x

Parasuraman, A. (2000). Technology readiness index (tri): A multiple-item scale to measure readiness to embrace new technologies. *Journal of Service Research, 2*, 307–320. doi:10.1177/109467050024001

Pardhasaradhi, Y., & Ahmed, S. (2007). Efficiency of Electronic Public Service Delivery in India: Public-Private Partnership as a Critical Factor. In *Proceedings of the 1st International Conference on Theory and Practice of Electronic Governance (ICEGOV 2007)*, Macao, China.

Pardo, T. A., & Burke, G. B. (2008). *Improving government interoperability: A capability framework for government managers.* Albany, NY: University at Albany, SUNY.

Park, R. (2008). Measuring Factors that Influence the Success of E-Government Initiatives'. In *Proceedings of the 41st Annual Hawaii International Conference on System Sciences (HICSS'08)*.

Park, C. (2006). Hedonic and utilitarian values of mobile internet in Korea. *International Journal of Mobile Communications, 4*, 497–508.

Park, J., Yang, S., & Lehto, X. (2007). Adoption of mobile technologies for Chinese consumers. *Journal of Electronic Commerce Research, 8*(3), 196–206.

Patton, M. Q. (1990). *Qualitative evaluation and research methods* (2nd ed.). Newbury Park, CA: Sage.

Paulraj, A., & Chen, I. (2007). Strategic buyer-supplier relationships, information technology and external logistics integration. *Journal of supply Chain Management, 4.*

Pavlou, P. (2003). Consumer acceptance of electronic commerce: Integrating trust and risk with the technology acceptance model. *International Journal of Electronic Commerce, 7*(3), 101–134.

Pavlou, P. A., & Chai, L. (2002). What drives electronic commerce across cultures? A cross-cultural empirical investigation of the theory of planned behavior. *Journal of Electronic Commerce Research, 3*(4), 240–253.

Pavlou, P. A., & Fygenson, M. (2006). Understanding and predicting electronic commerce adoption: an extension of the theory of planned behavior. *Management Information Systems Quarterly, 30*, 115–143.

Pedersen, P., Methlie, L., & Thorbjornsen, H. (2002). Understanding mobile commerce end-user adoption: A triangulation perspective and suggestions for an exploratory service evaluation framework. In *Proceedings of the 35th Hawaii International Conference on System Sciences* (p. 8).

Pelet, J.-É., & Papadopoulou, P. (2009, September 25-27). The effects of colors of e-commerce websites on mood, memorization and buying intention. Paper presented at the 4th Mediterranean Conference on Information Systems, Greece.

Pelet, J.-E., & Papadopoulou, P. (2010). Consumer responses to colors of e-commerce websites: An empirical investigation. In Kang, K. (Ed.), *E-Commerce*. Rijeka, Croatia: In-Tech. doi:10.5772/8897

Pelet, J.-E., & Papadopoulou, P. (2010). The effect of e-commerce websites colors on consumer trust. *International Journal of E-Business Research.*

Pellegrine, T. (2009). *A comparative study on approaches to social software and the semantic web.* Retrieved from http://www.semanticweb.at/file_upload/1_tmpphpvu-VU1T.pdf

Pelly, P. K., & Sia, S. K. (2007). Challenges in delivering cross-agency integrated e-services: The OBLS project'. *Journal of Information Technology, 22*(4).

Pentafronimos, G., Papastergiou, S., & Polemi, N. (2008). Interoperability testing for e-government web services. In *Proceedings of the 2nd International Conference on Theory and Practice of Electronic Governance (ICEGOV 2008)*, Egypt (Vol. 351, pp. 316-321).

Peristeras, V., Mentzas, G., Tarabanis, K. A., & Abecker, A. R. (2009). Transforming E-government and E-participation through IT. *IEEE Intelligent Systems, 24*(5), 14–19. doi:10.1109/MIS.2009.103

Peristeras, V., & Tarabanis, K. (2006). The connection, communication, consolidation, collaboration interoperability framework (C4IF) for information systems interoperability. *International Journal of Interoperability in Business Information Systems, 1*(1), 61–72.

Pervin, L. A. (1994). A critical analysis of current trait theory. *Psychological Inquiry, 5*, 103–113. doi:10.1207/s15327965pli0502_1

Phang, C. W., Li, Y., Sutanto, J., & Kankanhalli, A. (2005). Senior Citizens' Adoption of E-Government: In Quest of the Antecedents of Perceived Usefulness. In *Proceedings of the 38th Annual Hawaii International Conference on System Sciences (HICSS'05).*

Picture. (2009). *Final results from the picture project.* Retrieved from http://www.picture-eu.org/

Pihlström, M. (2007). Committed to content provider or mobile channel? Determinants of continuous mobile multimedia service use. *Journal of Information Technology Theory and Application, 9*, 1–24.

Pikkarainen, T., Pikkarainen, K., Karjaluoto, H., & Pahnila, S. (2004). Consumer acceptance of online banking: An extension of the technology acceptance model. *Internet Research, 14*(3), 224–235. doi:10.1108/10662240410542652

Pinho, J. C., & Macedo, I. M. (2008). Examining the antecedents and consequences of online satisfaction within the public sector: The case of taxation services. *Transforming Government: People, Process and Policy, 2*(3).

Polančič, G., Heričko, M., & Rozman, I. (2010). An empirical examination of application frameworks success based on technology acceptance model. *Journal of Systems and Software, 83*(4), 574–584. doi:10.1016/j.jss.2009.10.036

Porter, M. E. (1985). Technology and competitive advantage. *The Journal of Business Strategy, 5*(3), 60–78. doi:10.1108/eb039075

Prahalad, C. K., & Krishnan, M. (1999). The meaning of quality in the information age. *Harvard Business Review, 77*(5), 15–21.

Prahinksi, C., & Benton, W. C. (2004). Supplier evaluations: Communication strategies to improve supplier performance. *Journal of Operations Management, 22*(1), 39–62. doi:10.1016/j.jom.2003.12.005

Prasolova-Forland, E. (2008). Analyzing place metaphors in 3D educational collaborative virtual environments. *Computers in Human Behavior, 24*(2), 185–204. doi:10.1016/j.chb.2007.01.009

PriceWaterhouseCoopers. (2002). *Gesamtschweizerische strategie zur dauerhaften archivierung von unterlagen aus elektronischen systemen (Strategiestudie) – appendix III.* Retrieved from http://www.vsa-aas.org/uploads/media/d_strategie_anh_3.pdf

Puiggali, J. (2007). *Voto electrónico.* Paper presented at the 2nd Jornadas de Comercio Electrónico y Administración Electrónica, Saragossa, Spain.

Puliafito, A., & Tomarchio, O. (1999). Advanced Network Management Functionalities through the use of Mobile Software Agents. In *Proceedings of 3rd International Workshop on Intelligent Agents for Telecommunication Applications (IATA'99)*, Stockholm, Sweden.

Qi, J., Li, L., Li, Y., & Shu, H. (2009). An extension of technology acceptance model: Analysis of the adoption of mobile data services in China. *Systems Research and Behavioral Science, 26*(3), 391–407. doi:10.1002/sres.964

Raju, P. S. (1980). Optimum stimulation level: Its relationship to personality, demographics and exploratory behavior. *The Journal of Consumer Research, 7*, 272–282. doi:10.1086/208815

Ralyte, J., Jeusfeld, M., Backlund, P., Kuhn, H., & Arni-Bloch, N. (2008). A knowledge-based approach to manage information systems interoperability. *Information Systems, 33*, 754–784. doi:doi:10.1016/j.is.2008.01.008

Rao, B., Angelov, B., & Nov, O. (2006). Fusion of disruptive technologies: Lessons from the Skype case. *European Management Journal, 24*(2-3), 174–188. doi:10.1016/j.emj.2006.03.007

Rao, S., Truong, D., Senecal, S., & Le, T. (2007). How buyers' expected benefits, perceived risks, and e-business readiness influence their e-marketplaces usage. *Industrial Marketing Management, 36*, 1035–1045. doi:10.1016/j.indmarman.2006.08.001

Ravichandran, T., & Lertwongsatien, C. (2005). Effect of information resources and capabilities on firm performance: A resource-based perspective. *Journal of Management Information Systems, 21*(4), 237–276.

Ravi, K., Anandarajan, M., & Igbaria, M. (1999). Linking IT applications with manufacturing strategy: An intelligent decision support system approach. *Decision Sciences*, *30*(4), 959–992. doi:10.1111/j.1540-5915.1999.tb00915.x

Ray, I., Ray, I., & Narasimhamurthi, N. (2001). An anonymous electronic voting protocol for voting over the Internet. In *Proceedings of the Third International Workshop on Advanced Issues of E-Commerce and Web-based Information Systems*.

Real Madrid. (2009). *El voto electrónico en la Asamblea.* Retrieved from http://www.realmadrid.com/cs/Satellite/es/1193040472656/1202766421991/noticia/Noticia/El_voto_electronico_en_la_Asamblea.htm

Rehan, M., & Koyuncu, M. (2009). *Towards e-government: A survey of Turkey's progress.* Ankara, Turkey: ICEGOV.

Reichheld, F. F., & Sasser, E. W. (1990). Zero defections: Quality comes to services. *Harvard Business Review*, *68*(5), 105–111.

Reichheld, F. F., & Schefter, P. (2000). E-loyalty: Your secret weapon on the web. *Harvard Business Review*, *78*(4), 105–113.

Reid, M., & Levy, Y. (2008). Integrating trust and computer self-efficacy with TAM: An empirical assessment of customers' acceptance of banking information systems (BIS) in Jamaica. *Journal of Internet Banking and Commerce*, *12*(3).

Reid, V., Bardzki, B., & McNamee, S. (2004). Communication and Culture: Designing a Knowledge-enabled Environment to Effect Local Government Reform. *Electronic. Journal of E-Government*, *2*(3), 197–206.

Rettie, R., Grandcolas, U., & Deakins, B. (2005). Text message advertising: Response rates and branding effects. *Journal of Targeting. Measurement and Analysis for Marketing*, *13*, 304–313. doi:10.1057/palgrave.jt.5740158

Rhodes, S. R. (1983). Age-related differences in work attitudes and behavior: A review and conceptual analysis. *Psychological Bulletin*, *93*(2), 328–367. doi:10.1037/0033-2909.93.2.328

Richey, R. G., Daugherty, P. J., & Roath, A. S. (2007). Firm technological readiness and complementarity: Capabilities impacting logistics service competency and performance. *Journal of Business Logistics*, *28*(1), 195–228. doi:10.1002/j.2158-1592.2007.tb00237.x

Richins, M. (1983). An analysis of consumer interaction styles in the marketplace. *The Journal of Consumer Research*, *10*, 73–82. doi:10.1086/208946

Riding, R. (2001). *Cognitive style analysis – research administration.* Birmingham, AL: Learning and Training Technology.

Ridings, C. M., & Gefen, D. (2004). Virtual community attraction: Why people hang out online. *Journal of Computer-Mediated Communication*, *10*, 4.

Riesman, D. (1950). *The lonely crowd.* New Haven, CT: Yale University Press.

Rivest, R. (2006). *The three ballot voting system.* Retrieved from http://theory.csail.mit.edu/~rivest/Rivest-TheThreeBallotVotingSystem.pdf

Roberton, T. S., & Myers, J. H. (1969). Personality correlates of opinion leadership and innovative buying behavior. *JMR, Journal of Marketing Research*, *6*, 164–168. doi:10.2307/3149667

Roberts, J. A. (1995). Profiling levels of socially responsible consumer behaviour: A cluster analytic approach and its implications for marketing. *Journal of Marketing Theory and Practice*, 97-117.

Rodin, J. (1990). Control by any other name: Definitions, concepts, and processes. In Rodin, J., Schooler, C., & Schaie, K. W. (Eds.), *Self-directedness: Causes and effects throughout the life course.* Mahwah, NJ: Lawrence Erlbaum.

Rogers, E. (1995). *Diffusion of Innovations.* New York: Free Press.

Rogers, E. M. (1983). *Diffusion of innovations* (3rd ed.). New York, NY: Free Press.

Rogers, E. M., & Shoemaker, F. F. (1971). *Communication of Innovations: A Cross-Cultural Approach.* New York: Free Press.

Rogers, M. (1995). *Diffusion of innovations*. New York, NY: Free Press.

Rook, D. W., & Fisher, R. J. (1995). Normative influences on impulsive buying behavior. *The Journal of Consumer Research*, *22*, 305–313. doi:10.1086/209452

Rose, J., & Fogarty, G. (2006). Determinants of perceived usefulness and perceived ease of use in the technology acceptance model: Senior consumers' adoption of self-service banking technologies. In *Proceedings of the 2nd Biennial Academy of World Business, Marketing & Management Development Conference* (Vol. 2, pp. 122-129).

Rose, J., Sæbø, O., Nyvang, T., & Sanford, C. (2007). *D14.3a: The role of social networking software in eparticipation*. Retrieved from http://www.demo-net.org/what-is-it-about/research-papers-reports-1/demo-net-deliverables/RoseEtAl2007b

Ross, C., Orr, E. S., Sisic, M., Arseneault, J. M., Simmering, M. G., & Orr, R. R. (2009). Personality and motivations associated with Facebook use. *Computers in Human Behavior*, *25*, 578–586. doi:10.1016/j.chb.2008.12.024

Rossel, P., Finger, M., & Misuraca, G. (2006). Mobile e-Government Options: Between Technology-driven and User-centric. *Electronic. Journal of E-Government*, *4*(2), 79–86.

Rowe, G., & Wright, G. (1999). The Delphi technique as a forecasting tool: Issues and analysis. *International Journal of Forecasting*, *15*, 353–375. doi:10.1016/S0169-2070(99)00018-7

Roy, J. (2009). E-government and integrated service delivery in Canada: the Province of Nova Scotia as a case study. *Int. J. of Electronic Governance*, *2*(2/3), 223–238. doi:10.1504/IJEG.2009.029131

Rust, R. T., & Donthu, N. (1995). Capturing geographically localized misspecification error in retail store choice models. *JMR, Journal of Marketing Research*, *32*(1), 103–110. doi:10.2307/3152115

Ruth, S., & Mercer, D. (2007). Voting from the home or office? Don't hold your breath. *IEEE Internet Computing*, *11*(4), 68–71. doi:10.1109/MIC.2007.94

Ryan, A. B., & Suresh, S. (2009). Intelligent Agent based Mobile Shopper. In *Proceedings of sixth IFIP/IEEE International conference in Wireless Optical Communications and Networks*, Cairo, Egypt.

Sacco, G. M. (2007). Interactive exploration and discovery of e-government services. In *Proceedings of the 8th annual international conference on Digital government research*, Philadelphia, PA (Vol. 228, pp. 190-197).

Sæbø, O., Rose, J., & Skiftenes Flak, L. (2008). The shape of eParticipation: Characterizing an emerging research area. *Government Information Quarterly*, *25*(3), 400–428. doi:10.1016/j.giq.2007.04.007

Sahin, F., & Robinson, P. (2002). Flow coordination and information sharing in supply chains: Review, implications, and directions for future research. *Decision Sciences*, *33*(4), 505–536. doi:10.1111/j.1540-5915.2002.tb01654.x

Sahin, F., & Robinson, P. (2005). Information sharing and coordination in make-to-order supply chains. *Journal of Operations Management*, *23*(6), 579–598. doi:10.1016/j.jom.2004.08.007

Sahu, G. P., & Gupta, M. P. (2007). Users' Acceptance of E-Government: A Study of Indian Central Excise. [IJEGR]. *International Journal of Electronic Government Research*, *3*(3), 1–21.

Saiman, A. (2009). Barriers to efficient virtual business transactions. *Journal of Virtual Worlds Research*, *2*(3), 1–14.

Salhofer, P., Tretter, G., Stadlhofer, B., & Joanneum, F. H. (2008). Goal-oriented service selection. In *Proceedings of the 2nd International Conference on Theory and Practice of Electronic Governance (ICEGOV2008)*, Egypt (Vol. 351, pp. 60-66).

Salhofer, P., Stadlhofer, B., & Tretter, G. (2009). Ontology Driven e-Government. *Electronic. Journal of E-Government*, *7*(4), 415–424.

Sanders, N. R. (2005). IT alignment in supply chain relationships: A study of supplier benefits. *Journal of Supply Chain Management*, *41*(2), 4–13. doi:10.1111/j.1055-6001.2005.04102001.x

Sanders, N. R., & Premus, R. (2002). IT applications in supply chain organizations: A link between competitive priorities and organizational benefits. *Journal of Business Logistics, 23*(1), 65–83. doi:10.1002/j.2158-1592.2002.tb00016.x

Santin, A. O., Costa, R. G., & Maziero, C. A. (2008). A three-ballot-based secure electronic voting system. *IEEE Security and Privacy, 6*(3).

Saprikis, V., Chouliara, A., & Vlachopoulou, M. (2010). Perceptions towards online shopping: Analyzing the Greek University students' attitude. *Communications of the IBIMA*, 1-13.

Sarantis, D., Charalabidis, Y., & Psarras, J. (2008). *Towards standardising interoperability levels for information systems of public administrations*. Electronic Journal for e-Commerce Tools and Applications.

Sarikas, O. D., & Weerakkody, V. (2007). Realizing integrated e-government services: a UK local government perspective. *Transforming Government: People, Process and Policy, 1*(2).

Sauvage, T. (2003). The relationship between technology and logistics third-party providers. *International Journal of Physical Distribution and Logistics Management, 33*(3), 236–253. doi:10.1108/09600030310471989

Scharl, A., Dickinger, A., & Murphy, J. (2005). Diffusion and success factors of mobile marketing. *Electronic Commerce Research and Applications, 4*, 159–173. doi:10.1016/j.elerap.2004.10.006

Schaupp, L. C., Carter, L., & Hobbs, J. (2009). E-File Adoption: A Study of U.S. Taxpayers' Intentions. In *Proceedings of the 41ˢᵗ Annual Hawaii International Conference on System Sciences (HICSS'09)*.

Scheier, M. F., & Carver, C. S. (1985). Optimism, coping, and health: Assessment and implications of generalized outcome expectancies. *Health Psychology, 4*, 219–247. doi:10.1037/0278-6133.4.3.219

Schellong, A. (2008). *Citizen relationship management – a study of CRM in government*. Frankfurt, Germany: Peter Lang Verlag.

Schellong, A. (2009). Calling 311: Citizen relationship management in Miami-Dade county improving access to government information and services. In Rizvi, G., & De Jong, J. (Eds.), *The state of access: Success and failure of democracies to create equal opportunities* (pp. 191–206). Washington, DC: Brookings Institution Press.

Schiffman, L. G., & Kanuk, L. L. (2007). *Consumer behavior*. Upper Saddle River, NJ: Prentice Hall.

Schoenmakers, B. (2000). Fully auditable electronic secret-ballot elections. *Internet Technology Magazine*, 5-11.

Schönherr, M. (2004). Enterprise architecture frameworks. In Aier, S., & Schönherr, M. (Eds.), *Enterprise application integration – serviceorientierung und nachhaltige architekturen* (pp. 3–48). Berlin, Germany: Gito Verlag.

Schumacker, R. E., & Lomax, R. G. (2004). *A beginner's guide to structural equation modeling*. Mahwah, NJ: Lawrence Erlbaum.

Scupola, A., Henten, A., & Nicolajsen, H. W. (2009). E-Services: Characteristics, Scope and Conceptual Strengths. [IJESMA]. *International Journal of E-Services and Mobile Applications, 1*(3), 1–16.

Segars, A. H., & Grover, V. (1993). Re-examining perceived ease of use and usefulness: A confirmatory factors analysis. *Management Information Systems Quarterly, 17*(4), 517–526. doi:10.2307/249590

Sehl, M., & Faouzi, B. (2009). Multi-agent based framework for e-government. *Electronic Government: An International Journal, 6*(2), 177–192. doi:10.1504/EG.2009.024441

Selviaridis, K., & Spring, M. (2007). Third party logistics: A literature review and research agenda. *International Journal of Logistics Management, 18*(1), 125–150. doi:10.1108/09574090710748207

SEMIC. (n. d.). The semantic interoperability centre. *Europe*. Retrieved from http://www.semic.eu/semic/view/snav/About_SEMIC.xhtml.

Seong, S. (2008). *VOIP in Japan and Korea*. London: Ovum.

Serenko, A., & Detlor, B. (2002). *Agent toolkits: A general overview of the market and an assessment of instructor satisfaction with utilizing toolkits in the classroom.* Retrieved from http://foba.lakeheadu.ca/serenko/Agent_Toolkits_Working_Paper.pdf

ServiceMix. (n. d.). *The software apache foundation.* Retrieved from http://servicemix.apache.org/home.html

Shachaf, P., & Oltmann, S. M. (2007). E-Quality and E-Service Equality. In *Proceedings of the 40th Annual Hawaii International Conference on System Sciences (HICSS'07).*

Sheila, S. (2000). Health promotion and health for all. *Caribbean Health Journal, 2*(4), 10–11.

Shih, H. (2004). Extended technology acceptance model of internet utilization behavior. *Information & Management, 41*(6), 719–729. doi:10.1016/j.im.2003.08.009

Shim, S., Eastlick, M. A., Lotz, S. L., & Warrington, P. (2001). An online prepurchase intentions model: The role of intention to search. *Journal of Retailing, 77*(3), 397–416. doi:10.1016/S0022-4359(01)00051-3

Shin, D. H. (2009). Towards an understanding of the consumer acceptance of mobile wallet. *Computers in Human Behavior, 25,* 1343–1354. doi:10.1016/j.chb.2009.06.001

Shi, W., Shambare, N., & Wang, J. (2008). The adoption of internet banking: An institutional theory perspective. *Journal of Financial Services Marketing, 12*(4), 272–286. doi:10.1057/palgrave.fsm.4760081

Shugan, S. M. (2004). The impact of advancing technology on marketing and academic research. *Marketing Science, 23,* 469–476. doi:10.1287/mksc.1040.0096

Shum, S. B. (2003). The roots of computer supported argument visualization. In Kirschner, P. A., Buckingham Shum, S. J., & Carr, C. S. (Eds.), *Visualizing argumentation: Software tools for collaborative and educational sense-making.* Berlin, Germany: Springer-Verlag.

Silvera, D. H., Lavack, A. M., & Kropp, F. (2008). Impulse buying: The role of affect, social influence, and subjective wellbeing. *Journal of Consumer Marketing, 25,* 23–33. doi:10.1108/07363760810845381

Simon, H. A. (1960). *The new science of management decision.* New York, NY: Harper and Row.

Simonson, I. (1999). The effect of product assortment on buyer preferences. *Journal of Retailing, 75*(3), 347–370. doi:10.1016/S0022-4359(99)00012-3

Singhapakdi, A., & La Tour, M. S. (1991). The link between social responsibility orientation, motive appeals, and voting intention: A case of an anti-littering campaign. *Journal of Public Policy & Marketing, 10*(2), 118–129.

Sinkovics, R. R., & Roath, A. S. (2004). Strategic orientation, capabilities, and performance in manufacturer-3PL relationships. *Journal of Business Logistics, 25*(2), 43–64. doi:10.1002/j.2158-1592.2004.tb00181.x

Siomkos, G., & Vrechopoulos, A. (2002). Strategic marketing planning for competitive advantage in electronic commerce. *International Journal of Services Technology and Management, 3*(1), 22–38. doi:10.1504/IJSTM.2002.001614

Skinner, E. A. (1996). A guide to constructs of control. *Journal of Personality and Social Psychology, 71,* 549–570. doi:10.1037/0022-3514.71.3.549

Slyke, C. V., Belanger, F., & Comunale, C. L. (2004). Factors influencing the adoption of web-based shopping: The impact of trust. *Databases for Advances in Information Systems, 35*(2), 32–49.

Spekman, R. E. (1988). Strategic supplier selection: Understanding ling-term buyer relationships. *Business Horizons, 31*(4), 75–81. doi:10.1016/0007-6813(88)90072-9

Spence, J. (2008). Demographics of virtual worlds. *Journal of Virtual Worlds Research, 1*(2), 1–45.

Spur, G., Mertins, K., & Jochem, R. (1996). *Integrated enterprise modelling.* Berlin, Germany: Springer-Verlag.

Stank, T. P., Keller, S. B., & Daugherty, P. J. (2001). Supply chain collaboration and logistics service performance. *Journal of Business Logistics, 22*(2), 29–47. doi:10.1002/j.2158-1592.2001.tb00158.x

Stoica, V., & Ilas, A. (2009). Romanian Urban e-Government. Digital Services and Digital Democracy in 165 Cities. *Electronic. Journal of E-Government, 7*(2), 171–182.

Strauss, J., & Frost, R. (2009). *E-marketing* (5th ed.). Upper Saddle River, NJ: Prentice-Hall.

Suh, B., & Han, I. (2002). Effect of trust on customer acceptance of Internet banking. *Electronic Commerce Research and Applications*, *1*(3-4), 247–263. doi:10.1016/S1567-4223(02)00017-0

Sulaiman, A., Jaafar, N. I., & Mohezar, S. (2007). An overview of mobile banking adoption among the urban community. *International Journal of Mobile Communications*, *5*(2), 157–168. doi:10.1504/IJMC.2007.011814

Sultan, F., & Rohm, A. J. (2008). How to market to generation m(obile). *MIT Sloan Management Review*, *49*, 35–41.

Sun Microsystems. (2009). *Java ME technology*. Retrieved from http://java.sun.com/javame/ technology/ index.jsp

Sun Microsystems. (2009). *MySQL*. Retrieved from http://www.mysql.com

Sun, Q., Wang, C., & Cao, H. (2009). An extended TAM for analyzing adoption behavior of mobile commerce. In *Proceedings of the Eighth International Conference on Mobile Business* (pp. 52-56).

Suresh, S. (2001). *Applications of Policy and Mobile Agents in the Management of Computer Networks*. Unpublished Master's thesis, Anna University, Chennai, India.

Suresh, S. (2006). *Studies in agent based IP traffic congestion management in DiffServ networks.* Unpublished doctoral dissertation, University of South Australia, Adelaide, Australia.

Sustainable Energy Europe Campaign. (2008). *Directorate-general for energy and transport.* Retrieved from http://www.sustenergy.org/

Sweeney, J. C., Soutar, G. N., & Johnson, L. W. (1999). The role of perceived risk in the quality-value relationship: A study in a retail environment. *Journal of Retailing*, *75*, 77–105. doi:10.1016/S0022-4359(99)80005-0

Tambouris, E., Liotas, N., & Tarabanis, K. (2007). A framework for assessing eparticipation projects and tools. In *Proceedings of the 40th Annual Hawaii International Conference on System Sciences* (p. 90).

Tambouris, E., Liotas, N., Kaliviotis, D., & Tarabanis, K. (2007). A framework for scoping eParticipation. In *Proceedings of the 8th Annual International Conference on Digital Government Research: Bridging Disciplines & Domains* (pp. 288-289).

Tan, C. W., Pan, S. L., & Lim, E. T. K. (2005). Towards the Restoration of Public Trust in Electronic Governments: A Case Study of the E-Filing System in Singapore. In *Proceedings of the 38th Annual Hawaii International Conference on System Sciences (HICSS'05)*.

Tang, L. (2006). Group Effectiveness: An Integral and Developmental Perspective. In *Proceedings of the Annual meeting of the International Communication Association*, Dresden, Germany.

Tate, R. (1998). *An introduction to modeling outcomes in the behavioral and social science* (2nd ed.). Minneapolis, MN: Burgess.

Taylor, S., & Todd, P. A. (1995). Assessing IT usage: The role of prior experience. *Management Information Systems Quarterly*, *19*(2), 561–570. doi:10.2307/249633

Taylor, S., & Todd, P. A. (1995). Understanding information technology usage: A test of competing models. *Information Systems Research*, *6*(2), 144–176. doi:10.1287/isre.6.2.144

Teo, T. S. H., & Pok, S. H. (2003). Adoption of WAP-enabled mobile phones among Internet users. *Omega. International Journal of Management Science*, *31*(6), 483–498.

The Open Group. (2004). *TOGAF 8.1: Certification for practitioners version 1.xx by architecting-the-enterprise*. Retrieved from http://www.opengroup.org/togaf/cert/protected/certuploads/6853.pdf

The Yankee Group Report. (2007). *How open are the new VOIP market? A global perspective*. Retrieved from http://www.yankeegroup.com

Theodoridis, P., & Chatzipanagiotou, K. (2009). Store image attributes and customer satisfaction across different customer profiles within the supermarket sector in Greece. *European Journal of Marketing*, *43*(5-6), 708–734. doi:10.1108/03090560910947016

Thomas, R., & Harold, K. (2003). *Handbook of Consumer Behavior*. Upper Saddle River, NJ: Prentice-Hall.

Tihon, A. (2006). The Informational Attractors: A Different Approach of Information and Knowledge Management.

Tobin, P. K. J., & Bidoli, M. (2005). Factors affecting the adoption of Voice over Internet Protocol (VoIP) and other converged IP services in South Africa. *South African Journal of Business Management, 37*(1), 31–39.

Tucker, L. R., Dolich, I. J., & Wilson, D. T. (1981). Profiling environmentally responsible consumer-citizens. *Journal of the Academy of Marketing Science, 9*(4), 454–478. doi:10.1007/BF02729884

Tung, L. L., & Rieck, O. (2005). Adoption of electronic government services among business organizations in Singapore. *The Journal of Strategic Information Systems, 14*(4). doi:10.1016/j.jsis.2005.06.001

Turk, T., Blazic, B., & Trkman, P. (2008). Factors and sustainable strategies fostering the adoption of broadband communications in an enlarged European Union. *Technological Forecasting and Social Change, 75*(7), 933–951. doi:10.1016/j.techfore.2007.08.004

Tzagarakis, M., Karousos, N., Karacapilidis, N., & Nousia, D. (2009). Unleashing argumentation support systems on the Web: The case of CoPe_it! *International Journal of Web-Based Learning and Teaching Technologies, 4*(3), 22–38.

UbiPOL. (2009). *Project Document 2009*. Retrieved from http://www.ideal-ist.net/Countries/UK/PS-UK-3051

UE ENERGY STAR. (2009). *Introduction au Programme européen ENERGY STAR.* Retrieved from http://www.eu-energystar.org/fr/index.html

Umemoto, K. (2004). Practicing and Researching Knowledge Management at JAIST. In *Proceedings of the Technology Creation Based on Knowledge Science: Theory and Practice JAIST Forum*, Japan.

United Nations. (2008). *eGovernment survey 2008: From eGovernment to connected governance.* Retrieved from http://unpan1.un.org/intradoc/groups/public/documents/UN/UNPAN028607.pdf

Uno, Y. (1999). Why the Concept of Trans-Cultural Refraction Necessary. *NEWSLETTER: Intercultural Communication,* (35).

Van Bruggen, J. M., Boshuizen, H. P. A., & Kirschner, P. A. (2003). A cognitive framework for cooperative problem solving with argument visualization. In Kirschner, P. A., Buckingham Shum, S. J., & Carr, C. S. (Eds.), *Visualizing argumentation: Software tools for collaborative and educational sense-making* (pp. 25–47). Berlin, Germany: Springer-Verlag.

Van Hoek, R. (2002). Using information technology to leverage transport and logistics service operations in the supply chain: An empirical assessment of the interrelation between technology and operation management. *International Journal of Information Technology and Management, 1*(1), 115–130. doi:10.1504/IJITM.2002.001191

Van Waarden, F. (1992). Dimensions and types of policy networks. *European Journal of Political Research, 21*(1-2), 29–52. doi:10.1111/j.1475-6765.1992.tb00287.x

Varnali, K., & Toker, A. (2010). Mobile marketing: The-state-of-the-art. *International Journal of Information Management, 30*, 144–151. doi:10.1016/j.ijinfomgt.2009.08.009

Vassilakis, C., Lepouras, G., Halatsis, C., & Lobo, T. P. (2005). An XML model for electronic services. *Electronic Government, an Int. J., 2*(1), 41-55.

Vazquez-Salceda, J. (2004). Normative agents in health care: Uses and challenges. *AI Communications-Agents Applied in Health Care, 18*(3), 175–189.

Vehovar, V. (2003*). Security concern and on-line shopping: International study of the credibility of consumer information on the Internet.* Retrieved from http://www.consumerwebwatch.org/pdfs/Slovenia.pdf

Velasquez, M. G., & Rostankowski, C. (1985). *Ethics: Theory and practice.* Upper Saddle River, NJ: Prentice-Hall.

Velsen, L. V., Geest, T. V. d., Hedde, M. t., & Derks, W. (2008). Engineering User Requirements for e-Government Services: A Dutch Case Study. In *Proceedings of the 7th International Conference (EGOV 2008)*, Torino, Italy (Vol. 5184).

Venkatesh, V. (1999). Creation of favourable user perceptions: Exploring the role of intrinsic motivation. *Management Information Systems Quarterly, 23*(2), 239–260. doi:10.2307/249753

Venkatesh, V. (2000). Determinants of perceived ease of use: Integrating control, intrinsic motivation, and emotion into the technology acceptance model. *Information Systems Research*, *11*(4), 342–365. doi:10.1287/isre.11.4.342.11872

Venkatesh, V., & Davis, F. D. (1996). A model of the antecedents of perceived ease of use: Development and test. *Decision Sciences*, *27*(3), 451–481. doi:10.1111/j.1540-5915.1996.tb01822.x

Venkatesh, V., & Davis, F. D. (2000). A theoretical extension of the technology acceptance model: Four longitudinal field studies. *Management Science*, *46*(2), 186–204. doi:10.1287/mnsc.46.2.186.11926

Venkatesh, V., & Morris, M. G. (2000). Why don't men ever stop to ask for directions? Gender, social influence, and their role in technology acceptance and usage behavior. *Management Information Systems Quarterly*, *24*(1), 115–139. doi:10.2307/3250981

Venkatesh, V., & Morris, M. G. (2000). Why don't men ever stop to ask for directions? Gender, social influence, and their role in technology acceptance and usage behavior. *Management Information Systems Quarterly*, *24*(1), 115–139. doi:10.2307/3250981

Venkatesh, V., Morris, M. G., & Ackerman, P. L. (2000). A longitudinal field investigation of gender differences in individual technology adoption decision making processes. *Organizational Behavior and Human Decision Processes*, *83*(1), 33–60. doi:10.1006/obhd.2000.2896

Venkatesh, V., Morris, M. G., Davis, G. B., & Davis, F. D. (2003). User acceptance of information technology: Toward a unified view. *Management Information Systems Quarterly*, *27*(3), 425–478.

Verdegem, P., & Hauttekeete, L. (2008). The user at the centre of the development of one-stop government. *Int. J. of Electronic Governance*, *1*(3), 258–274. doi:10.1504/IJEG.2008.020449

Vicdan, H., & Ulusoy, E. (2008). Symbolic and experiential consumption of body in virtual worlds: From (Dis) embodiment to symembodiment. *Journal of Virtual Worlds Research*, *1*(2), 1–22.

Vickery, S. K., Calantone, R., & Droge, C. (1999). Supply chain flexibility: An empirical study. *Journal of Supply Chain Management*, *35*(3), 16–24. doi:10.1111/j.1745-493X.1999.tb00058.x

Vinoski, S. (2005). Towards integration – java business integration. *IEEE Internet Computing*, *9*(4), 89–91. doi:10.1109/MIC.2005.86

Vitvar, T., Kerrigan, M., Overeem, A., Peristeras, V., & Tarabanis, K. (2006, March). Infrastructure for the semantic Pan-European e-government. In *Proceedings of the Semantic Web and E-Government Conference*.

Vrechopoulos, P. A., O' Keefe, M. R., & Doukidis, I. G. (2000, June 19-21). Virtual store atmosphere in internet retailing. In *Proceedings of the 13th International Conference on Bled Electronic Commerce*, Slovenia.

Vrechopoulos, P. A., Siomkos, G., & Doukidis, G. (2001). Internet shopping adoption by Greek consumers. [f]. *European Journal of Innovation Management*, *4*(3), 142–152. doi:10.1108/14601060110399306

Wais, J. S., & Clemons, E. K. (2008). Understanding and implementing mobile social advertising. *International Journal of Mobile Marketing*, *3*, 12–18.

Walczuch, R., Lemmink, J., & Streukens, S. (2007). The effect of service employees' technology readiness on technology acceptance. *Information & Management*, *44*(2), 206–215. doi:10.1016/j.im.2006.12.005

Walser, K. (2008). Umrisse eines e-government-prozess-referenzmodells. *eGov-Präsenz*, *1*, 61-63.

Walser, K., & Riedl, R. (2009). Skizzierung transorganisationaler modularer e-government-geschäftsarchitekturen. In *Proceedings der 9 Internationalen Tagung für Wirtschaftsinformatik Business Services: Konzepte, Technologien, Anwendungen* (pp. 565-574).

Walser, K., & Riedl, R. (2010, July 1-2). Outline of a generic e-government architecture for political administrations – based on the policy cycle concept. In *Proceedings of the 4th International Conference on Methodologies, Technologies, and Tools Enabling e-Government, Olten, Switzerland* (pp. 1-10).

Walser, K. (2006). *Auswirkungen des CRM auf die IT-integration*. Lohmar, Germany: Eul-Verlag.

Walsham, G. (1995). Interpretive case studies in IS research: nature and method. *European Journal of Information Systems, 4*, 74–81. doi:10.1057/ejis.1995.9

Wang, L., Bretschneider, S., & Gant, J. (2005). Evaluating Web-Based E-Government Services with a Citizen-Centric Approach. In *Proceedings of the 38th Annual Hawaii International Conference on System Sciences (HICSS'05)*.

Wang, C., Lo, S., & Fang, W. (2008). Extending the technology acceptance model to mobile telecommunication innovation: The existence of network externalities. *Journal of Consumer Behaviour, 7*(2), 101–110. doi:10.1002/cb.240

Wang, Y. S., Wang, Y. M., Lin, Y. M., & Tang, T. I. (2003). Determinants of user acceptance of internet banking: An empirical study. *International Journal of Service Industry Management, 14*(5), 501–519. doi:10.1108/09564230310500192

Wankel, C., & DeFillippi, R. (2006). *New Visions of Graduate Management Education*. CT: Information Age Publishing.

Watson, R. T., Pitt, L. F., Berthon, P., & Zinkhan, G. M. (2002). U-commerce: Expanding the universe of marketing. *Journal of the Academy of Marketing Science, 30*, 333–348. doi:10.1177/009207002236909

Watson, R., Pitt, F., Berthon, P., & Zinkhan, G. (2002). U-Commerce: Expanding the universe of marketing. *Journal of the Academy of Marketing Science, 30*(4), 333–347. doi:10.1177/009207002236909

WAVE. (2009). *WAVE project portal*. Retrieved from http://wave-project.eu/

Webb, D. J., Mohr, L. A., & Harris, K. E. (2008). A re-examination of socially responsible consumption and its measurement. *Journal of Business Research, 68*, 91–98. doi:10.1016/j.jbusres.2007.05.007

Webber, C. A., Roberson, J. A., McWhinney, M. C., Brown, R. E., Pinckard, M. J., & Busch, J. F. (2006). After-hours power status of office equipment in the USA. *Energy, 31*, 2823–2838. doi:10.1016/j.energy.2005.11.007

Webster, F. E. (1975). Determining the characteristics of the socially conscious consumer. *The Journal of Consumer Research, 2*(3), 188–196. doi:10.1086/208631

Webster, J., & Martocchio, J. (1992). Microcomputer playfulness: Development of a measure with workplace implication. *Management Information Systems Quarterly, 16*, 201–225. doi:10.2307/249576

Webster, J., & Watson, R. (2002). Analyzing the past to prepare for the future. *Management Information Systems Quarterly, 26*(2).

Wei, T., Marthandan, G., Chong, A., Ooi, K., & Arumugam, S. (2009). What drives Malaysian m-commerce adoption? An empirical analysis. *Industrial Management & Data Systems, 109*(3), 370–388. doi:10.1108/02635570910939399

Welch, E. W., & Pandey, S. (2007). Multiple Measures of Website Effectiveness and their Association with Service Quality in Health and Human Service Agencies. In *Proceedings of the 40th Annual Hawaii International Conference on System Sciences (HICSS'07)*.

Wendy, O., & Leela, D. (2007). Citizen Participation and engagement in the Design of e-Government Services: The Missing Link in Effective ICT Design and Delivery. *Journal of the Association for Information Systems, 8*(9).

Wertheimer, M. (2004). *Trusted agent report: Diebold AccuVote-TS voting*. Columbia, MD: RABA Technologies.

Westbrook, R. A. (1980). A rating scale for measuring product/service satisfaction. *Journal of Marketing, 44*, 68–72. doi:10.2307/1251232

West, D. M. (2007). *Global E-Government*. Providence, RI: Center for Public Policy, Brown University.

Western Balkans Network for Inclusive eGovernment. (2008). *Roadmap for inclusive eGovernment in the Western Balkans: Building e-services accessible to all*. Retrieved from http://e-society.org.mk/portal/download/Roadmap-for-inclusive-eGovernment-in-the-Western-Balkans.pdf

Whitehead, A. N. (1967). *Science and the Modern World*. New York: The Press.

WHO. (2005). *Information technology in support of health care*. Retrieved from http://www.who.int/eht/en/InformationTech.pdf

Wimmer, M., Codagnone, C., & Janssen, M. (2008). Future of e-Government Research: 13 research themes identified in the eGovRTD2020 project. In *Proceeding of the 41st Hawaii International Conference on System Sciences*.

Wind, Y., & Mahajan, V. (2002). Convergence marketing. *Journal of Interactive Marketing, 16*, 64–79. doi:10.1002/dir.10009

Wood, R., & Bandura, A. (1989). Impact of conceptions of ability on self-regulatory mechanisms and complex decision making. *Journal of Personality and Social Psychology, 56*, 407–415. doi:10.1037/0022-3514.56.3.407

Wood, S. L., & Swait, J. (2002). Psychological indicator of innovation adoption: Cross-classification based on need for cognition and need for change. *Journal of Consumer Psychology, 12*(1), 1–13. doi:10.1207/S15327663JCP1201_01

Wooldridge, M., & Jennings, N. (1995). Intelligent Agents: Theory and Practice. *The Knowledge Engineering Review, 10*(2). doi:10.1017/S0269888900008122

World, I. S. T. (2008). *SEEMP: Single European employment market-place.* Retrieved from http://www.ist-world.com/ProjectDetails.aspx?ProjectId=dba1d15b009e48eca d153f131c29120c&SourceDatabaseId=8a6f60ff9d0d443 9b415324812479c31

Wright, S. L., Bailey, I. L., Tuan, K.-M., & Wacker, R. T. (1999). Resolution and legibility: A comparison of TFT-LCDs and CRTs. *Journal of the Society for Information Display, 7*, 253–256. doi:10.1889/1.1985290

Wu, I. L., & Chen, J. L. (2005). An extension of trust and TAM model with TPB in the initial adoption of on-line tax: An empirical study. *International Journal of Human-Computer Studies, 62*(6), 784–808. doi:10.1016/j.ijhcs.2005.03.003

Wu, J. H., & Wang, S. C. (2005). what drives mobile commerce? An empirical evaluation of the revised technology acceptance model. *Information & Management, 42*, 719–729. doi:10.1016/j.im.2004.07.001

Wu, J., & Wang, S. (2005). What drives mobile commerce? An empirical evaluation of the revised technology acceptance model. *Information & Management, 42*(5), 719–729. doi:10.1016/j.im.2004.07.001

Xanthidis, D., & Nicholas, D. (2007). Consumer preferences and attitudes towards eCommerce activities. Case study: Greece. In *Proceedings of the 6th WSEAS International Conference on E-ACTIVITIES* (pp. 134-139).

Yager, S. E., Kappelman, L. A., Maples, G. A., & Prybutok, V. R. (1997). Microcomputer playfulness: Stable or dynamic trait? *The Data Base for Advances in Information Systems, 28*, 43–52.

Yang, J., & Paul, S. (2005). E-government application at local level: issues and challenges: an empirical study. *Electronic Government, an Int. J., 2*(1), 56-76.

Yang, K. C. (2007). Exploring factors affecting consumer intention to use mobile advertising in Taiwan. *Journal of International Consumer Marketing, 20*, 33–49. doi:10.1300/J046v20n01_04

Yener, E. (2008). Adoption of Mobile Technologies by the Turkish Public Sector. Plenary Talk. In *Proceedings of the Third International Conference & Exhibitions on Mobile Government (mLife 2008)*, Antalya, Turkey.

Yildirim, G., Medeni, T., Aktaş, M., Kutluoğlu, U., & Kahramaner, Y. (2010). *M-Government as an Extension of E-Government Gateway: A Case Study*. Antalya, Turkey: ICEGEG.

Yi, M., Jackson, J., Park, J., & Probst, J. (2006). Understanding information technology acceptance by individual professionals: Toward an integrative view. *Information & Management, 43*(3), 350–363. doi:10.1016/j.im.2005.08.006

Yin, R. K. (1994). *Case study research design and methods.* Newbury Park, CA: Sage.

Yu, C. C. (2008). Building a Value-Centric e-Government Service Framework Based on a Business Model Perspective. In *Proceedings of the 7th International Conference (EGOV 2008)*, Torino, Italy (Vol. 5184).

Yunos, H. M., Gao, J. Z., & Shim, S. (2003). Wireless advertising's challenges and opportunities. *IEEE Computer, 36*(5), 30–37.

Zachman, J. A. (1987). A framework for information systems architecture. *IBM Systems Journal, 26*(3). doi:10.1147/sj.263.0276

Zaiem, I. (2005). Le Comportement Ecologique du Consommateur: Modélisation des Relations et Déterminants. *La Revue des Sciences de Gestion: Direction et Gestion, 40*, 75–88. doi:10.1051/larsg:2005032

Zeithaml, V. A. (1988). Consumer perceptions of price, quality, and value: A means-end model and synthesis of evidence. *Journal of Marketing, 52,* 2–22. doi:10.2307/1251446

Zeithaml, V. A. (2000). Service quality, profitability and the economic worth of customers: What we know and what we need to learn. *Journal of the Academy of Marketing Science, 28*(1), 67–85. doi:10.1177/0092070300281007

Zhang, J., & Mao, E. (2008). Understanding the acceptance of mobile SMS advertising among young Chinese consumers. *Psychology and Marketing, 25,* 787–805. doi:10.1002/mar.20239

Zhenyu, H. (2007). A comprehensive analysis of U.S. counties' e-Government portals: development status and functionalities. *European Journal of Information Systems, 16*(2).

Zhou, L., Dai, L., & Zhang, D. (2007). Online shopping acceptance model- A critical survey of consumer factors in online shopping. *Journal of Electronic Commerce Research, 8*(1), 41–62.

Zhu, S., Abraham, J., Paul, S. A., Reddy, M., Yen, J., Pfaff, M., & DeFlitch, C. (2007). *R-CAST-MED: Applying intelligent agents to support emergency medical decision-making teams.* Retrieved from http://agentlab.psu.edu/lab/publications/Zhu_AIME07.pdf

Zorotheos, A., & Kafeza, E. (2009). Users' perceptions on privacy and their intention to transact online: a study on Greek Internet users. *Direct Marketing: An International Journal, 3*(2), 139–153. doi:10.1108/17505930910964795

About the Contributors

Ada Scupola is a Senior Associate Professor at the Department of Communication, Business and Information Technologies, Roskilde University, Denmark. She holds a Ph.D. in social sciences from Roskilde University, an MBA from the University of Maryland at College Park, USA and a M.Sc. from the University of Bari, Italy. She is the editor-in-chief of *The International Journal of E-Services and Mobile Applications*. Her main research interests are user driven innovation, e-services, outsourcing, ICT in supply chain, adoption and diffusion of e-commerce and e-services in SMEs, and ICTs in clusters of companies. She is collaborating and has collaborated to several national and international research projects on the above subjects. Her research has been published in several international journals, among which include *International Journal of E-Business Research, Journal of Facilities Management, Journal of Business and Industrial Marketing, Library Management, Technological Forecasting and Social Change, The Journal of Information Science, International Journal of E-Services and Mobile Applications, The Information Society, Journal of Enterprise Information Management, Journal of Electronic Commerce in Organizations,, The Journal of Global Information Technology Management, Scandinavian Journal of Information Systems, The Journal of Electronic Commerce in Developing Countries,* and in numerous book chapters and international conferences.

* * *

Dimitrios Askounis is Associate Professor of Management Information and Decision Support Systems in the School of Electrical and Computer Engineering of the National Technical University of Athens (NTUA). He obtained his Diploma in Electrical Engineering and his PhD in Production Control Systems. He has been involved in numerous IT research projects funded by the EU (ESPRIT, BRITE-EURAM, FP5, FP6) in the thematic areas of e-business and e-government, business and data modeling, interoperability (e.g. GENESIS, LEX-IS, WEB-DEP), decision support, knowledge management, quality management, computer integrated manufacturing, enterprise resource planning, etc. He has also participated in several other management training projects funded by the EU within the EUROPAID framework in CEEC, NIS and MEDA countries and been involved in the monitoring and evaluation of large projects. Prof. Askounis has published over 80 papers in scientific journals and international conference proceedings.

Tyrone Edwards holds Master's degree in computer Science with Distinction from University of WestIndies, Mona, in 2010. He also holds Bachelor's degree in computer Science with Distinction from University of Technology, Jamaica in 2006. He is presently working as Lecturer in the School of

Computing and Information Technology in University of Technology, Jamaica since 2009 .Prior to that he has worked as Quality Assurance Analyst in Fiscal Services during 2007-2009. He presently got a publication to his credit in ACM proceedings. His research interests are mainly in Mobile Computing, Web and Intelligent Agents.

Michail N. Giannakos received his B.Sc in Informatics from the Ionian University, Corfu, Greece. He is currently working toward the Ph.D. degree at Ionian University. He has presented papers in several international conferences, such as International Conference on Information Communication Technologies in Education, International Conference on Computer Supported Education, MindTrek and Special Interesting Group for Information Technology Education. His main research interests are in the area of education technologies, social networks, learning systems' adoption, computer-assisted instruction, evaluation methodologies, teaching/learning strategies and instructional design. Mr. Giannakos is a member of the Greek Computer Society (GCS).

Ioannis G. Krasonikolakis holds a Master in Information Systems and a B.Sc. in Informatics both issued from Athens University of Economics and Business (AUEB). He is currently a Ph.D. student in the Department of Management Science and Technology at AUEB. He is a member of Eltrun (IMES) and ISTLab (IRIS) research groups operated at the Research Center at AUEB. He is currently working in the Technological Education Institute of Piraeus as an adjunct coordinator in the Department of Accounting and in the Department of Business Administration and has teaching experience in sections such as Object-Oriented programming, Marketing and Economics. His research interests evolve around interactive marketing, virtual reality retailing, consumer behavior in ubiquitous environments and store atmosphere in virtual reality. His works have been published in several conference proceedings.

Panagiota Papadopoulou holds a BSc (Hons) in Informatics from the National and Kapodistrian University of Athens, an MSc (Distinction) in Distributed and Multimedia Information Systems from Heriot-Watt University and a PhD in Information Systems from the National and Kapodistrian University of Athens. She is a research fellow in the Department of Informatics and Telecommunications at the National and Kapodistrian University of Athens. She has extensive, university-level teaching experience, as an adjunct faculty member at the University of Athens, the University of Pireaus, the University of Peloponnese, the University of Central Greece and other educational institutions in Greece. Dr. Papadopoulou has also actively participated in a number of European Community and National research projects. She has published more than 30 papers in international journals and conferences, with her current research interests focusing on online trust, e-commerce, interface design, web-based information systems and social computing.

Pier Paolo Carrus is Professor of management studies at the Cagliari University, Italy. His teaching and research is centered on management aspects of firms working in contexts with characteristics of relational and complex, dynamic changes. His research interests include cultural, strategic and organizational conditions for quality improvement in the performance of public service care firms, the analysis of the needs for innovative technological solutions for small and medium firms. He has published several papers, books as well as chapter in books. Pier Paolo Carrus can be contact at the Cagliari University (Italy), Via S. Ignazio 74, 09100. E-mail: ppcarrus@unica.it.

Yannis Charalabidis is Assistant Professor in the University of Aegean, in the area of eGovernance Information Systems, while also heading ICT research in the Decision Support Systems Laboratory of National Technical University of Athens (NTUA), coordinating policy making, research and pilot application projects for governments and enterprises worldwide. A computer engineer with a PhD in complex information systems, he has been employed for several years as an executive director in Singular IT Group, leading software development and company expansion in Europe, India and the US. He writes and teaches on Service Systems, Enterprise Interoperability, Government Transformation and Citizen Participation in NTUA and the University of Aegean. He has published more the 100 papers in international journals and conferences. He is Best Paper Award winner of the EGOV 2008 Conference, Best eGovernment Paper Nominee in the 42nd HICSS Conference and 1st Prize Nominee in the 2009 European eGovernment Awards.

Daniel González-Morales received the BS degree in Computer Science in University of Las Palmas of G.C in 1991 and the Ph.D. degree in 2003 in the University of La Laguna. Since 1991 he is an assistant professor in the department of Statistics and Computer Science in the University of La Laguna. He has over 15 years of experience as director technology in different projects related to e-Government and Public Administration. He has worked in several departments: Employment and Social Affairs, Canary Service of Employment and Canary Agency of Research, Innovation and Information Society. His research interests include e-Government, modernization of government information systems, software engineering, PMI, and BPMN.

Deirdre Lee is a Research Associate in the eGovernment Unit at the Digital Enterprise Research Institute (DERI), National University of Ireland, Galway (NUIG). Her research interests include eGovernment, eParticipation, data interoperability, the Semantic Web, collaborative environments, context-aware mobile and pervasive computing systems, and web services. She has worked on European Commission projects in the Sixth Framework Programme, the Competitiveness and Innovation Framework Programme (CIP), eParticipation Preparatory Action, and the Lifelong Learning Programme. Lee received a B.A. in Information and Communication Technology (ICT) and a M.Sc. in Computer Science from the University of Dublin, Trinity College, Ireland in 2004 and 2007 respectively. She has also worked as a research assistant in the IBM Research Lab, Zurich, Switzerland from 2004 – 2006.

Ourania Markaki is a Ph.D. candidate in the field of e-government in the Department of Electrical and Computer Engineering of the National Technical University of Athens (NTUA). She received her MBA in Techno-Economic Systems in 2008 and her Diploma in Electrical Engineering from NTUA in 2006. She has worked as a Research Assistant in the Telecommunications Laboratory of the Institute of Communication and Computer Systems (ICCS-NTUA); and since 2008, she is a Researcher in the Decision Support Systems Laboratory of NTUA, where she has been involved in several EU-funded projects, including the FP7 Greek Interoperability Centre. Her research interests include e-government interoperability, e-government evaluation, e-participation and multicriteria decision-making.

Ilias O. Pappas received a B.Sc Degree in Informatics (specialization in Humanities Informatics) from the Ionian University, Corfu, Greece in 2009. He is currently a M.Sc student of the Department of Informatics (specialization in Information Systems) at the Ionian University. His current research interests are in the areas of electronic commerce, mobile commerce, information technology adoption usage and strategy, innovation adoption, consumer behavior in online environments, customer relationship management (CRM) and Internet Marketing.

Adamantia G. Pateli is Lecturer of Information Systems at the Department of Informatics of the Ionian University, Corfu, Greece. She holds a B.Sc. Degree in Informatics (specialization in Information Systems) from the Athens University of Economics and Business (AUEB), a Masters Degree in Electronic Commerce from the University of Manchester, and a PhD degree from the Department of Management Science and Technology of the Athens University of Economics and Business (AUEB). Since 1997, Dr. Pateli has participated in a number of national and European-funded research projects in the areas of Electronic Commerce and Information Systems. She has published more than 30 research articles in leading academic journals, such as the *European Journal of Information Systems, Journal of Organizational Change Management, International Journal of Technology Management, Electronic Markets, and Management Decision*, as well as in several European and international peer-reviewed conferences. Her current research interests lie in the areas of information systems management, mobile and wireless services, e-government, electronic Human Resource management, technology-induced innovation and strategic technology alliances.

Jean-Eric Pelet owns a doctorate in marketing with distinction (Nantes University, France) and a MBA in information systems with distinction (Laval University, Quebec, Canada). He works as an assistant professor at SupAgro Montpellier (France) on problematic dedicated to the interface and to the consumer behaviour facing a website or any information system (e-learning, knowledge management, e-commerce platforms). Its main interest lies on the variables that enhance the navigation in order to help people to be more efficient on these. He works as a visiting professor in several places in France thanks to its Knowledge Management and Content Management System platform (kmcms.net) in Design School (Nantes), Business Schools (Paris, Reims), and Universities (Paris Dauphine – Nantes), on lectures focused on e-marketing, ergonomic, usability, and consumer behaviour. Dr. Pelet has also actively participated in a number of European Community and National research projects. His current research interests focus on web-based information systems, social networks, interface design and usability.

Carlos Peña-Dorta is the CTO of the company Arte Consultores Tecnologicos S.L.. He received the BS degree in Computer Science in the University of La Laguna in 1999. He has over ten years of experience working on projects for Public Administration (PA), particularly related to interoperability. Some of the most relevant projects which he has participated are SISPE (the national employment information system PA), SISPECAN (the PA regional information system for the Canary Islands Community) and PLATINO (platform interoperability Canary Islands Government). In recent years, he has collaborated with the University of La Laguna in R & D projects related to the adaptation of new technologies in information systems of Public Administration.

Vassilios Peristeras is the eGovernment Unit Leader at the Digital Enterprise Research Institute (DERI), National University of Ireland, Galway (NUIG). He is also Adjunct Lecturer at the University of Macedonia, Greece. He has studied Political Science, has postgraduate studies in public administration, masters in Information Systems, and holds a PhD in eGovernment. Vassilios has worked as scientific coordinator at the Greek National Centre for Public Administration, researcher at CERTH/ITI, IT Consultant in the United Nations. Since 1998 he has initiated, participated and coordinated several R&D projects at the national/international level (e.g. EU).

Roberta Pinna is Professor of Organization and Human Resource Management at the Cagliari University, Italy. Her research interests include supply chain management and integration of manufacturing and logistics strategies and structures, motivation and creativity, e-government and digital divide. This is reflected in several papers and articles published in national and international publications as well as chapter in books and books. She is also reviewer for an international journal. Roberta Pinna can be contact at the Cagliari University (Italy), Via S. Ignazio 74, 09100. E-mail: pinnar@unica.it.

Athanasia (Nancy) Pouloudi is an Associate Professor in the Department of Management Science and Technology at the Athens University of Economics and Business (AUEB), Greece. She holds a first degree in Informatics (AUEB, Greece), and an MSc and PhD degree in Information Systems (London School of Economics, UK). She has acted as scientific coordinator for AUEB in a number European Projects on e-government, e-business and e-inclusion. She is Senior Associate Editor of the European Journal of Information Systems and Associate Editor of IT & People and member of the Editorial Boards of Information & Management and the International Journal of Society, Information, Communication and Ethics. She has taught information systems at Brunel University (as lecturer) and the London School of Economics (as teaching assistant) and held visiting positions at Erasmus University (The Netherlands), the University of Hawaii (USA) and the Athens Laboratory of Business Administration (Greece). Nancy is an active member of the Association of Information Systems (AIS) and serves as Region 2 (Europe-Middle East-Africa) Representative of the AIS from July 2010 to June 2012.

David Price is a co-founder of the cloud-based web service Debategraph, which helps groups collaborate in thinking through complex issues by building and sharing interactive maps of domains of knowledge from multiple perspectives. Debategraph's collaborative partners have included: the UK Prime Minister's Office, the White House Office of Science and Technology Policy, and CNN. David has a Ph.D. from the University of Cambridge in organisational learning and environmental policy, and a B.Sc. in Business Administration from the University of Bath.

Yojana Priya Menda is a Masters Student in the eGovernment Unit at the Digital Enterprise Research Institute (DERI), National University of Ireland, Galway (NUIG). Her research interests include eGovernment, eParticipation, Semantic Web, Web development, Computational Linguistics, Content Management Systems and Business Information Systems. Yojana received a Bachelor of Technology in Information Technology (IT) from Indian Institute of Information Technology-Allahabad, India (2005-2009) and is currently pursuing her M.Sc. in Business Information Systems from Cairnes Business School, NUI-Galway, Ireland (2009-2011). She has also been parallel working on European Union eParticipation Projects.

Vaggelis Saprikis is a PhD Candidate at the University of Macedonia, Department of Applied Informatics, Thessaloniki, Greece. He is currently a Laboratory Assistant in the Technological Educational Institute of Western Macedonia. His research interests include e-business models, e-commerce, e-marketplaces and m-commerce. He has published papers in international journals and conferences and he is a member of Hellenic and international associations.

Elena Sánchez-Nielsen received the BS degree in Computer Science in University of Las Palmas of G.C in 1995 and the MS and Ph.D. degree in 1999 and 2003 respectively in Computer Science and Artificial Intelligence from the University of La Laguna. Since 1995 she is an assistant professor in the department of Statistics and Computer Science in the University of La Laguna. She has over 10 years of experience as research director and software architect in different projects related to e-Government and eParticipation at the Parliament of Canary Islands. Since 2005, she has been working at the Public Administration with projects related to smart information systems and interoperability issues. Her research interests include intelligent systems, information integration, semantic technologies, eGovernment, modernization of government information systems, e-Consultation technologies, audiovisual contents and social media.

Suresh Sankaranarayanan holds a PhD degree (2006) in Electrical Engineering with specialization in Networking from the University of South Australia. Later he has worked as a Postdoctoral Research Fellow and then as a Lecturer in the University of Technology, Sydney and at the University of Sydney, respectively. He is the recipient of University of South Australia President Scholarship, towards pursuing the PhD degree programme and has also bagged the IEEE travel award in 2005. Presently he is working as a Lecturer in the Department of Computing and leads the Intelligent Networking Research Group, in the University of West Indies, Kingston, Jamaica, since 2008. He has supervised fifteen research students leading to ME, M.S, M.phil and M.sc degrees. He has got to his credit, as on date, 37 research papers published in the Proceedings of major IEEE international conferences, as Book Chapters and in Journals. He is also a Reviewer and Technical Committee member for a number of IEEE Conferences and Journals. He has also given Keynote talks in IEEE conferences too. In additions he has conducted many tutorials, workshops and also given Guest Lectures in Networking and Agent Applications in various Universities, Colleges and Research Institutes. Presently he manages a collaborative research programme with Oakland University, Rochester, USA. Also received a research grant from University of WestIndies towards Wireless Sensor Network project towards patient Health Monitoring. His current research interests are mainly towards 'Intelligent Agents and their applications in Wireless Sensor based Mesh networks' used in the Health and Engineering sectors; Applications in mobile commerce; security aspects in the applications of Intelligent Agents. Further details can be obtained from his homepage: www.mona.uwi.edu/dmcs/staff/suresh/webpage-suresh.htm.

B.Senthil Arasu is currently involved in research and consultancy in the area of stock market efficiency, investor behavior, Equity Research and Risk management. He has a doctoral degree from Madurai Kamaraj University Madurai, India. In the past five years his research work has been published in various journals (both national and international) and his work has been presented in various International Conferences on Business and Management. He is presently working as an assistant professor in National Institute of Technology Tiruchirappalli India.

M.Sivagnanasundaram is currently pursuing his doctoral programme in National Institute of Technology, Tiruchirappalli, India. He is working in the area of consumer adoption of self service Technologies. His work has been published in Journal of Internet Banking Commerce, The Electronic Journal of Information Systems in Developing Countries.

N.Thamaraiselvan has been involved in research, consultancy and training in the area of brand management, marketing metrics, services marketing and strategic marketing. He was awarded a doctoral degree and MBA from National Institute of Technology Trichy, India. He has published various empirical papers in journals and awarded best paper for his empirical research on brand extension in International Conference on Business and Information 2006 held in Singapore. He is presently working as an associate professor in National Institute of Technology Tiruchirappalli India.

Kaan Varnalı, Ph.D. is an Assistant Professor at Department of Advertising, Istanbul Bilgi University, Turkey, where he teaches courses on various topics related with marketing through new media and consumer behavior. His prior research has appeared in various highly regarded scholarly outlets as journal articles, case studies, and book chapters. He also authored a book entitled "Mobile Marketing: Fundamentals & Strategy" published by McGraw-Hill, New York. His current research interests include consumer behavior in digital contexts, implications of social media on the universe of marketing, and mobile marketing strategy.

Maro Vlachopoulou is a Professor at the Department of Applied Informatics, University of Macedonia, Greece. Her professional expertise, research, and teaching interests include: marketing information systems, electronic commerce, e-business, e-marketing, internet marketing, enterprise resource planning (ERP) systems, customer relationship management (CRM) systems, e-supply chain management, e-logistics and healthcare information systems.

Adam P. Vrechopoulos is Assistant Professor at the Athens University of Economics and Business (AUEB), Department of Management Science and Technology. Since 2003, he is the Scientific Coordinator of the "Interactive Marketing and Electronic Services" (IMES) Research Group of the "ELTRUN" E-Business Research Center (www.eltrun.gr) at AUEB. He holds a Ph.D. in "Consumer Behaviour in Electronic Retailing" from Brunel University, Department of Information Systems and Computing, UK, an M.B.A. from the Athens Laboratory of Business Administration (ALBA), and a B.Sc. in Information Systems from the AUEB, Department of Informatics. He has published more than 90 papers in peer reviewed journals (among others in Journal of Retailing, Information Systems Journal, Journal of the Academy of Marketing Science, European Journal of Marketing, International Journal of Information Management), edited volumes and international conferences and 3 books. He has acted as a guest editor, reviewer, program scientific committee member and session chair for several international journals and conferences. He is Associate Editor of the European Journal of Information Systems and member of the Editorial Board of the Electronic Markets – The International Journal.

Theodora Zarmpou is an Electronic & Computer Engineer with a Master Degree in Management of Information Systems. Currently, she is a PhD Student in the Department of Applied Informatics, University of Macedonia, Thessaloniki, Greece. She is, also, a Laboratory Assistant in the Technological Educational Institute of Western Macedonia. Her research area is around mobile marketing, mobile government, mobile information systems, and mobile business. She has participated in international conferences and has published papers in international journals.

Index